ASPECTS OF COMPUTATION AND AUTOMATA THEORY WITH APPLICATIONS

LECTURE NOTES SERIES
Institute for Mathematical Sciences, National University of Singapore

Series Editors: Chi Tat Chong and Alexandre Thiery
Institute for Mathematical Sciences
National University of Singapore

ISSN: 1793-0758

Published

Vol. 42 *Aspects of Computation and Automata Theory with Applications*
edited by Noam Greenberg, Sanjay Jain, Keng Meng Ng, Sven Schewe, Frank Stephan, Guohua Wu & Yue Yang

Vol. 41 *Density Functionals for Many-Particle Systems: Mathematical Theory and Physical Applications of Effective Equations*
edited by Berthold-Georg Englert, Heinz Siedentop & Martin-Isbjörn Trappe

Vol. 40 *Modeling and Simulation for Collective Dynamics*
edited by Weizhu Bao, Peter A Markowich, Benoit Perthame & Eitan Tadmor

Vol. 39 *Models and Methods for Quantum Condensation and Fluids*
edited by Weizhu Bao, Yongyong Cai, Ionut Danaila & Peter A Markowich

Vol. 38 *Genealogies of Interacting Particle Systems*
edited by Matthias Birkner, Rongfeng Sun & Jan M Swart

Vol. 37 *Mathematics of Shapes and Applications*
edited by Sergey Kushnarev, Anqi Qiu & Laurent Younes

Vol. 36 *The Geometry, Topology and Physics of Moduli Spaces of Higgs Bundles*
edited by Richard Wentworth & Graeme Wilkin

Vol. 35 *Combinatorial and Toric Homotopy: Introductory Lectures*
edited by Alastair Darby, Jelena Grbić, Zhi Lü & Jie Wu

Vol. 34 *White Noise Analysis and Quantum Information*
edited by Luigi Accardi, Louis H Y Chen, Takeyuki Hida, Masanori Ohya, Si Si & Noboru Watanabe

Vol. 33 *Sets and Computations*
edited by Sy-David Friedman, Dilip Raghavan & Yue Yang

Vol. 32 *Mathemusical Conversations: Mathematics and Computation in Music Performance and Composition*
edited by Jordan B L Smith, Elaine Chew & Gérard Assayag

Vol. 31 *Geometric Analysis Around Scalar Curvatures*
edited by Fei Han, Xingwang Xu & Weiping Zhang

For the complete list of titles in this series, please go to
https://www.worldscientific.com/series/LNIMSNUS

Lecture Notes Series, Institute for Mathematical Sciences,
National University of Singapore
Vol. 42

ASPECTS OF COMPUTATION AND AUTOMATA THEORY WITH APPLICATIONS

Editors

Noam Greenberg
Victoria University of Wellington, New Zealand

Sanjay Jain
National University of Singapore, Singapore

Keng Meng Ng
Nanyang Technological University, Singapore

Sven Schewe
University of Liverpool, UK

Frank Stephan
National University of Singapore, Singapore

Guohua Wu
Nanyang Technological University, Singapore

Yue Yang
National University of Singapore, Singapore

World Scientific

NEW JERSEY • LONDON • SINGAPORE • BEIJING • SHANGHAI • HONG KONG • TAIPEI • CHENNAI • TOKYO

Published by

World Scientific Publishing Co. Pte. Ltd.
5 Toh Tuck Link, Singapore 596224
USA office: 27 Warren Street, Suite 401-402, Hackensack, NJ 07601
UK office: 57 Shelton Street, Covent Garden, London WC2H 9HE

Library of Congress Cataloging-in-Publication Data
Names: Greenberg, Noam, editor. | Jain, Sanjay, 1965 February 22– editor. | Ng, Keng Meng, editor. | Schewe, Sven, editor. | Stephan, Frank (Frank Christian), editor. | Wu, Guohua (Mathematician), editor. | Yang, Yue, 1964– editor.
Title: Aspects of computation and automata theory with applications / editors Noam Greenberg (Victoria University of Wellington, New Zealand), Sanjay Jain (National University of Singapore, Singapore), Keng Meng Ng (National University of Singapore, Singapore), Sven Schewe (University of Liverpool, UK), Frank Stephan (National University of Singapore, Singapore), Guohua Wu (National University of Singapore, Singapore), Yue Yang (National University of Singapore, Singapore).
Description: New Jersey : World Scientific, [2024] | Series: Lecture notes series, Institute for Mathematical Sciences, National University of Singapore, 1793-0758 ; vol. 42 | Includes bibliographical references.
Identifiers: LCCN 2023024536 | ISBN 9789811278624 (hardcover) | ISBN 9789811278631 (ebook for institutions) | ISBN 9789811278648 (ebook for individuals)
Subjects: LCSH: Machine theory. | Computational complexity. | Recursion theory.
Classification: LCC QA267 .A87 2024 | DDC 005.13/1--dc23/eng20230911
LC record available at https://lccn.loc.gov/2023024536

British Library Cataloguing-in-Publication Data
A catalogue record for this book is available from the British Library.

For any available supplementary material, please visit
https://www.worldscientific.com/worldscibooks/10.1142/13479#t=suppl

https://doi.org/10.1142/9789811278631_fmatter

Contents

https://doi.org/10.1142/9789811278631_fmatter

Foreword

The Institute for Mathematical Sciences (IMS) organizes thematic programs of longer duration as well as shorter workshops and public lectures. The themes are selected from among areas at the forefront of current research in the mathematical sciences and their applications. Each volume of the *IMS Lecture Notes Series* is a compendium of papers based on lectures and tutorials delivered at the IMS. The aim is to make original papers and expository articles on a subject of current interest available to the international research community. These volumes also serve as a record of activities that took place at the IMS. We hope that through regular publication of these Lecture Notes the Institute will achieve, in part, its objective of reaching out to the community of scholars in the promotion of research in the mathematical sciences.

Chi Tat Chong
Alexandre Thiery
Series Editors

case of recursion theory, but because the tools of computational complexity depend heavily on the relativised world one introduces (to model external influences on computations like the availability of knowledge about the physical world). For the full absence of such a world (that is, in the case of the empty set as oracle for the world), many fundamental questions are unknown, the most prominent of these being the question whether every nondeterministic polynomial time computation could be replaced by an equivalent deterministic one. Parameterised complexity is based on the insight that many natural problems are described by parameters and that they are in principle polynomial time solvable if one fixes some of these parameters to some constants. The field has, similarly to computational complexity, some basic assumptions and hierarchies and many results are based on the assumptions that the corresponding hierarchies do not collapse.

These two programmes were accompanied by two public lectures where leading experts gave a talk accessible to people who know the basics of mathematics and computer science, but who are not experts in the domain of the speakers. These lectures were part of the "Ng Kong Beng Public Lecture Series" of the Institute of Mathematical Sciences. On 14 September 2017, Denis Hirschfeldt spoke about "Waking Up from Leibniz' Dream: On the Unmechanizability of Truth" and on 24 September 2021, Jeffrey Shallit spoke about "Adventures in Automata with a Theorem-Prover". These two talks highlighted two different aspects of research in theorem proving: On one hand, Denis Hirschfeldt explained why there is no perfect theorem prover and why the basic principle of the truth of mathematical theorems is undecidable; on the other hand, Jeffrey Shallit spoke about successes in automata theory using a theorem prover which corrected some of the "findings" human researchers had made and published before. So the situation is a bit similar to that of chess playing: The technology is far from perfect and in theorem proving perfection is impossible, but it is better than what humans are able to do. In chess, for example, computers have not yet reached the state where their play is perfect and leads to the optimal result (either always win for one of the colours black and white or always draw for both colours, in spite of much research and fast computers, one does not even know which of these two cases will eventually come out, if at all) while in the year 1996 the chess computer "Deep Blue" defeated World Chess Champion Garry Kasparov and computer chess has made much progress since then so that the new generation machines easily defeat their predecessors from some years ago.

This volume contains 17 contributions reflecting the current state of art in the fields of the two programmes. Thirteen of these chapters were contributed to "Aspects of Computation", while four of these chapters were contributed to "Automata Theory and Applications".

The editors would like to express gratitude to the Institute of Mathematical Science, to its staff, its advisory board for funding, organising and providing an excellent support for the two programmes. Furthermore, we would like to thank the authors for contributing to this book.

We wish the readers that they find many new insights with this book and that the book helps also former participants to recall aspects of the two programmes. The organisers:

Noam Greenberg, Keng Meng Ng, Guohua Wu and Yue Yang (for "Aspects of Computation");

Sanjay Jain, Sven Schewe and Frank Stephan (for "Automata Theory and Applications: Games, Learning and Structures").

that, in the process of generating the c.e. degrees, the meet operator can be neglected, i.e., that any set which generates the c.e. degrees generates the nonzero c.e. degrees under join.

This conjecture was refuted in Ambos-Spies, Lempp and Slaman [6] where a generator is constructed which is not a join generator. Despite this negative answer, it is reasonable to ask whether there are some natural substantial parts of **R** which are generated under join by any generator. Our main result shows that this is indeed the case: any generator generates the high c.e. degrees under join (Corollary 2.9). In order to obtain this result we introduce and study the strongly meet inaccessible (s.m.i.) degrees. Here a c.e. degree **a** is s.m.i. if any set of c.e. degrees which generates **a** under join and meet generates **a** under join. So, in order to get our main result, it suffices to show that any high c.e. degree can be split into two s.m.i. degrees (Theorem 2.8).

Strong meet inaccessibility is a refinement of meet inaccessibility introduced in Ambos-Spies [3]. A c.e. degree **a** is meet inaccessible if no set **A** of c.e. degrees which does not intersect the lower cone of **a** generates **a**. So, in order to show that any generator intersects any nontrivial initial segment of **R**, it suffices to show that any nonzero c.e. degree bounds a meet inaccessible degree. In fact, Ambos-Spies [3] shows that any nonzero degree can be split into two meet inaccessible degrees, and this result was subsequently strengthened by Zhang [22] and Ding [8] who showed that the set **MI** of the meet inaccessible degrees is dense in **R**. As we show here the corresponding results fail for strong meet inaccessibility (see Lemmas 2.4 and 2.6 below). So the construction of strongly meet inaccessible degrees is considerably more delicate than that of meet inaccessible degrees.

We demonstrate the usefulness of the s.m.i. degrees by some further results on generators related to their ranks. Ambos-Spies [3] defines the rank of a generator **G** to be the least number $n \geq 1$ such that any c.e. degree can be generated by a subset of **G** of cardinality at most n, and he defines the rank of **G** to be ω (infinite) if no such number n exists. The join rank of a join generator is defined similarly. Now Ambos-Spies [3] has shown that there are join generators of infinite rank and definable join generators of rank n for any $n \geq 1$. He raised the question whether the corresponding results for generators and their ranks in place of join generators and their join ranks hold (see Problem 3 in [3]). Here, by refining the approach in [3] and by proving an appropriate existence result for s.m.i. degrees (Theorem 2.12), we affirmatively answer this question. In fact we show that there is a *definable* generator of rank ω (Theorem 2.11) and that for, any $n \geq 1$, there are definable generators which have rank and join rank n (Theorem 2.10).

The outline of this chapter is as follows. In Sec. 2, we present our main results. Here we first introduce strong meet inaccessibility and study some of the basic properties of this notion and its relation to the meet inaccessibility notion introduced in [3]. In particular, we give more handy characterizations of, and sufficient conditions for, strong meet inaccessibility (which subsequently are used in the constructions of s.m.i. degrees). Then we present our main results and state the existence theorems for s.m.i. degrees needed for the proofs. The proofs of these existence theorems are deferred to the subsequent sections. In Sec. 3, we explain the ideas underlying our constructions of s.m.i. degrees by constructing a low s.m.i. degree. Then in Secs. 4 and 5, this construction is refined in order to prove the required high-splitting theorem and bounding theorem for s.m.i. degrees which we need for our results on generators. Finally, in Sec. 6 we pose some open problems.

We conclude this section by giving some notation. Although we use the term "c.e." here sometimes for emphasis, all degrees considered in this paper are c.e. So boldface lower case letters denote c.e. degrees and boldface capital letters denote sets of c.e. degrees. We let $\mathbf{R}(\leq \mathbf{a}) = \{\mathbf{b} : \mathbf{b} \leq \mathbf{a}\}$ be the lower cone of $\mathbf{a}$. Similarly, $\mathbf{R}(\not\leq \mathbf{a}) = \{\mathbf{b} : \mathbf{b} \not\leq \mathbf{a}\}$ denotes the complement of the lower cone of $\mathbf{a}$ (etc.).

Related to generators we will use the following notation. For any set $\mathbf{A}$ of c.e. degrees we let $\mathrm{CL}(\mathbf{A})$ denote the closure of $\mathbf{A}$ (under join and meet), i.e., $\mathrm{CL}(\mathbf{A})$ is the least set $\mathbf{B}$ of c.e. degrees such that

$$\mathbf{A} \subseteq \mathbf{B} \tag{1.1}$$

$$\forall\, n \geq 0\, \forall\, \mathbf{a}_0, ..., \mathbf{a}_n \in \mathbf{B}\ (\mathbf{a}_0 \vee ... \vee \mathbf{a}_n \in \mathbf{B}) \tag{1.2}$$

and

$$\forall\, n \geq 0\, \forall\, \mathbf{a}_0, ..., \mathbf{a}_n \in \mathbf{B}\ (\mathbf{a}_0 \wedge ... \wedge \mathbf{a}_n \downarrow \Rightarrow\ \mathbf{a}_0 \wedge ... \wedge \mathbf{a}_n \in \mathbf{B}). \tag{1.3}$$

The closure of $\mathbf{A}$ under join is obtained by omitting clause (1.3) and is denoted by $\mathrm{CL}_j(\mathbf{A})$; the closure of $\mathbf{A}$ under meet is obtained by omitting clause (1.2) and is denoted by $\mathrm{CL}_m(\mathbf{A})$. We say that $\mathbf{A}$ generates $\mathbf{B}$ if $\mathbf{B}$ is contained in the closure of $\mathbf{A}$, $\mathbf{A}$ generates $\mathbf{B}$ under join if $\mathbf{B}$ is contained in the closure of $\mathbf{A}$ under join, and $\mathbf{A}$ generates $\mathbf{B}$ under meet if $\mathbf{B}$ is contained in the closure of $\mathbf{A}$ under meet. Similarly, we say that $\mathbf{A}$ generates $\mathbf{b}$ (under join, under meet) if $\mathbf{A}$ generates $\{\mathbf{b}\}$ (under join, under meet). Finally, we call $\mathbf{A}$ a generator if $\mathbf{A}$ generates $\mathbf{R}$ and we call $\mathbf{A}$ a join generator if $\mathbf{A}$ generates $\mathbf{R} \setminus \{\mathbf{0}\}$ under join. (Note that, by the existence of minimal pairs, $\mathbf{A}$ is a generator iff $\mathbf{A}$ generates $\mathbf{R} \setminus \{\mathbf{0}\}$.)

Since $\mathbf{R}$ is an upper semi-lattice, the above definitions will not change if we consider in (1.2) the binary case only, i.e., fix $n = 1$. A similar change in (1.3), however, is not admissible (see Ambos-Spies [3]).

For any subset $\mathbf{A}$ of $\mathbf{R}$, it is clear that $\mathbf{B} = \mathbf{R}$ satisfies conditions (1.1) to (1.3) and further, the intersection of any non-empty family of subsets of $\mathbf{R}$, each of which satisfies these three conditions, again satisfies the three conditions, so for any subset $\mathbf{A}$ of $\mathbf{R}$, $\mathrm{CL}(\mathbf{A})$ exists, and similarly, $\mathrm{CL}_j(\mathbf{A})$ and $\mathrm{CL}_m(\mathbf{A})$ exist. It is useful, however, to have "bottom-up" descriptions of these closures. It is easy to see that for any subset $\mathbf{A}$ of $\mathbf{R}$, we have

$$\mathrm{CL}_j(\mathbf{A}) = \{\mathbf{a}_0 \vee \cdots \vee \mathbf{a}_n : n \geq 0 \;\&\; \mathbf{a}_0, ..., \mathbf{a}_n \in \mathbf{A}\}$$

and

$$\mathrm{CL}_m(\mathbf{A}) = \{\mathbf{a}_0 \wedge \cdots \wedge \mathbf{a}_n : n \geq 0 \;\&\; \mathbf{a}_0, ..., \mathbf{a}_n \in \mathbf{A} \;\&\; \mathbf{a}_0 \wedge \cdots \wedge \mathbf{a}_n \downarrow\}.$$

To give a bottom-up description of the closure of a subset $\mathbf{A}$ of $\mathbf{R}$, we define $\mathrm{CL}^0(\mathbf{A}) = \mathbf{A}$, and, for $k \geq 0$, $\mathrm{CL}^{2k+1}(\mathbf{A}) = \mathrm{CL}_j(\mathrm{CL}^{2k}(\mathbf{A}))$, and $\mathrm{CL}^{2k+2}(\mathrm{A}) = \mathrm{CL}_m(\mathrm{CL}^{2k+1}(\mathbf{A}))$. Then we have

$$\mathbf{A} = \mathrm{CL}^0(\mathbf{A}) \subseteq \mathrm{CL}^1(\mathbf{A}) \subseteq \mathrm{CL}^2(\mathbf{A}) \subseteq \cdots$$

and

$$\mathrm{CL}(\mathbf{A}) = \bigcup_{i \geq 0} \mathrm{CL}^i(\mathbf{A}).$$

2. Main Section

Here we present our new results on generators. They are derived from theorems on the distribution of the strongly meet inaccessible degrees. The proofs of the latter, which from a technical point of view form the core of this chapter, are given in the subsequent three sections. We first introduce strong meet inaccesibility and give some basic properties of this new notion.

Definition 2.1: A c.e. degree $\mathbf{a}$ is *strongly meet inaccessible* (s.m.i.) if

$$\forall \mathbf{A} \subseteq \mathbf{R}(\mathbf{a} \in \mathrm{CL}(\mathbf{A}) \Rightarrow \mathbf{a} \in \mathrm{CL}_j(\mathbf{A})). \tag{2.1}$$

The set of the s.m.i. degrees is denoted by $\mathbf{SMI}$.

We obtain an alternative characterization of the s.m.i. degrees $\mathbf{a}$ by looking at the relation between the set of the degrees generated by the complement of the lower cone of $\mathbf{a}$ and the set of the degrees which cup to $\mathbf{a}$. Here we call a degree $\mathbf{b}$ $\mathbf{a}$-*cuppable* if $\mathbf{b} \leq \mathbf{a}$ and there is a degree $\mathbf{c} < \mathbf{a}$ such that $\mathbf{a} = \mathbf{b} \vee \mathbf{c}$ and we let

$$\mathrm{Cu}(\mathbf{a}) = \{\mathbf{b} \leq \mathbf{a} : \ \mathbf{b} \text{ is } \mathbf{a}\text{-cuppable}\} \ \text{ and } \ \mathrm{NCu}(\mathbf{a}) = \mathbf{R}(\leq \mathbf{a}) \setminus \mathrm{Cu}(\mathbf{a}).$$

Lemma 2.2: *Let* $\mathbf{a}$ *be a c.e. degree such that* $\mathbf{a} > \mathbf{0}$. *The following are equivalent.*

(i) $\mathbf{a}$ *is strongly meet inaccessible.*
(ii) $\mathbf{R}(\not\leq \mathbf{a}) \cup \mathrm{NCu}(\mathbf{a})$ *is closed under join and meet.*
(iii) $\mathrm{CL}(\mathbf{R}(\not\leq \mathbf{a})) \cap \mathrm{Cu}(\mathbf{a}) = \emptyset$.
(iv) $\mathrm{CL}_m(\mathbf{R}(\not\leq \mathbf{a})) \cap \mathrm{Cu}(\mathbf{a}) = \emptyset$.

Proof: The proof of the implication $(i) \Rightarrow (ii)$ is by contraposition. Assume that $\mathbf{R}(\not\leq \mathbf{a}) \cup \mathrm{NCu}(\mathbf{a})$ is not closed under join and meet. Fix $\mathbf{b}$ such that $\mathbf{b} \in \mathrm{CL}(\mathbf{R}(\not\leq \mathbf{a}) \cup \mathrm{NCu}(\mathbf{a}))$ and $\mathbf{b} \in \mathrm{Cu}(\mathbf{a})$. By the latter, take $\mathbf{c} < \mathbf{a}$ such that $\mathbf{a} = \mathbf{b} \vee \mathbf{c}$. Then, for $\mathbf{A} = \mathbf{R}(\not\leq \mathbf{a}) \cup \mathrm{NCu}(\mathbf{a}) \cup \{\mathbf{c}\}$, $\mathbf{a}$ is in the closure of $\mathbf{A}$. But since $\mathrm{NCu}(\mathbf{a})$ is an ideal, $\mathbf{a}$ is not in the join closure of $\mathbf{A}$.

The implications $(ii) \Rightarrow (iii)$ and $(iii) \Rightarrow (iv)$ are immediate.

Finally, for a proof of $(iv) \Rightarrow (i)$ assume

$$\mathrm{CL}_m(\mathbf{R}(\not\leq \mathbf{a})) \cap \mathrm{Cu}(\mathbf{a}) = \emptyset. \tag{2.2}$$

Since, by $\mathbf{a} > \mathbf{0}$, $\mathbf{a}$ is $\mathbf{a}$-cuppable, it follows that $\mathbf{a}$ cannot be expressed as a meet $\mathbf{c_0} \wedge \cdots \wedge \mathbf{c_n}$ with each $\mathbf{c_i} > \mathbf{a}$, i.e., $\mathbf{a}$ is nonbranching.

Now, given $\mathbf{A} \subseteq \mathbf{R}$ and $\mathbf{a} \in \mathrm{CL}(\mathbf{A})$, we have to show that $\mathbf{a} \in \mathrm{CL}_j(\mathbf{A})$. Take p minimal with $\mathbf{a} \in \mathrm{CL}^p(\mathbf{A})$. By eliminating the other possibilities, we show that $p \leq 1$, and hence $\mathbf{a} \in \mathrm{CL}_j(\mathbf{A})$.

If $p = 2k+2$ for $k \geq 0$, then $\mathbf{a} = \mathbf{c_0} \wedge \cdots \wedge \mathbf{c_n}$ with each $\mathbf{c_i} \in \mathrm{CL}^{2k+1}(\mathbf{A})$. By minimality of p, none of the $\mathbf{c_i}$ is equal to $\mathbf{a}$. This contradicts the fact that $\mathbf{a}$ is nonbranching.

If $p = 2k + 1$ with $k > 0$, then $\mathbf{a} = \mathbf{b_0} \vee \cdots \vee \mathbf{b_n}$ with each $\mathbf{b_i} \in \mathrm{CL}^{2k}(\mathbf{A})$. Taking n minimal, each $\mathbf{b_i}$ is in $\mathrm{Cu}(\mathbf{a})$. Since $k > 0$, each $\mathbf{b_i}$ can be expressed as a meet of elements in $\mathrm{CL}^{2k-1}(\mathbf{A})$. By (2.2), for each i, at least one of these elements, say $\mathbf{c_i}$, is less than or equal to $\mathbf{a}$. But then we have $\mathbf{a} = \mathbf{c_0} \vee \cdots \vee \mathbf{c_n}$ with each $\mathbf{c_i}$ in $\mathrm{CL}^{2k-1}(\mathbf{A})$, which implies that $\mathbf{a} \in \mathrm{CL}^{2k-1}(\mathbf{A})$, contradicting minimality of p. □

The following lemma, which provides a more handy sufficient condition for strong meet inaccessibility, gives the method we will use to construct s.m.i. degrees.

Lemma 2.3: *Let* $\mathbf{a}$ *be a c.e. degree such that* $\mathbf{a} > \mathbf{0}$ *and*

$$\forall \mathbf{b} \in \mathrm{Cu}(\mathbf{a}) \forall \mathbf{c_0}, \mathbf{c_1} > \mathbf{b}(\mathbf{c_0}, \mathbf{c_1} \not\leq \mathbf{a} \Rightarrow \exists \mathbf{d} \leq \mathbf{c_0}, \mathbf{c_1}(\mathbf{d} \not\leq \mathbf{a})). \tag{2.3}$$

Then $\mathbf{a}$ *is strongly meet inaccessible.*

Proof: By Lemma 2.2, it suffices to show that (2.2) holds. For a contradiction, suppose that $\mathbf{b} \in \mathrm{Cu}(\mathbf{a})$ and $\mathbf{b} = \mathbf{c_0} \wedge \cdots \wedge \mathbf{c_n}$ with each $\mathbf{c_i} \not\leq \mathbf{a}$. We assume that n is minimal and rule out all possibilities. Since $\mathbf{b} \leq \mathbf{a}$, $n = 0$ is impossible. Our assumption (2.3) rules out $n = 1$. If $n > 1$, then by (2.3), there is $\mathbf{d}$ with $\mathbf{d} \leq \mathbf{c_0}, \mathbf{c_1}$ and $\mathbf{d} \not\leq \mathbf{a}$. Letting $\mathbf{d_1} = \mathbf{d} \vee \mathbf{a}$, we have $\mathbf{d_1} \not\leq \mathbf{a}$ and $\mathbf{b} = \mathbf{d_1} \wedge \mathbf{c_2} \wedge \cdots \wedge \mathbf{c_n}$, contradicting minimality of n. □

Next, we look at some necessary (but not sufficient) conditions for strong meet inaccessibility. Obviously, any strongly meet inaccessible degree is nonbranching. In fact, strong meet inaccessibility is a proper strengthening of meet inaccessibility introduced in Ambos-Spies [3]. A c.e. degree $\mathbf{a}$ is *meet accessible* if there is a set $\mathbf{A}$ of c.e. degrees generating $\mathbf{a}$ such that $\mathbf{A}$ does not intersect the lower cone of $\mathbf{a}$; and $\mathbf{a}$ is *meet inaccessible* otherwise. As shown in [3], a c.e. degree $\mathbf{a}$ is meet inaccessible iff $\mathbf{a} > \mathbf{0}$ and $\mathbf{a} \notin \mathrm{CL}(\mathbf{R}(\not\leq \mathbf{a}))$. So, it is immediate by Lemma 2.2 that any s.m.i. degree is meet inaccessible. In [3], meet inaccessible degrees are used in order to show that, for any c.e. degree $\mathbf{a} > \mathbf{0}$ and any generator $\mathbf{G}$, $\mathbf{G}$ intersects the lower cone of $\mathbf{a}$. For this sake, it is observed that, for any degree $\mathbf{a}$ which bounds a meet inaccessible degree, the complement $\mathbf{R}(\not\leq \mathbf{a})$ of the lower cone of $\mathbf{a}$ does not generate $\mathbf{R}$, and it is shown that any nonzero degree bounds a meet inaccessible degree. In fact, Ambos-Spies [3] shows that any nonzero c.e. degree is the join of two meet inaccessible degrees whence the set $\mathbf{MI}$ of the meet inaccessible degrees is a join generator. Later on, the latter has been strengthened by Zhang [22] and Ding [8] who showed that $\mathbf{MI}$ is dense in $\mathbf{R}$. In contrast, it easily follows from results in the literature that the set of the strongly meet inaccessible degrees is not dense in $\mathbf{R}$. In fact, the following is true.

Lemma 2.4: *For any c.e. degree* $\mathbf{a} > \mathbf{0}$ *there are c.e. degrees* $\mathbf{b}$ *and* $\mathbf{c}$ *such that* $\mathbf{c} < \mathbf{b} \leq \mathbf{a}$ *and no c.e. degree in the closed interval* $[\mathbf{c}, \mathbf{b}]$ *is strongly meet inaccessible.*

Proof: Fix $\mathbf{a} > \mathbf{0}$. Since any nonzero c.e. degree bounds a contiguous degree (Ladner and Sasso [13]), w.l.o.g. we may assume that $\mathbf{a}$ is contiguous. Moreover, since any countable distributive lattice can be embedded into any nontrivial principal ideal of $\mathbf{R}$ by a map which preserves the greatest element (Ambos-Spies, Ding, Fejer [5]), we may fix pairwise incomparable c.e. degrees $\mathbf{a_0}, \mathbf{a_1}, \mathbf{a_2}, \mathbf{a_3}$ which generate (as atoms) the four-atom Boolean algebra in $\mathbf{R}(\leq \mathbf{a})$ with greatest element $\mathbf{a} = \mathbf{a_0} \vee \mathbf{a_1} \vee \mathbf{a_2} \vee \mathbf{a_3}$. So, since contiguous degrees have the covering property (below we come back to this in more detail when we discuss the possible ranks of generators), for any c.e. degree $\mathbf{d} \leq \mathbf{a}$ there are c.e. degrees $\mathbf{d_i} \leq \mathbf{a_i}$ $(i \leq 3)$ such that $\mathbf{d} = \mathbf{d_0} \vee \mathbf{d_1} \vee \mathbf{d_2} \vee \mathbf{d_3}$.

Now, let $\mathbf{b} = \mathbf{a_0} \vee \mathbf{a_1}$ and $\mathbf{c} = \mathbf{a_0}$. Then, given $\mathbf{d}$ such that $\mathbf{c} \leq \mathbf{d} \leq \mathbf{b}$, we have to show that $\mathbf{d}$ is not s.m.i. Note that $\mathbf{a_0} \leq \mathbf{d}$ and $\mathbf{a_0} = (\mathbf{a_0} \vee \mathbf{a_2}) \wedge (\mathbf{a_0} \vee \mathbf{a_3})$ where, by $\mathbf{d} \leq \mathbf{a_0} \vee \mathbf{a_1}$, it holds that $\mathbf{a_0} \vee \mathbf{a_2} \not\leq \mathbf{d}$ and $\mathbf{a_0} \vee \mathbf{a_3} \not\leq \mathbf{d}$. So, by Lemma 2.2, it suffices to show that $\mathbf{a_0}$ is $\mathbf{d}$-cuppable. Fix degrees $\mathbf{d_i} \leq \mathbf{a_i}$ $(i \leq 3)$ such that $\mathbf{d} = \mathbf{d_0} \vee \mathbf{d_1} \vee \mathbf{d_2} \vee \mathbf{d_3}$. Since $\mathbf{d} \leq \mathbf{a_0} \vee \mathbf{a_1}$, it follows that, for $j = 2, 3$, $\mathbf{d_j} \leq (\mathbf{a_0} \vee \mathbf{a_1}) \wedge \mathbf{a_j}$ hence $\mathbf{d_j} \leq \mathbf{a_1}$ by choice of the degrees $\mathbf{a_0}, \mathbf{a_1}, \mathbf{a_2}, \mathbf{a_3}$. So, for $\hat{\mathbf{d}} = \mathbf{d_1} \vee \mathbf{d_2} \vee \mathbf{d_3}$, $\hat{\mathbf{d}} \leq \mathbf{a_1}$ hence $\hat{\mathbf{d}} < \mathbf{d}$. Since, by $\mathbf{d_0} \leq \mathbf{a_0} \leq \mathbf{d}$, $\mathbf{d} = \mathbf{a_0} \vee \hat{\mathbf{d}}$, this implies the claim. □

The existence of generators which are not join generators, proven by Ambos-Spies, Lempp and Slaman [6], implies that the set of strongly meet inaccessible degrees is not a join generator. In order to show this, we use the following

Proposition 2.5: *Let* $\mathbf{A}$ *be a set of degrees generated by the set* $\mathbf{SMI}$ *of the strongly meet inaccessible degrees under join. Then, for any generator* $\mathbf{G}$, $\mathbf{G}$ *generates* $\mathbf{A}$ *under join.*

Proof: For a generator $\mathbf{G}$, $\mathbf{SMI} \subseteq \mathrm{CL}(\mathbf{G})$ hence $\mathbf{SMI} \subseteq \mathrm{CL}_j(\mathbf{G})$ by (2.1). Since $\mathbf{A} \subseteq \mathrm{CL}_j(\mathbf{SMI})$, it follows that $\mathbf{A} \subseteq \mathrm{CL}_j(\mathbf{G})$. □

Lemma 2.6: *The set* $\mathbf{SMI}$ *of the strongly meet inaccessible degrees is not a join generator.*

Proof: By Ambos-Spies, Lempp and Slaman [6], let $\mathbf{G}$ be a generator which does not generate $\mathbf{R} \setminus \{\mathbf{0}\}$ under join. Then, by Proposition 2.5, $\mathbf{SMI}$ does not generate $\mathbf{R} \setminus \{\mathbf{0}\}$ under join. □

Having given some basic properties of the s.m.i. degrees, we now turn to our main results. We first look at the question which degrees $\mathbf{a}$ are generated under join by *any* generator. By the existence of generators which are not join generators [6], there are degrees $\mathbf{a}$ which do not have this property. In fact, any join generator contains such a degree. So, for instance, there are a low degree $\mathbf{a}$ and a generator $\mathbf{G}$ such that $\mathbf{G}$ does not generate $\mathbf{a}$ under join. On the other hand, any strongly meet inaccessible $\mathbf{a}$ is generated under join by any generator. In fact, Proposition 2.5 shows that any degree which is generated under join by s.m.i. degrees has this property. A trivial example of a degree which is generated under join by any generator is the complete degree $\mathbf{0}'$ which, obviously is s.m.i. Our first technical main theorem gives a first nontrivial example of such a degree, namely a low degree.

Theorem 2.7: *There is a low c.e. degree* $\mathbf{a}$ *which is strongly meet inaccessible.*

The proof of Theorem 2.7 — which explains the basic construction of nontrivial s.m.i. degrees and which will be the basis for the other constructions of s.m.i. degrees — is given in Sec. 3.

Our main result in this direction, namely that any high degree is generated under join by any generator, is obtained along the same lines by proving the following.

Theorem 2.8: *Every high c.e. degree* $\mathbf{a}$ *is the join of two strongly meet inaccessible degrees.*

The proof of Theorem 2.8 is given in Sec. 4.

Corollary 2.9: *If* $\mathbf{G}$ *is a generator then* $\mathbf{G}$ *generates the set* $\mathbf{H}$ *of the high c.e. degrees under join.*

Proof: This is immediate by Theorem 2.8 and Proposition 2.5. □

In the remainder of this section we apply the strongly meet inaccessible degrees to some other questions on generators. We solve some open problems on the rank of (definable) generators raised in Ambos-Spies [3]. First, we recall the definition of the rank and join rank given in [3].

If a degree $\mathbf{a}$ is in the closure of a class $\mathbf{A}$ then

$$\mathrm{rank}_{\mathbf{A}}(\mathbf{a}) = \min\{n : \exists\, \mathbf{a}_1, \ldots, \mathbf{a}_n \in \mathbf{A}\ (\mathbf{a} \in \mathrm{CL}(\{\mathbf{a}_1, \ldots, \mathbf{a}_n\}))\},$$

and, similarly, if $\mathbf{a}$ is in the closure of $\mathbf{A}$ under join then

$$\text{j-rank}_{\mathbf{A}}(\mathbf{a}) = \min\{n : \exists\, \mathbf{a}_1, \ldots, \mathbf{a}_n \in \mathbf{A}\ (\mathbf{a} \in \mathrm{CL}_j(\{\mathbf{a}_1, \ldots, \mathbf{a}_n\}))\}.$$

Then the *rank* of a generator $\mathbf{G}$, rank($\mathbf{G}$) for short, is the least number $n \geq 1$ such that $\text{rank}_{\mathbf{G}}(\mathbf{a}) \leq n$ for all $\mathbf{a} \in \mathbf{R}$ if such an n exists, and rank($\mathbf{G}$) $= \omega$ otherwise. Similarly, the *join rank* (*j-rank* for short) of a join-generator $\mathbf{J}$, j-rank($\mathbf{J}$), is the least number $n \geq 1$ such that $\text{j-rank}_{\mathbf{J}}(\mathbf{a}) \leq n$ for all $\mathbf{a} \in \mathbf{R}$ if such an n exists, and j-rank($\mathbf{J}$) $= \omega$ otherwise. In [3] it is shown that, for any $n \geq 1$, there is a definable join generator $\mathbf{J}_n$ of join rank n and that there is a join-generator $\mathbf{J}_\omega$ of infinite join rank, and the question is raised whether these results on the join rank of join generators carry over to ranks and generators (see Problem 3 in [3]). Here, we answer this question affirmatively. Moreover, we show that there is a *definable* generator of infinite rank and that the required generators can be chosen so that they are join generators and that their ranks and join ranks agree:

Theorem 2.10: *For any $n \geq 1$ there is a definable join generator $\mathbf{G}_n$ of rank n and join rank n.*

Theorem 2.11: *There is a definable join generator $\mathbf{G}_\omega$ of infinite rank (hence infinite join rank).*

For the proofs of Theorems 2.10 and 2.11, we need the following existence result for s.m.i. degrees which is proved in Sec. 5, as well as a simple observation relating the rank of a s.m.i. degree to its join rank.

Theorem 2.12: *Let $\mathbf{b}_0, \ldots, \mathbf{b}_{n-1}$ ($n \geq 1$) be nonzero c.e. degrees. There are c.e. degrees $\mathbf{a}_0, \ldots, \mathbf{a}_{n-1}$ such that $\mathbf{0} < \mathbf{a}_i \leq \mathbf{b}_i$ for $i < n$ and $\mathbf{a}_0 \vee \cdots \vee \mathbf{a}_{n-1}$ is s.m.i.*

Proposition 2.13: *Assume that $\mathbf{a}$ is s.m.i. and $\mathbf{a} \in \text{CL}(\mathbf{A})$. Then $\mathbf{a} \in \text{CL}_j(\mathbf{A})$ and*

$$\text{rank}_{\mathbf{A}}(\mathbf{a}) = \text{j-rank}_{\mathbf{A}}(\mathbf{a}).$$

Proof: The first part of the claim is immediate by strong meet inaccessibility of $\mathbf{a}$. So, since, for $\mathbf{a} \in \text{CL}_j(\mathbf{A})$, $\text{rank}_{\mathbf{A}}(\mathbf{a}) \leq \text{j-rank}_{\mathbf{A}}(\mathbf{a})$, it suffices to show that $\text{j-rank}_{\mathbf{A}}(\mathbf{a}) \leq \text{rank}_{\mathbf{A}}(\mathbf{a})$. Assume that $\text{rank}_{\mathbf{A}}(\mathbf{a}) = n$ and fix $\mathbf{a}_0, \ldots, \mathbf{a}_{n-1} \in \mathbf{A}$ such that $\mathbf{a} \in \text{CL}(\mathbf{a}_0, \ldots, \mathbf{a}_{n-1})$. Then, by strong meet inaccessibility of $\mathbf{a}$, $\mathbf{a} \in \text{CL}_j(\mathbf{a}_0, \ldots, \mathbf{a}_{n-1})$ hence, by $\{\mathbf{a}_0, \ldots, \mathbf{a}_{n-1}\} \subseteq \mathbf{A}$, $\text{j-rank}_{\mathbf{A}}(\mathbf{a}) \leq n$. □

Just as the proof of the related results in [3], the proofs of Theorems 2.10 and 2.11 exploit some results on contiguous and nonbounding degrees in the literature, which we summarize first.

https://doi.org/10.1142/9789811278631_0001

Generators of the C.E. Degrees and Strongly Meet Inaccessible Degrees

Klaus Ambos-Spies

Institut für Informatik
University of Heidelberg
Im Neuenheimer Feld 205
D-69120 Heidelberg, Germany
ambos@math.uni-heidelberg.de

Decheng Ding

Department of Mathematics
Nanjing University
Nanjing, Jiangsu Province 210093, P. R. of China
dcding@nju.edu.cn

Peter Fejer

Department of Computer Science
University of Massachusetts Boston
Boston, MA 02125, USA
peter.fejer@umb.edu

We show that any set of computably enumerable (c.e.) degrees which generates the c.e. degrees under join and meet generates the high c.e. degrees under join. This result is obtained by showing that any high c.e. degree is the join of two strongly meet inaccessible (s.m.i.) degrees. Here a c.e. degree **a** is s.m.i. if any set of c.e. degrees which generates **a** under join and meet generates **a** under join. Moreover, by a further existence result for the s.m.i. degrees, we answer a question on the possible ranks of (definable) generators raised in Ambos-Spies [3] (where the rank of a generator is the supremum of the number of elements needed for generating any c.e. degree): there are definable generators of infinite rank and of rank n for any number $n \geq 1$.

Contents

1. Introduction

The partial ordering $(\mathbf{R}, \leq)$ of the computably enumerable degrees is an upper semilattice — i.e., for any c.e. degrees $\mathbf{a}$ and $\mathbf{b}$, their join $\mathbf{a} \vee \mathbf{b}$ exists — but not a lattice. In fact, as Lachlan [11] and Yates [21] have shown, for an incomparable pair of c.e. degrees, the meet may or may not exist. This asymmetry between joins and meets is further demonstrated by the fact that, by Sacks's splitting theorem [15], any nonzero c.e. degree is join-reducible (splits), i.e., the join of two lesser c.e. degrees, whereas, as shown by Lachlan [11] and Yates [21], an incomplete c.e. degree may be meet-reducible (branching), i.e., the meet of two greater c.e. degrees, or not. Presence and failure of meets are homogeneously distributed in the c.e. degrees: as Slaman [17] has shown, any nonempty interval of c.e. degrees contains an incomparable pair possessing a meet, hence a branching degree, whereas, by Ambos-Spies [2] and Fejer [10], respectively, any nonempty interval contains a pair without meet and a nonbranching degree. Despite this homogeneity, other results indicate that the lack of meets is more typical than their existence. So, as Ambos-Spies [2] and, independently, Harrington (unpublished) have shown, there is a degree $\mathbf{a} \neq \mathbf{0}, \mathbf{0}'$ such that $\mathbf{a}$ is not half of an incomparable pair with meet whereas any degree $\mathbf{a} \neq \mathbf{0}, \mathbf{0}'$ is half of an incomparable pair without meet.

The study of generators of the c.e. degrees sheds more light on the meet operator. Here a set $\mathbf{A}$ of c.e. degrees is a generator (of $\mathbf{R}$) if any c.e. degree is in the closure of $\mathbf{A}$ under join and meet, and $\mathbf{A}$ is a join (meet) generator if any nonzero (incomplete) c.e. degree is in the closure of $\mathbf{A}$ under join (meet). By Fejer's density theorem for the nonbranching degrees, any meet generator is dense whereas, by Sacks's splitting theorem, the set $\mathbf{L}$ of the low c.e. degrees is a join generator. Moreover, Ambos-Spies [3] has shown that any generator intersects any nonempty initial segment. So, for example, the set of the nonlow c.e. degrees is not a generator. These results together with the negative result on meets listed above, led Ambos-Spies to conjecture

https://doi.org/10.1142/9789811278631_fmatter

Preface

This volume of proceedings results from the IMS programmes "Aspects of Computation — in Celebration of the Research Work of Professor Rod Downey" (21 August to 15 September 2017) and "Automata Theory and Applications: Games, Learning and Structures" (20 to 24 September 2021); both programmes took place at the Institute for Mathematical Sciences at the National University of Singapore. The programme "Aspects of Computation" was dedicated to the research work of Rodney G. Downey, in celebration of his 60th birthday.

These aspects were footed in two research areas, first in Recursion Theory where Rodney G. Downey is a leading expert, and second in Parameterised Complexity, for which Rodney G. Downey is one of the founders. The Aspects of Computation also linked up with related areas, for example, Algorithmic Randomness which investigates the recursion-theoretic properties of randomly generated sequences. Furthermore, the programme also included the modelling of mathematical structures with recursion theory.

Set Theory turned out to be an important tool to explain why many open questions in other branches remained unresolved until today — they turned out to be independent of ZFC — even ZFC being the strongest general accepted system of axioms did not allow to decide these questions. In contrast to that, Recursion Theory produced insights of a different flavour, namely that many problems which one wanted to solve do not have any algorithm which can find the solution, that is, that these problems are undecidable. The experts in the field then explored further questions. Set Theory studies many additional properties and investigates how they relate with each other; Recursion Theory studied Turing degrees after Emil Post in 1944 introduced many fundamental reducibilities which are popular in the field until today. Post furthermore asked the important question whether there are recursively enumerable sets which are neither decidable nor Turing complete (the answer is "yes") and how the structure of Turing degrees looks like (this is heavily investigated but the research is still ongoing, as many details are still unknown). Reverse Mathematics carried ideas of Set Theory over to Recursion Theory and formulated additional axioms — many of them equivalent to certain mathematical theorems — and studied the interdependence of these axioms in certain fragments of the second order theory of the model of natural numbers (or a sufficiently close model with less induction axioms). Automata Theory is investigating the other end of computation, namely the computation with finite automata, and the intermediate level of languages in the Chomsky hierarchy (like context-free and context-sensitive languages). Finite automata are used to abstract the structure of computer programs by just considering the general skeleton of line numbers and investigating the possible runs in the programs provided that one restricts the variables to constant-sized ranges and reads the inputs symbol by symbol from some source. Researchers have in the course of years generalised the initial task of deciding membership in sets of words to the handling of mathematical structures and the analysis of games where two players alternately move a marker in a finite graph and where infinite runs are decided by looking at the set of infinitely often visited nodes. A special emphasis is on parity games where each node has a value and the winner depends on the question whether the largest such value is even or odd. Complexity Theory, and in particular Parameterised Complexity, sits in the middle of Automata Theory at the one end and Recursion Theory on the other hand. It differs from these two ends by the fact that important questions remained unresolved — not because of independence as in the case of set theory or because of non-existence of algorithms as in the

Recall that a c.e. degree is *contiguous* if it contains only one c.e. wtt-degree. Ladner and Sasso [13] have shown that any nonzero c.e. degree bounds a nonzero contiguous degree, and this has been extended by Ambos-Spies [1] as follows.

Let $\mathbf{c}_0, \ldots, \mathbf{c}_{n-1}$ be nonzero c.e. degrees ($n \geq 1$). There are c.e. degrees $\mathbf{b}_0, \ldots, \mathbf{b}_{n-1}$ such that, for $i < n$, $\mathbf{0} < \mathbf{b}_i \leq \mathbf{c}_i$, and, for $\emptyset \subset \alpha \subseteq \{0, \ldots, n-1\}$, $\bigvee_{i \in \alpha} \mathbf{b}_i$ is contiguous.

(2.4)

(This is a special case of Theorem 4.2 in [1].) As Stob [19] has shown, contiguous degrees have a local distributivity property. To make this more precise, call a c.e. degree $\mathbf{b}$ *n-covering* ($n \geq 1$) if, for any c.e. degrees $\mathbf{a}, \mathbf{b}_0, \ldots, \mathbf{b}_{n-1}$,

$$\mathbf{a} \leq \mathbf{b}_0 \vee \ldots \vee \mathbf{b}_{n-1} = \mathbf{b} \Rightarrow \exists\, \mathbf{a}_0 \leq \mathbf{b}_0, \ldots, \mathbf{a}_{n-1} \leq \mathbf{b}_{n-1}\ [\mathbf{a} = \mathbf{a}_0 \vee \ldots \vee \mathbf{a}_{n-1}] \tag{2.5}$$

holds. Then Stob [19] has shown that contiguous degrees are 2-covering and, as observed in [1] (see Corollary 1.5 there), this can be easily extended to show

Let $\mathbf{b}$ be contiguous. Then $\mathbf{b}$ is n-covering for all $n \geq 1$. (2.6)

The existence of nonzero *nonbounding* degrees, i.e., nonzero c.e. degrees which do not bound a minimal pair of c.e. degrees, has been established by Lachlan [12], and this result has been strengthened by Ambos-Spies and Soare [4], who have shown that there is an infinite sequence of nonbounding degrees which pairwise form minimal pairs.

There are c.e. degrees $\mathbf{c}_n > \mathbf{0}$ such that, for any $n \geq 0$, $\mathbf{c}_n$ is nonbounding, and, for any $n \neq n' \geq 0$, $\mathbf{c}_n \wedge \mathbf{c}_{n'} = \mathbf{0}$.

(2.7)

Now, following Ambos-Spies [3], say that a c.e. degree $\mathbf{b}$ is of *type* $n \geq 2$ *via* $\mathbf{b}_0, \ldots, \mathbf{b}_{n-1}$ if the degrees $\mathbf{b}_0, \ldots, \mathbf{b}_{n-1}$ are nonbounding and there is an embedding of the n-atom Boolean algebra $\mathcal{B}_n = (2^{\{1,\ldots,n\}}, \cup, \cap)$ into the initial segment $\mathbf{R}(\leq \mathbf{b})$ which preserves joins, meets and the least and greatest elements and which maps the atoms of $\mathcal{B}_n$ to the degrees $\mathbf{b}_0, \ldots, \mathbf{b}_{n-1}$; and say that $\mathbf{b}$ is of *type* n if $\mathbf{b}$ is of type n via some degrees $\mathbf{b}_0, \ldots, \mathbf{b}_{n-1}$.

Examples of degrees of type n are provided by the following fact.

Let $\mathbf{b}, \mathbf{b}_0, \ldots, \mathbf{b}_{n-1}$ ($n \geq 2$) be c.e. degrees such that $\mathbf{b} = \mathbf{b}_0 \vee \cdots \vee \mathbf{b}_{n-1}$, $\mathbf{b}_0, \ldots, \mathbf{b}_{n-1}$ are nonbounding and pairwise minimal pairs, and, for any α s.t. $\emptyset \subset \alpha \subseteq \{0, \ldots, n-1\}$, $\bigvee_{i \in \alpha} \mathbf{b}_i$ is contiguous. Then $\mathbf{b}$ is of type n via $\mathbf{b}_0, \ldots, \mathbf{b}_{n-1}$.

(2.8)

(Since the proof of (2.8), which exploits (2.6), is quite straightforward and implicit in the proof of Corollary 5.5 in [1], we omit it here.) Note that, by (2.4) and (2.7), for any $n \geq 2$ there are degrees $\mathbf{b}, \mathbf{b}_0, \ldots, \mathbf{b}_{n-1}$ as in the hypothesis of (2.8). So (by (2.6))

For $n \geq 2$ there is a contiguous (hence n-covering) degree $\mathbf{b}$ of type n

(2.9)

holds. Finally, we observe that

Let $\mathbf{b}$ be n-covering and of type n via $\mathbf{b}_0, \ldots, \mathbf{b}_{n-1}$ ($n \geq 2$). Then, for any c.e. degree $\mathbf{a} \leq \mathbf{b}$, $\mathbf{a}$ is nonbounding iff $\mathbf{a} \leq \mathbf{b}_i$ for some $i < n$

(2.10)

holds. (Namely, for a proof of the nontrivial implication, let $\mathbf{a} \leq \mathbf{b}$ be nonbounding. Then, since $\mathbf{b}$ is n-covering and $\mathbf{b} = \mathbf{b}_0 \vee \ldots \mathbf{b}_{n-1}$, there are c.e. degrees $\mathbf{a}_i \leq \mathbf{b}_i$ ($i < n$) such that $\mathbf{a} = \mathbf{a}_0 \vee \cdots \vee \mathbf{a}_{n-1}$. But, since the degrees $\mathbf{b}_i$ are pairwise minimal pairs and since $\mathbf{a}$ is nonbounding, this implies that, for some $i < n$, $\mathbf{a} = \mathbf{a}_i$ hence $\mathbf{a} \leq \mathbf{b}_i$.)

Some of the above results together with Theorem 2.12 imply the following sufficient condition for a generator to have rank $\geq n$ (for given $n \geq 2$).

Lemma 2.14: *Let $\mathbf{b}, \mathbf{b}_0, \ldots, \mathbf{b}_{n-1}$ be c.e. degrees such that $\mathbf{b}$ is of type n via $\mathbf{b}_0, \ldots, \mathbf{b}_{n-1}$ ($n \geq 2$), and let $\mathbf{G}$ be a generator such that*

$$\mathbf{G} \cap \mathbf{R}(\leq \mathbf{b}) \subseteq \mathbf{R}(\leq \mathbf{b}_0) \cup \cdots \cup \mathbf{R}(\leq \mathbf{b}_{n-1}) \tag{2.11}$$

holds. Then $\text{rank}(\mathbf{G}) \geq n$.

Proof: By Theorem 2.12, fix c.e. degrees $\mathbf{a}, \mathbf{a}_0, \ldots, \mathbf{a}_{n-1}$ such that $\mathbf{a}$ is s.m.i., $\mathbf{0} < \mathbf{a}_i \leq \mathbf{b}_i$ for $i < n$, and $\mathbf{a} = \mathbf{a}_0 \vee \cdots \vee \mathbf{a}_{n-1}$. Then, by Proposition 2.13, it suffices to show that $\text{j-rank}_{\mathbf{G}}(\mathbf{a}) \geq n$.

For a contradiction, assume that $\text{j-rank}_{\mathbf{G}}(\mathbf{a}) = p < n$ and fix $\mathbf{c}_0, \ldots, \mathbf{c}_{p-1} \in \mathbf{G}$ such that $\mathbf{a} = \mathbf{c}_0 \vee \cdots \vee \mathbf{c}_{p-1}$. Since $\mathbf{a} \leq \mathbf{b}$, it follows by (2.11) and by $p < n$ that there is a number $i < n$ such that

$$\mathbf{a} \leq \bigvee_{j \in \{0, \ldots, n-1\} \setminus \{i\}} \mathbf{b}_j.$$

On the other hand, by $\mathbf{b}$ having type n via $\mathbf{b}_0, \ldots, \mathbf{b}_{n-1}$,

$$\left(\bigvee_{j \in \{0,\ldots,n-1\} \setminus \{i\}} \mathbf{b}_j \right) \wedge \mathbf{b}_i = \mathbf{0}.$$

So $\mathbf{a} \wedge \mathbf{b}_i = \mathbf{0}$. But this is impossible since, by choice of $\mathbf{a}_i$, $\mathbf{a}_i \leq \mathbf{a}, \mathbf{b}_i$ and $\mathbf{a}_i \neq \mathbf{0}$. □

The preceding observations suffice to prove Theorem 2.10.

Proof: (of Theorem 2.10) Since $\mathbf{R}$ is a join generator of join rank 1 (hence rank 1), w.l.o.g. we may assume that $n \geq 2$. Let

$$\mathbf{G}_n = \{\mathbf{a} : \text{ there is no } \mathbf{b} \geq \mathbf{a} \text{ which is both } n\text{-covering and of type } n\}$$
$$\cup \{\mathbf{a} : \ \mathbf{a} \text{ is nonbounding}\}.$$

Obviously, $\mathbf{G}_n$ is definable, and in [3] it has been shown that $\mathbf{G}_n$ is a join generator of join rank n. So it suffices to show that $\text{rank}(\mathbf{G}_n) \geq n$.

By (2.9), fix c.e. degrees $\mathbf{b}, \mathbf{b}_0, \ldots, \mathbf{b}_{n-1}$ such that $\mathbf{b}$ is n-covering and $\mathbf{b}$ is of type n via $\mathbf{b}_0, \ldots, \mathbf{b}_{n-1}$. Then, by definition of $\mathbf{G}_n$,

$$\mathbf{G}_n \cap \mathbf{R}(\leq \mathbf{b}) = \{\mathbf{a} \leq \mathbf{b} : \ \mathbf{a} \text{ is nonbounding}\}.$$

So, by (2.10), (2.11) holds for $\mathbf{G} = \mathbf{G}_n$. Hence, $\text{rank}(\mathbf{G}_n) \geq n$ by Lemma 2.14. □

For the proof of Theorem 2.11, in addition to the above observations we need two definability results: First, Downey and Lempp [9] have shown that the class of the contiguous degrees is definable (namely, a c.e. degree $\mathbf{a}$ is contiguous if and only if $\mathbf{a}$ is 2-covering). Second, Nies [14] has shown that, for any definable class $\mathbf{D}$ of c.e. degrees, the ideal generated by $\mathbf{D}$ is definable too. By inspecting the proof, one immediately gets the corresponding result for the join closure.

Lemma 2.15: *(Nies) For any definable class* $\mathbf{D}$ *of c.e. degrees, the join closure* $\text{CL}_j(\mathbf{D})$ *of* $\mathbf{D}$ *is definable too.*

Proof: (of Theorem 2.11) Let

$$\mathbf{G}_\omega = \overline{\{\mathbf{a} : \exists\, \mathbf{b} \geq \mathbf{a}\, (\mathbf{b} \in \text{CL}_j(\mathbf{NB}) \cap \mathbf{CONT})\}} \cup \mathbf{NB}$$

where $\mathbf{NB}$ and $\mathbf{CONT}$ are the classes of the nonbounding degrees and contiguous degrees, respectively. Since, obviously, $\mathbf{NB}$ is definable hence

$\mathrm{CL}_j(\mathbf{NB})$ is definable by Lemma 2.15, it follows by definability of **CONT** (Downey and Lempp [9]) that $\mathbf{G}_\omega$ is definable.

To show that $\mathbf{G}_\omega$ is a join generator, given $\mathbf{a} \in \mathbf{R} \setminus \mathbf{G}_\omega$, it suffices to show that $\mathbf{a} \in \mathrm{CL}_j(\mathbf{G}_\omega)$. By choice of $\mathbf{a}$ and by definition of $\mathbf{G}_\omega$, $\mathbf{a} \notin \mathbf{NB}$ and there is a degree $\mathbf{b} \in \mathrm{CL}_j(\mathbf{NB}) \cap \mathbf{CONT}$ such that $\mathbf{a} \leq \mathbf{b}$. So, by downward closure of **NB** and by (2.6), there is a number $n \geq 2$ and nonzero nonbounding degrees $\mathbf{a}_i$ $(i < n)$ such that $\mathbf{a} = \mathbf{a}_0 \vee \cdots \vee \mathbf{a}_{n-1}$. So, in particular, $\mathbf{a} \in \mathrm{CL}_j(\mathbf{NB})$ which, by $\mathbf{NB} \subseteq \mathbf{G}_\omega$, implies the claim.

It remains to show that $\mathrm{rank}(\mathbf{G}_\omega) \geq n$ for all numbers $n \geq 2$. Fix $n \geq 2$, and, by (2.9), fix c.e. degrees $\mathbf{b}, \mathbf{b}_0, \ldots, \mathbf{b}_{n-1}$ such that $\mathbf{b}$ is contiguous and $\mathbf{b}$ is of type n via $\mathbf{b}_0, \ldots, \mathbf{b}_{n-1}$. Then, by Lemma 2.14, it suffices to show

$$\mathbf{G}_\omega \cap \mathbf{R}(\leq \mathbf{b}) \subseteq \mathbf{R}(\leq \mathbf{b}_0) \cup \cdots \cup \mathbf{R}(\leq \mathbf{b}_{n-1}).$$

So fix $\mathbf{a} \in \mathbf{G}_\omega \cap \mathbf{R}(\leq \mathbf{b})$. Since $\mathbf{a} \leq \mathbf{b}$ and $\mathbf{b} \in \mathrm{CL}_j(\mathbf{NB}) \cap \mathbf{CONT}$, it follows by definition of $\mathbf{G}_\omega$ that $\mathbf{a}$ is nonbounding. So, by (2.10), $\mathbf{a} \leq \mathbf{b}_i$ for some $i < n$, which completes the proof. □

3. Proof of Theorem 2.7

Before we give the proof of Theorem 2.7, we give some notation and facts used not only in this proof but also in the proofs of Theorems 2.8 and 2.12 given in the following two sections.

Let $(\varphi_e)_{e\geq 0}$, $(W_e)_{e\geq 0}$ and $(\{e\}^\sigma_e)_{e\geq 0}$ be computable numberings of the unary partial computable functions, c.e. sets, and Turing functionals, respectively, where W_e is the domain of φ_e, and fix uniformly computable enumerations $(\varphi_{e,s})_{s\geq 0}$, $(W_{e,s})_{s\geq 0}$ and $(\{e\}^\sigma_{e,s})_{s\geq 0}$ of φ_e, W_e and $\{e\}^\sigma$, respectively, such that

$$\varphi_{e,s}(x)\downarrow \text{ (i.e. } x \in W_{e,s}) \Rightarrow e, x < s, \tag{3.1}$$

$$\{e\}^X_s(x)\downarrow \Rightarrow e, x, u(X; e, x, s) < s \tag{3.2}$$

and

$$\{e\}^X_s(x)\downarrow \Rightarrow x < u(X; e, x, s) \tag{3.3}$$

hold, where $u(X; e, x)$ and $u(X; e, x, s)$ denote the use of $\{e\}^X_e(x)$ and $\{e\}^{X_s}_{e,s}(x)$, respectively.

For modelling some of the required Turing reductions, we use (standard) markers. The standard marker of W_e (with respect to the given enumeration $(W_{e,s})_{s\geq 0}$) is denoted by γ_e and defined by

$$\gamma_{e,0}(x) \quad = \langle x, 0\rangle$$

$$\gamma_{e,s+1}(x) = \begin{cases} \langle x, s+1\rangle & \text{if } W_{e,s+1} \upharpoonright x \neq W_{e,s} \upharpoonright x, \\ \gamma_{e,s}(x) & \text{otherwise.} \end{cases}$$

Note that $\gamma_e(x) = \lim_{s\to\omega} \gamma_{e,s}(x)$ exists, that $\gamma_e(x)$ is computable in W_e and that

$$\gamma_{e,s}(x) \leq \gamma_{e,s+1}(x) \leq \gamma_e(x) \tag{3.4}$$

for all numbers e, x, s. The following observation on the growth of the standard markers will be crucial.

Lemma 3.1: *(Standard Marker Lemma) Let A be any set and e be any index such that $W_e \not\leq_T A$. Then, for any infinite A-c.e. set D and any partial A-computable function ψ such that D is contained in the domain of ψ, there is an infinite A-computable subset $\hat{D}$ of D such that $\psi(x) < \gamma_e(x)$ for all $x \in \hat{D}$.*

This lemma is implicit already in Lachlan [11] and follows from the lemma to Theorem 2 of Chapter 18 of Shoenfield [16].

General format of the proof

We now turn to the proof of Theorem 2.7. By Lemma 2.3, it suffices to show that there is a low c.e. degree $\mathbf{a}$ such that $\mathbf{a} > \mathbf{0}$ and

$$\forall\, \mathbf{b_0}, \mathbf{b_1}\, [\mathbf{b_0} \vee \mathbf{b_1} = \mathbf{a} \Rightarrow \mathbf{a} \leq \mathbf{b_0} \quad \text{or} \quad \forall\, \mathbf{c_0}, \mathbf{c_1} \not\leq \mathbf{a}\, \exists\, \mathbf{d}\, (\mathbf{d} \leq \mathbf{b_1} \vee \mathbf{c_0}, \mathbf{b_1} \vee \mathbf{c_1}\ \&\ \mathbf{d} \not\leq \mathbf{a})] \tag{3.5}$$

holds.

By a finite injury priority argument, we construct a c.e. set A such that $\mathbf{a} = deg(A)$ has the required properties. Together with A, we enumerate auxiliary c.e. sets C_n $(n \geq 0)$. We let A_s and $C_{n,s}$ denote the finite parts of A and C_n, respectively, enumerated by the end of stage $s \geq 0$ of the construction. Moreover, we let $A_0 = C_{n,0} = \emptyset$, i.e., stage 0 of the construction is vacuous.

There is an infinite list of requirements R_e, $e \geq 0$, to be met. The requirements are of three different types, *noncomputability requirements* P_e, *lowness requirements* N_e, and *inaccessibility requirements* Q_e. The priority ordering is determined by $\mathrm{P}_e = \mathrm{R}_{3e}$, $\mathrm{N}_e = \mathrm{R}_{3e+1}$, and $\mathrm{Q}_e = \mathrm{R}_{3e+2}$.

At any stage $s+1$ of the construction, any requirement P_e, N_e and Q_e with $e \leq s$ may *require attention*, and the highest priority requirement which requires attention *becomes active* (or, in other words, *receives attention*). If R_e acts at stage $s+1$ then all lower priority requirements $\mathrm{R}_{e'}$, $e < e'$, are *initialized.* Moreover, if requirement R_e acts at stage $s+1$ then it may be declared to be *satisfied* at this stage. In this case, R_e remains satisfied unless it becomes initialized, i.e., if R_e is satisfied at stage s and R_e is not initialized at stage $s+1$ then R_e is satisfied at stage $s+1$ too. We say that R_e is *permanently satisfied at stage s* if R_e is satisfied at stage s and not initialized later, and R_e is *permanently satisfied* if there is a stage at which it is permanently satisfied.

We will argue that any requirement requires attention at most finitely often. So any requirement acts only finitely often and is initialized only finitely often.

Requirements

The requirements are as follows. In order to ensure that $\mathbf{a} > \mathbf{0}$ and $\mathbf{a}$ is low, we meet the *Noncomputability Requirements*

$$\mathrm{P}_e : A \neq \{e\}$$

and the *Lowness Requirements*

$$\mathrm{N}_e : \text{If } \{e\}_s^{A_s}(e)\downarrow \text{ for infinitely many } s \text{ then } \{e\}^A(e)\downarrow$$

respectively. In order to guarantee (3.5), for any numbers $n, m \geq 0$ and for the corresponding unique numbers $i_0, i_1, j_0, j_1, j_2, k_0, k_1$ and e such that $n = \langle i_0, i_1, j_0, j_1, j_2, k_0, k_1 \rangle$ and $e = \langle n, m \rangle$ we ensure

$$C_n \leq_{\mathrm{T}} W_{i_1} \oplus W_{k_0}, W_{i_1} \oplus W_{k_1} \tag{3.6}$$

and meet the *Inaccessibility Requirements*

$$\mathrm{Q}_e : \text{If } \quad (i)\ W_{i_0} = \{j_0\}^A \ \&\ W_{i_1} = \{j_1\}^A \ \&\ A = \{j_2\}^{W_{i_0} \oplus W_{i_1}},$$

$$(ii)\ W_{k_0}, W_{k_1} \not\leq_{\mathrm{T}} A, \text{ and}$$

$$(iii)\ A \not\leq_{\mathrm{T}} W_{i_0}$$

$$\text{then } (*)\ C_n \neq \{m\}^A.$$

To see that the Q_e requirements, together with (3.6) ensure that (3.5) holds, suppose that $\mathbf{b_0} \vee \mathbf{b_1} = \mathbf{a}$, $\mathbf{a} \not\leq \mathbf{b_0}$, and $\mathbf{c_0}, \mathbf{c_1} \not\leq \mathbf{a}$. To show that (3.5) is met, we need to show that there is $\mathbf{d}$ with $\mathbf{d} \leq \mathbf{b_1} \vee \mathbf{c_0}, \mathbf{b_1} \vee \mathbf{c_1}$ and $\mathbf{d} \not\leq \mathbf{a}$. We may choose $i_0, i_1, j_0, j_1, j_2, k_1, k_2$ such that $W_{i_0} \in \mathbf{b_0}$, $W_{i_1} \in \mathbf{b_1}$, $W_{i_0} = \{j_0\}^A$, $W_{i_1} = \{j_1\}^A$, $A = \{j_2\}^{W_{i_0} \oplus W_{i_1}}$, $W_{k_0} \in \mathbf{c_0}$, and $W_{k_1} \in \mathbf{c_1}$. Let $n = \langle i_0, i_1, j_0, j_1, j_2, k_0, k_1 \rangle$ and define $\mathbf{d} = deg(C_n)$. Then, (3.6) implies $\mathbf{d} \leq \mathbf{b_1} \vee \mathbf{c_0}, \mathbf{b_1} \vee \mathbf{c_1}$. Assuming that we meet Q_e for all $e = \langle n, m \rangle$, then, since for such e all the hypotheses $(i), (ii), (iii)$ of Q_e hold, we must have $C_n \neq \{m\}^A$ for all m, which ensures that $\mathbf{d} \not\leq \mathbf{a}$.

The strategies for satisfying condition (3.6) and meeting the requirements R_e are described next. There — just as in the following in general — when discussing condition (3.6) or an inaccessibility requirement Q_e, we always tacitly assume that $n = \langle i_0, i_1, j_0, j_1, j_2, k_0, k_1 \rangle$ and $e = \langle n, m \rangle$.

Strategy for satisfying condition (3.6)

Condition (3.6) is satisfied by marker permitting. To be more precise, the construction guarantees

$$x \in C_{n,s+1} \setminus C_{n,s} \Rightarrow \exists\, z < \min\{\gamma_{k_0,s}(x), \gamma_{k_1,s}(x)\}\ (z \in W_{i_1,s+1} \setminus W_{i_1,s}) \tag{3.7}$$

where γ_{k_p} is the *standard marker* of W_{k_p} $(p = 0, 1)$. (Note that, by (3.4), (3.7) implies that a number x cannot enter C_n after a stage s such that $W_{i_1,s} \upharpoonright \gamma_{k_p}(x) = W_{i_1} \upharpoonright \gamma_{k_p}(x)$ (for $p = 0, 1$). Since γ_{k_p} is W_{k_p}-computable it follows that (3.7) ensures that (3.6) holds.)

Strategy for meeting P_e

The strategy for meeting the noncomputability requirements is standard. It uses *followers*.

Requirement P_e *requires attention* at stage $s + 1$ if $e \leq s$, P_e is not satisfied at (the end of) stage s, and one of the following holds.

$$P_e \text{ has no follower at the end of stage } s. \tag{3.8}$$

$$P_e \text{ has follower } y \text{ at the end of stage } s,\ y \text{ is } \textit{realized} \text{ at stage } s, \text{ i.e., } \{e\}_s(y) = 0, \text{ and } y \notin A_s. \tag{3.9}$$

The corresponding action (if P_e receives attention at stage $s + 1$) is as follows. If (3.8) holds then appoint $y = s + 1$ as follower of P_e; and if (3.9) holds then put the follower y of P_e into A and declare P_e to be *satisfied*.

At any stage s there is at most one follower of P_e. We denote this follower by $y(e,s)$ and write $y(e,s)\uparrow$ if such a follower does not exist.

If P_e is *initialized* at stage $s+1$ then the follower of P_e at the end of stage s (if any) is cancelled (hence $y(e,s+1)\uparrow$). A follower which is never cancelled (i.e. which is appointed at a stage after which P_e is not initialized anymore) is called *permanent*. We will argue that there will be a (unique) permanent follower $y(e)$ of P_e and that this follower $y(e)$ witnesses that P_e is met, i.e., $A(y(e)) \neq \{e\}(y(e))$.

The noncomputability requirements are the only requirements which enumerate numbers into A. Since at any stage $s+1$ at most one requirement becomes active, it follows that

$$A_{s+1} \neq A_s \;\Rightarrow\; \exists!\, e \leq s\ (y(e,s)\downarrow \;\&\; A_{s+1} \setminus A_s = \{y(e,s)\}) \tag{3.10}$$

holds.

In the actual construction, the strategy for meeting the inaccessibility requirements $\mathrm{Q}_{e'}$ will require that the followers of the noncomputability requirements P_e have certain properties. For this sake, below we have to add another clause for P_e requiring attention.

Strategy for meeting N_e

The strategy for meeting the lowness requirements is standard too.

Requirement N_e *requires attention* at stage $s+1$ if $e \leq s$, N_e is not satisfied at (the end of) stage s, and the following holds.

$$\{e\}_s^{A_s}(e)\downarrow. \tag{3.11}$$

The corresponding action (if N_e receives attention at stage $s+1$) is as follows. Declare N_e to be *satisfied*.

Note that, by N_e becoming active, all followers of lower priority noncomputability requirements are cancelled. Since any follower appointed later will be greater than $s+1$ hence, by our convention (3.2), greater than $u(A_s; e, e, s)$, it follows by (3.10) that $\{e\}^A(e) = \{e\}_s^{A_s}(e)\downarrow$ and therefore that N_e is met unless N_e becomes initialized later. So if N_e is permanently satisfied then N_e is met.

Strategy for meeting Q_e

For the discussion of the strategy for meeting Q_e, fix e and the numbers n, m, i_0, i_1, j_0, j_1, j_2, k_0, k_1, such that $e = \langle n, m\rangle$ and $n = \langle i_0, i_1, j_0, j_1, j_2, k_0, k_1\rangle$.

There are two strategies for meeting Q_e. The first strategy attempts to refute hypothesis (i) of the requirement (see (3.12) and (3.13) below). This strategy is purely A-negative: it waits for a disagreement in one of the equations in (i) which can be preserved by putting a restraint on A. The second strategy attempts to meet the requirement by satisfying (*) (see (3.14) below). The latter strategy is A-negative too (in order to preserve a computation with oracle A) and, in addition, C_n-positive.

Requirement Q_e *requires attention* at stage $s+1$ if $e \leq s$, Q_e is not satisfied at (the end of) stage s, and one of the following holds.

$$\exists\, x \left(\begin{array}{l} A_s(x) \neq \{j_2\}_s^{W_{i_0,s} \oplus W_{i_1,s}}(x)\downarrow, \\ \text{and, for } u = u(W_{i_0,s} \oplus W_{i_1,s}; j_2, x, s) \text{ and for } p \leq 1, \\ W_{i_p,s} \restriction u = \{j_p\}_s^{A_s} \restriction u \end{array} \right) \tag{3.12}$$

$$\exists\, x\, \exists\, p \leq 1\, (W_{i_p,s}(x) = 1 \neq \{j_p\}_s^{A_s}(x)\downarrow) \tag{3.13}$$

$$\exists\, x \in \omega^{[e]} \tag{3.14}$$
$$\left(\{m\}_s^{A_s}(x) = 0 \;\&\; \exists\, z < \min\{\gamma_{k_0,s}(x), \gamma_{k_1,s}(x)\}(z \in W_{i_1,s+1} \setminus W_{i_1,s})\right).$$

(Note that, by our convention (3.2), the quantifier $\exists\, x$ in (3.12)–(3.14) can be replaced by the bounded quantifier $\exists\, x \leq s$ whence it is decidable whether these conditions hold or not.)

The corresponding action (if Q_e receives attention at stage $s+1$) is as follows. Declare Q_e to be *satisfied*. Moreover if (3.14) holds then, for the least x as there, put x into C_n. Note that the latter action is consistent with (3.7). (Also note that the set $\omega^{[e]}$ is reserved for Q_e, i.e., numbers from $\omega^{[e]}$ can be enumerated into C_n only by the Q_e-strategy.)

If Q_e is permanently satisfied then Q_e is met. This is immediate by the following claim since, if Q_e becomes active at stage $s+1$ and is not initialized later, then $A \restriction s+1 = A_s \restriction s+1$.

Claim 3.2: Assume that Q_e requires and receives attention at stage $s+1$ and that $A \upharpoonright s+1 = A_s \upharpoonright s+1$. Then either clause (i) in Q_e fails or clause $(*)$ in Q_e holds. So, in particular, Q_e is met.

Proof: If (3.12) holds, fix the least x as there and the corresponding u. Then, by (3.2), $x < s$ and $u(A_s; j_p, y, s) < s$ for all $y < u$ and $p \leq 1$. So, by $A \upharpoonright s+1 = A_s \upharpoonright s+1$, $A(x) = A_s(x)$ and, for $p \leq 1$, $\{j_p\}^A \upharpoonright u = \{j_p\}_s^{A_s} \upharpoonright u$. It follows, by choice of x and u, that $A(x) \neq \{j_2\}^{W_{i_0} \oplus W_{i_1}}(x)$ or, for some $p \leq 1$, $W_{i_p} \upharpoonright u \neq \{j_p\}^A \upharpoonright u$. In either case, this implies that hypothesis (i) of Q_e fails, hence Q_e is met.

If (3.13) or (3.14) holds then the argument that Q_e is met is similar. If (3.13) holds then, for any x and $p \leq 1$ as there, $1 = W_{i_p}(x) \neq \{j_p\}^A(x)$. So, again, hypothesis (i) of Q_e fails. Finally, if (3.14) holds then, for the least x as there, $C_n(x) = 1 \neq 0 = \{m\}^A(x)$. So the conclusion $(*)$ of Q_e holds. □

The above finitary strategy attempts to refute hypothesis (i) of Q_e by diagonalization or to guarantee the conclusion $(*)$ by diagonalization. It does not suffice to meet Q_e, and an additional infinitary strategy is needed which is based on the assumption that hypotheses (i) and (ii) of Q_e hold. Very roughly speaking, the idea of this strategy is as follows. By (i), $A = \{j_2\}^{W_{i_0} \oplus W_{i_1}}$. So a number y which enters A sufficiently late triggers a change of W_{i_0} or W_{i_1}. Now any follower y of a lower priority coding requirement which does not enter A sufficiently promptly is chosen so that a change of W_{i_1} gives the necessary permission to make an attack on Q_e via clause (3.14). So if the finitary strategy fails then we may argue that any "critical" number which enters A is "permitted" by W_{i_0}, hence $A \leq_T W_{i_0}$ (so hypothesis (iii) of Q_e fails and the requirement is met).

This infinitary strategy interferes with the strategies of the noncomputability requirements but the impact on a fixed requirement is finitary (hence compatible with the finite injury framework). To implement the strategy, we use the following e-eligibility notion.

A number y is *e-eligible* ($e = \langle n, m \rangle$, $n = \langle i_0, i_1, j_0, j_1, j_2, k_0, k_1 \rangle$) *via* numbers x, u, v and r at stage $s+1$ if the following hold.

$$x \in \omega^{[e]} \tag{3.15}$$

$$\{m\}_s^{A_s}(x) = 0 \tag{3.16}$$

$$u = u(A_s; m, x, s) \tag{3.17}$$

$$y \geq u \tag{3.18}$$

$$y \notin A_s \tag{3.19}$$

$$A_s \upharpoonright y+1 = \{j_2\}_s^{W_{i_0,s} \oplus W_{i_1,s}} \upharpoonright y+1 \tag{3.20}$$

$$v = \max\{u(W_{i_0,s} \oplus W_{i_1,s}; j_2, z, s) : z \leq y\} \tag{3.21}$$

$$\forall\, p \leq 1\ (W_{i_p,s} \upharpoonright v+1 = \{j_p\}_s^{A_s} \upharpoonright v+1) \tag{3.22}$$

$$r = \max\{u(A_s; j_p, z, s) : p \leq 1 \ \&\ z \leq v\} \tag{3.23}$$

and

$$v < \gamma_{k_0,s}(x), \gamma_{k_1,s}(x). \tag{3.24}$$

Call y *e-preeligible via* x, u, v and r at stage $s+1$ if (3.15)–(3.23) (but not necessarily (3.24)) hold. Call y *e-(pre)eligible* at stage $s+1$ if there are numbers x, u, v, and r such that y is *e-(pre)eligible* via x, u, v, and r at stage $s+1$. Note that, if y is e-(pre)eligible via x, u, v and r at stage $s+1$, then, by our conventions (3.2) and (3.3),

$$x < u \leq y < v < r < s. \tag{3.25}$$

(Also note that if y is e-(pre)eligible via x, u, v and r at stage $s+1$ then the numbers u, v and r are determined by y and x. So, in particular, since $x < y$, it is decidable whether y is e-(pre)eligible at stage $s+1$.)

The following claim explains how e-eligibility is used in meeting requirement Q_e.

Claim 3.3: Assume that Q_e requires attention only finitely often and is satisfied only finitely often. If there is an infinite computable set S of stages such that, for any $s \in S$,

$$\forall\, y, s'\ ([y \leq s \leq s' \ \&\ y \in A_{s'+1} \setminus A_{s'}] \Rightarrow y \text{ is e-eligible at stage } s'+1) \tag{3.26}$$

holds, then requirement Q_e is met.

Proof: Fix an infinite computable set S such that (3.26) holds for all $s \in S$. In order to show that Q_e is met, w.l.o.g. we may assume that hypothesis (i) in Q_e holds, and it suffices to show that

$$A \leq_T W_{i_0} \tag{3.27}$$

holds (hence hypothesis (iii) of Q_e fails).

For a proof of (3.27), it suffices to show that there is a function $s(z)$ ($z \geq 0$) such that

$$A \upharpoonright z+1 = A_{s(z)} \upharpoonright z+1 \tag{3.28}$$

and such that $s(z)$ can be uniformly computed from $W_{i_0} \oplus (A \upharpoonright z)$. Obviously, this gives an inductive procedure for uniformly computing $A(z)$ from W_{i_0} for any $z \geq 0$.

The definition of $s(z)$ is as follows. Fix $s_0 \geq e$ such that Q_e neither requires attention nor is satisfied after stage s_0. Note that, by (i), for any z there are numbers v_z and r_z such that, for all sufficiently large stages s and for $v(z,s) = v_z$ and $r(z,s) = r_z$, the following hold.

$$s > \max(z+1, s_0) \tag{3.29}$$

$$A \upharpoonright z = A_s \upharpoonright z \tag{3.30}$$

$$A_s \upharpoonright z+1 = \{j_2\}_s^{W_{i_0,s} \oplus W_{i_1,s}} \upharpoonright z+1 \tag{3.31}$$

$$v(z,s) = \max\{u(W_{i_0,s} \oplus W_{i_1,s}; j_2, x, s) : x \leq z\} \tag{3.32}$$

$$W_{i_0} \upharpoonright v(z,s) = W_{i_0,s} \upharpoonright v(z,s) \tag{3.33}$$

$$\forall\, p \leq 1\ (W_{i_p,s} \upharpoonright v(z,s) = \{j_p\}_s^{A_s} \upharpoonright v(z,s)) \tag{3.34}$$

$$r(z,s) = \max(\{u(A_s; j_p, x, s) : p \leq 1 \;\&\; x < v(z,s)\} \cup \{z+1\}). \tag{3.35}$$

(Note that, for given z and s, if there are numbers $v(z,s)$ and $r(z,s)$ such that (3.29)–(3.35) hold then these numbers are uniquely determined by z and s via (3.32) and (3.35).) So there are infinitely many stages s as above with $s \in S$. Moreover, using $W_{i_0} \oplus (A \upharpoonright z)$, as an oracle, it is decidable whether, for a stage s, there are numbers $v(z,s)$ and $r(z,s)$ such that (3.29)–(3.35) hold. So, if we let $s(z)$ be the least stage s such that $s \in S$ (hence (3.26) holds) and, for some $v(z,s)$ and $r(z,s)$, (3.29)–(3.35) hold then $s(z)$ can be uniformly computed from $W_{i_0} \oplus (A \upharpoonright z)$.

It remains to show that (3.28) holds. Fix z, let $s = s(z)$ and, for a contradiction, assume that $A \upharpoonright z+1 \neq A_s \upharpoonright z+1$. Then, by (3.30),

$$z \in A \setminus A_s. \tag{3.36}$$

Since, by (i), $A(z) = \{j_2\}^{W_{i_0} \oplus W_{i_1}}(z)$, it follows by (3.31), (3.32) and (3.33) that

$$W_{i_1} \upharpoonright v(z,s) \neq W_{i_1,s} \upharpoonright v(z,s).$$

So we may fix the unique $t \geq s$ such that

$$W_{i_1,t+1} \upharpoonright v(z,s) \neq W_{i_1,t} \upharpoonright v(z,s) = W_{i_1,s} \upharpoonright v(z,s). \tag{3.37}$$

On the other hand, since (by (3.35)) $z < r(z,s)$, it follows from (3.36) that there is a stage $s' \geq s$ such that

$$A_{s'+1} \upharpoonright r(z,s) \neq A_{s'} \upharpoonright r(z,s). \tag{3.38}$$

Note that $s' \leq t$ for the least such stage s' since otherwise, by (3.37), by (3.34) and by (3.35),

$$\begin{aligned} W_{i_1,t+1} \upharpoonright v(z,s) &\neq W_{i_1,s} \upharpoonright v(z,s) \\ &= \{j_1\}_s^{A_s} \upharpoonright v(z,s) \\ &= \{j_1\}_{t+1}^{A_{t+1}} \upharpoonright v(z,s). \end{aligned}$$

But this implies that Q_e requires attention via clause (3.13) at stage $t+2$ which (by (3.29)) contradicts the choice of s_0.

So for the remainder of the argument, we may fix s' maximal such that $s \leq s' \leq t$ and (3.38) holds. We will show that Q_e requires attention via clause (3.14) at stage $t+1$ contrary to choice of s_0.

Fix the (unique) number y which enters A at stage $s'+1$. Then

$$z \leq y < r(z,s) < s \tag{3.39}$$

(where the first inequality holds by (3.30) and the last inequality holds by choice of $r(z,s)$ and s and our convention (3.2)) and, by maximality of s',

$$(A_{t+1} \upharpoonright r(z,s)) \setminus (A_{s'} \upharpoonright r(z,s)) = \{y\}. \tag{3.40}$$

Moreover, by $s \in S$ and (3.39), y is e-eligible at stage $s'+1$, say via x, u, v and r. So $x \in \omega^{[e]}$, $\{m\}_{s'}^{A_{s'}}(x) = 0$ and $x < u = u(A_{s'}; m, x, s') \leq y$ which, by (3.40), implies $\{m\}_t^{A_t}(x) = 0$. So, in order to argue that Q_e requires attention via clause (3.14) at stage $t+1$, it suffices to show

$$W_{i_1,t+1} \upharpoonright \min\{\gamma_{k_0,t}(x), \gamma_{k_1,t}(x)\} \neq W_{i_1,t} \upharpoonright \min\{\gamma_{k_0,t}(x), \gamma_{k_1,t}(x)\}. \tag{3.41}$$

For a proof of (3.41), first note that, by e-eligibility of y via x, u, v and r at stage $s'+1$,

$$v = \max\{u(W_{i_0,s'} \oplus W_{i_1,s'}; j_2, x', s') : x' \leq y\} < \gamma_{k_0,s'}(x), \gamma_{k_1,s'}(x).$$

So, since the standard markers are nondecreasing in the stages and since $s' \leq t$, by (the inequality in) (3.37), it suffices to show that $v(z, s) \leq v$. But this follows from $z \leq y$ and $s \leq s' \leq t$, since, by (the equality in) (3.37), by (3.33), and by definition of $v(z, s)$,

$$\begin{aligned} v(z,s) &= \max\{u(W_{i_0,s} \oplus W_{i_1,s}; j_2, x', s) : x' \leq z\} \\ &= \max\{u(W_{i_0,s'} \oplus W_{i_1,s'}; j_2, x', s') : x' \leq z\}. \end{aligned}$$

This completes the proof of (3.41) and the proof of Claim 3.3. □

The following two claims provide some more facts on e-eligibility to be used in the construction: The first claim shows when we can find e-eligible numbers while the second claim tells us how e-eligibility of a number can be preserved.

Claim 3.4: Assume that clauses (i) and (ii) in Q_e hold, and that $\{m\}^A$ is total and, for almost all numbers $x \in \omega^{[e]}$, $\{m\}^A(x) = 0$. Then, for any infinite A-computable subset E of $\overline{A}$ there is an infinite subset F of E such that, for any $y \in F$, there are numbers x_y, u_y, v_y and r_y such that y is e-eligible via x_y, u_y, v_y and r_y at almost all stages.

Proof: Fix x_0 minimal such that $\{m\}^A(x) = 0$ for all $x \geq x_0$ in $\omega^{[e]}$. By (i), by totality of $\{m\}^A$, and by choice of E, for any number $x \geq 0$ there are unique numbers $u(x)$, $y(x)$, $v(x)$, $r(x)$ and $s(x)$ such that

$$\begin{aligned} u(x) &= u(A; m, x), \\ y(x) &= \mu y \geq u(x)\ (y \in E), \\ v(x) &= \max\{u(W_{i_0} \oplus W_{i_1}; j_2, z) : z \leq y(x)\}, \\ r(x) &= \max\{u(A; j_p, z) : p \leq 1\ \&\ z \leq v(x)\}, \end{aligned}$$

and $s(x)$ is the least stage $s > r(x)$ such that $\{m\}_s^{A_s}(x)\downarrow$, $u(A_s; m, x, s) = u(x)$, $A_s(z) = \{j_2\}^{W_{i_0,s} \oplus W_{i_1,s}}(z)$ for all $z \leq y(x)$, $v(x) = \max\{u(W_{i_0,s} \oplus W_{i_1,s}; j_2, z, s) : z \leq y(x)\}$, $W_{i_p,s}(z) = \{j_p\}_s^{A_s}(z)$ for all $p \leq 1$ and $z \leq v(x)$, $r(x) = \max\{u(A_s; j_p, z, s) : p \leq 1\ \&\ z \leq v(x)\}$, and $A \upharpoonright r(x) = A_s \upharpoonright r(x)$.

Obviously, $u(x)$, $y(x)$, $v(x)$, $r(x)$ can be computed by A, $x < u(x) \leq y(x) < v(x) < r(x) < s(x)$, $y(x) \in E$, and, for any $x \geq x_0$ in $\omega^{[e]}$, $y(x)$ is e-preeligible via x, $u(x)$, $v(x)$ and $r(x)$ at all stages $s + 1 > s(x)$. So it suffices to show that there are infinitely many numbers $x \in \omega^{[e]}$ such that $v(x) < \gamma_{k_0}(x)$ and $\gamma_{k_1}(x)$. But, by A-computability of $v(x)$, this is immediate by the Standard Marker Lemma. (Namely, a first application yields an infinite A-computable subset $\hat{D}$ of $\{x : x \in \omega^{[e]}\ \&\ x \geq x_0\}$

such that $v(x) < \gamma_{k_0}(x)$, and a second application yields an infinite A-computable subset $\hat{\hat{D}}$ of $\hat{D}$ such that $v(x) < \gamma_{k_1}(x)$.) □

Claim 3.5: Let e, y, x, u, v, r, s and s' be given such that $s \leq s'$, $e \leq s'$, y is e-(pre)eligible via x, u, v and r at stage $s+1$, $A_{s'} \upharpoonright r = A_s \upharpoonright r$, and Q_e is neither satisfied at stage s' nor requires attention at stage $s'+1$. Then y is e-(pre)eligible via x, u, v and r at stage $s'+1$.

Proof: Note that if (3.24) holds at stage s then it holds at all stages $s' \geq s$ since the standard markers are nondecreasing in the stages. So it suffices to consider the case of preeligibility.

By assumption, (3.15)–(3.23) and (3.25) hold, and it suffices to show that (3.16), (3.17), (3.19), (3.20), (3.21), (3.22) and (3.23) remain true if we replace s by s'. Since, by (3.25), $x < u \leq y < v < r < s$ and since, by assumption, $A_{s'} \upharpoonright r = A_s \upharpoonright r$, this is immediate for (3.16), (3.17), (3.19), and (3.23) where the last equation holds since, for $z \leq v$, the use of the computations $\{j_p\}_s^{A_s}(z) \downarrow$ is bounded by r hence these computations are preserved through the end of stage s'. The latter implies that (by $A_{s'} \upharpoonright y+1 = A_s \upharpoonright y+1$) the remaining conditions (3.20), (3.21) and (3.22) are preserved too, unless, for some $p \leq 1$, $W_{i_p,s'} \upharpoonright v+1 \neq W_{i_p,s} \upharpoonright v+1$. If the latter happens, however, then there is a number x' such that

$$1 = W_{i_p,s'}(x') \neq W_{i_p,s}(x') = \{j_p\}_s^{A_s}(x') \downarrow = \{j_p\}_{s'}^{A_{s'}}(x') \downarrow .$$

So, since, by assumption, Q_e is not satisfied at stage s' and since $e \leq s'$, it follows that Q_e requires attention at stage $s'+1$ via clause (3.13) contrary to assumption. So the clauses (3.20)–(3.22) hold at stage s' too, which completes the proof of the claim. □

Claim 3.3 leads to the following strategy for meeting Q_e: if the follower $y(e'', s)$ of a noncomputability requirement $\mathrm{P}_{e''}$ of lower priority than Q_e has not yet been enumerated into A, $y(e'', s)$ is not e-eligible, and there is a number $y > y(e'', s)$ such that $y \notin A_s$ and y is e-eligible at stage $s+1$ then we replace $y(e'', s)$ by (the least such) y. Then, by Claim 3.4 and Claim 3.5 we will argue that if Q_e is not satisfied for some trivial reasons then (almost all) permanent followers which do not enter A will become e-eligible and from this we conclude that the premise of Claim 3.3 will be satisfied.

To synchronize the attempts to make followers e-eligible for the higher priority inaccessibility requirements Q_e, we introduce e-states.

The *e-state* of a number y at the end of stage s is the (binary) string $\sigma(e,y,s)$ of length e defined by

$$\forall\, e' < e \Big(\sigma(e,y,s)(e') = 0 \Leftrightarrow$$
$$y \text{ is } e'\text{-eligible at stage } s+1 \text{ or } \mathrm{Q}_{e'} \text{ is satisfied at the end of stage } s \Big).$$

e-states are ordered by the standard lexicographical ordering, i.e., $\sigma < \sigma'$ if there is a number $e' < e$ such that $\sigma(e') \neq \sigma'(e')$ and, for the least such e', $\sigma(e') < \sigma'(e')$; and $\sigma \leq \sigma'$ if $\sigma < \sigma'$ or $\sigma = \sigma'$. In addition, we use the partial ordering $\preceq$ on $\{0,1\}^e$, where $\sigma \preceq \sigma'$ if $\sigma(e') \leq \sigma'(e')$ for all $e' < e$; and we write $\sigma \prec \sigma'$ if $\sigma \preceq \sigma'$ and $\sigma \neq \sigma'$. Note that $\sigma \prec \sigma'$ implies $\sigma < \sigma'$, but the converse in general fails. In order to distinguish between $<$ ($\leq$) and $\prec$ ($\preceq$), we say that σ is *(weakly) less* than σ' if $\sigma < \sigma'$ ($\sigma \leq \sigma'$) holds, and we say that σ *(weakly) precedes* σ' if $\sigma \prec \sigma'$ ($\sigma \preceq \sigma'$) holds. If we talk about the *least* e-state with a certain property, we always mean the least e-state with this property with respect to the lexicographical ordering $<$.

Strategy for meeting P_e revisited

In the spirit of the above comments, in order to make the action of the noncomputability requirements compatible with the strategy for meeting the inaccessibility requirements, we ensure that the e-state of the followers of P_e becomes as small as possible. Roughly speaking, this is ensured by replacing a follower y of P_e which is not yet in A and not yet realized by a larger number y' which is not yet in A and which has smaller e-state than y, whenever this is possible. To be more precise, if $y = y(e,s)$ is the follower of P_e at the end of stage s then a *declared* e-state denoted by $\hat{\sigma}(e,s)$ is attached to P_e. We ensure that the true e-state of y, $\sigma(e,y,s)$ weakly precedes the declared e-state $\hat{\sigma}(e,s)$ as long as y is not enumerated into A and not cancelled (if y is cancelled then the declared e-state is cancelled too), and if there is a number $y' > y$ which is not yet in A and which has an e-state less than the declared e-state of P_e then we replace the follower y by such a number y' and let the declared e-state be the true e-state of this number at the current stage.

The above is implemented by adding to (3.8) and (3.9) the following third clause causing P_e to require attention at stage $s+1$.

P_e has follower y at the end of stage s, y is not realized at stage s, and there is a number y' such that $y < y' \leq s$, $y' \notin A_s$, and $\sigma(e,y',s) < \hat{\sigma}(e,s)$. (3.42)

If this clause applies and P_e receives attention then the follower y is replaced by the least such y' and the declared e-state of P_e at stage $s+1$ is defined to be $\sigma(e, y', s)$. Moreover, if P_e receives attention via clause (3.8) at stage $s+1$ and a new follower is appointed then the maximum e-state 1^e becomes the declared e-state of P_e at stage $s+1$, i.e., $\hat{\sigma}(e, s+1) = 1^e$.

(Note that, by clause (3.42), a P_e-follower y may be replaced at stage $s+1$ by a number $y' \leq s$. So, for given e, y and s, it will not be decidable anymore whether y will be a P_e follower at a stage $\geq s$.)

The construction

The construction is immediate by the above. For clarity of presentation we explicitly state it here using the previously given definition of requiring attention. Parameters depending on the stage which are attached to the requirements persist unless explicitly stated otherwise. If any requirement R_e becomes initialized at stage $s+1$ and it was satisfied at the end of stage s then it becomes unsatisfied. Moreover, if a noncomputability requirement P_e becomes initialized at stage $s+1$ and it has a follower $y = y(e, s)$ at the end of stage s then this follower is cancelled as well as the declared e-state (i.e., $y(e, s+1) \uparrow$ and $\hat{\sigma}(e, s+1) \uparrow$).

At stage 0, no requirement is satisfied and no noncomputability requirement has a follower (i.e., $y(e, 0) \uparrow$ and $\hat{\sigma}(e, 0) \uparrow$ for $e \geq 0$), and $A_0 = C_{n,0} = \emptyset$ for all $n \geq 0$. Stage $s+1 > 0$ is as follows.

> *Stage* $s+1$. Fix k minimal such that R_k requires attention at stage $s+1$, fix e and $i \leq 2$ such that $k = 3e+i$, and distinguish the following three cases according to the type of requirement R_k.
>
> *Case 1: $k = 3e$, i.e., $\mathrm{R}_k = \mathrm{P}_e$.* If (3.8) holds then appoint $y = s+1$ as follower of P_e (i.e., let $y(e, s+1) = s+1$) and let $\hat{\sigma}(e, s+1) = 1^e$; if (3.9) holds then put the follower $y = y(e, s)$ of P_e into A and declare P_e to be *satisfied*; and if (3.42) holds then replace the follower $y = y(e, s)$ of P_e by the least number y' as there, i.e., let $y(e, s+1) = y'$ and let $\hat{\sigma}(e, s+1) = \sigma(e, y', s)$.
>
> *Case 2: $k = 3e+1$, i.e., $\mathrm{R}_k = \mathrm{N}_e$.* Declare N_e to be *satisfied.*
>
> *Case 3: $k = 3e+2$, i.e., $\mathrm{R}_k = \mathrm{Q}_e$.* Then declare Q_e to be *satisfied.* Moreover if (3.14) holds then, for the least x as there, put x into C_n (where $e = \langle n, m \rangle$ for some m).

In any case, declare that requirement R_k *receives attention* and is *active* at stage $s+1$, and initialize all requirements $R_{k'}$ with $k' > k$.

Verification

Obviously the construction satisfies (3.7). Hence (3.6) holds. So it suffices to show that all requirements are met.

We start with some observations on the followers and the declared states of the noncomputability requirements.

Obviously, for any e, P_e has no follower at the end of stage 0 and, for any $s \geq 0$, there is at most one follower of P_e at the end of stage $s+1$. So we may let $y(e,s)$ denote the follower of P_e at the end of stage s and let $y(e,s)\uparrow$ indicate that there is no such follower. Since at any stage at most one requirement acts and only the noncomputability requirements enumerate numbers into A, (3.10) is immediate. Moreover, by a straightforward induction on stage s, the indices e for which $y(e,s)$ is defined is an initial segment of ω of size $\leq s$, the followers (when defined) are nondecreasing in the stage and strictly increasing in the index, and a follower is bounded by its index from below and by the stage from above, i.e.,

$$\exists\, e \leq s\ (y(e',s)\downarrow \Leftrightarrow\ e' < e) \tag{3.43}$$

$$s < s'\ \&\ y(e,s)\downarrow\ \&\ y(e,s')\downarrow \Rightarrow\ y(e,s) \leq y(e,s') \tag{3.44}$$

$$e < e'\ \&\ y(e',s)\downarrow \Rightarrow\ y(e,s) < y(e',s) \tag{3.45}$$

$$y(e,s)\downarrow \Rightarrow\ e < y(e,s) \leq s \tag{3.46}$$

hold. Also note that if P_e has no follower at some stage then any of its later followers is greater than this stage, i.e.,

$$s < s'\ \&\ y(e,s)\uparrow\ \&\ y(e,s')\downarrow \Rightarrow\ y(e,s') > s, \tag{3.47}$$

and if P_e becomes active then the lower priority followers are cancelled hence

$$e < e'\ \&\ y(e,s) \neq y(e,s+1) \Rightarrow\ y(e',s+1)\uparrow. \tag{3.48}$$

Moreover, if the follower y is assigned to P_e at stage $s+1$ and P_e neither is initialized nor becomes active by stage $s' > s$ then, by cancellation of the lower priority followers at stage $s+1$, no number $\leq s+1$ enters A after stage s and prior to stage $s'+1$ whence

$$s < s' \;\&\; y(e,s) \neq y(e,s+1)\downarrow = y(e,s') \notin A_{s'} \;\Rightarrow\; A_{s'} \restriction s+2 = A_s \restriction s+2. \tag{3.49}$$

The declared state of a noncomputability requirement is defined at the end of a stage if and only if the requirement has a follower at the end of this stage, and the declared state changes if and only if the follower changes. Moreover, if a follower is appointed then the declared state assumes its maximum possible value (hence is weakly preceded by the e-state of the new follower at the previous stage), and if a follower is replaced with another number then the declared state is decreased and coincides with the true e-state of the new follower at the previous stage.

$$\hat{\sigma}(e,s)\downarrow \Leftrightarrow y(e,s)\downarrow \tag{3.50}$$

$$\hat{\sigma}(e,s) \neq \hat{\sigma}(e,s+1) \;\Leftrightarrow\; y(e,s) \neq y(e,s+1) \tag{3.51}$$

$$\hat{\sigma}(e,s) \neq \hat{\sigma}(e,s+1)\downarrow \Rightarrow \sigma(e,y(e,s+1),s) \preceq \hat{\sigma}(e,s+1) \tag{3.52}$$

$$\hat{\sigma}(e,s)\downarrow \neq \hat{\sigma}(e,s+1)\downarrow \Rightarrow \hat{\sigma}(e,s) > \hat{\sigma}(e,s+1) = \sigma(e,y(e,s+1),s). \tag{3.53}$$

Finally, observe that the declared states satisfy the following weak monotonicity property.

$$e < e' \;\&\; y(e,s) \notin A_s \;\&\; y(e',s)\downarrow \Rightarrow \hat{\sigma}(e,s) \leq \hat{\sigma}(e',s) \restriction e. \tag{3.54}$$

Namely, for a contradiction, assume that $e < e'$, $y(e,s) \notin A_s$ and $y(e',s)\downarrow$ hold but $\hat{\sigma}(e',s) \restriction e < \hat{\sigma}(e,s)$. Fix $s' < s$ minimal such that $y(e',t) = y(e',s)$ for all t with $s'+1 \leq t \leq s$. Then $y(e',s') \neq y(e',s'+1)\downarrow$ and (by (3.51) and by assumption) $\hat{\sigma}(e',s'+1) = \hat{\sigma}(e',s) < 1^{e'}$. So $\mathrm{P}_{e'}$ receives attention at stage $s'+1$ according to clause (3.42) whence, for $y = y(e',s'+1)$, $y(e',s') < y \leq s'$, $y \notin A_{s'}$ and $\sigma(e',y,s') = \hat{\sigma}(e',s'+1)$ hence $\sigma(e',y,s') \restriction e < \hat{\sigma}(e,s)$. On the other hand (by (3.48) and (3.51)) $y(e,s') = y(e,s)$ and $\hat{\sigma}(e,s') = \hat{\sigma}(e,s)$. Since (by (3.45)) $y(e,s') < y(e',s')$ and since $y(e,s') = y(e,s) \notin A_{s'}$, it follows that P_e requires attention at stage $s'+1$ according to clause (3.42) (or clause (3.9)) via y, which gives the desired contradiction.

Call a follower y of P_e *permanent* if $y = y(e,s)$ for all sufficiently large s and call an e-state σ the *permanent declared e-state* of P_e if $\sigma = \hat{\sigma}(e,s)$ for all sufficiently large s. Call requirement R_e *permanently satisfied at stage s* if R_e is satisfied at stage s and not initialized later (hence satisfied at all stages $s' \geq s$), and call R_e *permanently satisfied* if it is permanently satisfied at some stage.

Claim 3.6: Any requirement R_n requires attention only finitely often. Moreover, any requirement P_e has a permanent follower.

Proof: The proof is by induction on n. Fix n and, by inductive hypothesis, fix a stage $s_0 > n$ such that no requirement $R_{n'}$ with $n' < n$ requires attention after stage s_0. Then R_n receives attention at any stage $s+1 > s_0$ at which it requires attention and R_n is not initialized after stage s_0. So if R_n is a lowness or inaccessibility requirement then R_n will require attention at most once after stage s_0, since if it requires attention then it receives attention and is satisfied from this stage on (hence does not require attention anymore).

So, for the remainder of the proof, we may assume that R_n is a noncomputability requirement, say $R_n = P_e$. If P_e has no follower at the end of stage s_0 then (by $s_0 > n \geq e$) it requires and receives attention via clause (3.8) at stage $s_0 + 1$. So, since P_e is not initialized after stage s_0, $y(e,s) \downarrow$ for all $s > s_0$. So P_e does not require attention via (3.8) after stage $s_0 + 1$, and P_e requires attention at most once via (3.9) at a stage $s \geq s_0 + 1$ since if this happens then P_e becomes permanently satisfied at this stage hence does not require attention later. Finally, since $y(e,s)$ is defined for all $s \geq s_0+1$ it follows, by (3.50), that $\hat{\sigma}(e,s) \downarrow$ for such s. It follows, by (3.53), that $\hat{\sigma}(e,s)$ is weakly decreasing after stage s_0 and $\hat{\sigma}(e,s+1) < \hat{\sigma}(e,s)$ if and only if P_e becomes active via clause (3.42) at stage $s+1$. So P_e can act (hence require attention) via (3.42) at most $2^e - 1$ times after stage $s_0 + 1$. So $\mathcal{P}_e$ requires attention only finitely often, and, for the follower $y(e,s+1)$ existing at the last stage $s+1$ at which P_e acts, $y(e,s') = y(e,s+1)$ for all $s' > s$ hence $y(e,s+1)$ is permanent. □

Note that, by Claim 3.6 and by (3.50) and (3.51), requirement P_e does not only have a permanent follower but also a permanent declared e-state. Let $y(e)$ be the permanent follower of requirement P_e and let $\hat{\sigma}(e)$ be the permanent declared e-state of P_e. Moreover, let t_e+1 be the stage at which the permanent follower $y(e)$ is appointed. Note that, by (3.46) and (3.45), $e < y(e) < y(e')$ for $e < e'$. Moreover, it is immediate by Claim 3.6 and construction, that

$$A(y(e)) \neq \{e\}(y(e)) \tag{3.55}$$

(for any $e \geq 0$). Indeed, suppose that $A(y(e)) = 0$. Then, P_e is not permanently satisfied and if $\{e\}(y(e)) = 0$, then P_e would require attention infinitely often via (3.9), contradicting Claim 3.6. Now suppose that $A(y(e)) = 1$. If P_e requires attention at stage $t_e + 1$ via (3.8), then $y(e) = t_e + 1$, and (3.46) and the construction imply that $y(e) \notin A_{t_e+1}$, while if P_e requires attention at stage $t_e + 1$ via (3.42), we again have $y(e) \notin A_{t_e+1}$. It follows from (3.45) that after stage $t_e + 1$, $y(e)$ cannot be the follower of any requirement other than P_e. Thus, $A(y(e)) = 1$ implies that P_e received attention at some stage after $t_e + 1$ via (3.9) and $y(e)$ is realized, so again $A(y(e)) \neq \{e\}(y(e))$.

This immediately implies that the noncomputability requirements are met, and Claim 3.6 easily implies that the lowness requirements are met too.

Claim 3.7: Any requirement P_e and N_e is met.

Proof: The first part of the claim is immediate by (3.55). The proof of the second part is as follows. If $\{e\}_s^{A_s}(e)\downarrow$ for infinitely many stages s then, for the least stage $s \geq e$ such that no requirement of higher priority than N_e requires attention after stage s and $\{e\}_s^{A_s}(e)\downarrow$, N_e becomes active at stage $s+1$ thereby cancelling all lower priority followers. So (by choice of s and by (3.47)) $A \upharpoonright s+1 = A_s \upharpoonright s+1$ which, by our convention (3.2), ensures that $\{e\}^A(e) = \{e\}_s^{A_s}(e)\downarrow$. □

It remains to show that the inaccessibility requirements Q_e are met. For the remainder of the proof, fix $e \geq 0$, and fix $n, m, i_0, i_1, j_0, j_1, j_2, k_0, k_1$ such that $e = \langle n, m\rangle$ and $n = \langle i_0, i_1, j_0, j_1, j_2, k_0, k_1\rangle$. Moreover, by Claim 3.6, fix the greatest stage s_0 such that Q_e or a higher priority requirement requires attention at stage s_0. Note that $s_0 \geq e+1$ (since P_e has higher priority than Q_e, P_e eventually requires attention, and P_e requires attention only at stages $s+1 > e$). Moreover, by initialization,

$$\forall\, e'' > e\ (y(e'', s_0)\uparrow\ \&\ \hat{\sigma}(e'', s_0)\uparrow). \tag{3.56}$$

The following observation will be useful for showing that Q_e is met.

Claim 3.8: (a) Let s, s' and y be given such that $s_0 \leq s < s'$ and $A_{s'} \upharpoonright s+2 = A_s \upharpoonright s+2$. Then $\sigma(e+1, y, s') \preceq \sigma(e+1, y, s)$.

(b) Let e'', s' and y be given such that $e < e''$, $s_0 \leq s'$, and $y = y(e'', s') \notin A_{s'}$. Then $\sigma(e+1, y, s') \preceq \hat{\sigma}(e'', s') \upharpoonright e+1$.

Proof: (a) Given $e' \leq e$ such that $\sigma(e+1,y,s)(e')=0$, it suffices to show that

$$\sigma(e+1,y,s')(e')=0. \tag{3.57}$$

If $Q_{e'}$ is satisfied at stage s then, by $s_0 \leq s$, $Q_{e'}$ is permanently satisfied at stage s_0 hence satisfied at stage s' whence (3.57) holds. Otherwise, there are numbers x, u, v and r such that y is e'-eligible via x, u, v and r at stage $s+1$, and it suffices to show that y is e'-eligible via x, u, v and r at stage $s'+1$ too. But, since $s_0 \leq s$ and $r < s$, this is immediate by Claim 3.5.

(b) Fix $s < s'$ minimal such that $y(e'', s+1) = y(e'', s') = y$. Then, by (3.49), $A_{s'} \restriction s+2 = A_s \restriction s+2$ and, for all t with $s+1 \leq t \leq s'$, $y(e'',t) = y(e'',s') = y$ whence $s_0 \leq s$ by (3.56). So, by part (a) of the claim, $\sigma(e+1,y,s') \preceq \sigma(e+1,y,s)$. On the other hand, by choice of s, it follows, by (3.52), that $\sigma(e+1,y,s) \preceq \hat{\sigma}(e'',s+1) \restriction e+1$ and, by (3.50) and (3.51), that $\hat{\sigma}(e'',s') = \hat{\sigma}(e'',s+1)$. So $\sigma(e+1,y,s') \preceq \hat{\sigma}(e'',s') \restriction e+1$. □

Claim 3.9: Q_e is met.

Proof: If Q_e is permanently satisfied then, for the greatest stage $s+1$ at which Q_e receives attention, Q_e is not initialized later whence $A \restriction s+2 = A_s \restriction s+2$. So Q_e is met by Claim 3.2. Hence, for the remainder of the proof we may assume that Q_e is not satisfied at any stage $s \geq s_0$. Moreover, w.l.o.g. we may assume that the clauses (i) and (ii) in Q_e hold while clause $(*)$ fails (since, otherwise, Q_e is obviously met). So, in particular, the assumptions made in Claim 3.3 and Claim 3.4 hold.

By the above and by Claim 3.3, in order to show that Q_e is met, it suffices to show that there is an infinite computable set S of stages s satisfying (3.26). For giving such a set S, we need some more notation and observations.

Let $e' \leq e$ and $s \geq s_0$. A number y is *truly e'-eligible via x, u, v, r at stage $s+1$* if y is e'-eligible via x, u, v, r at stage $s+1$ and $A \restriction r = A_s \restriction r$; y is *truly e'-eligible at stage $s+1$* if y is truly e'-eligible via some numbers x, u, v and r at stage $s+1$; and the $(e+1)$-state of y *truly precedes* the $(e+1)$-state σ at stage s ($\sigma(e+1,y,s) \preceq_t \sigma$ for short) if, for any $e' \leq e$ such that $\sigma(e')=0$ and $Q_{e'}$ is not satisfied at the end of stage s_0, y is truly e'-eligible at stage $s+1$.

Note that if $e' \leq e$, $s \geq s_0$, $Q_{e'}$ is not satisfied at stage s_0 and y is truly e'-eligible at stage $s+1$ then (by (3.25)) $y \notin A$ and, by choice of s_0 and

Claim 3.5, y is (truly) e'-eligible at stage $s'+1$ for all stages $s' \geq s$. It follows that, for any number y, any stages s and s' such that $s' \geq s \geq s_0$ and any $(e+1)$-states σ and σ' such that $\sigma \preceq \sigma'$ and such that the $(e+1)$-state of y truly precedes σ at stage s, it holds that the $(e+1)$-state of y at stage s' truly precedes σ', i.e.,

$$\forall\, s, s', y \geq 0\, \forall\, \sigma, \sigma' \in \{0,1\}^{e+1} \\ ([s_0 \leq s \leq s' \,\&\, \sigma(e+1, y, s) \preceq_t \sigma \,\&\, \sigma \preceq \sigma'] \Rightarrow \sigma(e+1, y, s') \preceq_t \sigma') \tag{3.58}$$

holds. Moreover, obviously, for $s \geq s_0$, $\sigma(e+1, y, s) \preceq_t \sigma$ implies $\sigma(e+1, y, s) \preceq \sigma$. Conversely, (by (3.25))

$$\forall\, s \geq s_0\, \forall\, y \geq 0\, \forall\, \sigma \in \{0,1\}^{e+1} \\ ([\sigma(e+1, y, s) \preceq \sigma \,\&\, A \upharpoonright s+2 = A_s \upharpoonright s+2] \Rightarrow \sigma(e+1, y, s) \preceq_t \sigma). \tag{3.59}$$

Now let σ^* be the least $(e+1)$-state σ (with respect to the ordering $\leq$) such that, for infinitely many numbers $y \notin A$, there is a stage $s \geq s_0$ at which the $(e+1)$-state of y truly precedes σ, i.e.,

$$\sigma^* = \mu\, \sigma \in \{0,1\}^{e+1} [\overset{\infty}{\exists}\, y \notin A\, \exists\, s \geq s_0\, (\sigma(e+1, y, s) \preceq_t \sigma)]$$

(note that σ^* exists since A is co-infinite and, for any y and $s \geq s_0$, the $(e+1)$-state of y truly precedes 1^{e+1} at stage s), and fix $e_0 > e$ minimal such that

$$y(e_0) \notin A \tag{3.60}$$

and, for any $y \geq e_0$ with $y \notin A$, there is no stage $s \geq s_0$ and no $\sigma < \sigma^*$ such that the $(e+1)$-state of y truly precedes σ at stage s, i.e.,

$$\forall\, y \geq e_0\, \forall\, s \geq s_0\, \forall\, \sigma < \sigma^*\, (y \notin A \;\Rightarrow\; \sigma(e+1, y, s) \not\preceq_t \sigma) \tag{3.61}$$

holds (note that e_0 exists since, by choice of σ^*, (3.61) holds for almost all numbers e_0 while, by (3.55), (3.60) holds for infinitely many e_0).

The following properties of σ^* will be crucial.

$$\forall\, e'' \geq e_0\, \forall\, s > t_{e_0}\, (\hat{\sigma}(e'', s) \downarrow \Rightarrow \sigma^* \leq \hat{\sigma}(e'', s) \upharpoonright e+1) \tag{3.62}$$

$$\forall\, e'' \geq e_0\, (y(e'') \notin A \;\Rightarrow\; \hat{\sigma}(e'') \upharpoonright e+1 = \sigma^*) \tag{3.63}$$

$$\sigma^*(e) = 0. \tag{3.64}$$

The proof of (3.62) is as follows. Since $y(e_0) \notin A$ and $\hat{\sigma}(e_0, s) = \hat{\sigma}(e_0, t_{e_0}+1)$ for all $s > t_{e_0}$, by (3.54) (and (3.50)), it suffices to show that $\sigma^* \leq \hat{\sigma}(e_0, t_{e_0}+1) \restriction e+1$. This is done as follows. By choice of t_{e_0} and by (3.60), P_{e_0} becomes active via clause (3.8) or (3.42) at stage $t_{e_0}+1$ — hence $\sigma(e_0, y(e_0), t_{e_0}) \preceq \hat{\sigma}(e_0, t_{e_0}+1)$ — and $A \restriction t_{e_0}+2 = A_{t_{e_0}} \restriction t_{e_0}+2$. So, by (3.59), $\sigma(e+1, y(e_0), t_{e_0}) \preceq_t \hat{\sigma}(e_0, t_{e_0}+1) \restriction e+1$. Since $e_0 \leq y(e_0)$ it follows by choice of e_0 that $\sigma^* \leq \hat{\sigma}(e_0, t_{e_0}+1) \restriction e+1$, which completes the proof of (3.62).

For a proof of (3.63), fix $e'' \geq e_0$ such that $y(e'') \notin A$, and, for a contradiction, assume that $\hat{\sigma}(e'') \restriction e+1 \neq \sigma^*$. Then, by (3.62), $\sigma^* < \hat{\sigma}(e'') \restriction e+1$. But, by choice of $\sigma^*$and (3.58) this implies that there is a number $y > y(e'')$ such that $y \notin A$ and the $(e+1)$-state of y strictly precedes $\hat{\sigma}(e'') \restriction e+1$ at all sufficiently large stages. So $\mathrm{P}_{e''}$ requires attention infinitely often via clause (3.42) contrary to Claim 3.6. (Note that, by $y(e'') \notin A$ and (3.55), the permanent follower $y(e'')$ is never realized.)

Finally, for a proof of (3.64), for a contradiction assume that $\sigma^*(e) = 1$. Then, by (3.61), there is no number y and no stage s such that

$$\begin{gathered} y \notin A \;\&\; e_0 \leq y \;\&\; s \geq s_0 \;\&\; \sigma(e+1, y, s) \preceq_t \sigma^* \;\& \\ y \text{ is truly } e\text{-eligible at stage } s+1. \end{gathered} \tag{3.65}$$

(Namely, for such y and s, $\sigma(e+1, y, s) \preceq_t \sigma$ for the string $\sigma < \sigma^*$ defined by $\sigma(e) = 0$ and $\sigma(e') = \sigma^*(e')$ for $e' < e$.) So, in order to get the desired contradiction, it suffices to show that there is a number y and a stage s satisfying (3.65). Let E be the A-computably enumerable set

$$E = \{y \notin A : \exists\, e'' \geq e_0 \;\exists\, s \geq s_0\; (y(e'', s) \neq y(e'', s+1) = y \;\&$$

$$A \restriction s+2 = A_s \restriction s+2 \;\&\; \hat{\sigma}(e'', s+1) \restriction e+1 = \sigma^*)\}.$$

Then, for any $y \in E$, $y \notin A$, $e_0 \leq y$ (by (3.46)), and, by (3.52), (3.59) and (3.58), $\sigma(e+1, y, s) \preceq_t \sigma^*$ for almost all stages s. So it only remains to show that there is a number $y \in E$ which is truly e-eligible at some stage $s \geq s_0$ hence at almost all stages s. But the existence of such a number y follows by Claim 3.4 since E is infinite — namely, for any number $e'' \geq e_0$ such that $y(e'') \notin A$, $y(e'') \in E$ since $y(e'', t_{e''}) \neq y(e'', t_{e''}+1) = y(e'')$, $A \restriction t_{e''}+2 = A_{t_{e''}} \restriction t_{e''}+2$ (by (3.49)), and $\hat{\sigma}(e'', t_{e''}+1) \restriction e+1 = \sigma^*$ (by (3.63)) — and since any infinite A-c.e. set contains an infinite A-computable subset. This completes the proof of (3.64).

Now S is defined by

$$\begin{gathered} S = \{s > t_{e_0} : \forall\, e'' > e_0\; (y(e'', s)\downarrow \;\& \\ y(e'', s) \notin A_s \Rightarrow \hat{\sigma}(e'', s+1) \restriction e+1 = \sigma^*)\}. \end{gathered} \tag{3.66}$$

Obviously, S is computable. Moreover, S is infinite. Namely, for any $e'' > e_0$ such that $y(e'') \notin A$, $y(e''', t_{e''} + 1) = y(e''')$ and $y(e''') \in A$ if and only if $y(e''') \in A_{t_{e''}+1}$ for all $e''' \leq e''$. So, for any such e'', $t_{e''} + 1 \in S$ by (3.63). Infinity of S follows since there are infinitely many e'' with $y(e'') \notin A$.

It remains to show that the elements s of S satisfy (3.26). So fix $s \in S$, a number $y \leq s$ and a stage $s' \geq s$ such that $y \in A_{s'+1} \setminus A_{s'}$. We have to show that y is e-eligible at stage $s'+1$. Fix e'' such that $y = y(e'', s')$. Since y enters A at stage $s'+1 > t_{e_0} + 1$ it follows that $e_0 < e''$. Moreover, since $y \leq s$, it follows by (3.47) that $y(e'', t)$ is defined for all t with $s \leq t \leq s'$. So, by (3.53) and (3.51) and by choice of s,

$$\hat{\sigma}(e'', s') \upharpoonright e+1 \leq \hat{\sigma}(e'', s) \upharpoonright e+1 = \sigma^*$$

hence, by (3.62), $\hat{\sigma}(e'', s') \upharpoonright e+1 = \sigma^*$. It follows by Claim 3.8(b) that $\sigma(e+1, y, s')$ precedes σ^*. But, by (3.64), this implies that y is e-eligible at stage $s'+1$.

This completes the proof of Claim 3.9. □

This completes the proof of Theorem 2.7.

4. Proof of Theorem 2.8

Given a high c.e. degree $\mathbf{a}$, it suffices to give c.e. degrees $\mathbf{a_0}$ and $\mathbf{a_1}$ such that

$$\mathbf{a} = \mathbf{a_0} \vee \mathbf{a_1} \tag{4.1}$$

and, for $i = 0, 1$,

$$\begin{gathered}\forall\, \mathbf{b_0}, \mathbf{b_1}\ [\mathbf{b_0} \vee \mathbf{b_1} = \mathbf{a_i} \Rightarrow \mathbf{a_i} \leq \mathbf{b_0} \ \text{ or} \\ \forall\, \mathbf{c_0}, \mathbf{c_1} \not\leq \mathbf{a_i}\ \exists\, \mathbf{d}\ (\mathbf{d} \leq \mathbf{b_1} \vee \mathbf{c_0}, \mathbf{b_1} \vee \mathbf{c_1}\ \&\ \mathbf{d} \not\leq \mathbf{a_i})]\end{gathered} \tag{4.2}$$

hold. (Note that (4.2) ensures that $\mathbf{a}_i$ is s.m.i. unless $\mathbf{a}_i = \mathbf{0}$. So, by $\mathbf{a} > \mathbf{0}$ and by (4.1), either $\mathbf{a_0}$ and $\mathbf{a_1}$ are s.m.i. or one of these degrees is $\mathbf{0}$ and $\mathbf{a}$ is s.m.i. In either case, this implies Theorem 2.8.)

The strategy for satisfying (4.2) is based on the corresponding strategy introduced in the proof of Theorem 2.7 above. For $i = 0, 1$ we enumerate c.e. sets A_i and auxiliary c.e. sets $C_{i,n}$ ($n \geq 0$) such that, for any numbers $n, m \geq 0$ and for the corresponding unique numbers $i_0, i_1, j_0, j_1, j_2, k_0, k_1$ and e such that $n = \langle i_0, i_1, j_0, j_1, j_2, k_0, k_1 \rangle$ and $e = \langle n, m \rangle$,

$$C_{i,n} \leq_T W_{i_1} \oplus W_{k_0}, W_{i_1} \oplus W_{k_1} \tag{4.3}$$

holds and the *Inaccessibility Requirements*

$$\mathrm{Q}^i_e : \text{If} \quad (i)\ W_{i_0} = \{j_0\}^A \ \&\ W_{i_1} = \{j_1\}^{A_i} \ \&\ A_i = \{j_2\}^{W_{i_0} \oplus W_{i_1}},$$

$$(ii)\ W_{k_0}, W_{k_1} \not\leq_{\mathrm{T}} A_i, \text{ and}$$

$$(iii)\ A_i \not\leq_{\mathrm{T}} W_{i_0}$$

$$\text{then } (*)\ C_{i,n} \neq \{m\}^{A_i}$$

are met. (In the following, when we consider one of the sets $C_{i,n}$ or one of the requirements Q^i_e, we tacitly assume that $n = \langle i_0, i_1, j_0, j_1, j_2, k_0, k_1\rangle$ and $e = \langle n, m\rangle$.)

Just as in the proof of Theorem 2.7, condition (4.3) is satisfied by guaranteeing

$$x \in C_{i,n,s+1} \setminus C_{i,n,s} \Rightarrow \exists\, z < \min\{\gamma_{k_0,s}(x), \gamma_{k_1,s}(x)\}\ (z \in W_{i_1,s+1} \setminus W_{i_1,s}), \tag{4.4}$$

and we use the finitary strategy for meeting requirement Q^i_e by diagonalization introduced there. So requirement Q^i_e *requires attention* at stage $s+1$ if $e \leq s$, Q^i_e is not satisfied at (the end of) stage s, and one of the following holds.

$$\exists\, x \left(\begin{array}{l} A_{i,s}(x) \neq \{j_2\}_s^{W_{i_0,s} \oplus W_{i_1,s}}(x)\downarrow, \\ \text{and, for } u = u(W_{i_0,s} \oplus W_{i_1,s}; j_2, x, s) \text{ and for } p \leq 1, \\ W_{i_p,s} \restriction u = \{j_p\}_s^{A_{i,s}} \restriction u \end{array} \right) \tag{4.5}$$

$$\exists\, x\, \exists\, p \leq 1\ (W_{i_p,s}(x) = 1 \neq \{j_p\}_s^{A_{i,s}}(x)\downarrow) \tag{4.6}$$

$$\exists\, x \in \omega^{[e]} \tag{4.7}$$
$$\left(\{m\}_s^{A_{i,s}}(x) = 0 \ \&\ \exists\, z < \min\{\gamma_{k_0,s}(x), \gamma_{k_1,s}(x)\}\ (z \in W_{i_1,s+1} \setminus W_{i_1,s})\right).$$

The corresponding action (if Q^i_e receives attention at stage $s+1$) is essentially as before: Declare Q^i_e to be *satisfied*. Moreover if (4.7) holds then, for the least x as there, put x into $C_{i,n}$. Note that the latter action is consistent with (4.4). Moreover, as in the proof of Theorem 2.7 (compare with Claim 3.2), the following holds (for $i = 0, 1$) implying that Q^i_e is met if it is permanently satisfied.

Claim 4.1: Assume that Q^i_e requires and receives attention at stage $s+1$ and that $A_i \upharpoonright s+1 = A_{i,s} \upharpoonright s+1$. Then either clause (i) in Q^i_e fails or clause $(*)$ in Q^i_e holds. So, in particular, Q^i_e is met.

While in the basic construction, however, we attempted to guarantee $A \upharpoonright s+1 = A_s \upharpoonright s+1$ by cancelling the lower priority followers of the noncomputability requirements when Q_e received attention, in the current setting (where the noncomputability requirements are replaced by some coding requirements, see below) cancellation is not possible. So, instead, when Q^i_e receives attention at stage $s+1$, then it imposes a restraint $r^i_Q(e, s+1) = s+1$ on A_i. Correspondingly, we initialize Q^i_e later if (and only if) this restraint becomes injured. More formally, Q^i_e is *initialized* at stage $s'+1$ if $A_{i,s'+1} \upharpoonright r^i_Q(e, s') \neq A_{i,s'} \upharpoonright r^i_Q(e, s')$ (as in the previous constructions we assume that any parameters stay unchanged during the construction unless explicitly stated otherwise). If initialized at stage $s'+1$ then Q^i_e becomes *unsatisfied* at stage $s'+1$ and the restraint is lifted by setting $r^i_Q(e, s'+1) = 0$. If Q^i_e is satisfied at stage $s+1$ and not initialized later then we say that Q^i_e is *permanently satisfied* (at stage $s+1$). (Q^i_e can become initialized only by the action of a coding requirement (to be defined below) where the coding requirement may have higher or lower priority than Q^i_e. But, since a coding requirement has a choice whether it puts a code into A_0 or into A_1, just as in the proof of the Sacks splitting theorem, by enforcing the highest priority restraint, eventually Q^i_e will not be initialized anymore.)

As in the proof of Theorem 2.7, the direct strategy for meeting the inaccessibility requirements does not suffice to guarantee that these requirements are met. Unless Q^i_e is permanently satisfied (hence met by Claim 4.1) or met for some trivial reason, in addition it has to be ensured that numbers which enter A_i "late" are "eligible". To make this more precise, define *i-e-(pre)eligibility* of a number y (via numbers x, u, v and r) at stage $s+1$ just as e-(pre)eligibility in the proof of Theorem 2.7 but with the set A_i in place of A (i.e., by replacing A_s by $A_{i,s}$ in (3.15)–(3.24)); and, similarly, the *i-e-state* $\sigma_i(e, y, s)$ of a number y at stage s is defined as the e-state $\sigma(e, y, s)$ in the basic construction with A_i, $Q^i_{e'}$ and i-e'-eligibility in place of A, $Q_{e'}$ and e'-eligibility, respectively. Then, as there (compare with Claim 3.3), one can observe that the following holds (for $i = 0, 1$).

Claim 4.2: Assume that Q^i_e requires attention only finitely often and is satisfied only finitely often. If there is an infinite computable set S of stages such that, for any $s \in S$,

$$\forall\, y, s'\ ([y \leq s \leq s'\ \&\ y \in A_{i,s'+1} \setminus A_{i,s'}] \Rightarrow y \text{ is } i\text{-}e\text{-eligible at stage } s'+1) \tag{4.8}$$

holds, then requirement Q^i_e is met.

Moreover, as in the proof of Theorem 2.7 we can prove the following two claims (corresponding to Claims 3.4 and 3.5) giving sufficient conditions for the existence of i-e-eligible numbers and for the preservation of i-e-eligibility ($i = 0, 1$).

Claim 4.3: Assume that clauses (i) and (ii) in Q^i_e hold, and that $\{m\}^{A_i}$ is total and, for almost all numbers $x \in \omega^{[e]}$, $\{m\}^{A_i}(x) = 0$. Then, for any infinite A_i-computable subset E of $\overline{A_i}$ there is an infinite subset F of E such that, for any $y \in F$, there are numbers x_y, u_y, v_y and r_y such that y is i-e-eligible via x_y, u_y, v_y and r_y at almost all stages.

Claim 4.4: Let e, y, x, u, v, r, s and s' be given such that $s \leq s'$, $e \leq s'$, y is i-e-(pre)eligible via x, u, v and r at stage $s+1$, $A_{i,s'} \upharpoonright r = A_{i,s} \upharpoonright r$, and Q^i_e is neither satisfied at stage s' nor requires attention at stage $s'+1$. Then y is i-e-(pre)eligible via x, u, v and r at stage $s'+1$.

Having reviewed the basic features of the inaccessibility strategy, we now turn to the strategy for guaranteeing that, for $\mathbf{a}_i = deg(A_i)$, (4.1) holds. This strategy is based on *Martin permitting*, a sort of "almost - always" permitting argument pertaining to high c.e. degrees (see e.g. Soare [18], Chapter XI.2 for more details).

We start with some notation. Given an infinite c.e. set A and a computable 1-1 function a enumerating A, let $A_s = \{a(0), \ldots, a(s-1)\}$, let $\overline{a}(x, s)$ be the $(x+1)$th element of $\overline{A_s}$ in order, let $\overline{a}(x)$ be the $(x+1)$th element of $\overline{A}$ in order, and let

$$c_{\overline{A}}(x) = \mu s\ (\overline{a}(x, s) = \overline{a}(x)).$$

Now, since the given degree $\mathbf{a}$ is high, the following holds (see e.g. Theorem XI.2.1 in [18]).

Fact 4.5: There is a c.e. set $A \in \mathbf{a}$ and a computable 1-1 function a enumerating A such that $c_{\overline{A}}$ dominates all computable functions.

For the remainder of the proof fix such a set A and the corresponding enumeration function a and enumeration $(A_s)_{s\geq 0}$. Say that a number y is

M(artin)-permitted (by A) at stage $s+1$ if $\overline{a}(y, s+1) \neq \overline{a}(y, s)$. Then, by Fact 4.5, for any given computable function $f(y)$, y is M-permitted at a stage $> f(y)$ for almost all y.

Now in order to guarantee $A =_{\mathrm{T}} A_0 \oplus A_1$ (hence (4.1)), we use a combined Martin and marker permitting. To be more precise, we define a partial computable marker function $\beta(e, s)$ (i.e., a partial computable function $\beta : \omega \times \omega \to \omega$ with computable domain) with the following properties (for any $e, e', s, s', y \geq 0$).

Initially, the marker is nowhere defined, i.e.,

$$(\beta_0)\ \forall\, e\ (\beta(e, 0)\uparrow).$$

At any stage $s+1 > 0$ either, for the least e such that $\beta(e, s)\uparrow$, $\beta(e, s+1)$ becomes defined while all other marker positions remain unchanged, or a marker $\beta(e)$ defined at stage s may be moved to a higher position or lifted in which case the markers below remain unchanged while the markers above are lifted. Moreover, if the marker $\beta(e)$ is put down on a new position at stage $s+1$ then we let $\beta(e, s+1) = \langle s+1, 2^{s+1}\rangle > s+1$. So, more formally, the following holds (for all $s \geq 0$).

$$(\beta_{00})\ \exists\, e \leq s$$
$$\Big[\forall\, e' < e\ (\beta(e', s) = \beta(e', s+1)\downarrow)\ \&$$
$$\beta(e, s) \neq \beta(e, s+1)\ \&$$
$$(\beta(e, s+1) = \langle s+1, 2^{s+1}\rangle \vee (\beta(e, s)\downarrow\ \&\ \beta(e, s+1)\uparrow))\ \&$$
$$\forall\, e'' > e\ (\beta(e'', s+1)\uparrow)\Big].$$

Here we assume that the pairing function $\langle\ ,\ \rangle$ is not only strictly increasing in both arguments but also satisfies $x \leq \langle x, 0\rangle$ and $\langle x, 2^x\rangle < \langle x+1, 0\rangle$ for all $x \geq 0$. (The latter property will be used below in order to keep the traces defined there in order.)

As one can easily check (by induction on s), (β_0) and (β_{00}) guarantee that, at any stage s, β is defined on an initial segment of ω of length $\leq s$,

$$(\beta_1)\ \exists\, e \leq s\ (\beta(e', s)\downarrow \Leftrightarrow\ e' < e),$$

that β is strictly increasing on this initial segment,

$$(\beta_2)\ [e < e'\ \&\ \beta(e', s)\downarrow]\ \Rightarrow\ \beta(e, s) < \beta(e', s),$$

that, whenever defined, $\beta(e,s)$ is nondecreasing in s, exceeds e, and is bounded by $\langle s+1, 2^{s+1}\rangle$,

$$(\beta_3) \quad [s \leq s' \,\&\, \beta(e,s)\downarrow \,\&\, \beta(e,s')\downarrow]$$

$$\Rightarrow e < \beta(e,s) \leq \beta(e,s') \leq \langle s', 2^{s'}\rangle,$$

that the value of $\beta(e)$ at a stage $s+1$ where it is put down or moved to a new value is greater than this stage,

$$(\beta_4)\ \beta(e,s) \neq \beta(e,s+1)\downarrow \Rightarrow \beta(e,s+1) = \langle s+1, 2^{s+1}\rangle > s+1,$$

and that the values of the marker on different arguments differ (when defined) independent of the stages, i.e.,

$$(\beta_5)\ [e \neq e' \,\&\, \beta(e,s)\downarrow \,\&\, \beta(e',s')\downarrow] \Rightarrow \beta(e,s) \neq \beta(e',s').$$

Now, in order to guarantee that A is Turing equivalent to $A_0 \oplus A_1$, we guarantee that, eventually, the marker $\beta(e)$ comes to a final position $\beta^*(e) \in \omega$, i.e.,

$$(\beta_6)\ \beta^*(e) = \lim_{s\to\omega} \beta(e,s) < \omega \text{ exists},$$

and that the moves of the marker are related to the given enumeration of the given set A and the constructed enumerations of the sets A_i $(i = 0, 1)$ under construction as follows.

If defined on e at stage s, the marker $\beta(e)$ is allowed to move or to be lifted at stage $s+1$ only if the old marker position is M-permitted, i.e.,

$$(\beta_7)\ \beta(e,s)\downarrow \neq \beta(e,s+1) \Rightarrow \overline{a}(\beta(e,s),s) \neq \overline{a}(\beta(e,s),s+1),$$

and any such move has to be made recognizable for $A_0 \oplus A_1$ by enumerating a new number y with $y < \beta(e,s)$ into A_0 or A_1 at stage $s+1$, i.e.,

$$(\beta_8)\ \beta(e,s)\downarrow \neq \beta(e,s+1) \Rightarrow \exists\, i \leq 1\, \exists\, y < \beta(e,s)(y \in A_{i,s+1} \setminus A_{i,s}).$$

Conversely, if a number y is enumerated into A_0 or A_1 at stage $s+1$ then this is witnessed by a corresponding move of β, i.e.,

$$(\beta_9) y \in A_{i,s+1} \setminus A_{i,s} \Rightarrow \exists\, e\, [\beta(e,s)\downarrow \,\&\, y \in [e, \beta(e,s)) \\ \&\, \beta(e,s) \neq \beta(e,s+1)].$$

Finally, for any element $a(s)$ of A,

$$(\beta_{10})\ \beta(a(s),s)\uparrow \text{ or } \beta(a(s),s+1)\uparrow.$$

Note that, by (β_4), (β_3), and (β_6), the last condition implies that, for $e \in A$, the final position $\beta^*(e)$ of the marker on e exceeds the stage at which the number e is enumerated into A, namely

$$(\beta_{10^*})\ \beta^*(a(s)) \geq s+1.$$

That the above rules guarantee $A =_{\mathrm{T}} A_0 \oplus A_1$ is shown as follows.

Claim 4.6: Assume that the partial computable marker β satisfies the conditions (β_1)–(β_{10}) for all e, e', s, s', y. Then $A \leq_{\mathrm{T}} \beta^*$, $\beta^* \leq_{\mathrm{T}} A$, $\beta^* \leq_{\mathrm{T}} A_0 \oplus A_1$, and $A_0 \oplus A_1 \leq_{\mathrm{T}} \beta^*$. So, in particular, $A =_{\mathrm{T}} A_0 \oplus A_1$.

Proof: Note that, by (β_6), $\beta^*(e) = \lim_{s\to\omega} \beta(e,s) \in \omega$ is well-defined and, by (β_3), $\beta^*(e) = \sup\{\beta(e,s) : s \geq 0\ \&\ \beta(e,s)\downarrow\}$. So $A \leq_{\mathrm{T}} \beta^*$ holds since, by (the above and) (β_{10^*}), a number e is in A iff $e \in A_{\beta^*(e)}$; $\beta^* \leq_{\mathrm{T}} A$ holds since, by (β_7), $\beta^*(e) = \beta(e,s)$ for the least s such that $\beta(e,s)\downarrow$ and $\overline{a}(\beta(e,s),s) = \overline{a}(\beta(e,s))$; $\beta^* \leq_{\mathrm{T}} A_0 \oplus A_1$ holds since, by (β_8), $\beta^*(e) = \beta(e,s)$ for the least s such that $\beta(e,s)$ is defined and $A_i \restriction \beta(e,s) = A_{i,s} \restriction \beta(e,s)$ for $i = 0,1$; and $A_0 \oplus A_1 \leq_{\mathrm{T}} \beta^*$ holds since, by (β_9), it holds that, for any $y \geq 0$ and $i \leq 1$, $A_i(y) = A_{i,s}(y)$ for the least stage s such that $\beta^*(e) = \beta(e,s)$ for all $e \leq y$. □

The task to guarantee the above rules for the eth marker $\beta(e)$, in particular to guarantee that $\beta(e)$ comes to a limit and that, for $e = a(s)$, (β_{10}) holds, is assigned to the eth *coding requirement* (for $e \geq 0$)

$$\mathrm{P}_e : \ \beta^*(e) \in \omega \text{ exists and if } e = a(s) \text{ and } \beta(e,s)\downarrow \text{ then } \beta(e,s+1)\uparrow.$$

The priority ordering of the requirements is given by $\mathrm{P}_e = \mathrm{R}_{3e}$ and $\mathrm{Q}^i_e = \mathrm{R}_{3e+i+1}$ $(i = 0,1,\ e \geq 0)$.

The basic strategy for meeting P_e is twofold. First, if e is minimal such that $\beta(e,s)$ is undefined (and there is no higher priority coding requirement which becomes active at stage $s+1$) then P_e puts down marker $\beta(e)$ (according to (β_4)) by letting $\beta(e,s+1) = \langle s+1, 2^{s+1}\rangle$. Second, if e enters A at stage $s+1$, i.e., if $e = a(s)$, and $\beta(e,s)\downarrow$ then P_e lifts the marker $\beta(e)$ and all markers $\beta(e'')$ with $e < e''$.

Note that this action is consistent with the rules (β_1)–(β_5) and guarantees (β_6) and (β_{10}) where, for the former, one should note that $\beta(e)$ changes only if P_e or a higher priority coding requirement becomes active and that this will happen only finitely often. Moreover (by (β_3)), the above action is consistent with (β_7) since e entering A at stage $s+1$ M-permits the lifting of $\beta(e)$ at stage $s+1$. In order to ensure (β_8), however, when $\beta(e)$ is lifted at stage $s+1$ (in order to guarantee (β_{10})) then, for some $i \leq 1$, a *trace* $y < \beta(e,s)$ has to be put into A_i at stage $s+1$. This action in turn has to be made compatible with (β_9) for which it suffices to choose the trace y so that $y \in [e, \beta(e,s)) \setminus A_{i,s}$ holds. So, when $\beta(e)$ is put down at stage $s+1$ then at the same stage *i-e-traces* $\beta^i(e,s+1) \in [e, \beta(e,s+1)) \setminus A_{i,s}$ $(i = 0,1)$ are put down too, and one of these traces is enumerated into its corresponding set A_i if $\beta(e)$ is lifted later in order to ensure (β_{10}). The decision, which of the traces is chosen, is based on the current restraints. Namely, in order to guarantee that the restraint imposed by a requirement is eventually obeyed, we have to guarantee that, if some higher priority restraints might be injured, then the highest priority restraint which may be injured by the enumeration of a trace is protected by choosing the trace going into the other side (we refer to this as the *splitting condition* in the following).

Just as in the basic construction, the positive requirements P_e have to cooperate with the higher priority inaccessibility requirements in order to make the strategy for meeting the latter work. This requires some amendments of the just described basic P_e-strategy. In particular, if the current i-e-trace $\beta^i(e,s)$ of the coding requirement P_e can be replaced by a larger unused number y having i-e-state σ less than the (declared) i-e-state of $\beta^i(e,s)$ then we have to do so. Since such a number y may be bigger than the current position of the marker $\beta(e)$, this requires moving $\beta(e)$ beyond y in order to ensure that $y \in [e, \beta(e))$. Now moving $\beta(e)$ requires M-permission. By the domination property of A, in almost all cases, eventually we get such a permission, but we have to ensure that, while waiting for this permission, the i-e-state σ of y is preserved. In order to achieve this, we have to impose a restraint on A_i once we see the candidate y (compare with Claim 4.4). Moreover, once permission is given to move the marker $\beta(e)$, in addition we have to put one of the current traces of P_e into A_0 or A_1. Of course, we cannot use the i-e-trace since enumeration of $\beta^i(e,s) < y$ into A_i may destroy the i-e-state σ of y. So we have to enumerate the $(1-i)$-e-trace $\beta^{1-i}(e,s)$ into A_{1-i}. The latter, however, may cause the following two problems. First, this might violate the splitting condition. In order to prevent this from happening, whenever one of the traces is less than the restraint imposed by the higher priority requirements on the corresponding side then, once permission is given to move $\beta(e)$, we lift the marker (simultaneously enumerating the trace satisfying the splitting condition in its corresponding set) and start all over again with the P_e-strategy. Second, by enumerating the $(1-i)$-e-trace into A_{1-i}, this trace is used up and we have to assign a new $(1-i)$-e-trace which may have a greater $(1-i)$-e-state than the previous trace. So improving the i-e-state of the i-e-trace may worsen the $(1-i)$-e-state of the $(1-i)$-e-trace. In order to overcome the latter problem, we do not assign a single i-e-trace $\beta^i(e)$ to P_e but a sequence $\vec{\beta}^i(e) = (\beta^i_0(e), \ldots, \beta^i_{2^e}(e))$ of 2^e+1 i-e-traces and assign a declared i-e-state to the sequence (as a whole) where $\beta^i_0(e) < \beta^i_1(e) < \cdots < \beta^i_{2^e}(e)$, the i-e-state of any of these traces precedes the declared i-e-state, and, for $k < 2^e$, the trace y_{k+1} (for ease of notation, we write y_k for $\beta^i_k(e)$) is chosen big enough such that enumerating y_{k+1} into A_i will not affect the i-e-state of the smaller members $y_0, \ldots, y_k$ in the sequence. Since the declared i-e-state can be improved at most $2^e - 1$ times, this guarantees that there are enough of the current $(1-i)$-e-traces left for the necessary updates of i-e-traces (and vice versa) and we can preserve the i-e-states of the unused traces by enumerating the traces of the given sequence in decreasing order. Correspondingly to the basic construction, the first sequence of i-e-traces for P_e is assigned the declared i-e-state 1^e, and the current sequence of i-e-traces with state σ is only replaced by a new sequence if the states of all members of this new sequence precede a state $\sigma' < \sigma$ (so that the declared state can be improved to σ' by replacing the old sequence with the new sequence).

For the implementation of the just described P_e-strategy, we use the following definition. Given a binary string σ of length e, call a sequence $\vec{y} = (y_0, \ldots, y_{2^e})$ (of numbers) of length 2^e+1 *i-σ-z-eligible* at stage $s+1$ if the following hold.

(1) $z < y_0 < y_1 < \cdots < y_{2^e} < s$.
(2) For $k \leq 2^e$, $y_k \notin A_{i,s}$.
(3) For $k \leq 2^e$, $\sigma_i(e, y_k, s) \preceq \sigma$. Moreover, for any $e' < e$ and $k \leq 2^e$ such that $\sigma(e') = 0$ and $\mathrm{Q}^i_{e'}$ is not satisfied at the end of stage s, there are numbers x_k, u_k, v_k and r_k such that y_k is i-e'-eligible via x_k, u_k, v_k and r_k at stage $s+1$ and $r_k < y_{k+1}$ (where we let $y_{2^e+1} = s$).

As explained above, if the replacement of the current sequence $\vec{\beta}^i(e, s)$ of i-e-traces by a sequence of larger unrestrained traces allows to improve the (declared) i-e-state then the P_e-strategy attempts to do this in the following two steps. First, by imposing an appropriate restraint on A_i, the state of the potential replacement is preserved; then, once M-permission is given, the actual replacement takes place. The first part of this action is formalized by the following definition of requiring attention (while the second part will be described in Case 1 of Step 1 of stage $s+1$ of the construction below). Whether P_e requires attention at stage $s+1$ depends on the restraint $R^i(e,s)$ on A_i to be respected by P_e at stage $s+1$ while the restraint $R^i(e,s)$ depends on which higher priority coding requirements $\mathrm{P}_{e'}$, $e' < e$, require attention at stage $s+1$. So these concepts are defined simultaneously by induction on e.

The restraint imposed on A_i by the requirements of higher priority than P_e ($i = 0,1$) (including the restraints which higher priority requirements which require attention at stage $s+1$ may want to impose) is denoted by $R^i(e,s)$ and is defined as follows. Let

$$R^i(e,s) = s+1$$

if there is a number $e' < e$ such that $\mathrm{Q}^i_{e'}$ requires attention at stage $s+1$ or $\mathrm{P}_{e'}$ requires attention via i at stage $s+1$, and let

$$R^i(e,s) = \max\{r^i(e',s) : e' < e\}$$

where

$$r^i(e',s) = \max\{r^i_P(e',s), r^i_Q(e',s)\}$$

otherwise.

If $\beta(e,s)\downarrow$, $r^0_P(e,s) = r^1_P(e,s) = 0$, and $\hat{\sigma}_0(e,s)\downarrow$ and $\hat{\sigma}_1(e,s)\downarrow$ are the declared 0-e-state and declared 1-e-state at the end of stage s, then requirement P_e *requires attention* via $(i,\sigma,\vec{y})$ at stage $s+1$ if $i \leq 1$, σ is a binary string of length e such that $\sigma < \hat{\sigma}_i(e,s)$, and $\vec{y}$ is an i-σ-$\max\{R^i(e,s),\beta(e,s)\}$-eligible sequence at stage $s+1$; P_e *requires attention via* i at stage $s+1$ if i is minimal such that P_e requires attention via $(i,\sigma,\vec{y})$ for some σ and $\vec{y}$; and P_e *requires attention* at stage $s+1$ if there is an $i \leq 1$ such that P_e requires attention via i at stage $s+1$.

If requirement P_e is *initialized* at stage s then $\beta(e,s)$ is lifted ($\beta(e,s)\uparrow$), and, for $i = 0,1$, the sequence of the i-e-traces is cancelled ($\vec{\beta}^i(e,s)\uparrow$), the declared i-e-state is cancelled ($\hat{\sigma}_i(e,s+1)\uparrow$), and the restraint imposed on A_i by P_e is cancelled ($r^i_P(e,s) = 0$).

Using the notation introduced above, the construction is as follows where at stage 0 all requirements are initialized.

Stage $s+1$. The stage consists of three steps. Either the first or the second step, however, is vacuous.

Step 1. Fix e minimal such that $\beta(e,s)\downarrow$, $\beta(e,s)$ is M-permitted at stage $s+1$, and one of the following holds.

(a) $e = a(s)$.
(b) For some $i \leq 1$, $\beta_0^i(e,s) < R^i(e,s)$.
(c) For some $i \leq 1$, $r_P^i(e,s) > 0$.

If there is no such e then proceed to Step 2. (Note that this happens only if $\beta(a(s),s)\uparrow$ since otherwise $\beta(a(s),s)$ is M-permitted at stage $s+1$.) Otherwise, distinguish the following two cases, perform the corresponding action, and afterwards move to Step 3 (skipping Step 2).

Case 1: Conditions (a) and (b) fail and, for some $i \leq 1$ such that $r_P^i(e,s) > 0$, there is a binary string σ of length e such that $\sigma < \hat{\sigma}_i(e,s)$ and such that there is an i-σ-$\max\{R^i(e,s),\beta(e,s)\}$-eligible sequence $\vec{y}$ at stage $s+1$.

Then, for the least such i, the least corresponding σ, and the least corresponding $\vec{y}$, let

$\beta(e,s+1) = \langle s+1, 2^{s+1}\rangle$,
$\vec{\beta^i}(e,s+1) = \vec{y}$,
$\hat{\sigma}_i(e,s+1) = \sigma$,
$r_P^i(e,s+1) = 0$, and
declare that P_e becomes *i-active* at stage $s+1$.

Moreover, if there is a number $k \leq 2^e$ such that $\beta_k^{1-i}(e,s) \notin A_{1-i,s}$ (below we will show that such a number exists) then, for the greatest such k, put $\beta_k^{1-i}(e,s)$ into $A_{1-i,s+1}$.

Case 2: Otherwise. For $i=0,1$, let n_i be the least number $n < 3e$ such that, for some $e' < e$,

$$[n = 3e' \ \&\ \beta_0^i(e,s) < r_P^i(e',s)]$$
$$\text{or}$$
$$[n = 3e'+1+i \ \&\ \beta_0^i(e,s) < r_Q^i(e',s)]$$
$$\text{or}$$
$$[n = 3e' \ \&\ \mathrm{P}_{e'} \text{ requires attention via } i \text{ at stage } s+1]$$
$$\text{or}$$
$$[n = 3e'+1+i \ \&\ \mathrm{Q}_{e'}^i \text{ requires attention at stage } s+1]$$

holds, and let $n_i = 3e$ if no such number n exists; fix i_0 such that either $n_{i_0} < n_{1-i_0}$ or $i_0 = 0$ and $n_0 = n_1$; and do the following.

(1) If $\beta_0^{1-i_0}(e,s) \notin A_{1-i_0,s}$ then put $\beta_0^{1-i_0}(e,s)$ into $A_{1-i_0,s+1}$ (below we will show that $\beta_0^{1-i_0}(e,s) \notin A_{1-i_0,s}$ holds), and
(2) initialize P_e.

In either case, declare that P_e *becomes active* at stage $s+1$, and initialize all requirements $P_{e'}$ with $e' > e$. Moreover, initialize all requirements $Q^i_{e'}$ such that $A_{i,s+1} \restriction r^i_Q(e',s) \neq A_{i,s} \restriction r^i_Q(e',s)$. (Note that numbers can go into A_i in the first step of a stage only. So $A_{i,s+1}$ is given by the end of this step.)

Step 2. Fix e minimal such that $\beta(e,s)\uparrow$ and do the following (for $i = 0,1$).

Let $\beta(e,s+1) = \langle s+1, 2^{s+1}\rangle$,
let $\vec{\beta}^i(e,s+1) = (\langle s+1,0\rangle,\ldots,\langle s+1,2^e\rangle)$,
let $\hat{\sigma}_i(e,s+1) = 1^e$,
let $r^i_P(e,s+1) = 0$, and
declare that P_e *becomes active* at stage $s+1$.
Moreover, initialize all requirements $P_{e'}$ with $e < e'$ (actually this action is vacuous since $\beta(e',s)\uparrow$ for $e' > e$).

Step 3. Fix e minimal as in Step 1 or — if Step 1 is vacuous — as in Step 2.
For any $e' < e$ such that $P_{e'}$ requires attention at stage $s+1$, fix the unique $i \leq 1$ such that $P_{e'}$ requires attention via i at stage $s+1$, let $r^i_P(e',s+1) = s+1$, and declare that $P_{e'}$ *receives attention (via i)* at stage $s+1$.

Finally, for any $e' < e$ and $i \leq 1$ such that $A_{i,s+1} \restriction s+1 = A_{i,s} \restriction s+1$ and $Q^i_{e'}$ requires attention at stage $s+1$, let $r^i_Q(e',s+1) = s+1$ and declare that $Q^i_{e'}$ *receives attention* at stage $s+1$. Moreover, if (4.7) holds (for e' in place of e) then, for the least x as there, put x into $C_{i,n}$ (where $e' = \langle n,m\rangle$ for some m).

This completes the construction.

Note that in this construction the terms *receiving attention* and *becoming active* have different meanings. A coding requirement P_e can become active only in Step 1 or 2 of a stage whereas it can receive attention only in Step 3 (and the latter, but not the former, only occurs if P_e requires attention). An inaccessibility requirement Q^i_e does not become active at all but it may receive attention in Step 3 of a stage at which it requires attention. The coding requirement P_e may enumerate a number into A_0 or A_1 only if it becomes active (and this happens (if and) only if it becomes active in Step 1 of the stage). If a (coding or inaccessibility) requirement receives attention then it imposes a restraint (on one of the sets A_0 or A_1) but does not initialize any (coding or inaccessibility) requirements. If P_e becomes active then it initializes the lower priority coding requirements as well as the inaccessibility requirements which are injured by the action of P_e. (The action of P_e may also injure the restraint imposed by some higher priority coding requirement $P_{e'}$, so that this restraint becomes useless. Still, if this happens, for technical convenience, we do not cancel the restraint. Similarly, we let such a higher priority requirement $P_{e'}$ receive attention if it requires attention no matter whether or not the restraint it wants to impose became injured in Step 1 of stage $s + 1$.)

Proof of Correctness.

Obviously the construction satisfies (4.4). Hence (4.3) holds. So, by Claim 4.6, it suffices to show that the marker β satisfies the conditions (β_1)–(β_{10}) (hence, in particular, the coding requirements P_e are met) and that the inaccessibility requirements Q^i_e are met.

We first show that the conditions (β_1)–(β_{10}) are satisfied where we proceed in three steps: First, following some basic observations on the moves of the marker β, we show that (β_1)–(β_5) are satisfied (Claim 4.7). Then, using some observations on the relations among the positions of the marker β on e and the corresponding traces and declared states together with the fact that the least i-e-trace does not enter A_i (Claim 4.8), we prove conditions (β_7)–(β_{10}) (Claim 4.9). Finally, for a proof of (β_6), we show that the coding requirements act and require attention only finitely often and the inaccessibility requirements require attention only finitely often, from which we deduce that not only the marker β on e but also the restraints, traces and declared states come to a limit (Claim 4.10).

Note that at any stage $s + 1 > 0$ there is a unique number e, in the following denoted by e_{s+1}, such that P_e becomes active at stage $s + 1$. Since all coding requirements are initialized at stage 0, since all coding requirements $P_{e'}$ with $e' > e_{s+1}$ are initialized at stage $s + 1$, and since — unless $\beta(e_{s+1}, s)\downarrow$ — e_{s+1} is the least e such that $\beta(e, s)\uparrow$, it follows by a straightforward induction on s that (β_0) and (for $e = e_{s+1}$) (β_{00}) hold. As observed above, this implies

Claim 4.7: Conditions (β_1)–(β_5) are satisfied.

To verify conditions (β_7)–(β_{10}) next (and for the remainder of the proof in general), we start with some observations on the relations between the marker $\beta(e,s)$ and the corresponding traces $\beta^i_k(e,s)$ and their declared *i-e*-states $\hat{\sigma}_i(e,s)$ and introduce some notation.

It is immediate by construction that (for $i \leq 1$ and $e, s \geq 0$)

$$\beta(e,s)\downarrow \Leftrightarrow \vec{\beta}^i(e,s)\downarrow \Leftrightarrow \hat{\sigma}_i(e,s)\downarrow \tag{4.9}$$

holds. (In the following we will tacitly use this fact.) So, if we call a stage s an *e-stage* if $\beta(e,s)$ is defined, then the parameters $\beta(e,s)$, $\vec{\beta}^i(e,s)$ and $\hat{\sigma}_i(e,s)$ attached to P_e are defined at stage s if and only if s is an e-stage. Also note that, by (β_1), any e-stage s is greater than e and any e-stage is an e'-stage for all $e' \leq e$. Moreover, for any e-stage s, $\vec{\beta}^i(e,s) = (\beta^i_0(e,s),\ldots,\beta^i_{2^e}(e,s))$ where $\beta^i_0(e,s) < \beta^i_1(e,s) < \cdots < \beta^i_{2^e}(e,s)$. When dealing with these traces, we sometimes identify the vector $\vec{\beta}^i(e,s)$ (if defined) with the corresponding finite set, i.e., let $\vec{\beta}^i(e,s) = \{\beta^i_0(e,s),\ldots,\beta^i_{2^e}(e,s)\}$, and we write $\vec{\beta}^i(e,s) < \vec{\beta}^i(e',s')$ if $\vec{\beta}^i(e,s)$ and $\vec{\beta}^i(e',s')$ are defined and $\beta^i_{2^e}(e,s) < \beta^i_0(e',s')$ (i.e., $\max \vec{\beta}^i(e,s) < \min \vec{\beta}^i(e',s')$) and, similarly, $\vec{\beta}^i(e,s) < y$ ($y < \vec{\beta}^i(e,s)$) if $\vec{\beta}^i(e,s)$ is defined and $\beta^i_{2^e}(e,s) < y$ ($y < \beta^i_0(e,s)$).

Moreover, the following notions will be useful. We say that $\vec{\beta}^i(e)$ is *newly appointed at stage* $s+1$ if P_e is active in Step 2 of stage $s+1$ (i.e., iff $\vec{\beta}^i(e,s)\uparrow \neq \vec{\beta}^i(e,s+1)\downarrow$), and we say that $\vec{\beta}^i(e)$ *becomes upgraded at stage* $s+1$ if P_e is i-active at stage $s+1$ (i.e., iff $\vec{\beta}^i(e,s)\downarrow \neq \vec{\beta}^i(e,s+1)\downarrow$). For any e-stage s, we let $s(e,s)$ be the greatest stage $\leq s$ at which $\vec{\beta}^i(e)$ is newly appointed (note that s does not depend on i since $\vec{\beta}^0(e)$ and $\vec{\beta}^1(e)$ are newly appointed at the same stages), and we let $s^i(e,s)$ be the greatest stage $\leq s$ at which $\vec{\beta}^i(e)$ is newly appointed or becomes upgraded. Moreover, we call e-stages s' and s'' *e-equivalent* ($s' \sim_e s''$) if any stage s with $\min(s',s'') \leq s \leq \max(s',s'')$ is an e-stage, and we say that $\vec{\beta}^i(e,s)$ is *upgraded* if s is an e-stage and $\vec{\beta}^i(e)$ becomes upgraded at stage $s^i(e,s)$. Note that, for an e-stage s, $s(e,s)$ is the least stage $s' \leq s$ with $s' \sim_e s$ and that $s(e,s) \leq s^i(e,s) \leq s$. Also note that

$$\beta(e,s)\downarrow \Rightarrow \vec{\beta}^i(e,s) = \vec{\beta}^i(e,s^i(e,s))\downarrow \tag{4.10}$$

since P_e neither becomes i-active nor is initialized at any stage t with $s^i(e,s) < t \leq s$. Moreover, if $\vec{\beta}^i(e)$ is newly appointed at stage $s+1$ then, by (β_1) and by construction, $e \leq s$, $\vec{\beta}^i(e, s+1) = (\langle s+1, 0\rangle, \ldots, \langle s+1, 2^e\rangle)$ and $\beta(e, s+1) = \langle s+1, 2^{s+1}\rangle$ whence

$$\begin{aligned}\beta(e,s)\downarrow \,\&\, s(e,s) = s^i(e,s) \Rightarrow e < s^i(e,s) &\leq \vec{\beta}^i(e, s^i(e,s)) \\ &= (\langle s^i(e,s), 0\rangle, \ldots, \langle s^i(e,s), 2^e\rangle) \\ &< \langle s^i(e,s), 2^{s^i(e,s)}\rangle = \beta(e, s^i(e,s))\end{aligned} \tag{4.11}$$

holds, while if $\vec{\beta}^i(e)$ is upgraded at stage $s+1$ then $\beta(e,s) < \vec{\beta}^i(e, s+1) < s$, and none of the new i-e-traces has been previously enumerated into A_i (and none is enumerated into A_i at stage $s+1$) whence

$$\begin{aligned}\beta(e,s)\downarrow \,\&\, s(e,s) < s^i(e,s) \Rightarrow\ & \beta(e, s^i(e,s)-1) < \vec{\beta}^i(e, s^i(e,s)) < s^i(e,s) \\ & < \langle s^i(e,s), 2^{s^i(e,s)}\rangle = \beta(e, s^i(e,s)) \\ & \text{and} \\ & \vec{\beta}^i(e, s^i(e,s)) \cap A_{i,s^i(e,s)} = \emptyset\end{aligned} \tag{4.12}$$

holds. Together with (β_3) the preceding three observations imply that the traces $\vec{\beta}^i(e,s)$ (if defined) are contained in the open interval $(e, \beta(e,s))$,

$$\beta(e,s)\downarrow \Rightarrow\ e < \vec{\beta}^i(e,s) < \beta(e,s) \leq \langle s, 2^s\rangle, \tag{4.13}$$

that the traces $\vec{\beta}^i(e,s')$ (if defined) exceed $\langle s, 2^s\rangle$ for any previous stage s at which the traces were undefined,

$$s < s' \,\&\, \beta(e,s)\uparrow \,\&\, \beta(e,s')\downarrow \Rightarrow\ \langle s, 2^s\rangle < \vec{\beta}^i(e,s') \tag{4.14}$$

and that the traces $\vec{\beta}^i(e,s)$ are nondecreasing in s (where defined),

$$s < s' \,\&\, \beta(e,s)\downarrow \,\&\, \beta(e,s')\downarrow \Rightarrow\ \vec{\beta}^i(e,s) = \vec{\beta}^i(e,s') \ \vee\ \vec{\beta}^i(e,s) < \vec{\beta}^i(e,s'). \tag{4.15}$$

(For the latter two facts, recall that the pairing function is strictly increasing in either argument and satisfies $\langle s, 2^s\rangle < \langle s+1, 0\rangle$ for all s, and observe that, for e-equivalent stages $s < s'$, either $\vec{\beta}^i(e,s) = \vec{\beta}^i(e,s') = \vec{\beta}^i(e, s^i(e,s))$ (namely if $s^i(e,s) = s^i(e,s')$) or $\vec{\beta}^i(e,s) < \beta(e, s^i(e,s)) < \vec{\beta}^i(e,s')$ (namely if $s^i(e,s) < s^i(e,s')$).) It follows that, at any stage s, the sets of i-traces of the coding requirement are pairwise different, in fact

$$e < e' \,\&\, \beta(e',s)\downarrow \Rightarrow\ \vec{\beta}^i(e,s) < \beta(e, s^i(e,s)) < \vec{\beta}^i(e',s) \tag{4.16}$$

holds. For a proof of (4.16), fix e, e', s such that $e < e'$ and s is an e'-stage. By (4.10) and (4.13), it suffices to show that $\beta(e, s^i(e,s)) < \vec{\beta}^i(e',s)$. But, since $\beta(e, s^i(e,s)) = \langle s^i(e,s), 2^{s^i(e,s)}\rangle$ and since $\mathrm{P}_{e'}$ is initialized at stage $s^i(e,s)$, this is immediate by (4.14).

Since a number y is enumerated into A_i at stage $s+1$ only if y is an i-e-trace (for some $e \geq 0$) at stage s, (4.13) implies

$$A_{i,s+1} \subseteq \omega \upharpoonright \langle s, 2^s \rangle. \tag{4.17}$$

So if $\vec{\beta}^i(e)$ is newly appointed at stage $s+1$ then, by (4.11), the newly appointed i-e-traces are not in $A_{i,s+1}$. Since the corresponding observation for traces becoming upgraded at stage $s+1$ trivially holds (compare (4.12)),

$$\beta(e,s) \downarrow \Rightarrow \vec{\beta}^i(e, s^i(e,s)) \cap A_{i,s^i(e,s)} = \emptyset \tag{4.18}$$

follows. Moreover,

$$\begin{aligned} s \sim_e s+1 \;\&\; A_{i,s+1} \upharpoonright \langle s^i(e,s), 2^{s^i(e,s)} \rangle \neq A_{i,s} \upharpoonright \langle s^i(e,s), 2^{s^i(e,s)} \rangle \;\Rightarrow \\ [\, \mathrm{P}_e \text{ is } (1-i)\text{-active at stage } s+1 \;\& \\ \exists\, k \leq 2^e \; (A_{i,s+1} \setminus A_{i,s} = \{\beta^i_k(e, s^i(e,s))\})\,] \,. \end{aligned} \tag{4.19}$$

Namely, since P_e becomes active at stage $s^i(e,s)$, all lower priority requirements $\mathrm{P}_{e''}$, $e < e''$, are initialized at stage $s^i(e,s)$ hence (by (4.14)) enumerate only numbers $> \langle s^i(e,s), 2^{s^i(e,s)} \rangle$ into A_i after stage $s^i(e,s)$, and, since $s+1$ is an e-stage, no higher priority requirement $\mathrm{P}_{e'}$, $e' < e$, may act at stage $s+1$. So P_e is the only requirement which may enumerate a number into A_i at stage $s+1$ and — since $s+1$ is an e-state — this happens only if P_e is $(1-i)$-active at stage $s+1$ whence the claim follows by (4.10).

Since the i-e-traces are enumerated into A_i in decreasing order, (4.18) and (4.19), together with (4.10), (4.13) and the fact that no number enters A_i at stage $s^i(e,s)$, imply (for $k \leq 2^e$)

$$\begin{aligned} \beta(e,s) \downarrow \;\&\; \beta^i_k(e, s^i(e,s)) \notin A_{i,s} \;\Rightarrow \\ A_{i,s} \upharpoonright \beta^i_{k+1}(e, s^i(e,s)) = A_{i,s^i(e,s)-1} \upharpoonright \beta^i_{k+1}(e, s^i(e,s)) \\ (\text{where we let } \beta^i_{2^e+1}(e, s^i(e,s)) = \beta(e, s^i(e,s))). \end{aligned} \tag{4.20}$$

Next we turn to the declared states. Note that, correspondingly to (4.10),

$$\beta(e,s) \downarrow \Rightarrow \hat{\sigma}_i(e,s) = \hat{\sigma}_i(e, s^i(e,s)) \downarrow \tag{4.21}$$

holds (hence $\vec{\beta}^i(e,s)$ is upgraded iff $\hat{\sigma}_i(e,s) < 1^e$). Moreover, if defined at stages s and $s+1$, then $\hat{\sigma}_i(e,s) = \hat{\sigma}_i(e,s+1)$ unless P_e becomes i-active at stage $s+1$, in which case $\hat{\sigma}_i(e,s+1) < \hat{\sigma}_i(e,s)$. Hence

$$s' \leq s'' \;\&\; s' \sim_e s'' \;\Rightarrow\; \hat{\sigma}_i(e,s'') \leq \hat{\sigma}_i(e,s') \downarrow \tag{4.22}$$

and

$$s \sim_e s+1 \Rightarrow (\hat{\sigma}_i(e, s+1) < \hat{\sigma}_i(e, s) \Leftrightarrow \text{P}_e \text{ is } i\text{-active at stage } s+1) \tag{4.23}$$

hold. Since the declared i-e-state can assume only 2^e different values hence can be decreased at most $2^e - 1$ times, (4.22) and (4.23) imply

$$\beta(e, s) \downarrow \Rightarrow \ |\{s' : s \sim_e s' \ \& \ \text{P}_e \text{ is } i\text{-active at stage } s'\}| < 2^e. \tag{4.24}$$

Since there are $2^e + 1$ i-e-traces (if defined) and since i-e-traces are enumerated into A_i in decreasing order (compare (4.20)) it follows (from (4.24) with $1 - i$ in place of i) by (4.18), (4.19) and (4.20) that

$$\beta(e, s) \downarrow \Rightarrow \quad \begin{array}{c} \beta_0^i(e, s^i(e, s)) \notin A_{i,s} \text{ and} \\ A_{i,s} \upharpoonright \beta_1^i(e, s^i(e, s)) = A_{i,s^i(e,s)-1} \upharpoonright \beta_1^i(e, s^i(e, s)). \end{array} \tag{4.25}$$

By (4.10), the latter implies

Claim 4.8: Let s be an e-stage. Then, for $i \leq 1$, $\vec{\beta}^i(e, s) = \vec{\beta}^i(e, s^i(e, s))$, $\beta_0^i(e, s) \notin A_{i,s}$, and $A_{i,s} \upharpoonright \beta_1^i(e, s) = A_{i,s^i(e,s)-1} \upharpoonright \beta_1^i(e, s)$.

Note that Claim 4.8 implies that if P_e is i-active at stage $s+1$ then there is a number $k \leq 2^e$ such that $\beta_k^{1-i}(e, s) \notin A_{1-i,s}$, and, for the greatest such k, the $(1-i)$-e-trace $\beta_k^{1-i}(e, s)$ is enumerated into A_{1-i} at stage $s+1$; and, similarly, if Case 2 of Step 1 applies to stage $s+1$ then $\beta_0^{1-i_0}(e, s)$ is not in $A_{1-i_0,s}$ and $\beta_0^{1-i_0}(e, s)$ is enumerated into A_{1-i_0} at stage $s+1$. So if P_e becomes active via Step 1 at stage $s+1$ then, for some $i \leq 1$, there is an i-e-trace $\beta_k^i(e, s)$ which is newly enumerated into A_i at stage $s+1$. Since, for the least e such that $\beta(e, s) \downarrow \neq \beta(e, s+1)$ (if any), P_e becomes active in Step 1 of stage $s+1$ it follows that

$$\begin{array}{c} \beta(e, s) \downarrow \neq \beta(e, s+1) \ \& \ \forall\, e' < e\, (\beta(e', s) = \beta(e', s+1)) \Rightarrow \\ \exists\, i \leq 1\, \exists\, k \leq 2^e\, (\beta_k^i(e, s) \in A_{i,s+1} \setminus A_{i,s}) \end{array} \tag{4.26}$$

holds. This observation is crucial for establishing the following claim.

Claim 4.9: Conditions (β_7)–(β_{10}) are satisfied.

Proof: For a proof of (β_7) note that, for the least e such that $\beta(e, s) \downarrow\neq \beta(e, s+1)$, P_e becomes active in Step 1 of stage $s+1$ whence $\beta(e, s)$ is M-permitted. By (β_2) this implies (β_7). (β_8) is immediate by (4.26), (4.13), and (β_2). For a proof of (β_9) assume $y \in A_{i,s+1} \setminus A_{i,s}$. Then $y = \beta_k^i(e, s)$ for some e and $k < 2^e$ where P_e becomes active in Step 1 of stage $s+1$ whence $\beta(e, s+1) \neq \beta(e, s) \downarrow$. So (β_9) follows from (4.13). Finally, for a proof of (β_{10}), w.l.o.g. assume that $\beta(a(s), s) \downarrow$. Then, since $a(s) < \beta(a(s), s)$, $\beta(a(s), s)$ is M-permitted at stage $s + 1$. So either $\mathrm{P}_{a(s)}$ becomes active according to Case 2 in Step 1 of stage $s+1$ or a higher priority requirement P_e, $e < a(s)$, becomes active. In either case, $\mathrm{P}_{a(s)}$ is initialized at stage $s + 1$, hence, $\beta(a(s), s + 1) \uparrow$. □

Next we show that all requirements require attention only finitely often, that the coding requirements become active only finitely often, and that the marker β reaches a final position $\beta^*(e) \in \omega$ for all $e \geq 0$ whence the marker condition (β_6) is satisfied too.

Claim 4.10: (a) For $e \geq 0$, the coding requirement P_e becomes active only finitely often, requires attention only finitely often and is initialized only finitely often. Moreover, $\beta^*(e) = \lim_s \beta(e, s) \in \omega$ whence, in particular, (β_6) is satisfied.

(b) For $e \geq 0$ and $i \leq 1$, the inaccessibility requirement Q^i_e requires attention only finitely often and is initialized only finitely often.

Proof: The two parts of the claim are proven simultaneously by induction on the index n of the corresponding requirements $\mathrm{R}_n = \mathrm{P}_e$ and $\mathrm{R}_n = \mathrm{Q}^i_e$.

Fix n and, by inductive hypothesis, fix a stage $s_0 > 0$ such that no requirement $\mathrm{R}_{n'}$ with $n' < n$ requires attention, becomes active, or is initialized after stage $s_0 - 1$. So any restraint imposed by such a higher priority requirement at stage s_0 is permanent and is bounded by s_0.

Next fix $s_1 \geq s_0$ such that, for any coding requirement $\mathrm{P}_{e''}$ which becomes active after stage s_1 and for any $i \leq 1$ and $s \geq s_1$ such that $\beta^i_0(e'', s)$ is defined, $s_0 < \beta^i_0(e'', s)$. The existence of such a stage s_1 is established as follows. First note that, by (4.13), $\beta^i_0(e'', s) \leq s_0$ implies that $e'' < s_0$. So, by (4.14), it suffices to show that any requirement $\mathrm{P}_{e''}$ which becomes active infinitely often is initialized infinitely often. But this is immediate by (4.24).

Note that, by choice of s_1, for any requirement $\mathrm{P}_{e''}$ which becomes active according to Case 2 in Step 1 of a stage $s+1 > s_1$, $n \leq n_i$ for the parameter n_i ($i = 0, 1$) defined there. Namely, by $s_1 \geq s_0$, $3e'' \geq n$, no requirement $\mathrm{R}_{n'}$ with $n' < n$ requires attention at stage $s+1$, and the restraint imposed by such a requirement $\mathrm{R}_{n'}$ is bounded by s_0 hence is less than $\beta^i_0(e'', s)$ by choice of s_1.

Now distinguish the following two cases according to the type of requirement R_n.

Case 1: $R_n = P_e$ *(i.e.,* $n = 3e$*).* It suffices to show that

$$\exists\, s\, \forall\, s' \geq s\, (\beta(e, s') \downarrow) \tag{4.27}$$

holds. Namely, fix s as in (4.27). Then, by (4.24), P_e becomes active only finitely often after stage s. Since P_e can be initialized only at a stage where some $P_{e'}$ with $e' \leq e$ becomes active, it follows that P_e is initialized only finitely often. Since $\beta(e, s') \downarrow \neq \beta(e, s'+1) \downarrow$ implies that P_e becomes active at stage $s'+1$, it follows that $\beta^*(e) = \lim_{s' \to \omega} \beta(e, s')$ exists and, by (4.27), $\beta^*(e) \in \omega$. Finally, in order to show that P_e requires attention only finitely often, let s' be the last stage at which some $P_{e'}$ with $e' \leq e$ becomes active. It suffices to observe that P_e receives attention at any stage $> s'$ at which it requires attention and that P_e receives attention at most once after stage s' since the restraint which is imposed then will never be cancelled.

Now, for a proof of (4.27), for a contradiction assume that $\beta(e, s) \uparrow$ for infinitely many s. Fix $s_2 > s_1$ such that $a(s) > e$ for all $s \geq s_2$ and such that $\beta(e, s_2) \uparrow$. By the latter, P_e becomes active in Step 2 of stage $s_2 + 1$ hence $\beta(e, s_2+1) \downarrow$ and $r^0_P(e, s_2+1) = r^1_P(e, s_2+1) = 0$. So, by assumption, we may fix $s_3 > s_2$ minimal such that $\beta(e, s_3+1) \uparrow$. Since no higher priority requirement may act at stage s_3+1 (by $s_3 \geq s_0$), it follows that P_e becomes active according to Case 2 in Step 1 of stage s_3+1. Moreover, since $e < a(s_3)$ (by $s_3 \geq s_2$) and since, for $i \leq 1$, $R^i(e, s_3) = R^i(e, s_0 - 1) \leq s_0 < \beta^i_0(e, s_3)$ (by choice of s_0 and by $s_3 \geq s_1$), the conditions (a) and (b) given there fail. So (c) must hold, i.e., there is a (unique) number $i \leq 1$ such that $r^i_P(e, s_3) > 0$. On the other hand, since P_e does not become active according to Case 1, there is no string $\sigma \in \{0,1\}^e$ and no vector $\vec{y}$ such that $\sigma < \hat{\sigma}_i(e, s_3)$ and $\vec{y}$ is i-σ-$\max\{R^i(e, s_3), \beta(e, s_3)\}$-eligible at stage $s_3 + 1$.

Now, by $r^i_P(e, s_2) = 0 < r^i_P(e, s_3)$, fix s maximal such that $s_2 \leq s < s_3$ and $r^i_P(e, s) = 0$. Then P_e requires and receives attention via i at stage $s+1$ whence there is a string $\sigma \in \{0,1\}^e$ and a vector $\vec{y}$ such that $\sigma < \hat{\sigma}_i(e, s)$ and $\vec{y}$ is i-σ-$\max\{R^i(e, s), \beta(e, s)\}$-eligible at stage $s+1$, and, by maximality of s, $r^i_P(e, s') = s+1$ for all s' with $s < s' \leq s_3$. By the latter, P_e does not become active at any such stage s' (since $r^i_P(e, s') = 0$ for any stage s' at which P_e is active) hence $\hat{\sigma}_i(e, s) = \hat{\sigma}_i(e, s_3)$ and $\beta(e, s) = \beta(e, s_3)$. So, since, by choice of s_0, $R^i(e, s) = R^i(e, s_3)$ too, in order to get the desired contradiction it suffices to show that $\vec{y}$ is i-σ-$\max\{R^i(e, s), \beta(e, s)\}$-eligible at stage $s_3 + 1$. For this sake, by definition of i-σ-$\max\{R^i(e, s), \beta(e, s)\}$-eligibility and by Claim 4.4, it suffices to show that $A_{i,s_3} \upharpoonright s + 1 = A_{i,s} \upharpoonright s + 1$ since, by choice of s_0, a requirement $Q^i_{e'}$ with $e' < e$ is satisfied at stage s iff it is satisfied at stage s_3, and does not require attention after stage s_0.

Now since neither P_e nor a higher priority requirement becomes active at a stage $s'+1$ where $s \leq s' < s_3$, for a proof of $A_{i,s_3} \restriction s+1 = A_{i,s} \restriction s+1$, it suffices to show that no lower priority requirement enumerates a trace $< s+1$ into A_i at such a stage $s'+1$. So fix $\mathrm{P}_{e''}$, $e < e''$, and, for a contradiction, assume that $\beta^i_k(e'', s') < s+1$ is in $A_{i,s'+1} \setminus A_{i,s'}$. Since, by choice of s', either P_e requires attention via i at stage $s'+1$ or $r^i_P(e, s') = s+1$, it follows that $\beta^i_0(e'', s') < s+1 \leq R^i(e'', s'+1)$. So $\mathrm{P}_{e''}$ becomes active according to Case 2 in Step 1 of stage $s'+1$. Moreover, for the parameters n_0 and n_1 defined there, $n_i \leq 3e = n$ (since P_e requires attention via i at stage $s'+1$ or $r^i_P(e, s') = s+1$). On the other hand, by $s' > s_1$, $n \leq n_{1-i}$ as observed above. But, since, for a stage s' such that P_e requires attention via i at stage $s'+1$ or $r^i_P(e, s') = s+1$, P_e does not require attention via $1-i$ at stage $s'+1$ and $r^{1-i}_P(e, s') = 0$, the latter implies that $n_{1-i} > n$ hence $n_i < n_{1-i}$. But, by construction, this implies that $\mathrm{P}_{e''}$ does not enumerate a trace in A_i at stage $s'+1$ which gives the desired contradiction and completes the proof of (4.27).

Case 2: $\mathrm{R}_n = \mathrm{Q}^i_e$ *(i.e.,* $n = 3e+1+i$*).* If Q^i_e does not require attention after stage s_1, then it can only be initialized once after stage s_1 and the result holds for the requirement, so we can assume that Q^i_e requires attention at stage $s+1 > s_1$ and it suffices to show that Q^i_e receives attention and is permanently satisfied at stage $s+1$.

In order to show that Q^i_e receives attention at stage $s+1$, it suffices to show that $A_{i,s+1} \restriction s+1 = A_{i,s} \restriction s+1$. For a contradiction, assume that this is not the case. Then, for some $e'' > e$, $\mathrm{P}_{e''}$ becomes active at stage $s+1$ and enumerates a trace $\beta^i_k(e'', s) \leq s$ into A_i. Since, by Q^i_e requiring attention, $R^i(e'', s) = s+1$ it follows that $\beta^i_0(e'', s) < R^i(e'', s)$. So $\mathrm{P}_{e''}$ acts according to Case 2 of Step 1 of stage s+1 and, for the parameters n_0 and n_1 defined there, $n_{1-i} \leq n_i$ since otherwise $\mathrm{P}_{e''}$ would not enumerate a trace into A_i. But, by Q^i_e requiring attention, $n_i \leq 3e+1+i = n$ while on the other hand, as observed above, $n \leq n_{1-i}$. Moreover, by definition, n_{1-i} cannot attain the value $3e+1+i$. So $n_i < n_{1-i}$ giving the desired contradiction.

The proof that Q^i_e is permanently satisfied at stage $s+1$ is similar. For a contradiction, fix $s' > s$ minimal such that Q^i_e is initialized at stage $s'+1$. Then $A_{i,s'} \restriction r^i_Q(e, s') \neq A_{i,s'+1} \restriction r^i_Q(e, s')$ where, by minimality of s', $r^i_Q(e, s') = r^i_Q(e, s+1) = s+1$. So (by the priority ordering of the requirements and by choice of s_0) a coding requirement $\mathrm{P}_{e''}$ with $e'' > e$ becomes active at stage $s'+1$ and enumerates a trace $\beta^i_k(e'', s') < r^i_Q(e, s')$

into A_i. Since $r^i_Q(e,s') \leq R^i(e'',s')$ and $\beta^i_0(e'',s') \leq \beta^i_k(e'',s')$, it follows that $\mathrm{P}_{e''}$ acts according to Case 2 in Step 1 of stage $s'+1$ and that $n_i \leq 3e+1+i = n$. Moreover, since $\mathrm{P}_{e''}$ enumerates $\beta^i_k(e'',s')$ into A_i, $n_{1-i} \leq n_i$. But this is impossible, since (as observed above) $n \leq n_{1-i}$ by choice of s_1 and $n_{1-i} \neq 3e+1+i$ by definition. □

By Claim 4.10, the parameters attached to the coding and inaccessibility requirements come to a limit. So we may let

$$\vec{\beta}^i(e) = (\beta^i_0(e), \ldots, \beta^i_{2^e}(e)) = \lim_{s\to\omega} \vec{\beta}^i(e,s) \in \omega^{2^e+1},$$

$$\hat{\sigma}_i(e) = \lim_{s\to\omega} \hat{\sigma}_i(e,s),$$

$$r^i_P(e) = \lim_{s\to\omega} r^i_P(e,s) < \omega,$$

and

$$r^i_Q(e) = \lim_{s\to\omega} r^i_Q(e,s) < \omega.$$

Moreover, we let

$$r_P(e) = \max\{r^0_P(e), r^1_P(e)\} \text{ and } r_P(e,s) = \max\{r^0_P(e,s), r^1_P(e,s)\},$$

and we let t_e+1 and t^i_e+1 be the last stages at which P_e becomes active, and at which the i-e-traces become newly defined or upgraded, respectively. Note that $t^i_e \leq t_e$ and, for $e < e'$, $t_e < t^i_{e'}$. Moreover, for all $s \geq t_e$, $\beta(e,s+1) = \beta(e,t_e+1) = \beta^*(e)$, while, for all stage $s \geq t^i_e$, $\vec{\beta}^i(e,s+1) = \vec{\beta}^i(e,t^i_e+1) = \vec{\beta}^i(e)$, $\hat{\sigma}^i(e,s+1) = \hat{\sigma}^i(e,t^i_e+1) = \hat{\sigma}^i(e)$, and $s^i(e,s+1) = t^i_e+1$ (hence, in particular, $s+1 \sim_e t^i_e+1$). Also note that, by (4.25), the latter imply

$$\beta^i_0(e) \notin A_i \text{ and } A_i \restriction \beta^i_1(e) = A_{i,t^i_e} \restriction \beta^i_1(e). \tag{4.28}$$

It remains to show that the inaccessibility requirements Q^i_e are met.

Claim 4.11: Q^i_e is met.

Proof: For the proof, fix $e \geq 0$ and $i \leq 1$ and fix the least stage s_0 such that no requirement of higher priority than P_{e+1} becomes active or requires attention or is initialized after stage s_0 (note that Q^0_e and Q^1_e, hence $\mathrm{Q}^i_{e'}$, have higher priority than P_{e+1}).

Note that, by choice of s_0,

$$e < e'' \ \& \ \beta(e'',s') \downarrow \ \& \ s_0 < s^i(e'',s') - 1 \ \& \ y = \beta^i_k(e'',s') \notin A_{i,s'} \ \Rightarrow \\ \sigma_i(e+1,y,s') \preceq \hat{\sigma}_i(e'',s') \restriction e+1 \tag{4.29}$$

holds. (Compare with Claim 3.8 in the proof of Theorem 2.7.) The proof is as follows. By (4.10), $\vec{\beta}^i(e'',s') = \vec{\beta}^i(e'',s^i(e'',s'))$ hence $y = \beta^i_k(e'',s^i(e'',s'))$. It follows, by construction, that $\sigma_i(e'',y,s^i(e'',s')-1) \preceq \hat{\sigma}_i(e'',s^i(e'',s'))$ and, for any $e' \leq e$ such that $\hat{\sigma}_i(e'',s^i(e'',s'))(e') = 0$ and $\mathrm{Q}^i_{e'}$ is not satisfied at stage $s^i(e'',s') - 1$, y is i-e'-eligible at stage $s^i(e'',s')$ via numbers x, u, v and r where $r < \beta^i_{k+1}(e'',s^i(e'',s'))$ (and where $\beta^i_{2^{e''}+1}(e'',s^i(e'',s')) = \langle s^i(e'',s'), 2^{s^i(e'',s')}\rangle$ by convention). Since $s_0 < s^i(e'',s') - 1$, it follows by (4.20) and by Claim 4.4, that, for any such e', y is i-e'-eligible at stage $s'+1$ too. Since, by $s_0 < s^i(e'',s')-1$, for $e' \leq e$, $\mathrm{Q}^i_{e'}$ is satisfied at stage s' iff $\mathrm{Q}^i_{e'}$ is satisfied at stage $s^i(e'',s') - 1$ it follows that $\sigma_i(e+1,y,s') \preceq \hat{\sigma}_i(e'',s^i(e'',s'))$. So the claim follows by (4.21).

The remainder of the proof is organized as the proof of the corresponding Claim 3.9 in the proof of Theorem 2.7. As there w.l.o.g. we may assume that Q^i_e is not satisfied at any stage $s \geq s_0$ and that the clauses (i) and (ii) in Q^i_e hold while clause $(*)$ fails. So, in particular, the assumptions made in Claim 4.2 and Claim 4.3 hold, and it suffices to show that there is an infinite computable set S of stages s satisfying (4.8).

For the definition of such a set S, first define *true i-e'-eligibility* and the i-$(e+1)$-state of y *truly preceding* the $(e+1)$-state σ at stage s ($\sigma_i(e+1,y,s) \preceq_t \sigma$) just as the corresponding notions in the proof of Theorem 2.7 with A_i, $\mathrm{Q}^i_{e'}$ and i-e'-eligibility in place of A, $\mathrm{Q}_{e'}$ and e'-eligibility, respectively. In addition, say that $\sigma_i(e+1,y,s)$ precedes σ at stage s *via* r if, for any $e' \leq e$ such that $\sigma(e') = 0$ and $\mathrm{Q}^i_{e'}$ is not satisfied at stage s_0, y is i-e'-eligible via numbers x, u, v and r' at stage $s+1$ such that $r' \leq r$, and say that $\sigma_i(e+1,y,s)$ truly precedes σ at stage s *via* r if $\sigma_i(e+1,y,s)$ precedes σ at stage s via r and $A_i \restriction r = A_{i,s} \restriction r$. Then, obviously, the analogs of (3.58) and (3.59) hold and, for the newly introduced notion, we obtain the following variant of (3.59):

$$\forall\, s \geq s_0 \ \forall\, y \geq 0 \ \forall\, \sigma \in \{0,1\}^{e+1} \\ ([\sigma_i(e+1,y,s) \preceq \sigma \text{ via } r \ \& \ A_i \restriction r = A_{i,s} \restriction r] \Rightarrow \sigma_i(e+1,y,s) \preceq_t \sigma \text{ (via } r)). \tag{4.30}$$

Then, based on these modified notions, define σ^* correspondingly too:

$$\sigma^* = \mu\, \sigma \in \{0,1\}^{e+1} [\overset{\infty}{\exists}\, y \notin A_i \ \exists\, s \geq s_0 \ (\sigma_i(e+1,y,s) \preceq_t \sigma)].$$

Finally, fix $e_0 > e$ minimal such that

$$s_0 < t^i_{e_0}, \tag{4.31}$$

$$r_P(e_0) = 0 \tag{4.32}$$

and

$$\forall\, y \geq e_0\, \forall\, s \geq s_0\, \forall\, \sigma < \sigma^*\, (y \notin A_i \;\Rightarrow\; \sigma_i(e+1, y, s) \not\leq_t \sigma) \tag{4.33}$$

hold. Since (4.31) and (4.33) hold for almost all numbers e_0, in order to show that such a number e_0 exists, it suffices to show that there are infinitely many $e'' > e$ such that $r_P(e'') = 0$. For a contradiction, assume that this is not the case and fix $e_1 > e$ such that $r_P(e'') > 0$ for all $e'' \geq e_1$. Then, for any coding requirement $\mathrm{P}_{e''}$ with $e'' \geq e_1$ and for any stage $s+1$ at which $\mathrm{P}_{e''}$ becomes active (hence $r_P(e'', s+1) = 0$) and $\beta(e'', s+1)\downarrow$, there is a stage $s' > s$ such that $\mathrm{P}_{e''}$ receives attention at stage $s'+1$. So the function f which assigns to any stage s with $\beta(e'', s) \neq \beta(e'', s+1)\downarrow$ for some $e'' \geq e_1$ the least stage $s' > s+1$ at which $r_P(e'', s') > 0$ (while $f(s) = s$ otherwise) is total and computable. Since, for such e'' and s, $s < \beta(e'', s+1)$ it follows that, for almost all such pairs (e'', s), $\beta(e'', s+1)$ is M-permitted at some stage $s'' > f(s)$. So, unless $\mathrm{P}_{e''}$ became active or initialized at a stage t with $s+1 < t \leq s''$, $\beta(e'', s'') = \beta(e'', s+1)$ and $r_P(e'', s'') > 0$ whence $\mathrm{P}_{e''}$ becomes active at stage $s''+1$ or becomes initialized by some higher priority requirement becoming active. It follows that, for almost all e'' either the marker $\beta(e'')$ is moved infinitely often or is eventually permanently undefined. But this contradicts (β_6).

The properties (3.62), (3.63) and (3.64) of σ^*, which were crucial in the proof of Theorem 2.7, here become as follows.

$$\forall\, e'' \geq e_0\, \forall\, s > t_{e_0}\, (\hat{\sigma}_i(e'', s)\downarrow \Rightarrow \sigma^* \leq \hat{\sigma}_i(e'', s) \restriction e+1) \tag{4.34}$$

$$\exists\, e_1 \geq e_0\, \forall\, e'' \geq e_1\, (\hat{\sigma}_i(e'') \restriction e+1 = \sigma^*) \tag{4.35}$$

$$\sigma^*(e) = 0. \tag{4.36}$$

For a proof of (4.34), for a contradiction, fix $e'' \geq e_0$ minimal such that $\hat{\sigma}_i(e'', s) \restriction e+1 < \sigma^*$ for some $s > t_{e_0}$ and let s be the least such stage. Distinguish the following two cases depending on whether $e'' = e_0$ or $e_0 < e''$.

Case 1: $e'' = e_0$. Then, by $s > t_{e_0}$, $\hat{\sigma}_i(e_0, s) = \hat{\sigma}_i(e_0, t^i_{e_0} + 1) = \hat{\sigma}_i(e_0)$ (and $\vec{\beta}^i(e_0) = \vec{\beta}^i(e_0, t^i_{e_0} + 1)$) hence $\hat{\sigma}_i(e_0) \restriction e + 1 < \sigma^*$. So P_{e_0} becomes i-active at stage $t^i_{e_0} + 1$. It follows, by construction, that $\vec{\beta}^i(e_0)$ is i-$\hat{\sigma}_i(e_0)$-$\max\{R^i(e_0, t^i_{e_0}), \beta(e_0, t^i_{e_0})\}$-eligible at stage $t^i_{e_0} + 1$ hence $\sigma_i(e + 1, \beta^i_0(e_0), t^i_{e_0}) \preceq \hat{\sigma}_i(e_0) \restriction e + 1$ via $\beta^i_1(e_0)$. Since, by (4.28), $\beta^i_0(e_0) \notin A_i$ and $A_i \restriction \beta^i_1(e_0) = A_{i,t^i_{e_0}} \restriction \beta^i_1(e_0)$, it follows by (4.30) that $v_i(e+1, \beta^i_0(e_0), t^i_{e_0}) \preceq_t \hat{\sigma}_i(e_0) \restriction e + 1$. So, since $e_0 < \beta^i_0(e_0)$, $\hat{\sigma}_i(e_0) \restriction e + 1 \not< \sigma^*$ by (4.33) contrary to assumption.

Case 2: $e_0 < e''$. Let $s' = s^i(e'', s) - 1$. Then $t_{e_0} < s'$ since, by P_{e_0} becoming active at stage $t_{e_0} + 1$, $P_{e''}$ is initialized at stage $t_{e_0} + 1$, and (since $\hat{\sigma}_i(e'', s) < 1^{e''}$ by assumption) $P_{e''}$ becomes i-active at stage $s' + 1$. So, by construction, there is an i-$\hat{\sigma}_i(e'', s)$-$\max\{R^i(e'', s'), \beta(e'', s')\}$-eligible sequence $\vec{y}$ at stage $s' + 1$. By $e_0 < e''$ it follows that the sequence $\vec{y} \restriction 2^{e_0} + 1$ is i-$\hat{\sigma}_i(e'', s) \restriction e_0$-$\max\{R^i(e_0, s'), \beta(e_0, s')\}$-eligible at stage $s' + 1$. Since, by minimality of e'', $\hat{\sigma}_i(e'', s) \restriction e + 1 < \hat{\sigma}_i(e_0, s') \restriction e + 1$ (hence $\hat{\sigma}_i(e'', s) \restriction e_0 < \hat{\sigma}_i(e_0, s')$) and since, by choice of e_0, $r_P(e_0, s') = 0$, it follows that P_{e_0} requires attention at stage $s' + 1$ hence $r_P(e_0, s' + 1) = s' + 1$. Since P_{e_0} neither is initialized nor becomes active after stage $t_0 + 1$, it follows that $r_P(e_0) > 0$ contrary to choice of e_0.

The proof of (4.35) is indirect too. For a contradiction, assume that (4.35) fails. Then, by (4.34), $\sigma^* < \hat{\sigma}_i(e'') \restriction e + 1$ — hence $\sigma^* < \hat{\sigma}_i(e'', t_{e''} + 1) \restriction e + 1$ — for infinitely many numbers $e'' \geq e_0$. On the other hand, by choice of σ^*, for any $e'' \geq e_0$ and z there is a sequence $\vec{y}$ and a stage s' such that $\vec{y}$ is i-$\sigma^* 1^{e''-e-1}$-z-eligible at all stages $s \geq s'$. It follows that, for any $e'' \geq e_0$ and any stage $s + 1 > s_0$ such that $P_{e''}$ becomes active at stage $s + 1$ and $\sigma^* < \hat{\sigma}_i(e'', s + 1) \restriction e + 1$, there will be a least stage $s' > s$ such that $P_{e''}$ requires attention at stage $s' + 1$ or is initialized at this stage. So, by M-permitting, we may argue that for almost all (e'', s) as above either $P_{e''}$ becomes active after stage $s + 1$ or is initialized later whence $t_{e''} \neq s$. But this contradicts the assumption.

Finally, for a proof of (4.36), for a contradiction, assume that $\sigma^*(e) = 1$. Define $\sigma \in 2^{e+1}$ by $\sigma \restriction e = \sigma^* \restriction e$ and $\sigma(e) = 0$. Then, $\sigma < \sigma^*$ and we will show

$$\overset{\infty}{\exists}\, y \notin A_i\ \exists s \geq s_0\ (\sigma_i(e + 1, y, s) \preceq_t \sigma) \tag{4.37}$$

which contradicts the definition of σ^*.

To establish (4.37), define

$$E = \{y \notin A_i : \exists\, s \geq s_0\, (\sigma_i(e+1, y, s) \preceq_t \sigma^*)\}.$$

Then, E is clearly computably enumerable in A_i, and we claim that E is infinite. To see this, fix $e_1 \geq e_0$ as in (4.35) and for any $e'' \geq e_1$, consider $y = \beta_0^i(e'')$ and $s = t_{e''}^i$. Then, $s \geq s_0$ and, by (4.28), $y \notin A_i$. By construction and (4.35), $\sigma_i(e{+}1, y, s) = \sigma_i(e'', y, s) \restriction e{+}1 \preceq \hat{\sigma}_i(e'', s{+}1) \restriction e{+}1 = \sigma^*$ and $\sigma_i(e{+}1, y, s)$ precedes σ^* via $\beta_1^i(e'')$, so by (4.28) and (4.30), $\sigma_i(e{+}1, y, s) \preceq_t \sigma^*$ and thus $y \in E$. By (4.16), for $e'' \neq e'''$, $\beta_0^i(e'') \neq \beta_0^i(e''')$, so E is infinite.

It follows that E has an infinite A_i-computable subset E', and, by Claim 4.3, there is an infinite subset F of E' such that for every $y \in F$, y is truly i-e-eligible at all sufficiently large stages. For any $y \in F$, we also have $\sigma_i(e+1, y, s) \preceq_t \sigma^*$ for all sufficiently large stages s, so for any $y \in F$, $\exists s \geq s_0(\sigma_i(e+1, y, s) \preceq_t \sigma)$, establishing (4.37), and hence (4.36) by contradiction.

Now S is defined by

$$S = \{s > t_{e_1} : \forall\, e'' > e_1\, (\hat{\sigma}_i(e'', s)\downarrow \Rightarrow\ \hat{\sigma}_i(e'', s) \restriction e+1 = \sigma^*)\} \qquad (4.38)$$

where e_1 is as in (4.35).

Obviously, S is computable. Moreover, S is infinite since, by (4.35), $t_{e''}^i + 1 \in S$ for all $e'' > e_1$. It remains to show that the elements s of S satisfy (4.8). So fix $s \in S$, a number $y \leq s$ and a stage $s' \geq s$ such that $y \in A_{i,s'+1} \setminus A_{i,s'}$. We have to show that y is i-e-eligible at stage $s'+1$. Fix e'' and k such that $y = \beta_k^i(e'', s')$. Since y enters A_i at stage $s'+1 > t_{e_1}+1$ it follows that $e_1 < e''$. Moreover, since $y \leq s$, it follows (by (4.14)) that s and s' are e''-equivalent whence by (4.22), by $s \in S$ and by (4.34),

$$\hat{\sigma}_i(e'', s') \restriction e+1 \leq \hat{\sigma}_i(e'', s) \restriction e+1 = \sigma^* \leq \hat{\sigma}_i(e'', s') \restriction e+1,$$

i.e., $\hat{\sigma}_i(e'', s') \restriction e{+}1 = \sigma^*$. Since, by (4.29), $\sigma_i(e{+}1, y, s') \preceq \hat{\sigma}_i(e'', s') \restriction e{+}1$, it follows that $\sigma_i(e+1, y, s')$ precedes σ^*. So y is i-e-eligible at stage $s'+1$ by (4.36).

This completes the proof of Claim 4.11. □

5. Proof of Theorem 2.12

The proof combines the basic construction of an s.m.i. degree with the permitting technique first used by Dekker in [7] and then formalized by

Yates in [20]. Given noncomputable c.e. sets $B_0, \dots, B_{n-1}$ $(n \geq 1)$, it suffices to construct pairwise disjoint c.e. sets $A_0, \dots, A_{n-1}$ such that, for $A = A_0 \cup \dots \cup A_{n-1}$ and $\mathbf{a} = deg(A)$,

$$A_i \text{ is noncomputable } (i \leq n-1), \tag{5.1}$$

$$A_i \leq_{\mathrm{T}} B_i \ (i \leq n-1) \tag{5.2}$$

and (3.5) hold. Since the construction is very similar to the basic construction of a low s.m.i. degree given in the proof of Theorem 2.7, we only point out the necessary changes.

Let $(B_{i,s})_{s \geq 0}$ be a computable enumeration of the given set B_i $(i \leq n-1)$. The finite part of A_i enumerated by the end of stage s is denoted by $A_{i,s}$, and we let $A_s = A_{0,s} \cup \dots \cup A_{n-1,s}$.

In order to ensure (3.5), just as in the proof of Theorem 2.7, we enumerate auxiliary c.e. sets C_n $(n \geq 0)$ and ensure that the requirements Q_e are met and condition (3.7) is satisfied. (The index n of C_n will (in the rest of this section only) be used implicitly in the index e of Q_e and should not be confused with the n used in the statement of the theorem. All future uses of n in this section will refer to this number n introduced above.) In order to ensure that A_i is noncomputable, i.e., in order to ensure (5.1), we meet the requirements

$$\mathrm{P}_{ne+i} : A_i \neq \{e\}$$

for $e \geq 0$ and $i \leq n-1$. Condition (5.2) is satisfied by permitting, i.e., for $i \leq n-1$ and all numbers x and stages s we ensure

$$x \in A_{i,s+1} \setminus A_{i,s} \Rightarrow B_{i,s+1} \upharpoonright x+1 \neq B_{i,s} \upharpoonright x+1. \tag{5.3}$$

The requirements are ordered by $\mathrm{R}_{2e} = \mathrm{P}_e$ and $\mathrm{R}_{2e+1} = \mathrm{Q}_e$. The strategy for meeting the inaccessibility requirements Q_e is exactly as in the proof of Theorem 2.7, and e-(pre)eligibility and e-states are defined as there. The strategy for meeting the noncomputability requirements P_e has to be refined, since now the enumeration of a (realized) follower into A_i requires permitting by B_i (according to condition (5.3)). So a single follower will not be sufficient, and — as usual for constructions by permitting — we appoint a new follower if all existing followers are realized and the requirement is not satisfied. Since, for the sake of the inaccessibility requirements, the e-states of the P_e-followers have to be optimized, this requires that a declared state is assigned to each follower (not just to the requirement as in the basic construction). We let $y_p(e,s)$ be the pth follower of P_e in order of magnitude at the end of stage s (and write $y_p(e,s)\uparrow$ if there is no such follower). Then, if $y_p(e,s)$ is defined, a declared e-state $\hat{\sigma}_p(e,s)$ is assigned to this follower at the end of stage s. This leads to the following revision of the definition for a noncomputability requirement requiring attention and the corresponding action.

For $e = ne' + i$, P_e *requires attention* at stage $s+1$ if $e \leq s$, P_e is not satisfied at (the end of) stage s, and one of the following holds.

$$\text{There is a realized follower } y \text{ of } \mathrm{P}_e \text{ at the end of stage } s \text{ (i.e., a follower } y \text{ such that } \{e'\}_s(y) = 0) \text{ and } B_{i,s+1} \restriction y+1 \neq B_{i,s} \restriction y+1. \tag{5.4}$$

$$\text{There is a follower } y \text{ of } \mathrm{P}_e \text{ at the end of stage } s, \text{ say } y = y_p(e,s), \text{ and there is a number } y' \text{ such that } y < y' \leq s,\ y' \notin A_s, \text{ and } \sigma(e,y',s) < \hat{\sigma}_p(e,s). \tag{5.5}$$

$$\text{All followers of } \mathrm{P}_e \text{ at the end of stage } s \text{ (if any) are realized.} \tag{5.6}$$

If P_e receives attention then P_e becomes active via the first of the above clauses which holds, and the action is as follows. If (5.4) holds then the least y as there is enumerated into A_i and the requirement P_e is declared to be satisfied. If (5.5) holds (and (5.4) does not hold), then for the least y as there, for the corresponding p and for the least corresponding y', y is replaced by y', the declared state of $y_p(e,s+1)$ is set to $\hat{\sigma}_p(e,s+1) = \sigma(e,y',s)$ and all P_e-followers $y'' > y$ which exist at the end of stage s are cancelled (note that, by $y < y'$, the latter ensures that $y' = y_p(e,s+1)$). Finally, if (5.6) holds (and (5.4) and (5.5) do not hold), then $y = s+1$ is appointed as follower and, for p such that $y = y_p(e,s+1)$, the declared state of y is set to $\hat{\sigma}_p(e,s+1) = 1^e$.

If P_e becomes active at stage $s+1$ then, just as in the basic construction, all requirements of lower priority are initialized, where, for a noncomputability requirement P_n, becoming initialized now means that all followers and their declared states are cancelled and the requirement is declared to be unsatisfied. Moreover, we say that P_e becomes *active via* p at stage $s+1$ if either P_e becomes active according to clause (5.4) or (5.5) and $y = y_p(e,s)$ for the least y as there or P_e becomes active according to clause (5.6) and $y_p(e,s+1) = s+1$ is appointed at stage $s+1$.

Up to this modification of P_e requiring attention and the corresponding changes in the activity of P_e the construction is the same as in the proof of

Theorem 2.7. The proof of correctness follows the lines of the corresponding part of the proof of Theorem 2.7 whence we only give a rough sketch in the following.

The observations on the relations among the followers for different requirements made there directly carry over. For the followers $y_0(e,s),\ldots,y_p(e,s)$ of a requirement P_e at the end of stage s at which the requirement is not satisfied, one should observe that for any follower $y_l(e,s)$ $(l \leq p)$, $y_l(e,s) \notin A_s$ and $y_l(e,s)$ is not a follower of any other requirement at stage s. Moreover, for $l < p$, the stage at which $y_l(e,s)$ is appointed (according to clause (5.5) or (5.6)) precedes the corresponding stage for $y_{l+1}(e,s)$ and $\hat{\sigma}_l(e,s) \leq \hat{\sigma}_{l+1}(e,s)$.

Since at any stage $s+1$ at most one follower is enumerated into at most one of the sets A_i, the above in particular implies that the sets A_i are pairwise disjoint.

The argument that any requirement requires attention only finitely often (see Claim 3.6) now becomes a bit more sophisticated for the noncomputability requirements: Given P_e $(e = ne'+i)$, by inductive hypothesis, fix a stage $s_0 > e$ such that no higher priority requirement requires attention after stage s_0, and, for a contradiction, assume that P_e requires attention infinitely often. Then P_e is not initialized after stage s_0, P_e is not satisfied after stage s_0, and P_e receives attention at any stage $s+1 > s_0$ at which it requires attention. Note that once $y_p(e,s)$ is appointed and not cancelled afterwards, P_e can become active via p at most 2^e times (since, whenever this happens, the declared state $\hat{\sigma}_p(e,s')$ is decreased). Since, by assumption, P_e acts infinitely often, it follows, by induction on p, that, for any p there is a permanent follower $y_p(e)$ with a permanent declared state $\hat{\sigma}_p(e)$ and $A_i(y_p(e)) = \{e'\}(y_p(e)) = 0$. Moreover, $y_p(e,s') = y_p(e)$ for all $s' \geq s$ where $s > s_0$ is minimal such that, for all $p' \leq p$, $y_{p'}(e,s) = y_{p'}(e)$ and $\hat{\sigma}_{p'}(e,s) = \hat{\sigma}_{p'}(e)$. Since $\hat{\sigma}_p(e)$ is nondecreasing in p and bounded by 1^e, $\hat{\sigma}_p(e)$ comes to a limit for all sufficiently large p hence is computable in p. It follows that the least stage $s(p) > s_0$ such that $y_p(e) = y_p(e,s)$ for $s \geq s(p)$ and $y_p(e)$ is realized at stage $s(p)$ can be computed. Since, by choice of s_0, (5.4) fails for all stages $s \geq s_0$ and since $y_p(e) \geq p$, it follows that $B_i \upharpoonright p = B_{i,s(p)} \upharpoonright p$. So B_i is computable contrary to assumption, giving the desired contradiction.

As in the proof of Theorem 2.7, the fact that all requirements require attention only finitely often easily implies that the noncomputability requirements are met (see Claim 3.7). So it only remains to show that the inaccessibility requirements Q_e are met too. Here the proof given in Theorem 2.7 (see Claim 3.9) directly carries over. Note that Claim 3.2, Claim 3.3, Claim 3.4 and Claim 3.5 proven there hold in the given context too. So, as in the proof of Claim 3.9, given a stage s_0 such that neither Q_e nor a higher priority requirement requires attention after stage s_0, it suffices to show that — assuming that Q_e is not satisfied at any stage $s \geq s_0$ and that the clauses (i) and (ii) in Q_e hold while clause $(*)$ fails — there is an infinite computable set S of stages s satisfying (3.26). Such a set S is obtained as in the proof Theorem 2.7 by making some obvious changes: in the definition of stage e_0 there, condition (3.60) has to be replaced by the condition that P_{e_0} is not permanently satisfied (note that in the old context (3.60) is equivalent to this fact), and, correspondingly, the set S now consists of the stages $s > t_{e_0}$ satisfying

$$\forall e'' > e_0 \, \forall p \geq 0 \, (y_p(e'', s) \downarrow \; \& \; P_{e''} \text{ is not satisfied at stage } s \\ \Rightarrow \hat{\sigma}_p(e'', s+1) \restriction e+1 = \sigma^*)$$

where now $t_{e_0} + 1$ is the last stage at which P_{e_0} becomes active.

6. Open Problems

We conclude the chapter with some open problems.

Open Problem 6.1: Can Theorem 2.8 be extended to show that any high degree can be split into two high s.m.i. degrees?

An affirmative answer to Open Problem 6.1 would show that any generator of the high degrees is a join generator of the high degrees.

The next open problem asks if Theorem 2.8 can be generalized in a different direction.

Open Problem 6.2: Can every high_2 or even every non-low_2 degree be split into two s.m.i. degrees?

An affirmative answer to Open Problem 6.2 would show that every generator of the c.e. degrees generates the high_2 (or even the non-low_2) degrees under join.

Open Problem 6.3: Is the class of s.m.i. degrees definable in $\mathbf{R}$?

Neither the definition of strong meet inaccessibility nor any of the three equivalent conditions given in Lemma 2.2 are obviously first-order.

Open Problem 6.4: Does every strongly meet inaccessible degree satisfy Condition (2.3)?

Since Condition (2.3) is clearly first-order, an affirmative answer to Open Problem 6.4 together with Lemma 2.3 would imply an affirmative answer to Open Problem 6.3.

References

1. Ambos-Spies, Klaus. Contiguous r.e. degrees. Computation and proof theory (Aachen, 1983), 1–37, Lecture Notes in Math., 1104, Springer, Berlin, 1984.
2. Ambos-Spies, Klaus. On pairs of recursively enumerable degrees. Trans. Amer. Math. Soc. 283 (1984), 507–531.
3. Ambos-Spies, Klaus. Generators of the recursively enumerable degrees. Recursion theory week (Oberwolfach, 1984), 1–28, Lecture Notes in Math., 1141, Springer, Berlin, 1985.
4. Ambos-Spies, Klaus; Soare, Robert I. The recursively enumerable degrees have infinitely many one-types. Ann. Pure Appl. Logic 44 (1989), 1–23.
5. Ambos-Spies, Klaus; Ding, Decheng; Fejer, Peter A. Embedding distributive lattices preserving 1 below a nonzero recursively enumerable Turing degree. Logical methods (Ithaca, NY, 1992), 92–129, Progr. Comput. Sci. Appl. Logic, 12, Birkhäuser, Boston, MA, 1993.
6. Ambos-Spies, Klaus; Lempp, Steffen; Slaman, Theodore A. Generating sets for the recursively enumerable Turing degrees. Computational prospects of infinity. Part II. Presented talks, 1–22, Lect. Notes Ser. Inst. Math. Sci. Natl. Univ. Singap., 15, World Sci. Publ., Hackensack, NJ, 2008.
7. Dekker, J. C. E. A theorem on hypersimple sets. Proc. Amer. Math. Soc. 5 (1954), 791–796.
8. Ding, Decheng. Every p-generic r.e. degree is meet-inaccessible. Chinese J. Contemp. Math. 14 (1993), 331–337.
9. Downey, Rodney G.; Lempp, Steffen. Contiguity and distributivity in the enumerable Turing degrees. J. Symbolic Logic 62 (1997), 1215–1240.
10. Fejer, Peter A. The density of the nonbranching degrees. Ann. Pure Appl. Logic 24 (1983), 113–130.
11. Lachlan, A. H. Lower bounds for pairs of recursively enumerable degrees. Proc. London Math. Soc. (3) 16 (1966), 537–569.
12. Lachlan, A. H. Bounding minimal pairs. J. Symbolic Logic 44 (1979), 626–642.
13. Ladner, Richard E.; Sasso, Leonard P., Jr. The weak truth table degrees of recursively enumerable sets. Ann. Math. Logic 8 (1975), 429–448.
14. Nies, André. Parameter definability in the recursively enumerable degrees. J. Math. Log. 3 (2003), 37–65.
15. Sacks, Gerald E. Degrees of unsolvability. Princeton University Press, Princeton, N.J. 1963 ix+174 pp.
16. Shoenfield, Joseph. Degrees of unsolvability. North-Holland Publishing Co., Amsterdam, 1971.

17. Slaman, Theodore A. The density of infima in the recursively enumerable degrees. International Symposium on Mathematical Logic and its Applications (Nagoya, 1988). Ann. Pure Appl. Logic 52 (1991), 155–179.
18. Soare, Robert I. Recursively enumerable sets and degrees. A study of computable functions and computably generated sets. Perspectives in Mathematical Logic. Springer-Verlag, Berlin, 1987. xviii+437 pp.
19. Stob, Michael. wtt-degrees and T-degrees of r.e. sets. J. Symbolic Logic 48 (1983), 921–930.
20. Yates C. E. M. Three theorems on the degrees of recursively enumerable sets. Duke Math. J. 32 (1965), 461–468.
21. Yates, C. E. M. A minimal pair of recursively enumerable degrees. J. Symbolic Logic 31 (1966), 159–168.
22. Zhang, Qing Long. The density of the meet-inaccessible r.e. degrees. J. Symbolic Logic 57 (1992), 585–596.

https://doi.org/10.1142/9789811278631_0002

Cuts in the ML Degrees

Katherine Arthur
School of Mathematics and Statistics
Victoria University of Wellington, Wellington, New Zealand

Noam Greenberg[a]
School of Mathematics and Statistics
Victoria University of Wellington, Wellington, New Zealand
greenberg@msor.vuw.ac.nz
http://homepages.mcs.vuw.ac.nz/~greenberg/

Joseph S. Miller[b]
Department of Mathematics, University of Wisconsin–Madison
480 Lincoln Dr., Madison, WI 53706, USA
jmiller@math.wisc.edu
http://www.math.wisc.edu/~jmiller/

We show that the cut defined by a real number $r \in [0,1]$ is realised in the hierarchy of p-bases in the ML degrees if and only if it is left-Π^0_3.

1. Introduction

In [1], the authors characterise the sets that are computable from some pair of relatively random sequences, or equivalently, from both halves of some ML-random sequence. There are only countably many such sets, they are all K-trivial, and the Turing degrees of these sets form an ideal. It turns out that this ideal is one among a hierarchy of ideals $\mathcal{B}_p$ in the K-trivial degrees, each indexed by rationals $p \in [0,1]$, with $p < q$ implying that

[a]Greenberg was supported by a Marsden Fund grant #17-VUW-090.
[b]Miller was supported by grant #358043 from the Simons Foundation.

$\mathcal{B}_p \subsetneq \mathcal{B}_q$. If $p = k/n$ with $k < n$ natural numbers, then $\mathcal{B}_p$ is the collection of sets A which for some random sequence Z (equivalently, for $Z = \Omega$ being any left-c.e. random sequence), A is computable from the join of any k of the n-columns of Z. Various similar characterizations of these ideals are known; for example, see [1, Prop. 5.1].

Since the $\mathcal{B}_p$ are a strictly ordered chain of ideals, it is natural to ask: which cuts are realised? Namely for which reals $r \in (0,1)$ is there a set A that is an element of $\mathcal{B}_p$ exactly for $p > r$? There are only countably many K-trivial sets, and so only countably many cuts are realised this way. In this chapter, we characterise these cuts:

Theorem 1.1: *The following are equivalent for a real number $r \in (0,1)$:*

(1) *There is a set A such that for all $p \in \mathbb{Q} \cap [0,1]$, $A \in \mathcal{B}_p \iff p > r$.*
(2) *r is right-Σ^0_3.*

By (2), we mean that the right cut $\{p \in \mathbb{Q} : p > r\}$ is Σ^0_3. We note that since each ideal $\mathcal{B}_p$ is characterised by being computable from a collection of random sequences, [2, Thm. 2.1] implies that we may take A to be c.e. in (1).

Remark 1.2: When $r \in (0,1)$ is rational, the conditions of Theorem 1.1 hold. However, in this case, one can also ask whether there is a set A with $A \in \mathcal{B}_p \iff p \geqslant r$. A positive answer follows from [2, Thm. 3.3]. Alternatively, the construction below can be modified to obtain such a set A.

The main tool used to explore the ideals $\mathcal{B}_p$ is *cost functions.* We recall some definitions. A *cost function* is a computable function $\mathbf{c}\colon \mathbb{N}^2 \to \mathbb{R}^{\geqslant 0}$. In this chapter, we only consider cost functions $\mathbf{c}$ with the following extra properties:

(i) *Monotonicity*: for all x and s, $\mathbf{c}(x,s) \leqslant \mathbf{c}(x,s+1)$ and $\mathbf{c}(x,s) \geqslant \mathbf{c}(x+1,s)$;
(ii) The *limit condition*: for all x, $\underline{\mathbf{c}}(x) = \lim_s \mathbf{c}(x,s)$ is finite and $\lim_{x\to\infty} \underline{\mathbf{c}}(x) = 0$;
(iii) For all x and s, $\mathbf{c}(x,s) \leqslant 1$;
(iv) For all $s < x$, $\mathbf{c}(x,s) = 0$.

The idea is that a cost function $\mathbf{c}$ measures, in an analytic way, the complexity of a computable approximation $\langle A_s \rangle$ of a Δ^0_2 set A. Intuitively, the fewer the mind-changes, the simpler A is. The number $\mathbf{c}(x,s)$ is the cost of changing A on x at stage s, namely of setting $A_s(x) \neq A_{s-1}(x)$. The monotonicity condition says that the cost of changing x goes up as time passes, and that at any given stage, it is cheaper to change A on larger numbers. The limit condition puts a restraint on the costs, ensuring they are not too onerous in the limit. The notion of obedience tells us which computable approximations are simple from $\mathbf{c}$'s point of view:

Definition 1.3: Let $\langle A_s \rangle$ be a computable approximation of a Δ^0_2 set A, and let $\mathbf{c}$ be a cost function. The *total* $\mathbf{c}$*-cost* of $\langle A_s \rangle$ is

$$\mathbf{c}\langle A_s \rangle = \sum_{s<\omega} \mathbf{c}_s(x) \; [\![x \text{ is least such that } A_s(x) \neq A_{s-1}(x)]\!].$$

We say that A *obeys* $\mathbf{c}$ if for some computable approximation $\langle A_s \rangle$ of A, $\mathbf{c}\langle A_s \rangle$ is finite.

In [1], it is shown that for all rational $p \in (0,1)$, $A \in \mathcal{B}_p$ if and only if A obeys the cost function $\mathbf{c}_{\Omega,p}$ defined by

$$\mathbf{c}_{\Omega,p}(x,s) = \begin{cases} (\Omega_s - \Omega_x)^p, & \text{if } x \geqslant s; \\ 0, & \text{if } x < s. \end{cases}$$

Here $\langle \Omega_s \rangle$ is some increasing computable approximation of a left-c.e. ML-random sequence Ω. This characterisation of the ideals $\mathcal{B}_p$ shows that Theorem 1.1 is really a theorem about cost functions. For two cost functions $\mathbf{c}$ and $\mathbf{c}'$, write $\mathbf{c} \ll \mathbf{c}'$ if:

- for all x and s, $\mathbf{c}(x,s) \leqslant \mathbf{c}'(x,s)$; and
- for every constant k, $\underline{\mathbf{c}}'(x) > k\underline{\mathbf{c}}(x)$ for all but finitely many x.

We prove:

Proposition 1.4: *Let $\{\mathbf{c}_p : p \in \mathbb{Q} \times (0,1)\}$ be a collection of uniformly computable cost functions, such that if $p < q$, then $\mathbf{c}_q \ll \mathbf{c}_p$. Then for any real number $r \in (0,1)$, the following are equivalent:*

(1) *There is a set A such that for all $p \in \mathbb{Q} \cap [0,1]$, A obeys $\mathbf{c}_p$ if and only if $p > r$.*
(2) *r is right-Σ^0_3.*

It is readily observed that $\mathbf{c}_{\Omega,q} \ll \mathbf{c}_{\Omega,p}$ whenever $p < q$, and so Proposition 1.4 implies Theorem 1.1.

2. Proof of Proposition 1.4

Before we prove Proposition 1.4, we introduce some notation and state a lemma. Suppose that $\langle A_s \rangle$ is a computable approximation of a set A. A *speed-up* of $\langle A_s \rangle$ is an approximation $\langle A_{h(s)} \rangle$ where $h\colon \mathbb{N} \to \mathbb{N}$ is computable and strictly increasing. For simplicity, we write $\langle A_h \rangle$ for $\langle A_{h(s)} \rangle$. It is not difficult to see that if $\langle A_h \rangle$ is a speed-up of $\langle A_s \rangle$, then for any cost function $\mathbf{c}$, $\mathbf{c}\langle A_h \rangle \leqslant \mathbf{c}\langle A_s \rangle$. In fact, there are several reasons that the cost on the left might be smaller. Suppose that x is the least such that $A_{h(s)}(x) \neq A_{h(s-1)}(x)$. So the step s contribution to $\mathbf{c}\langle A_h \rangle$ is $\mathbf{c}(x, s)$. In contrast, the step $h(s)$ contribution to $\mathbf{c}\langle A_s \rangle$ is at least $\mathbf{c}(x, h(s))$, which by monotonicity is at least $\mathbf{c}(x, s)$. It may be more, since it is possible that there is some $y < x$ such that $A_{h(s)}(y) \neq A_{h(s)-1}(y)$, but it just happens that $A_{h(s)}(y) = A_{h(s-1)}(y)$. And of course, relative to $\langle A_s \rangle$, $\mathbf{c}\langle A_h \rangle$ only counts some of the stages, namely those in the range of h. We will make use of the following, which is well-known, and follows from the techniques in [3]:

Lemma 2.1: *A Δ^0_2 set A obeys a cost function $\mathbf{c}$ if and only if every computable approximation $\langle A_s \rangle$ of A has a speed-up $\langle A_h \rangle$ with $\mathbf{c}\langle A_h \rangle < \infty$.*

We fix an effective listing $\langle h_e \rangle$ of partial "speed-up" functions. That is:

- $\langle h_e \rangle$ are uniformly partial computable;
- Each h_e is either total, or its domain is a finite initial segment of ω;
- Each h_e is strictly increasing on its domain;
- Every strictly increasing computable function is h_e for some e.

Further, for every e and s, let $n_{e,s} = \max \operatorname{dom} h_{e,s}$; by withholding convergences, we may assume that:

- $\operatorname{dom} h_{e,s}$ is an initial segment of ω; and
- $h_{e,s}(n_{e,s}) < s$.

For any cost function $\mathbf{c}$, we can define

$$\mathbf{c}\langle A_{h_e} \rangle[s] = \sum_{m \leqslant n_{e,s}} \mathbf{c}(x, m) \, [\![x \text{ is least such that } A_{h_e(m)} \neq A_{h_e(m-1)}]\!].$$

The value $\mathbf{c}\langle A_{h_e} \rangle[s]$ is computable, uniformly in e, s and in a computable index for $\mathbf{c}$. And if h_e is total, then $\mathbf{c}\langle A_{h_e} \rangle = \lim_s \mathbf{c}\langle A_{h_e} \rangle[s]$.

(1) $\Longrightarrow$ (2) of Proposition 1.4 is essentially [3, Fact 2.13], which is uniform. We are given a Δ^0_2 set A; we fix a computable approximation $\langle A_s \rangle$ for A. By Lemma 2.1, A obeys $\mathbf{c}_p$ if and only if there are some e and M such that h_e is total and for all s, $\mathbf{c}_p\langle A_{h_e} \rangle[s] \leqslant M$. This is a Σ^0_3 predicate

of p. Note that the collection of p such that A obeys $\mathbf{c}_p$ must be a right cut (a final segment of $\mathbb{Q} \cap (0,1)$); this follows from the assumption that $\mathbf{c}_q \leqslant \mathbf{c}_p$ for $p < q$.

Before we give the details, we briefly discuss the proof of (2) $\Longrightarrow$ (1). We are given a right-Σ_3^0 real $r \in (0,1)$, and define a computable approximation $\langle A_s \rangle$ of the desired set A. The value of r can be guessed by the true path on a tree of strategies: one duty of the strategies is to guess, given $p \in \mathbb{Q} \cap (0,1)$, whether $p > r$ or not; locally the behaviour of the true path is Σ_2^0/Π_2^0, so to approximate the Σ_3^0 predicate $p > r$, we need to keep trying different existential witnesses for the outermost quantifier.

Suppose that a strategy τ works with some rational number $p = p^\tau$. There are two possibilities. The infinite outcome $\tau\hat{\ }\infty$ believes that it has proof that $p > r$, and so it is $\tau\hat{\ }\infty$'s responsibility to ensure that A obeys $\mathbf{c}_p$. This is both done passively, by initialisations, and more actively, by setting strict bounds on the action of weaker requirements. The speed-up of $\langle A_s \rangle$ which witnesses that A obeys $\mathbf{c}_p$ is the restriction of our approximation to the $\tau\hat{\ }\infty$-stages. There are two kinds of nodes σ that may change A, and thus increase the cost measured by $\tau\hat{\ }\infty$: nodes to the right of $\tau\hat{\ }\infty$, and nodes extending $\tau\hat{\ }\infty$. For each node σ, we assign a bound δ^σ on the amount of cost that σ's action may cause to nodes (strategies) strictly above it (nodes that σ extends). We distribute the bounds δ^σ so that the total damage caused by all nodes extending $\tau\hat{\ }\infty$ is finite. The nodes to the right of $\tau\hat{\ }\infty$ (including the finite outcome $\tau\hat{\ }\mathtt{fin}$) contribute *nothing* to $\tau\hat{\ }\infty$'s cost. This is the result of initialisations and our speed-up: at the mth $\tau\hat{\ }\infty$ stage, nodes to the right only change A on numbers greater than m, and we measure the $\mathbf{c}_p$-cost of these changes at stage m. We use the assumption (iii) above, that if $s < x$ then $\mathbf{c}_p(x, s) = 0$.

Now consider the Σ_2^0 outcome $\tau\hat{\ }\mathtt{fin}$. This outcome believes that $p \leqslant r$, and so tries to ensure that A does not obey $\mathbf{c}_p$. By Lemma 2.1, it suffices to check all speed-ups of our base approximation $\langle A_s \rangle$. We make use of the following strengthening of Lemma 2.1, which is Fact 2.2 of [3]:

Lemma 2.2: *Suppose that $\langle A_s \rangle$ is a computable approximation of a set A that obeys a cost function $\mathbf{c}$. Then for any $\varepsilon > 0$, there is a speed-up of $\langle A_s \rangle$ with total cost bounded by ε.*

Thus, in order to show that A does not obey $\mathbf{c}_p$, it suffices to ensure that for all e, $\mathbf{c}_p\langle A_{h_e} \rangle \geqslant 1$. The node τ will be assigned one e. It needs to change A on numbers x so that the cost $\mathbf{c}_p\langle A_{h_e}\rangle$ increases. The node τ faces two difficulties:

- Some nodes above τ restrain τ from adding more than δ^τ to their cost; and δ^τ is much smaller than 1.
- The speed-up function h_e is revealed to τ very slowly.

The second difficulty is technical: we see $h_e(m)$ converge to some value t only at some stage s much later than t. Thus, τ discovers that it had to change A_t on some value; but A_t was already defined at stage t. This is addressed easily by giving τ an infinite collection (which we denote by $\omega^{[\tau]}$) of potential inputs for x to play with; for a suitable $x \in \omega^{[\tau]}$, the node τ keeps $A_r(x) \neq A_t(x)$ for stages $r \geqslant s$ until we see a value of h_e greater than s.

The first difficulty is fundamental: this is where we use the assumptions on the relative growth-rate of the cost functions $\mathbf{c}_p$. Take some node τ working to increase $\mathbf{c}_q\langle A_{h_e}\rangle$ for some e and q, and let ρ be some node above τ that is concerned about incurring cost from τ's action. The node ρ only cares if it is trying to keep costs low; that is, if $\rho\hat{\ }\infty \preccurlyeq \tau$. Let $p = p^\rho$ be the rational number that ρ is working with; it is trying to keep the $\mathbf{c}_p$-cost of some approximation finite. Now the outcome $\rho\hat{\ }\infty$, and therefore τ, believe that they have proof that $p > r$. The node τ is working with the assumption that $q \leqslant r$. Thus, we can arrange that $q < p$. The assumption $\mathbf{c}_q \ll \mathbf{c}_p$ now means that τ can change A to make the $\mathbf{c}_q$-cost large while keeping the $\mathbf{c}_p$-damage very small: smaller than δ^τ.

We now give the details. Let $r \in (0,1)$ be right-Σ^0_3. There are uniformly computable, non-decreasing sequences $\langle \ell^{p,e}_s\rangle_{s<\omega}$ (of natural numbers) for $p \in \mathbb{Q}$ and $e < \omega$ such that for all such p, $p > r$ if and only if for some $e < \omega$, $\langle \ell^{p,e}_s\rangle$ is unbounded.

We define a computable approximation $\langle A_s\rangle$ of a Δ^0_2 set A. We will meet two types of requirements. The first type of requirements are indexed by $p \in \mathbb{Q} \cap (0,1)$:

$$N_p : \text{If } p > r, \text{ then } A \text{ obeys } \mathbf{c}_p.$$

Requirements of the second type are indexed by $p \in \mathbb{Q} \cap (0,1)$ and $e < \omega$:

$$R_{p,e} : \text{If } p < r \text{ and } h_e \text{ is total, then } \mathbf{c}_p\langle A_{h_e}\rangle \geqslant 1.$$

As discussed above, meeting these requirements suffices to ensure (1) of the proposition.

Approximating r

We work with a full binary tree of strategies. The strategies are the finite sequences of the symbols ∞ and `fin`.

By recursion on the length $|\sigma|$ of a node σ on the tree, we define:

- $p^\sigma \in \mathbb{Q} \cap (0,1)$ and $e^\sigma \in \omega$; the node σ will attempt to meet either N_{p^σ} or R_{p^σ,e^σ};
- a rational number $r^\sigma > p^\sigma$; this is an upper bound on the value of r believed by σ.

The meaning of the outcome ∞ is that we believe that $p^\sigma > r$, and so we meet N_{p^σ} by defining a suitable speed-up of our approximation for A. The meaning of the outcome `fin` is that we believe that $p^\sigma \leqslant r$, and so we meet R_{p^σ,e^σ}.

We use an effective ω-ordering of all the pairs $(p,e) \in (\mathbb{Q}\cap(0,1))\times\omega$. We start with the root of the tree, which is the empty sequence $\langle\rangle$, by letting $(p^{\langle\rangle}, e^{\langle\rangle})$ be the least pair in our ordering; we let $r^{\langle\rangle} = 1$.

Suppose that σ is on the tree and that we have already defined p^σ, e^σ and r^σ. We then define these parameters for the children $\sigma\hat{}\infty$ and $\sigma\hat{}\mathtt{fin}$. We start with the latter:

(a) $r^{\sigma\hat{}\infty} = p^\sigma$.
(b) $r^{\sigma\hat{}\mathtt{fin}} = r^\sigma$.

Then, for both children τ of σ, we let (p^τ, e^τ) be the next pair (p,e) on our list after (p^σ, e^σ) such that $p < r^\tau$.

For brevity, for any node σ, we write:

- ℓ_s^σ for $\ell_s^{p^\sigma,e^\sigma}$.
- h^σ for h_{e^σ} (and similarly h_s^σ for $h_{e^\sigma,s}$).

Allocating capital to nodes

Computably, we assign to each node σ a positive rational number δ^σ such that

$$\sum \delta^\sigma \leqslant 1$$

(where the sum ranges over all strategies σ). The idea of the parameter δ^σ

is that σ promises any τ with $\tau\hat{\ }\infty \preccurlyeq \sigma$ that it will not add more than δ^σ to the cost accrued by τ.[c]

Construction

At stage s, we define the path of accessible nodes by recursion. If a strategy σ is accessible at stage s, then we say that s is a *σ-stage.*

We start with $A_0 = 0^\infty$.

The root is always accessible. Suppose that a node σ is accessible at stage s. If $|\sigma| = s$, we halt the stage. We also initialise all nodes weaker than σ.

Suppose that $|\sigma| < s$.

First, let $t < s$ be the last $\sigma\hat{\ }\infty$-stage before stage s; $t = 0$ if there was no such stage. If $\ell_s^\sigma > t$, then we let $\sigma\hat{\ }\infty$ be the next accessible node.

Suppose that $\ell_s^\sigma \leqslant t$. We will define the notion of a *σ-action* stage. Let w be the last σ-action stage prior to stage s; $w = 0$ if there was no such stage. Let $n = \max \operatorname{dom} h_s^\sigma$; let s^* be the last stage prior to stage s at which $\sigma\hat{\ }\mathtt{fin}$ was initialised. If:

(i) $\mathbf{c}_{p^\sigma}(\langle A_{h^\sigma}\rangle)[s] < 1$;

(ii) $n > w$; and

(iii) there is a number $x > s^*$, $x \in \omega^{[\sigma]}$ satisfying[d]

$$\mathbf{c}_{r^\sigma}(x, s) \leqslant \mathbf{c}_{p^\sigma}(x, n) \cdot \delta^\sigma,$$

then we choose the least such x, set $A_{s+1}(x) = 1 - A_s(x)$, and call s a σ-action stage. Otherwise, σ makes no change to A at stage s. In either case, we let $\sigma\hat{\ }\mathtt{fin}$ be the next accessible node.

2.1. *Verification*

Let δ^* denote the true path. Because we never terminate a stage s before we get to a node of length s, and the strategy tree is binary splitting, the true path is infinite.

---

[c]Actually, it will be $2\delta^\sigma$, for a truly unimportant reason. The last σ-action may add to the cost σ is measuring a quantity close to 1, making the total cost close to 2; from τ's point of view, the increase is then close to $2\delta^\sigma$.

More importantly, note that the value δ^σ does not depend on the stage number. A reasonable approach would be to shrink δ^σ each time σ is initialised. We do not need to do this, because even when σ is initialised, the amount that it previously added to the total cost it is monitoring has not gone away, and so it does not need to start afresh.

[d]Recall that $\omega^{[\rho]}$, for $\rho \in \{\infty, \mathtt{fin}\}^{<\omega}$, is a partition of ω into pairwise disjoint, infinite computable sets.

Toward verifying that the requirements are met, we show that the true path approximates r correctly. For the first part of the next lemma, note that if τ extends σ, then $r^\sigma \geqslant r^\tau$, so $\inf_{\sigma\in\delta^*} r^\sigma = \lim_{\sigma\in\delta^*} r^\sigma$.

Lemma 2.3:

(a) $r = \inf_{\sigma\in\delta^*} r^\sigma$.
(b) *For all rational* $p \in (0, r)$, *for all* e, *there is some* $\sigma \in \delta^*$ *with* $(p^\sigma, e^\sigma) = (p, e)$.

Proof: First, by induction on the length of $\sigma \in \delta^*$, we verify that $r^\sigma > r$. For the root this is clear since $r < 1$. If $\sigma \in \delta^*$ and $r^\sigma > r$, there are two cases. If $\sigma\hat{}\infty \in \delta^*$ then $\langle \ell_s^\sigma \rangle$ is unbounded, which implies that $p^\sigma = r^{\sigma\hat{}\infty} > r$. Otherwise $\sigma\hat{}\mathtt{fin} \in \delta^*$ and $r^{\sigma\hat{}\mathtt{fin}} = r^\sigma > r$.

Let $\tilde{r} = \inf_{\sigma\in\delta^*} r^\sigma$. Let $p \in (0, \tilde{r})$ be rational and let $e < \omega$. For all $\tau \in \delta^*$, $r^\tau > p$. Thus, we never skip over the pair (p, e) when assigning pairs to the nodes on the true path. It follows that there is some $\sigma \in \delta^*$ with $(p^\sigma, e^\sigma) = (p, e)$. This verifies (b).

Suppose, for a contradiction, that $\tilde{r} > r$. Let $p \in (r, \tilde{r})$ be rational, and let e witness that $p > r$, that is, $\langle \ell_s^{p,e} \rangle$ is unbounded. Let $\sigma \in \delta^*$ with $(p^\sigma, e^\sigma) = (p, e)$. Then $\sigma\hat{}\infty \in \delta^*$ and $r^{\sigma\hat{}\infty} = p < \tilde{r}$, which is a contradiction. □

The next lemma shows that action by a node does increase the total cost it is monitoring. Let σ be any node, and let s be a σ-action stage. We let

- $n_s^\sigma = \max \operatorname{dom} h_s^\sigma$;
- y_s^σ be the number acted upon by σ, that is, the unique number $y \in \omega^{[\sigma]}$ such that $A_{s+1}(y) \neq A_s(y)$.

Lemma 2.4: *Let* σ *be any node, and suppose that* $s < s'$ *are two* σ*-action stages. Then*

$$\mathbf{c}_{p^\sigma}(\langle A_{h^\sigma} \rangle)[s'] \geqslant \mathbf{c}_{p^\sigma}(\langle A_{h^\sigma} \rangle)[s] + \mathbf{c}_{p^\sigma}(y_s^\sigma, n_s^\sigma).$$

Proof: Let $y = y_s^\sigma$. We may assume that $s' = s^+$ is the next σ-action stage after stage s. Also let s^- be the previous σ-action stage prior to stage s ($s^- = 0$ if there was no such stage). Since σ does not act between stages s^- and s, and between stages s and s^+,

- $A_t(y)$ is constant for $t \in (s^-, s]$; and
- $A_t(y)$ is constant for $t \in (s, s^+]$.

The point is that no other node can change A on an element of $\omega^{[\sigma]}$. Note that

$$s^- < n^\sigma_s \leqslant h^\sigma(n^\sigma_s) < s < n^\sigma_{s^+} \leqslant h^\sigma(n^\sigma_{s^+}) < s^+.$$

Thus there is some $m \in (n^\sigma_s, n^\sigma_{s^+}]$ such that $s^- < h^\sigma(m-1) \leqslant s < h^\sigma(m) \leqslant s^+$. Then $A_{h^\sigma(m-1)}(y) \neq A_{h^\sigma(m)}(y)$. This shows that stage m of the approximation A_{h^σ} contributes at least $\mathbf{c}_{p^\sigma}(y,m) \geqslant \mathbf{c}_{p^\sigma}(y,n^\sigma_s)$ to $\mathbf{c}_{p^\sigma}(\langle A_{h^\sigma}\rangle)[s^+]$, and this was not seen at stage s. □

Lemma 2.5: *Let τ be any node. Then*

$$\sum\{\mathbf{c}_{p^\tau}(y^\tau_s, n^\tau_s) \ : \ s \textit{ is a } \tau\textit{-action stage}\} < 2.$$

Proof: For $t \leqslant \omega$, let

$$S^\tau_t = \sum\{\mathbf{c}_{p^\tau}(y^\tau_s, n^\tau_s) \ : \ s \text{ is a } \tau\text{-action stage } \ \& \ s < t\}.$$

Then Lemma 2.4 implies that for every τ-action stage s,

$$S^\tau_s \leqslant \mathbf{c}_{p^\tau}(\langle A_{h^\tau}\rangle)[s] < 1.$$

If there are infinitely many τ-action stages then $S^\tau_\omega \leqslant 1$. Otherwise, let s be the last τ-action stage. As $\mathbf{c}_{p^\tau}(y^\tau_s, n^\tau_s) \leqslant 1$, we have

$$S^\tau_\omega = S^\tau_s + \mathbf{c}_{p^\tau}(y^\tau_s, n^\tau_s) < 2.$$ □

Lemma 2.6: *Let σ be a node and suppose that $\sigma\hat{\ }\mathtt{fin} \in \delta^*$. Then there are only finitely many σ-action stages. If h^σ is total then $\mathbf{c}_{p^\sigma}\langle A_{h^\sigma}\rangle \geqslant 1$.*

Proof: If h^σ is partial, then there cannot be more than one σ-action stage after stage $\max \operatorname{dom} h^\sigma$. Suppose that h^σ is total. We will show that eventually, $\mathbf{c}_{p^\sigma}(\langle A_{h^\sigma}\rangle)[s] \geqslant 1$, which will also imply that there are only finitely many σ-action stages. Suppose, for a contradiction, that for all s, $\mathbf{c}_{p^\sigma}(\langle A_{h^\sigma}\rangle)[s] < 1$.

Let s^* be the last stage at which $\sigma\hat{\ }\mathtt{fin}$ is initialised. Since $r^\sigma > p^\sigma$, we know that for all but finitely many x,

$$\underline{\mathbf{c}}_{r^\sigma}(x) < \underline{\mathbf{c}}_{p^\sigma}(x) \cdot \delta^\sigma.$$

Let x^* be the least $x > s^*$, $x \in \omega^{[\sigma]}$ satisfying this inequality. Then for all but finitely many stages t, for all s,

$$\mathbf{c}_{r^\sigma}(x^*, s) < \mathbf{c}_{p^\sigma}(x^*, t) \cdot \delta^\sigma.$$

For sufficiently late stages s, we have $n = \max \operatorname{dom} h^\sigma_s > x^*$ and $\mathbf{c}_{r^\sigma}(x^*, s) < \mathbf{c}_{p^\sigma}(x^*, n) \cdot \delta^\sigma$. This shows that there are infinitely many σ-action stages. Let t^* be a late σ-action stage; let $\varepsilon^* = \mathbf{c}_{p^\sigma}(x^*, t^*)$, which is positive. For every σ-action stage $s > t^*$, by minimality of y^σ_s, we have $y^\sigma_s \leqslant x^*$, and as $n^\sigma_s > t^*$, monotonicity of $\mathbf{c}_{p^\sigma}$ implies that $\mathbf{c}_{p^\sigma}(y^\sigma_s, n^\sigma_s) \geqslant \varepsilon^*$. Thus by Lemma 2.4, between any two σ-action stages, the partial cost $\mathbf{c}_p(\langle A_{h_e}\rangle)[s]$ grows by at least ε^*, so eventually grows beyond 1, which is a contradiction. □

Now fix some $p \in \mathbb{Q} \cap (0, 1)$.

Lemma 2.7: *Suppose that $p < r$. Then for all e, the requirement $R_{p,e}$ is met.*

Proof: By Lemma 2.3(b), let $\sigma \in \delta^*$ such that $(p^\sigma, e^\sigma) = (p, e)$. Since $p < r$, $\sigma\hat{\ }\mathtt{fin} \in \delta^*$. Then Lemma 2.6 implies that $R_{p,e}$ is met. □

Lemma 2.8: *Suppose that $p > r$. Then the requirement N_p is met.*

Proof: Let σ be the longest node on the true path such that $r^\sigma > p$. So $p^\sigma \leqslant p$ and $\sigma\hat{\ }\infty \in \delta^*$. Let s^* be sufficiently late so that:

- σ is not initialised after stage s^*; and
- For every τ such that $\tau\hat{\ }\mathtt{fin} \preccurlyeq \sigma$, there are no τ-action stages after stage s^*;

the latter uses Lemma 2.6. Let $s_0 < s_1 < s_2 < \ldots$ be the increasing enumeration of the $\sigma\hat{\ }\infty$-stages after stage s^*. We show that $\mathbf{c}_{p^\sigma}\langle A_{s_k}\rangle$ is finite, which suffices since $p^\sigma \leqslant p$.

Let $k \geqslant 1$; let x_k be the least such that $A_{s_k}(x_k) \neq A_{s_{k-1}}(x_k)$. Let τ_k be the node such that $x_k \in \omega^{[\tau_k]}$. So there is some τ_k-action stage $t_k \in [s_{k-1}, s_k)$ such that $x_k = y^{\tau_k}_{t_k}$. Since $t_k > s^*$, we know that $\tau_k\hat{\ }\mathtt{fin}$ lies to the right of $\sigma\hat{\ }\infty$, or τ_k extends $\sigma\hat{\ }\infty$. In the first case (which includes the case $\tau_k = \sigma$), $\tau_k\hat{\ }\mathtt{fin}$ is initialised at stage s_{k-1}, and so $x_k > s_{k-1} \geqslant k$, which implies that $\mathbf{c}_{p^\sigma}(x_k, k) = 0$; so stage k contributes no cost to the total cost $\mathbf{c}_{p^\sigma}\langle A_{s_k}\rangle$.

Suppose that τ_k extends $\sigma\hat{\ }\infty$. Then $t_k = s_{k-1}$, and more importantly, $r^{\tau_k} \leqslant r^{\sigma\hat{\ }\infty} = p^\sigma$. Thus

$$\mathbf{c}_{p^\sigma}(x_k, k) \leqslant \mathbf{c}_{r^\tau}(x_k, s_{k-1}) \leqslant \mathbf{c}_{p^\tau}(x_k, n^\tau_{s_{k-1}}) \cdot \delta^\tau.$$

It follows that

$$\mathbf{c}_{p^\sigma}\langle A_{s_k}\rangle = \sum_k \mathbf{c}_{p^\sigma}(x_k, k) \leqslant \sum_{\tau \geqslant \sigma\hat{}\infty} \delta^\tau \cdot \sum \{\mathbf{c}_{p^\tau}(y_s^\tau, n_s^\tau) \;:\; s \text{ a } \tau\text{-action stage}\} \leqslant \sum_\tau 2\delta^\tau \leqslant 2$$

(using Lemma 2.5), and so is finite as required. □

References

1. Noam Greenberg, Joseph S. Miller, and André Nies. Computing from projections of random points: a dense hierarchy of subideals of the K-trivial degrees. *Journal of Mathematical Logic*, vol. 20 (2020), 41 pages. World Scientific Publishing Company.
2. Noam Greenberg, Joseph S. Miller, André Nies, and Daniel Turetsky. Martin-Löf reducibility and cost functions. *Israel Journal of Mathematics*, to appear.
3. André Nies. Calculus of cost functions. In *The Incomputable*, Theory Appl. Comput., pages 183–216. Springer, Cham, 2017.

https://doi.org/10.1142/9789811278631_0003

Cantor–Bendixson Ranks for Almost Prime Models

Nikolay Bazhenov

Sobolev Institute of Mathematics
4 Acad. Koptyug Ave., Novosibirsk, 630090, Russia
and
Novosibirsk State University
2 Pirogova St., Novosibirsk, 630090, Russia
bazhenov@math.nsc.ru

Margarita Marchuk[a]

Sobolev Institute of Mathematics
4 Acad. Koptyug Ave., Novosibirsk, 630090, Russia
and
Novosibirsk State University
2 Pirogova St., Novosibirsk, 630090, Russia
margaretmarchuk@gmail.com

For a complete theory T, the set of n-types of T admits a natural topology. This chapter studies connections between topological spaces of types and the algorithmic complexity of isomorphisms among decidable structures. Let $\mathbf{d}$ be a Turing degree. A decidable structure is decidably $\mathbf{d}$-categorical if it has a unique decidable copy, up to $\mathbf{d}$-computable isomorphisms. It is known that if T has a decidable prime model $\mathcal{N}$, then $\mathcal{N}$ is decidably $\mathbf{0}'$-categorical. In other words, if the Cantor–Bendixson ranks of all types realized in a decidable model $\mathcal{N} \models T$ are equal to zero, then $\mathcal{N}$ is unique, up to computable isomorphisms. We obtain the following result: if all types realized in a countable model $\mathcal{M} \models T$ are

[a]The work of N. Bazhenov was supported by the program of fundamental scientific researches of the SB RAS № I.1.1, project № 0314-2019-0002. The reported study of M. Marchuk was funded by RFBR according to the research project № 17-01-00247. The work was done partially while the authors were visiting the Institute for Mathematical Sciences, National University of Singapore in 2017. The visit was supported by the Institute. The authors are grateful to D. Belanger and S. Goncharov for fruitful discussions.

ranked and the supremum of the Cantor–Bendixson ranks of these types, denoted by $CB(\mathcal{M})$, is a successor ordinal, then $\mathcal{M}$ is an almost prime model (i.e. there is a finite tuple $\bar{c}$ such that $(\mathcal{M}, \bar{c})$ is a prime model of its own first-order theory). As a corollary, we show that for any decidable model $\mathcal{M} \models T$, if $CB(\mathcal{M})$ is a finite ordinal, then $\mathcal{M}$ is decidably $\mathbf{0}'$-categorical. This solves a problem of Belanger. Furthermore, we prove that for any $n \in \omega$, there is an almost prime model $\mathcal{M}$ with $CB(\mathcal{M}) = n$.

1. Introduction

This chapter studies algorithmic complexity of isomorphisms among decidable algebraic structures. A computable structure $\mathcal{S}$ is *computably categorical* if for any computable copy $\mathcal{A}$ of $\mathcal{S}$, there is a computable isomorphism from $\mathcal{A}$ onto $\mathcal{S}$. In other words, the structure $\mathcal{S}$ is computably categorical if it has a unique computable presentation, up to computable isomorphisms. The systematic study of computable categoricity goes back to the works of Fröhlich and Shepherdson [1], and Mal'tsev [2, 3]. A particularly simple example of a computably categorical structure is provided by the following fact: every finitely generated, computable structure is computably categorical [2]. In [1], the authors constructed two isomorphic computable fields F and K such that F has a splitting algorithm, but K has no splitting algorithm. This implies that the field F is not computably categorical. Mal'tsev [3] showed that the abelian group $R_\infty = \bigoplus_{i\in\omega} \mathbb{Q}$ has a computable copy with no c.e. basis, and hence, R_∞ is not computably categorical.

In subsequent years, the notion of computable categoricity has been relativized (e.g., to the levels of the hyperarithmetical hierarchy [4, 5] or the Ershov hierarchy [6, 7]). The problems related to computable categoricity have been receiving much attention in the literature, see Sec. 2.3 for a more detailed discussion. In some familiar classes of structures, the property of "being computably categorical" admits a simple syntactic description: e.g., a computable Boolean algebra is computably categorical if and only if it has only finitely many atoms [8, 9]. Nevertheless, the groundbreaking result of Downey, Kach, Lempp, Lewis-Pye, Montalbán, and Turetsky [10] shows that in general, there is no simple syntactic characterization of computable categoricity: it was proved that the index set of computably categorical structures is Π^1_1-complete. Informally speaking, this means that the simplest possible description of, say, computably categorical graphs is provided by the *verbatim* definition of a computably categorical structure.

The situation changes quite drastically if one starts considering only *decidable structures*. Recall that a computable L-structure $\mathcal{S}$ is *decidable* if the complete diagram of $\mathcal{S}$ is computable, i.e. given a first-order L-formula $\psi(\bar{x})$ and a tuple $\bar{a}$ from $\mathcal{S}$, one can effectively check whether $\psi(\bar{a})$ is true in $\mathcal{S}$.

Let $\mathbf{d}$ be a Turing degree. A structure $\mathcal{S}$ is *decidably* $\mathbf{d}$*-categorical* (or $\mathbf{d}$*-autostable relative to strong constructivizations* [11]) if $\mathcal{S}$ has a decidable copy, and for any decidable structures $\mathcal{A}$ and $\mathcal{B}$ isomorphic to $\mathcal{S}$, there is a $\mathbf{d}$-computable isomorphism from $\mathcal{A}$ onto $\mathcal{B}$. If $\mathbf{d} = \mathbf{0}$, then we say that $\mathcal{S}$ is *decidably categorical.* Informally, one can say that the notion of decidable categoricity is a counterpart of computable categoricity in the realm of decidable structures.

Nurtazin [12] obtained a complete characterization of decidably categorical structures (see Theorem 2.1). This result has the following noteworthy consequences:

(1) Every decidably categorical structure $\mathcal{S}$ is an *almost prime model,* i.e. there is a finite tuple $\bar{c} \in \mathcal{S}$ such that the structure $(\mathcal{S}, \bar{c})$ is a prime model of the theory $Th(\mathcal{S}, \bar{c})$.
(2) The property of "being decidably categorical" can be described by a computable $\Sigma_{\omega+2}$ formula.

Hence, decidable categoricity *does* admit a pretty tame syntactic characterization, which contrasts with the result of [10] mentioned above. Goncharov and Marchuk [13, 14] showed that the complexity bound from the second item is sharp: the index set of decidably categorical graphs is $\Sigma^0_{\omega+2}$-complete. Quite surprisingly, in [15] it was proved that the index set of decidably categorical Boolean algebras is also $\Sigma^0_{\omega+2}$-complete. Note that the index set of computably categorical Boolean algebras is just Σ^0_3-complete.

On the other hand, in [16] (Corollary 5.6) it was shown that the index set of $\mathbf{0}'$-decidably categorical, decidable structures is Π^1_1-complete. Hence, the result of Nurtazin does not have a reasonable generalization for the case of $\mathbf{0}'$-decidable categoricity.

In view of the results above, it is natural to ask whether one can obtain simple *sufficient* conditions for being $\mathbf{0}'$-decidably categorical. Since decidable categoricity is closely connected to prime models (i.e. models which realize only principal types), one can consider models which realize types of a bounded Cantor–Bendixson rank (to be explained in Sec. 2.2). At the workshop "Aspects of Computation" (Singapore, August 21–September 15, 2017), Belanger asked the following question:

Problem 1.1: (Belanger) *Let n be a non-zero natural number. Suppose* *that $\mathcal{S}$ is a decidable structure such that every type realized in $\mathcal{S}$ has Cantor–Bendixson rank at most n. Is $\mathcal{S}$ always $\mathbf{0}'$-decidably categorical?*

Here we give a positive answer to the problem. The outline of this chapter is as follows. Section 2 contains the necessary preliminaries and a discussion of related work. In Sec. 3, we prove the main result (Theorem 3.1): if all types from a countable structure $\mathcal{S}$ are ranked and the supremum of the Cantor–Bendixson ranks of these types, denoted by $CB(\mathcal{S})$, is a successor ordinal, then $\mathcal{S}$ is an almost prime model. The answer to Problem 1.1 is a simple consequence of this result. The rest of the chapter discusses the surroundings of the main result. In Sec. 3.1, we give analysis of Cantor–Bendixson ranks for well-orders. In Sec. 3.2, we prove that for any natural number n, there is an almost prime model $\mathcal{S}$ with $CB(\mathcal{S}) = n$ (Theorem 3.7).

2. Preliminaries

We consider only computable languages. We assume that for every considered countable structure, its domain is a subset of ω.

2.1. *Prime models and decidable categoricity*

Let T be a complete theory in a language L. A structure $\mathcal{M} \models T$ is a *prime model* of T if $\mathcal{M}$ is elementarily embeddable into any model of the theory T. If we say that $\mathcal{M}$ is a prime model and we do not specify the theory T, then this means that $\mathcal{M}$ is a prime model of its own first-order theory $Th(\mathcal{M})$. A structure $\mathcal{M}$ is an *almost prime model* if there is a finite tuple $\bar{c}$ from $\mathcal{M}$ such that $(\mathcal{M}, \bar{c})$ is a prime model.

An L-formula $\psi(\bar{x})$ is a *complete formula* of the theory T if it satisfies the following conditions:

(a) $T \vdash \exists \bar{x} \psi(\bar{x})$.
(b) For any L-formula $\xi(\bar{x})$, we have either $T \vdash \forall \bar{x}[\psi(\bar{x}) \to \xi(\bar{x})]$ or $T \vdash \forall \bar{x}[\psi(\bar{x}) \to \neg\xi(\bar{x})]$.

Note that the set of complete formulas of T is co-c.e. in T.

A structure $\mathcal{M} \models T$ is an *atomic model* of T if every tuple from $\mathcal{M}$ satisfies some complete formula of T. Vaught proved that $\mathcal{M}$ is a prime model if and only if $\mathcal{M}$ is a countable atomic model (see, e.g., [17]).

Theorem 2.1: (Nurtazin [12]) *Let $\mathcal{M}$ be a decidable L-structure. Then $\mathcal{M}$*

is decidably categorical if and only if there exists a finite tuple $\bar{c}$ from $\mathcal{M}$ with the following properties:

(a) *The structure $(\mathcal{M}, \bar{c})$ is a prime model of the theory $Th(\mathcal{M}, \bar{c})$.*
(b) *Given an $(L \cup \{\bar{c}\})$-formula $\psi(\bar{x})$, one can effectively, uniformly in ψ, determine whether ψ is a complete formula of $Th(\mathcal{M}, \bar{c})$.*

Goncharov (see Theorem 2 in [11]) obtained the following sufficient condition for decidable **d**-categoricity.

Theorem 2.2: (Goncharov [11]) *Let* **d** *be a Turing degree, and $\mathcal{M}$ be a decidable L-structure. Suppose that $\bar{c}$ is a finite tuple from $\mathcal{M}$ such that the following conditions hold:*

(a) *The structure $(\mathcal{M}, \bar{c})$ is a prime model.*
(b) *Given an $(L \cup \{\bar{c}\})$-formula $\psi(\bar{x})$, one can effectively relative to* **d** *(and uniformly in ψ) determine whether ψ is a complete formula of the theory $Th(\mathcal{M}, \bar{c})$.*

Then $\mathcal{M}$ is decidably **d***-categorical.*

Corollary 2.3: *If a decidable structure $\mathcal{M}$ is an almost prime model, then $\mathcal{M}$ is decidably* $\mathbf{0}'$*-categorical.*

Proof: Choose a tuple $\bar{c}$ such that $(\mathcal{M}, \bar{c})$ is a prime model. It is easy to show that the theory $Th(\mathcal{M}, \bar{c})$ is decidable. Hence, the set of complete formulas of $Th(\mathcal{M}, \bar{c})$ is co-c.e. Therefore, one can apply Theorem 2.2 and obtain that $\mathcal{M}$ is decidably $\mathbf{0}'$-categorical. □

2.2. *Types and Cantor–Bendixson ranks*

Let T be a complete theory in a language L, and $p = p(x_1, \ldots, x_n)$ be a set of L-formulas with free variables $x_1, \ldots, x_n$. The set p is called a (complete) *n-type* of T if p has the following properties:

(a) the set $p \cup T$ is satisfiable, and
(b) for any L-formula $\varphi(x_1, \ldots, x_n)$, we have either $\varphi \in p$ or $\neg\varphi \in p$.

By $S_n(T)$ we denote the set of all n-types of T.

A type $p \in S_n(T)$ is *principal* if it contains a complete formula of T. Note that a countable structure $\mathcal{M} \models T$ is a prime model of T if and only if all types realized in $\mathcal{M}$ are principal.

Let φ be an L-formula in free variables $x_1, \dots, x_n$ such that the set $T \cup \{\varphi\}$ is satisfiable. We define a set of n-types

$$[\varphi] = \{p \in S_n(T) : \varphi \in p\}.$$

The set $S_n(T)$ is endowed with *Stone topology* (or Mal'tsev topology), which is generated by taking the sets $[\varphi]$ as basic open sets.

Let X be a compact Hausdorff topological space. By X' we denote the set of all limit points of X. For an ordinal α, the *α-th Cantor–Bendixson derivative* of X is defined by transfinite induction as follows.

- $X^0 = X$;
- $X^{\alpha+1} = (X^\alpha)'$;
- If λ is a limit ordinal, then $X^\lambda = \bigcap_{\beta<\lambda} X^\beta$.

The smallest ordinal α with $X^{\alpha+1} = X^\alpha$ is called the *Cantor–Bendixson rank* (or *CB-rank* for short) of the space X.

Definition 2.4: Let $p \in S_n(T)$ be a type. The *CB-rank* of the type p, denoted by $CB(p)$, is defined as follows.

(a) If $p \in S_n(T)^\alpha \setminus S_n(T)^{\alpha+1}$ for some ordinal α, then we say that the type p is *ranked* and $CB(p) = \alpha$.
(b) If $p \in S_n(T)^\beta$ for all ordinals β, then p is *unranked* and $CB(p) = \infty$.

Note that $CB(p) = \alpha < \infty$ if and only if p is an isolated point in $S_n(T)^\alpha$. In particular, $CB(p) = 0$ iff p is a principal type of the theory T.

Let $\mathcal{M} \models T$, and let $\bar{c} = c_1, \dots, c_n$ be a tuple from $\mathcal{M}$. By $tp^{\mathcal{M}}(\bar{c})$ we denote the type realized by the tuple $\bar{c}$ in $\mathcal{M}$, i.e.

$$tp^{\mathcal{M}}(\bar{c}) = \{\varphi(x_1, \dots, x_n) : \mathcal{M} \models \varphi(\bar{c})\}.$$

Definition 2.5: Let T be a complete theory in a language L, and let $\mathcal{M} \models T$ be an L-structure. Suppose that n is a non-zero natural number. We define the following parameters:

(1) $CB_n(\mathcal{M}) = \sup\big\{CB(tp^{\mathcal{M}}(\bar{c})) : \bar{c} \in dom(\mathcal{M})^n,\ CB(tp^{\mathcal{M}}(\bar{c})) < \infty\big\}$.
(2) $CB(\mathcal{M}) = \sup\{CB_n(\mathcal{M}) : 1 \le n < \omega\}$.

In other words, $CB_n(\mathcal{M})$ is the supremum of the CB-ranks of all ranked n-types realized in $\mathcal{M}$, and $CB(\mathcal{M})$ is the supremum of the CB-ranks of all ranked types in $\mathcal{M}$.

2.3. *Related work*

The problems related to computable categoricity have received a lot of attention in the literature. The reader is referred to, e.g., the surveys [18, 19] for a detailed discussion and bibliographic references. Here we give only a brief overview of the results, which consider the Turing degrees of isomorphisms among decidable structures.

The *decidable categoricity spectrum* of a decidable structure $\mathcal{S}$ is the set

$$DecCatSpec(\mathcal{S}) = \{\mathbf{d} : \mathcal{S} \text{ is decidably } \mathbf{d}\text{-categorical}\}.$$

A degree $\mathbf{c}$ is the *degree of decidable categoricity* for $\mathcal{S}$ if $\mathbf{c}$ is the least degree in the spectrum $DecCatSpec(\mathcal{S})$.

The notions of decidable categoricity spectrum and degree of decidable categoricity were introduced by Goncharov [11]. Note that these notions are analogues of *categoricity spectrum* and *degree of categoricity*, respectively: e.g., the categoricity spectrum of a computable structure $\mathcal{A}$ contains all Turing degrees which are capable of computing isomorphisms among arbitrary computable copies of $\mathcal{A}$. The systematic study of categoricity spectra was initiated by Fokina, Kalimullin, and Miller [20]. To name only a few, the following results on categoricity spectra are known. For a computable non-limit ordinal α, every Turing degree $\mathbf{d}$, which is d.c.e. in and above $\mathbf{0}^{(\alpha)}$, is a degree of categoricity [20, 21]. There is a Σ^0_2 degree which is not a degree of categoricity [22]. There is a computable algebraic field with no degree of categoricity [23]. For more results on categoricity spectra, we refer the reader to, e.g., [18, 19, 24, 25].

Goncharov proved that any c.e. Turing degree is the degree of decidable categoricity for a prime model (Theorem 3 in [11]). In [26], it was shown that there is a decidable prime model with no degree of decidable categoricity. In [27], it was proved that for any computable ordinal α, every degree, which is c.e. in and above $\mathbf{0}^{(\alpha+1)}$, is the degree of decidable categoricity for a decidable structure. Furthermore, if α is infinite, then a similar result can be obtained for degrees $\mathbf{d}$ which are *d.c.e.* in and above $\mathbf{0}^{(\alpha+1)}$ [19]. For more results on decidable categoricity spectra, the reader is referred to [19, 28, 29].

We note that the recursion-theoretic properties of Cantor–Bendixson ranks for types were studied in the papers of Sacks [30] and Chong [31]. Their approach is different from ours: For a complete and substructure complete theory T, they consider spaces of 1-types, which are defined as follows. Let $\mathcal{A}$ be a countable substructure of a model of T. Then the space $S\mathcal{A}$ contains all 1-types over $dom(\mathcal{A})$.

3. Cantor–Bendixson Ranks and Almost Primeness

The first result of the section gives a positive answer to the question stated in Problem 1.1.

Theorem 3.1: *Let $\mathcal{M}$ be a countable L-structure with the following properties:*

(i) All types $tp^{\mathcal{M}}(\bar{c})$, $\bar{c} \in \mathcal{M}$, are ranked.
(ii) $CB(\mathcal{M})$ is a successor ordinal.

Then $\mathcal{M}$ is an almost prime model. Consequently, by Corollary 2.3, if $\mathcal{M}$ is decidable, then $\mathcal{M}$ is decidably $\mathbf{0}'$-categorical.

Before the proof, we explain why Theorem 3.1 provides us with a solution to Problem 1.1. As in Problem 1.1, we take a number $n \in \omega$ and a decidable structure $\mathcal{S}$ such that for any tuple $\bar{c}$, we have $CB(tp^{\mathcal{S}}(\bar{c})) \leq n$. Then it is easy to see that $CB(\mathcal{S})$ is a finite ordinal, and every type realized in $\mathcal{S}$ is ranked. Thus, Theorem 3.1 shows that $\mathcal{S}$ is an almost prime model.

Proof: First, we obtain the following auxiliary fact.

Lemma 3.2: *Let $\mathcal{M}$ be a countable L-structure, and let $\bar{c}$ be a finite tuple from $\mathcal{M}$. Suppose that α is an ordinal. If $CB(tp^{\mathcal{M}}(\bar{c})) \geq \alpha$, then for any tuple $\bar{a}$ from $\mathcal{M}$, we have $CB(tp^{\mathcal{M}}(\bar{c}, \bar{a})) \geq \alpha$.*

Proof: The proof employs induction on α. The case $\alpha = 0$ is obvious.

Suppose that the statement of the lemma holds for an ordinal α. Let $\bar{c}$ be a tuple such that $CB(tp^{\mathcal{M}}(\bar{c})) \geq \alpha + 1$, and let $\bar{a}$ be an arbitrary tuple from $\mathcal{M}$. By the induction hypothesis, we obtain that $CB(tp^{\mathcal{M}}(\bar{c}, \bar{a})) \geq \alpha$.

Towards contradiction, assume that $CB(tp^{\mathcal{M}}(\bar{c}, \bar{a})) = \alpha$. Then one can choose a formula $\psi(\bar{x}, \bar{y})$ with the following properties:

(a) $\mathcal{M} \models \psi(\bar{c}, \bar{a})$.
(b) For any tuple $\bar{c}_1, \bar{a}_1$ from $\mathcal{M}$, the following holds: If $CB(tp^{\mathcal{M}}(\bar{c}_1, \bar{a}_1)) \geq \alpha$, i.e. the type $tp^{\mathcal{M}}(\bar{c}_1, \bar{a}_1)$ belongs to the α-th Cantor–Bendixson derivative, then

$$\mathcal{M} \models \psi(\bar{c}_1, \bar{a}_1) \Rightarrow tp^{\mathcal{M}}(\bar{c}_1, \bar{a}_1) = tp^{\mathcal{M}}(\bar{c}, \bar{a}).$$

Definition 3.3: A formula $\psi(\bar{x}, \bar{y})$ with the properties (a) and (b) above is called an *α-complete formula* for the tuple $\bar{c}, \bar{a}$ (with respect to $Th(\mathcal{M})$).

We define a new formula $\xi(\bar{x}) = \exists \bar{y} \psi(\bar{x}, \bar{y})$. Note that ξ satisfies the following:

(1) $\mathcal{M} \models \xi(\bar{c})$.
(2) Let $\bar{d}$ be a tuple such that $CB(tp^{\mathcal{M}}(\bar{d})) \geq \alpha$ and $\mathcal{M} \models \xi(\bar{d})$. Fix a tuple $\bar{b}$ with $\mathcal{M} \models \psi(\bar{d}, \bar{b})$. By the induction hypothesis, we have $CB(tp^{\mathcal{M}}(\bar{d}, \bar{b})) \geq \alpha$. Since the formula ψ is α-complete for $\bar{c}, \bar{a}$, we have $tp^{\mathcal{M}}(\bar{d}, \bar{b}) = tp^{\mathcal{M}}(\bar{c}, \bar{a})$ and $tp^{\mathcal{M}}(\bar{d}) = tp^{\mathcal{M}}(\bar{c})$.

Therefore, the formula ξ is α-complete for the tuple $\bar{c}$, and $CB(tp^{\mathcal{M}}(\bar{c})) \leq \alpha$, which contradicts the choice of $\bar{c}$. Hence, we have $CB(tp^{\mathcal{M}}(\bar{c}, \bar{a})) \geq \alpha + 1$.

Now suppose that δ is a limit ordinal, and $CB(tp^{\mathcal{M}}(\bar{c})) \geq \delta$. For any tuple $\bar{a}$ and any ordinal $\beta < \delta$, by the induction hypothesis, we have $CB(tp^{\mathcal{M}}(\bar{c}, \bar{a})) \geq \beta$. Thus, $CB(tp^{\mathcal{M}}(\bar{c}, \bar{a})) \geq \delta$. Lemma 3.2 is proved. □

Let $CB(\mathcal{M}) = \alpha$. If $\alpha = 0$, then the structure $\mathcal{M}$ realizes only principal types and hence, $\mathcal{M}$ is a prime model. Thus, we may assume that $\alpha > 0$.

Since α is a successor ordinal, there is a number n and an n-tuple $\bar{c}$ such that $CB(tp^{\mathcal{M}}(\bar{c})) = \alpha$. We prove that the structure $(\mathcal{M}, \bar{c})$ is a prime model.

Let $\bar{a}$ be an arbitrary tuple from $\mathcal{M}$. By Lemma 3.2, we have $CB(tp^{\mathcal{M}}(\bar{c}, \bar{a})) \geq \alpha$. Since every type realized in $\mathcal{M}$ is ranked, we deduce that $CB(tp^{\mathcal{M}}(\bar{c}, \bar{a})) = \alpha$.

Let $\psi(\bar{x}, \bar{y})$ be an α-complete formula for the tuple $\bar{c}, \bar{a}$ (with respect to the theory $Th(\mathcal{M})$). It is sufficient to show that the $(L \cup \{\bar{c}\})$-formula $\psi(\bar{c}, \bar{y})$ is a complete formula of the theory $Th(\mathcal{M}, \bar{c})$.

Assume that $\bar{a}_1$ is an arbitrary tuple such that $\mathcal{M} \models \psi(\bar{c}, \bar{a}_1)$. By Lemma 3.2, we have $CB(tp^{\mathcal{M}}(\bar{c}, \bar{a}_1)) = \alpha$. Since the formula ψ is α-complete, we obtain that $tp^{\mathcal{M}}(\bar{c}, \bar{a}_1) = tp^{\mathcal{M}}(\bar{c}, \bar{a})$. Therefore, for any $(L \cup \{\bar{c}\})$-formula $\xi(\bar{c}, \bar{y})$, we have:

- If $\mathcal{M} \models \xi(\bar{c}, \bar{a})$, then $Th(\mathcal{M}, \bar{c}) \vdash \forall \bar{y}[\psi(\bar{c}, \bar{y}) \to \xi(\bar{c}, \bar{y})]$.
- If $\mathcal{M} \not\models \xi(\bar{c}, \bar{a})$, then $Th(\mathcal{M}, \bar{c}) \vdash \forall \bar{y}[\psi(\bar{c}, \bar{y}) \to \neg\xi(\bar{c}, \bar{y})]$.

This shows that the formula $\psi(\bar{c}, \bar{y})$ is complete. Hence, $(\mathcal{M}, \bar{c})$ is a prime model. Theorem 3.1 is proved. □

3.1. *Well-orders and Cantor–Bendixson ranks*

The goal of this subsection is to show that in general, one cannot omit the first requirement of Theorem 3.1. In order to obtain this, here we work with countable well-orders. First, we give the necessary preliminaries.

The reader is referred to Chapter 6 in [32] for the formal definition of infinitary formulas, and their properties. Let n be a natural number. By

Σ_n^{inf} we denote the class of infinitary Σ_n formulas, and Π_n^{inf} denotes the class of infinitary Π_n formulas.

Let $\mathcal{A}$ and $\mathcal{B}$ be L-structures. We say that $\mathcal{A} \leq_n \mathcal{B}$ if every Π_n^{inf} sentence true in $\mathcal{A}$ is also true in $\mathcal{B}$. We write $\mathcal{A} \equiv_n \mathcal{B}$ if $\mathcal{A} \leq_n \mathcal{B}$ and $\mathcal{B} \leq_n \mathcal{A}$. We will use the following results on the relations $\leq_n$ for countable linear orders.

Lemma 3.4: (Lemma 15.7 in [32]) *Suppose that $\mathcal{A}$ and $\mathcal{B}$ are linear orders, $a \in \mathcal{A}$, and $b \in \mathcal{B}$. Suppose also that $\mathcal{A}_0$, $\mathcal{A}_1$, $\mathcal{B}_0$, $\mathcal{B}_1$ are non-empty intervals such that*

$$\mathcal{A} = \mathcal{A}_0 + \{a\} + \mathcal{A}_1, \quad \mathcal{B} = \mathcal{B}_0 + \{b\} + \mathcal{B}_1.$$

Then the condition $(\mathcal{A}, a) \leq_n (\mathcal{B}, b)$ holds if and only if $\mathcal{A}_0 \leq_n \mathcal{B}_0$ and $\mathcal{A}_1 \leq_n \mathcal{B}_1$.

Lemma 3.5: (Lemma 15.10 in [32]) *Suppose that $\alpha = \omega^{n+1} \cdot \gamma + \delta$, where $\gamma \neq 0$, $n < \omega$, and $\delta < \omega^n$. Consider the ordinal $\beta = \omega^{n+1} \cdot (\gamma + \gamma_1) + \delta$, where $\gamma_1 \geq 0$. Then we have $\alpha \equiv_{2n+1} \beta$.*

The next result shows the following. Suppose that one relaxes the conditions of Theorem 3.1 as follows: now we do not require that *all* types realized in the structure $\mathcal{M}$ are ranked. Then both outcomes are possible: sometimes, $\mathcal{M}$ *is* an almost prime model; and sometimes, it *is not*.

Proposition 3.6: *Consider a countable ordinal $\alpha = \omega^\omega \cdot \alpha_0$, where $\alpha_0 \geq 2$.*

(a) The well-order α is an almost prime model if and only if $\alpha_0 < \omega$.
(b) For any tuple $\bar{c}$ from α, we have $CB(tp^\alpha(\bar{c})) \in \{0, \infty\}$. In particular, this implies that $CB(\alpha) = 0$.

Proof: (a) This fact was obtained in Corollary 1 of [33].

(b) Fix a copy $\mathcal{L}$ of α. Let 0 be the least element inside $\mathcal{L}$, and let d_0 be the element such that the interval $[0; d_0)_\mathcal{L}$ is isomorphic to ω^ω. Let $\bar{c} = c_0 <_\mathcal{L} c_1 <_\mathcal{L} \cdots <_\mathcal{L} c_n$ be a tuple from $\mathcal{L}$.

First, suppose that $c_n <_\mathcal{L} d_0$. Then one can apply the analysis from Sec. 2 of [33], and show that $CB(tp^\mathcal{L}(\bar{c})) = 0$, i.e. the tuple $\bar{c}$ realizes a principal type. We omit the technical details here, but a typical example looks like follows. Suppose that $\bar{c} = (c_0, c_1, c_2, c_3)$, where $c_0 = 0$, $[c_0; c_1)_\mathcal{L} \cong \omega$, $[c_0; c_2)_\mathcal{L} \cong \omega^2$, and c_3 is a successor of c_2. Then a complete formula

$\psi(x_0, x_1, x_2, x_3)$ for the tuple $\bar{c}$ can be recovered as follows.

$$Succ_1(y,z) = (y < z) \,\&\, \neg\exists w[y < w < z];$$
$$Lim_1(z) = \neg\exists y Succ_1(y,z);$$
$$Succ_2(y,z) = (y < z) \,\&\, Lim_1(y) \,\&\, Lim_1(z) \,\&\, \neg\exists w[Lim_1(w) \,\&\, (y < w < z)];$$
$$Lim_2(z) = Lim_1(z) \,\&\, \neg\exists y Succ_2(y,z);$$
$$Succ_3(y,z) = (y < z) \,\&\, Lim_2(y) \,\&\, Lim_2(z) \,\&\, \neg\exists w[Lim_2(w) \,\&\, (y < w < z)];$$
$$\psi(\bar{x}) = \forall y(x_0 \le y) \,\&\, Succ_2(x_0,x_1) \,\&\, Succ_3(x_0,x_2) \,\&\, Succ_1(x_2,x_3).$$

Now we need to prove the following: if $c_n \ge_{\mathcal{L}} d_0$, then we have $CB(tp^{\mathcal{L}}(\bar{c})) = \infty$. Lemma 3.2 implies that it is sufficient to obtain that $CB(tp^{\mathcal{L}}(c_n)) = \infty$. Hence, we will denote c_n by c, and consider only the 1-type of c.

We prove that $CB(tp^{\mathcal{L}}(c)) \ge \alpha$ for any ordinal α. For $\alpha = 0$, this is obvious.

Suppose that we already showed the following: if $c \ge_{\mathcal{L}} d_0$, then $CB(tp^{\mathcal{L}}(c)) \ge \alpha$. Towards contradiction, assume that $CB(tp^{\mathcal{L}}(c)) = \alpha$. Choose an α-complete formula $\psi(x)$ for the element c. Suppose that ψ is a first-order Σ_n formula. Note that the formula ψ satisfies the following:

- $\mathcal{L} \models \psi(c)$.
- Let e be an arbitrary element with $e \ge_{\mathcal{L}} d_0$. By the induction hypothesis, we have $CB(tp^{\mathcal{L}}(e)) \ge \alpha$. Therefore, if $\mathcal{L} \models \psi(e)$, then $tp^{\mathcal{L}}(e) = tp^{\mathcal{L}}(c)$.

By choosing a large enough number n, w.l.o.g., we may assume that the interval $[0;c)_{\mathcal{L}}$ is isomorphic to $\omega^{n+1} \cdot \gamma + \delta$, where $\gamma \ge \omega^\omega$ and $\delta < \omega^n$. Choose an element $e \in \mathcal{L}$ such that the interval $[0;e)_{\mathcal{L}}$ is a copy of $\omega^{n+1} \cdot (\gamma + 1) + \delta$. By Lemma 3.5, we have $[0;e)_{\mathcal{L}} \equiv_{2n+1} [0;c)_{\mathcal{L}}$. Hence, we apply Lemma 3.4 and obtain that $(\mathcal{L}, e) \equiv_{2n+1} (\mathcal{L}, c)$. Since ψ is a Σ_n formula, this implies that $\mathcal{L} \models \psi(e)$ and $tp^{\mathcal{L}}(e) = tp^{\mathcal{L}}(c)$. On the other hand, the analysis of formulas from Sec. 2 of [33] shows that the types $tp^{\mathcal{L}}(e)$ and $tp^{\mathcal{L}}(c)$ are different. We illustrate this by a typical example. Assume that $[0;c)_{\mathcal{L}} \cong \omega^\omega + \omega^2 + 3$. Then, w.l.o.g., one can choose $\gamma = \omega^\omega$ and $n = 3$. This implies that the interval $[0;e)_{\mathcal{L}}$ is isomorphic to $\omega^\omega + \omega^4 + \omega^2 + 3$. Therefore, there is an element d such that $[d;e)_{\mathcal{L}}$ is isomorphic to the well-order $\omega^4 + \omega^2 + 3$, and this property is witnessed by a first-order formula. On the other hand, there is no such element for the original c. Therefore, we obtained a contradiction, and we deduce that $CB(tp^{\mathcal{L}}(c)) \ge \alpha + 1$.

Let δ be a limit ordinal. It is clear that if $CB(tp^{\mathcal{L}}(c)) \geq \beta$ for all $\beta < \delta$, then $CB(tp^{\mathcal{L}}(c)) \geq \delta$.

Hence, we obtained that $CB(tp^{\mathcal{L}}(c)) = \infty$. Proposition 3.6 is proved. □

3.2. *Finite Cantor–Bendixson ranks*

In this subsection, we show that any natural number n can be the Cantor–Bendixson rank $CB(\mathcal{M})$ for an almost prime model $\mathcal{M}$. We leave open whether this result can be extended to infinite ordinals.

Theorem 3.7: *Let n be a natural number. There exists an almost prime model $\mathcal{M}_n$ with the following properties:*

- *All types $tp^{\mathcal{M}_n}(\bar{c})$, $\bar{c} \in \mathcal{M}_n$, are ranked.*
- *$CB(\mathcal{M}_n) = n$.*

Proof: If $n = 0$, then any prime model $\mathcal{M}_0$ satisfies the conditions of the theorem.

Further proof employs induction on n. Suppose that $n = 1$. Consider the language $L_1 = \{P_{1,i} : i \in \omega\}$, where every $P_{1,i}$ is a unary predicate. We define a countable L_1-structure $\mathcal{M}_1$ as follows. First, we put an element $a_1^* \in \mathcal{M}_1$ with $\mathcal{M}_1 \models \bigwedge\!\!\!\bigwedge_{i\in\omega} \neg P_{1,i}(a_1^*)$. By CF we denote the following set of infinitary formulas:

$$CF = \left\{ \mathop{\&}_{j\leq k} \neg P_{1,j}(x) \,\&\, \bigwedge\!\!\!\bigwedge_{i>k} P_{1,i}(x) : k \in \omega \right\}.$$

For each formula $\xi(x)$ from CF, we add to $\mathcal{M}_1$ a unique element b such that $\xi(b)$ holds.

Now we compute the CB-ranks of tuples from $\mathcal{M}_1$. For every $b \in \mathcal{M}_1$ with $b \neq a_1^*$, there exists a complete formula

$$\psi_b^{\mathcal{M}_1}(x) = \mathop{\&}_{j\leq k} \neg P_{1,j}(x) \,\&\, P_{1,k+1}(x),$$

which is satisfied by the element b. If $\bar{b} = b_1, b_2, \dots, b_m$ is a tuple from $\mathcal{M}_1 \setminus \{a_1^*\}$, then one can define a complete formula

$$\psi_{\bar{b}}^{\mathcal{M}_1}(x_1, x_2, \dots, x_m) = \mathop{\&}_{1\leq i\leq m} \psi_{b_i}^{\mathcal{M}_1}(x_i).$$

Note that $\mathcal{M}_1 \models \psi_{\bar{b}}(\bar{b})$. Hence, we have $CB(tp^{\mathcal{M}_1}(\bar{b})) = 0$ for all $\bar{b}$ from $\mathcal{M}_1 \setminus \{a_1^*\}$. Furthermore, it is not hard to show that any tuple $\bar{a} \in \mathcal{M}_1$ satisfies the following: if $a_1^* \notin \bar{a}$, then $CB(tp^{\mathcal{M}_1}(\bar{a})) = 0$; otherwise, we have

$CB(tp^{\mathcal{M}_1}(\bar{a})) = 1$. Using this fact, it is easy to obtain that $(\mathcal{M}_1, a_1^*)$ is a prime model.

Suppose that $n = m + 1$, where $m \geq 1$. Assume that we already constructed the desired L_m-structure $\mathcal{M}_m$ with $CB(\mathcal{M}_m) = m$, and there is a unique element a_m^* such that $CB(tp^{\mathcal{M}_m}(a_m^*)) = m$ and $\mathcal{M}_m \models \bigwedge\!\!\!\bigwedge_{i\in\omega} \neg P_{m,i}(a_m^*)$. We also assume that $(\mathcal{M}_m, a_m^*)$ is a prime model.

We define a new language

$$L_{m+1} = L_m \cup \{E_m, R_m\} \cup \{P_{m+1,i} : i \in \omega\},$$

where $P_{m+1,i}$, $i \in \omega$, are unary predicates, E_m and R_m are binary predicates. A countable L_{m+1}-structure $\mathcal{M}_{m+1}$ is built as follows. The L_m-reduct of $\mathcal{M}_{m+1}$ is a disjoint union of infinitely many copies of $\mathcal{M}_m$. The predicate E_m is an equivalence relation on $\mathcal{M}_{m+1}$, which picks out the domain of each copy of $\mathcal{M}_m$. An E_m-*class* is an E_m-equivalence class inside $\mathcal{M}_{m+1}$. We often identify an E_m-class $[a]_{E_m}$ and the corresponding L_m-structure on the domain $[a]_{E_m}$. For $a, b \in \mathcal{M}_{m+1}$, we set $\mathcal{M}_{m+1} \models R_m(a, b)$ if and only if a and b belong to different E_m-classes and $tp^{\mathcal{M}_m}(a) = tp^{\mathcal{M}_m}(b)$. We fix an E_m-class, denoted by $\mathcal{M}_m^*$, such that for all $a \in \mathcal{M}_m^*$, we have $\mathcal{M}_{m+1} \models \bigwedge\!\!\!\bigwedge_{i\in\omega} \neg P_{m+1,i}(a)$. By CF' we denote the following set of infinitary formulas:

$$CF' = \left\{ \mathop{\&}_{j\leq k} \neg P_{m+1,j}(x) \,\&\, \bigwedge\!\!\!\bigwedge_{i>k} P_{m+1,i}(x) : k \in \omega \right\}.$$

For each formula $\xi(x) \in CF'$, we choose an E_m-class $\mathcal{M}_{m,\xi} \neq \mathcal{M}_m^*$, and for all $a \in \mathcal{M}_{m,\xi}$, we set $\mathcal{M}_{m+1} \models \xi(a)$. As per usual, here we assume that every E_m-class inside $\mathcal{M}_{m+1}$ either is equal to $\mathcal{M}_m^*$, or corresponds to a unique formula $\xi \in CF'$. By a_{m+1}^* we denote the copy of a_m^* inside the E_m-class $\mathcal{M}_m^*$, i.e. the unique element from $\mathcal{M}_m^*$ with $CB(tp^{\mathcal{M}_m^*}(a_{m+1}^*)) = m$ and $\mathcal{M}_{m+1} \models \bigwedge\!\!\!\bigwedge_{i\in\omega} \neg P_{m,i}(a_{m+1}^*)$.

We sketch the computations of the CB-ranks for tuples from $\mathcal{M}_{m+1}$. Let $\bar{b}$ be a tuple from $\mathcal{M}_m$ such that $CB(tp^{\mathcal{M}_m}(\bar{b})) = k$. Then by $\psi_{\bar{b}}^{\mathcal{M}_m}(\bar{x})$ we denote a k-complete formula for $\bar{b}$ in $\mathcal{M}_m$.

Let a be an element from $\mathcal{M}_{m+1}$. Since a lies in some E_m-class, one can slightly abuse the notations and treat a as an element from $\mathcal{M}_m$. Thus, it makes perfect sense to consider the type $tp^{\mathcal{M}_m}(a)$. Let $l = CB(tp^{\mathcal{M}_m}(a))$. W.l.o.g., we may assume that $l \neq 0$.

First, suppose that $a \notin \mathcal{M}_m^*$ and

$$\mathcal{M}_{m+1} \models \mathop{\&}_{j\leq k} \neg P_{m+1,j}(a) \,\&\, P_{m+1,k+1}(a).$$

Then we claim that the formula

$$\psi_a^{\mathcal{M}_{m+1}}(x) := \psi_a^{\mathcal{M}_m}(x) \,\&\, \mathop{\&}_{j\leq k} \neg P_{m+1,j}(x) \,\&\, P_{m+1,k+1}(x)$$

is an l-complete formula inside $\mathcal{M}_{m+1}$ for the element a. Indeed, let b be an arbitrary element from $\mathcal{M}_{m+1}$ with $CB(tp^{\mathcal{M}_{m+1}}(b)) \geq l$ and $\mathcal{M}_{m+1} \models \psi_a^{\mathcal{M}_{m+1}}(b)$. Notice that b and a belong to the same E_m-class $[a]_{E_m}$. Moreover, note the following: if a sequence of 1-types $(p_i)_{i\in\omega}$ converges to the type $q := tp^{\mathcal{M}_{m+1}}(b)$ (in the corresponding Mal'tsev topological space), then one can assume that every type p_i is realized by an element from the E_m-class $[a]_{E_m}$ (just consider the neighbourhood of q induced by the formula $\neg P_{m+1,0}(x) \& \ldots \& \neg P_{m+1,k}(x) \& P_{m+1,k+1}(x)$). Using this observation, it is not hard to deduce that the CB-rank of the type $tp^{\mathcal{M}_m}(b)$, i.e. the type inside $\mathcal{M}_m$, is at least l. Since the formula $\psi_a^{\mathcal{M}_m}$ is l-complete inside $\mathcal{M}_m$, we obtain that $tp^{\mathcal{M}_m}(b) = tp^{\mathcal{M}_m}(a)$. This implies that $tp^{\mathcal{M}_{m+1}}(b) = tp^{\mathcal{M}_{m+1}}(a)$ and hence, we have $CB(tp^{\mathcal{M}_{m+1}}(a)) = l$. In particular, this shows that $CB(\mathcal{M}_{m+1}) \geq m$.

Now assume that $a \in \mathcal{M}_m^*$. We claim that the formula

$$\psi_a^{\mathcal{M}_{m+1}}(x) := \psi_a^{\mathcal{M}_m}(x) \,\&\, \neg P_{m+1,0}(x)$$

is an $(l+1)$-complete formula in $\mathcal{M}_{m+1}$. Indeed, consider 1-types

$$q_i := tp^{\mathcal{M}_{m+1}}(b_i),\ i \in \omega,$$

where b_i is an isomorphic copy of a inside the E_m-class $[b_i]_{E_m}$ and $\mathcal{M}_{m+1} \models \neg P_{m+1,0}(b_i) \& \ldots \& \neg P_{m+1,i}(b_i) \& P_{m+1,i+1}(b_i)$. The sequence $(q_i)_{i\in\omega}$ converges to the type $tp^{\mathcal{M}_{m+1}}(a)$, and the argument above shows that for every i, $CB(q_i) = l$. Hence, we obtain that $CB(tp^{\mathcal{M}_{m+1}}(a)) \geq l+1$. On the other hand, let c be an arbitrary element from $\mathcal{M}_{m+1}$ such that $CB(tp^{\mathcal{M}_{m+1}}(c)) \geq l+1$ and $\mathcal{M}_{m+1} \models \psi_a^{\mathcal{M}_{m+1}}(c)$. If c does not belong to $\mathcal{M}_m^*$, then by applying the argument similar to that one above, one can deduce that $CB(tp^{\mathcal{M}_m}(c)) \geq l+1$ and $tp^{\mathcal{M}_{m+1}}(c)$ is equal to some q_i, $i \in \omega$. Since $CB(q_i) = l < l+1$, this contradicts the choice of the element c. Therefore, c must belong to $\mathcal{M}_m^*$. Recall that $\psi_a^{\mathcal{M}_m}$ is an l-complete formula inside $\mathcal{M}_m^*$. Thus, it is not hard to obtain that $tp^{\mathcal{M}_{m+1}}(c) = tp^{\mathcal{M}_{m+1}}(a)$. Hence, $CB(tp^{\mathcal{M}_{m+1}}(a)) = l+1$, and this implies that $CB(\mathcal{M}) \geq m+1$, and a_{m+1}^* is the unique element from $\mathcal{M}_{m+1}$ such that $CB(tp^{\mathcal{M}_{m+1}}(a_{m+1}^*)) = m+1$ and $\mathcal{M}_{m+1} \models \bigwedge\!\!\!\bigwedge_{i\in\omega} \neg P_{m+1,i}(a_{m+1}^*)$.

In order to show that $CB(\mathcal{M}) \leq m+1$, we need to consider the CB-ranks for tuples of arbitrary length from $\mathcal{M}_{m+1}$. Roughly speaking, if we want to obtain an appropriate l-complete (or $(l+1)$-complete) formula for a tuple $\bar{a}$ from $\mathcal{M}_{m+1}$, then it is necessary to specify the actions of the predicates E_m and R_m on $\bar{a}$. We illustrate this by a typical example. Consider a tuple $\bar{a} = a_0, a_1, a_2$ from $\mathcal{M}_{m+1}$ such that $a_2 = a^*_{m+1}$, a_1 is a copy of a^*_m inside the E_m-class $[a_0]_{E_m}$, and

$$\mathcal{M}_{m+1} \models E_m(a_0, a_1) \,\&\, \neg P_{m+1,0}(a_0) \,\&\, P_{m+1,1}(a_0).$$

Then one can take

$$\begin{aligned}\psi_{\bar{a}}^{\mathcal{M}_{m+1}}(x_0, x_1, x_2) = \psi_{a_0,a_1}^{\mathcal{M}_m}(x_0, x_1) \,\&\, E_m(x_0, x_1) \& \neg E_m(x_1, x_2)\\ \& \neg R_m(x_0, x_1) \& R_m(x_1, x_2) \& \neg P_{m+1,0}(x_0)\\ \& P_{m+1,1}(x_0) \& \neg P_{m+1,0}(x_2) \& \psi_{a_2}^{\mathcal{M}_m}(x_2)\end{aligned}$$

as an $(m+1)$-complete formula for the tuple $\bar{a}$. In general, we have

$$\begin{aligned}CB(tp^{\mathcal{M}_{m+1}}(\bar{a})) = \sup(\{CB(tp^{\mathcal{M}_m}(a)) : a \in \mathcal{M}_{m+1} \setminus \mathcal{M}^*_m,\ a \in \bar{a}\} \cup\\ \{CB(tp^{\mathcal{M}_m}(a)) + 1 : a \in \mathcal{M}^*_m,\ a \in \bar{a}\}).\end{aligned}$$

Finally, we need to show that $(\mathcal{M}_{m+1}, a^*_{m+1})$ is a prime model. By $\varphi_{\bar{a}}^{(\mathcal{M}_m, a^*_m)}(a^*_m, \bar{x})$ we denote a complete formula for $\bar{a}$ in $(\mathcal{M}_m, a^*_m)$. If $a \in \mathcal{M}_{m+1} \setminus \mathcal{M}^*_m$ and $\mathcal{M}_{m+1} \models \underset{i \le k}{\&} \neg P_{m+1,i}(a) \& P_{m+1,k+1}(a)$, then the formula

$$\begin{aligned}\varphi_a^{(\mathcal{M}_{m+1}, a^*_{m+1})}(x) := \exists y [R_m(y, a^*_{m+1}) \& \underset{i \le k}{\&} \neg P_{m+1,i}(y) \& P_{m+1,k+1}(y)\\ \& E_m(x, y) \& \varphi_a^{(\mathcal{M}_m, a^*_m)}(y, x)]\end{aligned}$$

is a complete formula for a in $(\mathcal{M}_{m+1}, a^*_{m+1})$. Indeed, if the element y from the formula above exists, then y must be the copy of a^*_m inside the E_m-class $[x]_{E_m}$. This implies that both types $tp^{(\mathcal{M}_m, y)}(a)$ and $tp^{(\mathcal{M}_{m+1}, a^*_{m+1})}(a)$ are principal.

If $b \in \mathcal{M}^*_m$, then

$$\varphi_b^{(\mathcal{M}_{m+1}, a^*_{m+1})}(x) = E_m(x, a^*_{m+1}) \& \varphi_b^{(\mathcal{M}_m, a^*_m)}(a^*_{m+1}, x)$$

is a complete formula for b in $(\mathcal{M}_{m+1}, a^*_{m+1})$. As per usual, if $\bar{a} = a_1, \dots, a_k$ is a tuple from $\mathcal{M}_{m+1}$, then one can define a complete formula

$$\varphi_{\bar{a}}^{(\mathcal{M}_{m+1}, a^*_{m+1})}(x_1, \dots, x_k) = \underset{1 \le i \le k}{\&} \varphi_{a_i}^{(\mathcal{M}_{m+1}, a^*_{m+1})}(x_i).$$

Consequently, we deduce that $(\mathcal{M}_{m+1}, a^*_{m+1})$ is a prime model. Theorem 3.7 is proved. □

References

1. A. Fröhlich and J. C. Shepherdson, Effective procedures in field theory, *Philos. Trans. Roy. Soc. London Ser. A.* **248** (1956), 407–432.
2. A. I. Mal'tsev, Constructive algebras I, *Russ. Math. Surv.* **16** (1961), 77–129.
3. A. I. Mal'tsev, On recursive abelian groups, *Sov. Math. Dokl.* **32** (1962), 1431–1434.
4. C. J. Ash, Recursive labelling systems and stability of recursive structures in hyperarithmetical degrees, *Trans. Am. Math. Soc.* **298** (1986), 497–514.
5. C. J. Ash, Categoricity in hyperarithmetical degrees, *Ann. Pure Appl. Logic.* **34** (1987), 1–14.
6. B. Khoussainov, F. Stephan, and Y. Yang, Computable categoricity and the Ershov hierarchy, *Ann. Pure Appl. Logic.* **156** (2008), 86–95.
7. D. Cenzer, G. LaForte, and J. Remmel, Equivalence structures and isomorphisms in the difference hierarchy, *J. Symb. Log.* **74** (2009), 535–556.
8. S. S. Goncharov and V. D. Dzgoev, Autostability of models, *Algebra Logic.* **19** (1980), 28–37.
9. J. B. Remmel, Recursive isomorphism types of recursive Boolean algebras, *J. Symb. Log.* **46** (1981), 572–594.
10. R. G. Downey, A. M. Kach, S. Lempp, A. E. M. Lewis-Pye, A. Montalbán, and D. D. Turetsky, The complexity of computable categoricity, *Adv. Math.* **268** (2015), 423–466.
11. S. S. Goncharov, Degrees of autostability relative to strong constructivizations, *Proc. Steklov Inst. Math.* **274** (2011), 105–115.
12. A. T. Nurtazin, Strong and weak constructivization and computable families, *Algebra Logic.* **13** (1974), 177–184.
13. S. S. Goncharov and M. I. Marchuk, Index sets of constructive models of bounded signature that are autostable relative to strong constructivizations, *Algebra Logic.* **54** (2015), 108–126.
14. S. S. Goncharov and M. I. Marchuk, Index sets of constructive models of finite and graph signatures that are autostable relative to strong constructivizations, *Algebra Logic.* **54** (2016), 428–439.
15. S. S. Goncharov, N. A. Bazhenov, and M. I. Marchuk, The index set of Boolean algebras autostable relative to strong constructivizations, *Sib. Math. J.* **56** (2015), 393–404.
16. N. Bazhenov, M. Harrison-Trainor, I. Kalimullin, A. Melnikov, and K. M. Ng, Automatic and polynomial-time algebraic structures, *J. Symb. Log.* **84** (2019), 1630–1669.
17. C. C. Chang and H. J. Keisler, *Model theory*. North-Holland, Amsterdam, 1973.
18. E. B. Fokina, V. Harizanov, and A. Melnikov, Computable model theory. In ed. R. Downey, *Turing's Legacy: Developments from Turing's Ideas in Logic*, vol. 42, *Lect. Notes Logic*, pp. 124–194. Cambridge University Press, Cambridge, 2014.
19. N. A. Bazhenov, Categoricity spectra of computable structures, *Journal of Mathematical Sciences* **256** (2021), 34–50.

20. E. B. Fokina, I. Kalimullin, and R. Miller, Degrees of categoricity of computable structures, *Arch. Math. Logic.* **49** (2010), 51–67.
21. B. F. Csima, J. N. Y. Franklin, and R. A. Shore, Degrees of categoricity and the hyperarithmetic hierarchy, *Notre Dame J. Formal Logic.* **54** (2013), 215–231.
22. B. A. Anderson and B. F. Csima, Degrees that are not degrees of categoricity, *Notre Dame J. Formal Logic.* **57** (2016), 389–398.
23. R. Miller, **d**-computable categoricity for algebraic fields, *J. Symb. Log.* **74** (2009), 1325–1351.
24. N. A. Bazhenov, I. S. Kalimullin, and M. M. Yamaleev, Degrees of categoricity and spectral dimension, *J. Symb. Log.* **83** (2018), 103–116.
25. B. F. Csima and J. Stephenson, Finite computable dimension and degrees of categoricity, *Ann. Pure Appl. Logic.* **170** (2019), 58–94.
26. N. Bazhenov, Prime model with no degree of autostability relative to strong constructivizations. In eds. A. Beckmann, V. Mitrana, and M. Soskova, *Evolving Computability*, vol. 9136, *Lect. Notes Comput. Sci.*, pp. 117–126. Springer, Cham, 2015.
27. N. Bazhenov, Autostability spectra for decidable structures, *Math. Struct. Comput. Sci.* **28** (2018), 392–411.
28. N. A. Bazhenov, Degrees of autostability relative to strong constructivizations for Boolean algebras, *Algebra Logic.* **55** (2016), 87–102.
29. N. A. Bazhenov and M. I. Marchuk, Degrees of autostability for prime Boolean algebras, *Algebra Logic.* **57** (2018), 98–114.
30. G. E. Sacks, Effective bounds on Morley rank, *Fund. Math.* **103** (1979), 111–121.
31. C. T. Chong, Degree-theoretic bounds on the Morley rank, *Arch. Math. Logik Grundlagenforsch.* **26** (1987), 137–145.
32. C. J. Ash and J. F. Knight, *Computable structures and the hyperarithmetical hierarchy.* vol. 144, *Stud. Logic Found. Math.*, Elsevier Science B.V., Amsterdam, 2000.
33. S. S. Goncharov, N. A. Bazhenov, and M. I. Marchuk, Index set of linear orderings that are autostable relative to strong constructivizations, *J. Math. Sci.* **221** (2017), 840–848.

https://doi.org/10.1142/9789811278631_0004

Isomorphism Types of Rogers Semilattices in the Analytical Hierarchy

Nikolay Bazhenov[a]

Sobolev Institute of Mathematics
4 Acad. Koptyug Ave., Novosibirsk, 630090, Russia
and
Department of Mathematics, School of Sciences and Humanities,
Nazarbayev University,
53 Qabanbaybatyr Ave., Astana, 010000, Kazakhstan
bazhenov@math.nsc.ru

Sergey Ospichev[b]

Sobolev Institute of Mathematics
4 Acad. Koptyug Ave., Novosibirsk, 630090, Russia
and
Novosibirsk State University
2 Pirogova St., Novosibirsk, 630090, Russia
ospichev@math.nsc.ru

Mars Yamaleev[c]

Kazan Federal University
18 Kremlyovskaya St., Kazan, 420008, Russia
mars.yamaleev@kpfu.ru

A numbering of a countable family S is a surjective map from the set of natural numbers ω onto S. A numbering ν is reducible to a numbering μ if there is an effective procedure which given a ν-index of an object from

[a]The work of N. Bazhenov was supported by Nazarbayev University Faculty Development Competitive Research Grants N090118FD5342.
[b]The work of S. Ospichev was funded by RFBR according to the research project No. 17-01-00247.
[c]The work of M. Yamaleev was supported by Russian Science Foundation, project No. 18-11-00028.

S, computes a μ-index for the same object. The reducibility between numberings gives rise to a class of upper semilattices, which are usually called Rogers semilattices. This chapter studies Rogers semilattices for families $S \subset P(\omega)$ belonging to various levels of the analytical hierarchy. We prove that for any non-zero natural numbers $m \neq n$, any non-trivial Rogers semilattice of a Π^1_m-computable family cannot be isomorphic to a Rogers semilattice of a Π^1_n-computable family. One of the key ingredients of the proof is an application of the result by Downey and Knight on degree spectra of linear orders.

1. Introduction

Uniform computations for families of mathematical objects constitute a classical line of research in recursion theory. An important approach to studying such computations is provided by the theory of numberings. The theory goes back to the ground-breaking work of Gödel [21], where an effective numbering of first-order formulae was used in the proof of the incompleteness theorems. One of the first results in the theory of computable numberings was obtained by Kleene [28]: he gave a construction of a universal partial computable function. After that, the foundations of the modern theory of numberings were developed by Kolmogorov and Uspenskii [30,43] and, independently, by Rogers [41].

Let $\mathcal{S}$ be a countable set. A *numbering* of $\mathcal{S}$ is a surjective map ν from the set of natural numbers ω onto $\mathcal{S}$. A standard tool for comparing the complexity of different numberings is provided by the notion of *reducibility* between numberings: A numbering ν is *reducible* to another numbering μ, denoted by $\nu \leq \mu$, if there is total computable function $f(x)$ such that $\nu(x) = \mu(f(x))$ for all $x \in \omega$. In other words, there is an effective procedure which, given a ν-index of an object from $\mathcal{S}$, computes a μ-index for the same object.

In this chapter, we consider only families $\mathcal{S}$ containing subsets of ω, i.e., we always assume that $\mathcal{S} \subseteq P(\omega)$ and $\mathcal{S}$ is countable.

Suppose that $\mathcal{S}$ is a family of computably enumerable (c.e.) sets. A numbering ν of the family $\mathcal{S}$ is *computable* if the set

$$G_\nu = \{\langle n, x\rangle : x \in \nu(n)\} \tag{1.1}$$

is c.e. A family $\mathcal{S}$ is *computable* if it has a computable numbering.

In a standard recursion-theoretic way, the notion of reducibility between numberings gives rise to the *Rogers upper semilattice* (or *Rogers semilattice* for short) of a family: For a computable family $\mathcal{S}$, this semilattice contains the degrees of all computable numberings of $\mathcal{S}$. As per usual, here two numberings have the same degree if they are reducible to each other.

There is a large body of literature on Rogers semilattices of computable families. To name only a few, computable numberings were studied by Badaev [5,6], Ershov [14,16], Friedberg [20], Goncharov [22,23], Khutoretskii [27], Lachlan [31,32], Mal'tsev [34], Pour-El [40], Selivanov [42], and many other researchers.

Goncharov and Sorbi [24] started developing the theory of generalized computable numberings. One of their approaches to generalized computations can be summarized as follows. Let Γ be a complexity class (e.g., Σ^0_1, d-Σ^0_1, Σ^0_n, or Π^1_n). A numbering ν of a family $\mathcal{S}$ is Γ-*computable* if the set G_ν from (1.1) belongs to the class Γ. We say that a family $\mathcal{S}$ is Γ-*computable* if it has a Γ-computable numbering. Note that the classical notion of a computable numbering becomes a synonym of a Σ^0_1-computable numbering.

In a similar way to computable numberings, one can introduce the notion of the Rogers semilattice of a Γ-computable family, see Sec. 2.1 for the details. One of natural questions in the area of generalized computable numberings can be formulated as follows.

Problem 1.1: *Given a complexity class Γ, study the isomorphism types of Rogers semilattices of Γ-computable families.*

We give a short overview of some results related to Problem 1.1, further related work is discussed in Sec. 2.2. We say that an upper semilattice is a *Rogers Γ-semilattice* if it is isomorphic to the Rogers semilattice of a Γ-computable family of sets. A semilattice is *non-trivial* if it contains more than one element.

Let n be a non-zero natural number. Badaev, Goncharov, and Sorbi [9] proved that for any $m \geq n+3$, any non-trivial Rogers Σ^0_m-semilattice is not isomorphic to a Rogers Σ^0_n-semilattice. Podzorov [39] obtained a generalization of this theorem: he showed that a similar result holds for $m \geq n+2$. It is still open whether this fact is true for $m = n+1$.

Badaev and Goncharov [2] extended the result of [9] to the hyperarithmetical hierarchy: They showed that for any computable ordinals $\alpha > 0$ and $\beta \geq \alpha + 3$, any non-trivial Rogers Σ^0_β-semilattice is not isomorphic to a Rogers Σ^0_α-semilattice.

We note that the situation in the Ershov hierarchy is quite the opposite. Recall that for a non-zero natural number n, Σ^{-1}_n denotes the class of all n-c.e. sets. The result of Herbert, Jain, Lempp, Mustafa, and Stephan (Theorem 2 in [25]) implies that every Rogers Σ^{-1}_n-semilattice is also a Rogers Σ^{-1}_{n+1}-semilattice.

In this chapter, we continue the investigations of [2, 9, 39] in the setting of the analytical hierarchy. We prove that for any natural numbers $m > n \geq 1$, a non-trivial Rogers Π^1_m-semilattice is not isomorphic to a Rogers Π^1_n-semilattice (Theorem 3.1). Quite unexpectedly, one of the main ingredients of the proof is an application of the result of Downey and Knight [13] on degree spectra of linear orders.

The structure of the chapter is as follows. Section 2 contains the necessary preliminaries and a discussion of related work. In Sec. 3, we prove Theorem 3.1: First, we give a complete outline of the proof including all key ideas. After that, the proofs of the auxiliary claims are given in separate subsections.

2. Preliminaries

We use lowercase bold Latin letters (e.g., $\mathbf{a}$, $\mathbf{b}$, $\mathbf{c}$) to denote m-degrees. Lowercase bold Latin letters with a subscript T (e.g., $\mathbf{x}_T$, $\mathbf{y}_T$, $\mathbf{z}_T$) denote Turing degrees.

We assume that for any considered countable structure, its universe is contained in ω. For a structure $\mathcal{A}$, $D(\mathcal{A})$ denotes the atomic diagram of $\mathcal{A}$.

We treat upper semilattices as structures in the language $L_{usl} = \{\leq, \vee\}$. If $\mathcal{A}$ is an upper semilattice and $a \leq_{\mathcal{A}} b$ are elements from $\mathcal{A}$, then the *interval* $[a; b]_{\mathcal{A}}$ is the semilattice

$$[a; b]_{\mathcal{A}} := (\{c \in \mathcal{A} : a \leq_{\mathcal{A}} c \leq_{\mathcal{A}} b\}; \leq_{\mathcal{A}}, \vee_{\mathcal{A}}).$$

For a complexity class Γ, we say that an L_{usl}-structure $\mathcal{A} = (\omega; \leq_A, \vee_A)$ is a Γ-*presentation* of an upper semilattice $\mathcal{M}$ if $\mathcal{A}$ satisfies the following conditions:

(1) the function $\vee_A$ is total computable,
(2) the relation $\sim_A := \leq_A \cap \geq_A$ is a congruence on $\mathcal{A}$,
(3) the relations $\leq_A$ and $\sim_A$ both belong to the class Γ,
(4) the quotient structure $\mathcal{A}/_{\sim_A}$ is isomorphic to $\mathcal{M}$.

2.1. *Numberings*

Suppose that ν is a numbering of a family $\mathcal{S}_0$, and μ is a numbering of a family $\mathcal{S}_1$. Note that the condition $\nu \leq \mu$ always implies that $\mathcal{S}_0 \subseteq \mathcal{S}_1$.

Numberings ν and μ are *equivalent* (denoted by $\nu \equiv \mu$) if $\nu \leq \mu$ and $\mu \leq \nu$. The numbering $\nu \oplus \mu$ of the family $\mathcal{S}_0 \cup \mathcal{S}_1$ is defined as follows:

$$(\nu \oplus \mu)(2x) = \nu(x), \quad (\nu \oplus \mu)(2x + 1) = \mu(x).$$

The following fact is well-known (see, e.g., p. 36 in [16]): If ξ is a numbering of a family $\mathcal{T}$, then

$$(\nu \leq \xi \,\&\, \mu \leq \xi) \ \Leftrightarrow\ (\nu \oplus \mu \leq \xi).$$

For further background on numberings, the reader is referred to, e.g., [7, 16, 17].

Let Γ be a complexity class with the following properties:

(a) If ν is a Γ-computable numbering and μ is a numbering such that $\mu \leq \nu$, then μ is also Γ-computable.
(b) If numberings ν and μ are both Γ-computable, then the numbering $\nu \oplus \mu$ is also Γ-computable.

For example, it is not hard to show that for any non-zero natural number n, each of the classes Σ^0_n, Σ^{-1}_n, and Π^1_n has these properties.

Consider a Γ-computable family $\mathcal{S}$. By $Com_\Gamma(\mathcal{S})$ we denote the set of all Γ-computable numberings of $\mathcal{S}$. Since the relation $\equiv$ is a congruence on the structure $(Com_\Gamma(\mathcal{S}); \leq, \oplus)$, we use the same symbols $\leq$ and $\oplus$ on numberings of $\mathcal{S}$ and on $\equiv$-equivalence classes of these numberings.

The quotient structure $\mathcal{R}_\Gamma(\mathcal{S}) := (Com_\Gamma(\mathcal{S})/{\equiv}; \leq, \oplus)$ is an upper semilattice. We say that $\mathcal{R}_\Gamma(\mathcal{S})$ is the *Rogers semilattice* of the Γ-computable family $\mathcal{S}$. For the sake of convenience, we use the following notation:

$$\mathcal{R}^1_n(\mathcal{S}) := \mathcal{R}_{\Pi^1_n}(\mathcal{S}).$$

2.2. *Related work*

The questions related to Problem 1.1 were extensively studied for Rogers Σ^0_n-semilattices. In particular, the following results on the number of isomorphism types of Rogers Σ^0_n-semilattices are known. Ershov and Lavrov [19] (see also p. 72 in [16], and [18]) showed that there are finite families $\mathcal{S}_i$, $i \in \omega$, of c.e. sets such that the semilattices $\mathcal{R}_{\Sigma^0_1}(\mathcal{S}_i)$ are pairwise non-isomorphic. In other words, there are infinitely many isomorphism types of Rogers Σ^0_1-semilattices. V'yugin [44] proved that there are infinitely many pairwise elementarily non-equivalent Rogers Σ^0_1-semilattices. Badaev, Goncharov, and Sorbi [8] proved that for any natural number $n \geq 2$, there are infinitely many pairwise elementarily non-equivalent Rogers Σ^0_n-semilattices. The reader is referred to, e.g., [3, 4, 10, 24] for further results on Rogers Σ^0_n-semilattices.

Rogers semilattices in the analytical hierarchy were previously studied by Dorzhieva [11, 12]. In particular, she showed that for any non-zero n

and any non-trivial Rogers Π^1_n-semilattice $\mathcal{R}$, the first-order theory of $\mathcal{R}$ is hereditarily undecidable (Theorem 3 in [12]).

Kalimullin, Puzarenko, and Faizrakhmanov [26] considered *computable* Π^1_1*-numberings*. A Π^1_1*-numbering* of a family $\mathcal{S}$ is a *partial* map ν acting from ω onto $\mathcal{S}$ such that the domain of ν is enumeration reducible to the Π^1_1-complete set $\mathcal{O}$. A Π^1_1-numbering ν is *computable* if the set

$$G^*_\nu = \{\langle n, x\rangle : n \in dom(\nu),\ x \in \nu(n)\}$$

is enumeration reducible to $\mathcal{O}$.

3. Main Result

Theorem 3.1: *Let $m > n$ be non-zero natural numbers. If $\mathcal{R}$ is a non-trivial Rogers Π^1_m-semilattice, then $\mathcal{R}$ is not isomorphic to a Rogers Π^1_n-semilattice.*

Proof: Here we give a complete outline of the proof. This includes all of its main ingredients, but the proofs of some auxiliary statements are given in separate subsections.

First, we give an upper bound on the complexity of (atomic diagrams of) arbitrary intervals in a Rogers Π^1_n-semilattice. Let $\mathcal{E}_n$ be the Σ^1_n-complete set.

Proposition 3.2: *Let $\mathcal{M}$ be a Rogers Π^1_n-semilattice. If $a \leq_{\mathcal{M}} b$ are elements from $\mathcal{M}$, then the interval $[a;b]_{\mathcal{M}}$ has a $\Sigma^0_2(\mathcal{E}_n)$-presentation.*

Second, we obtain a family of upper semilattices which are realizable as intervals in every non-trivial Rogers Π^1_k-semilattice.

Let $\mathbf{D}_m$ denote the semilattice of m-degrees. As per usual, we assume that $\mathbf{D}_m$ does not contain the degrees $\deg_m(\emptyset)$ and $\deg_m(\omega)$. For a non-empty set $A \subset \omega$, let $\mathbf{D}_m(\leq A)$ be the principal ideal $\{\mathbf{d} : \mathbf{d} \leq \deg_m(A)\}$ in the semilattice $\mathbf{D}_m$.

Suppose that ν and μ are Π^1_k-computable numberings of a family $\mathcal{S}$. We say that ν is Δ^1_k*-reducible* to μ, denoted by $\nu \leq_{\Delta^1_k} \mu$, if there is a total Δ^1_k function $f(x)$ such that $\nu(x) = \mu(f(x))$, for all $x \in \omega$. The numberings ν and μ are Δ^1_k*-equivalent*, denoted by $\nu \equiv_{\Delta^1_k} \mu$, if $\nu \leq_{\Delta^1_k} \mu$ and $\mu \leq_{\Delta^1_k} \nu$.

Proposition 3.3: *Let k be a non-zero natural number, and $\mathcal{S}$ be a Π^1_k-computable family such that $\mathcal{S}$ contains at least two elements. Suppose that a set $U \subseteq \omega$ is immune and Δ^1_k. Then for any Π^1_k-computable numbering α of* *the family $\mathcal{S}$, there is a Π^1_k-computable numbering β of $\mathcal{S}$ with the following properties:*

(a) $\beta \equiv_{\Delta^1_k} \alpha$.
(b) If $\mathcal{S}$ is finite, then the principal ideal induced by β inside $\mathcal{R}^1_k(\mathcal{S})$ (we denote this ideal by $[\widehat{\beta}]^1_k$) is isomorphic to the structure $\mathbf{D}_m(\leq U)$.
(c) If $\mathcal{S}$ is infinite, then the ideal $[\widehat{\beta}]^1_k$ is isomorphic to the structure $\mathbf{D}_m(\leq U) \setminus \{\mathbf{0}\}$, i.e. to the ideal $\mathbf{D}_m(\leq U)$ with the least element omitted.

Proposition 3.3 allows us to find plenty of linearly ordered intervals inside an arbitrary non-trivial Rogers Π^1_k-semilattice. This is obtained via the following result:

Proposition 3.4: *Suppose that $\mathcal{L}$ is a countable linear order with least and greatest elements. Then there exists a set $A \subseteq \omega$ with the following properties:*

(1) the principal ideal $\mathbf{D}_m(\leq A)$ is isomorphic to $\mathcal{L}$,
(2) A is immune, and
(3) A is c.e. in $(D(\mathcal{L}) \oplus \emptyset^{(2)})$.

Now we are ready to introduce the last key ingredient, and to finish the proof. The last ingredient uses the result of Downey and Knight [13] on the degree spectra of linear orders.

Suppose that $\mathcal{M}$ is a countable infinite structure, and α is a computable ordinal. A Turing degree $\mathbf{x}_T$ is the *αth jump degree* of $\mathcal{M}$ if it is the least degree in the set

$$\{\deg_T(D(\mathcal{A}))^{(\alpha)} : \mathcal{A} \cong \mathcal{M}, \text{ and the domain of } \mathcal{A} \text{ is } \omega\}.$$

A degree $\mathbf{x}_T$ is *proper αth jump degree* of $\mathcal{M}$ if $\mathbf{x}_T$ is the αth jump degree of $\mathcal{M}$ and for any computable ordinal $\beta < \alpha$, $\mathcal{M}$ has no βth jump degree.

Theorem 3.5: (Downey and Knight [13], see also [1, 29]) *For any computable ordinal $\alpha \geq 2$ and any Turing degree $\mathbf{x}_T \geq \mathbf{0}^{(\alpha)}$, there is a linear order having proper αth jump degree $\mathbf{x}_T$.*

Corollary 3.6: *For any Turing degree $\mathbf{x}_T$, there exists a linear order $\mathcal{L}$ with the following properties: $\deg_T(D(\mathcal{L})) \leq \mathbf{x}_T^{(3)}$, and there is no copy $\mathcal{A} \cong \mathcal{L}$ with $\deg_T(D(\mathcal{A})) \leq \mathbf{x}_T$.*

In Corollary 3.6, the desired structure $\mathcal{L}$ is the linear order from Theorem 3.5 having proper second jump degree $\mathbf{x}_T^{(3)}$.

Recall that $m > n \geq 1$. Towards contradiction, assume that $\mathcal{R}$ is a non-trivial Rogers Π^1_m-semilattice which is also a Rogers Π^1_n-semilattice.

Let $\mathbf{e}_T$ be the Turing degree of the set $\mathcal{E}_n$. By applying Corollary 3.6 to the degree $\mathbf{x}_T = \mathbf{e}_T^{(2)}$, one can choose a linear order $\mathcal{L}$ such that $\mathcal{L}$ has least and greatest elements, $\deg_T(D(\mathcal{L})) \leq \mathbf{e}_T^{(5)}$, and there is no copy $\mathcal{A}$ of $\mathcal{L}$ with $\deg_T(D(\mathcal{A})) \leq \mathbf{e}_T^{(2)}$.

By Proposition 3.4, one can find an immune set $A \subset \omega$ such that $\mathbf{D}_m(\leq A) \cong \mathcal{L}$ and A is c.e. in $(D(\mathcal{L}) \oplus \mathbf{0}^{(2)})$. In particular, this implies that the set A is $\Sigma^0_1(\mathbf{e}_T^{(5)})$. Since the set $\mathcal{E}_n$ is Δ^1_m, the set A is also Δ^1_m.

Since the Rogers Π^1_m-semilattice $\mathcal{R}$ is non-trivial, the corresponding Π^1_m-computable family $\mathcal{S}$ contains more than one element. We apply Proposition 3.3 to $\mathcal{S}$ and the immune Δ^1_m set A, and find an interval $[a;b]_{\mathcal{R}}$ which is isomorphic either to $\mathcal{L}$, or to the structure $\mathcal{L} \setminus \{0\}$ (i.e. the order $\mathcal{L}$ with the least element omitted). W.l.o.g., one may assume that $[a;b]_{\mathcal{R}}$ is a copy of $\mathcal{L}$.

Recall that we assumed that $\mathcal{R}$ is a Rogers Π^1_n-semilattice. Thus, by Proposition 3.2, the interval $[a;b]_{\mathcal{R}}$ has a $\Sigma^0_2(\mathcal{E}_n)$-presentation $\mathcal{P}$. Using the presentation $\mathcal{P}$, it is straightforward to build a linear order $\mathcal{A}_0 \cong \mathcal{L}$ such that $D(\mathcal{A}_0) \leq_T \mathbf{e}_T^{(2)}$. This contradicts the choice of the order $\mathcal{L}$. Therefore, $\mathcal{R}$ cannot be a Rogers Π^1_n-semilattice.

This concludes the (outline of the) proof of Theorem 3.1. The proofs of Propositions 3.2–3.4 are given below. □

3.1. *Proof of Proposition 3.2*

We follow the lines of the proof of Lemma 7 in [39]. For the sake of self-completeness, here we give a detailed exposition.

Let ν be a numbering of a family $\mathcal{S}$. The operator

$$\Psi(\nu;\cdot)\colon \{W \subseteq \omega : W \text{ is c.e., } W \neq \emptyset\} \to \{[\mu]_{\equiv} : \mu \leq \nu\}$$

is defined as follows. Let W be a non-empty c.e. set. Choose a total computable function $f(x)$ such that $range(f) = W$, and define $\Psi(\nu;W)$ as the $\equiv$-class of the numbering $\nu \circ f$. It is not hard to show that the operator Ψ is well-defined, i.e. the value $\Psi(\nu;W)$ does not depend on the choice of a function f. Moreover, if μ is a numbering from the class $\Psi(\nu;W)$, then $range(\mu) = \nu(W) = \{\nu(x) : x \in W\}$.

The operator Ψ has the following properties (see p. 1120 in [39]):

(a) For any numbering μ, the condition $\mu \leq \nu$ holds iff there is a c.e. set W with $[\mu]_{\equiv} = \Psi(\nu;W)$.

(b) For any non-empty c.e. sets U and V, we have $\Psi(\nu; U \cup V) = \Psi(\nu; U) \vee \Psi(\nu; V)$.
(c) $\Psi(\nu; U) \leq \Psi(\nu; V)$ if and only if $\nu(U) \subseteq \nu(V)$ and there is a partial computable function $\theta(x)$ such that $U \subseteq dom(\theta)$, $\theta(U) \subseteq V$, and for every $x \in U$, we have $\nu(x) = \nu(\theta(x))$.

Recall that $\mathcal{M}$ is a Rogers Π^1_n-semilattice. Hence, one can choose a Π^1_n-computable family $\mathcal{S}$, and identify an arbitrary element $c \in \mathcal{M}$ with a class $[\xi]_\equiv$, where ξ is a Π^1_n-computable numbering of the family $\mathcal{S}$.

Choose a numbering ν of $\mathcal{S}$ such that $b = [\nu]_\equiv$. By the property (a) of the operator Ψ, there is a non-empty c.e. set U such that $a = \Psi(\nu; U)$. We define an L_{usl}-structure $\mathcal{A} = (\omega; \leq_A, \vee_A)$ as follows:

(1) $\vee_A$ is a total computable function such that $W_{x \vee_A y} = W_x \cup W_y$ for all x and y;
(2) $x \leq_A y$ iff $\Psi(\nu; U \cup W_x) \leq \Psi(\nu; U \cup W_y)$.

The properties of Ψ imply that the relation $\sim_A := (\leq_A \cap \geq_A)$ is a congruence on $\mathcal{A}$, and the function $F\colon [e]_{\sim_A} \mapsto \Psi(\nu; U \cup W_e)$ is an isomorphism from $\mathcal{A}/{\sim_A}$ onto $[a; b]_{\mathcal{M}}$.

In order to finish the proof, now it is sufficient to show that the relation $\leq_A$ is $\Sigma^0_2(\mathcal{E}_n)$. By the property (c) of the operator Ψ, the condition $x \leq_A y$ is equivalent to

$$(\exists i \in \omega)[U \cup W_x \subseteq dom(\varphi_i) \ \&\ \varphi_i(U \cup W_x) \subseteq U \cup W_y \ \&\ (\forall z \in U \cup W_x)(\nu(z) = \nu(\varphi_i(z)))]. \quad (3.1)$$

Since the condition $\nu(z) = \nu(v)$ can be rewritten as

$$(\forall k \in \omega)[k \in \nu(z) \leftrightarrow k \in \nu(v)]$$

and ν is a Π^1_n-computable numbering, Eq. (3.1) describes a $\Sigma^0_2(\mathcal{E}_n)$ condition. Therefore, the semilattice $\mathcal{M}$ has a $\Sigma^0_2(\mathcal{E}_n)$-presentation. Proposition 3.2 is proved.

3.2. *Proof sketch for Proposition 3.3*

The proof of this result is entirely based on Theorem 2 from the paper by Podzorov [38].

Let m be a non-zero natural number. A numbering ν is Δ^0_m-*reducible* to a numbering μ, denoted by $\nu \leq_{\Delta^0_m} \mu$, if there is a total Δ^0_m function $f(x)$ such that $\nu(x) = \mu(f(x))$, for every x. We write $\nu \equiv_{\Delta^0_m} \mu$ if $\nu \leq_{\Delta^0_m} \mu$ and $\mu \leq_{\Delta^0_m} \nu$.

If ν is a Σ^0_m-computable numbering of a family S, then by $[\widehat{\nu}]^0_m$ we denote the principal ideal induced by ν inside the Rogers Σ^0_m-semilattice of S (see [38] for more details).

Theorem 3.7: (Podzorov [38]) *Suppose that m and n are natural numbers such that $2 \leq m \leq n$. Suppose also that a set $U \subseteq \omega$ is immune and Δ^0_m. Let S be a Σ^0_n-computable family such that S contains at least two elements. Then for any Σ^0_n-computable numbering α of the family S, there is a Σ^0_n-computable numbering β of S with the following properties:*

(a) $\beta \equiv_{\Delta^0_m} \alpha$.
(b) If S is finite, then $[\widehat{\beta}]^0_n$ is isomorphic to the structure $\mathbf{D}_m(\leq U)$.
(c) If S is infinite, then the ideal $[\widehat{\beta}]^0_n$ is isomorphic to the structure $\mathbf{D}_m(\leq U) \setminus \{\mathbf{0}\}$.

A careful step-by-step analysis of the proof of Theorem 3.7 shows that this result works well for analytical sets: Roughly speaking, the construction described in [38] depends only on the complexity of the set U, and the desired numbering β can be considered as a Δ^0_m permutation (or in our case, as a Δ^1_k permutation) of the numbering α. Thus, one can recover the proof of Proposition 3.3 from the Podzorov's construction just by using some basic properties of the analytical hierarchy and the Tarski–Kuratowski algorithm.

3.3. *Proof of Proposition 3.4*

Before giving the proof, we briefly discuss some known related results. The results of Lachlan [33] imply that for any countable linear order $\mathcal{L}$ with least element, there exists an initial segment of $\mathbf{D}_m$ isomorphic to $\mathcal{L}$. Ershov [15] gave a characterization of the semilattice $\mathbf{D}_m$ up to isomorphism, as a c-universal upper semilattice of cardinality continuum, see also [16,37].

Let $\mathbf{ID}_m$ denote the upper semilattice of immune m-degrees. Following [35], here we assume that the m-degree of a non-empty finite set also belongs to $\mathbf{ID}_m$. Using the characterization of Ershov [15], Mal'tsev [35] proved that the structure $\mathbf{ID}_m$ is isomorphic to $\mathbf{D}_m$. Therefore, one can obtain the following:

Lemma 3.8: (essentially Mal'tsev [35]) *For any countable linear order $\mathcal{L}$ with least element, there is an initial segment of $\mathbf{ID}_m$ isomorphic to $\mathcal{L}$.*

In a way, Proposition 3.4 can be treated as a refinement of Lemma 3.8.

Proof: We follow the Odifreddi's exposition (see § 6.2 in [36]) of the results of Lachlan [33]. Suppose that $U \subseteq \omega$. For a non-empty c.e. set W, choose a total computable function f such that $range(f) = W$. Define

$$\Psi_U(W) = f^{-1}(U).$$

Note that in a way, the operator $\Psi(\nu;\cdot)$ from the proof of Proposition 3.2 is a counterpart of the operator Ψ_U in the realm of numberings.

We recall some basic properties of the operator Ψ_U (see pp. 561–562 in [36] and p. 388 in [16]):

(a) The m-degree of the set $\Psi_U(W)$ does not depend on the choice of f.
(b) $\Psi_U(W) \leq_m U$.
(c) If $W_1 \subseteq W_2$ are c.e. sets, then $\Psi_U(W_1) \leq_m \Psi_U(W_2)$.
(d) If W_1 is a c.e. set and $W_1 =^* W_2$, then $\Psi_U(W_1) \equiv_m \Psi_U(W_2)$.
(e) For any set $X \leq_m U$, there is a c.e. set W_0 such that $\Psi_U(W_0) \equiv_m X$.

The properties above imply that Ψ_U induces an epimorphism of upper semilattices, acting from $\mathcal{E}^*$ onto the ideal $\mathbf{D}_m(\leq U)$. We slightly abuse the notation and identify Ψ_U with the induced epimorphism.

W.l.o.g., we may assume that $\mathcal{L}$ is an infinite linear order, and the domain of $\mathcal{L}$ is ω. Moreover, we assume that 0 is the least element of $\mathcal{L}$, and 1 is the greatest element of $\mathcal{L}$. Fix a strongly $D(\mathcal{L})$-computable sequence of finite linear orders $\{L_s\}_{s\in\omega}$ such that $L_0 = \{0 < 1\}$, $\bigcup_{s\in\omega} L_s = \mathcal{L}$, $L_s \subseteq L_{s+1}$, $L_{3t} = L_{3t+1} = L_{3t+2}$, and L_{3t} contains exactly $(t+2)$ elements, for any s and t. For an element $a \in \mathcal{L}$, let $l(a) = \min\{s : a \in L_s\}$.

At a stage s, we construct infinite disjoint c.e. sets $P_{a,s}$, $a \in L_s$, and a computable equivalence relation E_s. We also build finite sets A_s and B_s. For $a \in L_s$, let

$$R_{a,s} := \bigcup_{b \leq_{L_s} a} P_{b,s}.$$

The sets have the following properties:

- $A_s \subseteq A_{s+1}$, $B_s \subseteq B_{s+1}$, $E_s \subseteq E_{s+1}$, and $A_s \cap B_s = \emptyset$,
- for any $a \in L_s$, $R_{a,s} \subseteq R_{a,s+1}$,
- $R_{1,s} \cup A_s \cup B_s = \omega$ and $R_{1,s} \cap (A_s \cup B_s) = \emptyset$,
- for any $x \in \omega$, the equivalence class $[x]_{E_s}$ is finite; moreover, for any s, the canonical index of $[x]_{E_s}$ can be computed uniformly in x;
- every $x \in \omega$ satisfies exactly one of the following:
  - there is $a \in L_s$ such that $[x]_{E_s} \subseteq P_{a,s}$,

  - $[x]_{E_s} \subseteq A_s$, or
  - $[x]_{E_s} \subseteq B_s$;
- if $a \in L_s$ and $x \in P_{a,s+1}$, then there is an element $y \in P_{a,s} \cap [x]_{E_{s+1}}$.

Note that the last property implies the following: for any $a \in \mathcal{L}$ and $s \geq l(a)$, the set $P_{a,s} \cap P_{a,l(a)}$ is infinite. At stage $s+1$, if we do not explicitly specify A_{s+1}, then we assume that $A_{s+1} = A_s$. The same applies to other sets.

Intuition. Before giving a formal construction, we discuss the intuition behind this construction. Recall that we want to construct an immune set A such that $\mathbf{D}_m(\leq A)$ is isomorphic to $\mathcal{L}$. With help of the functional Ψ_A we transfer our problem to the semilattice $\mathcal{E}^*$. Thus, we will build c.e. sets $R_{a,l(a)}$ which will correspond to the elements $a \in \mathcal{L}$. In order to obtain the desired isomorphism, we must ensure that other c.e. sets (i.e. sets W not equal to our $R_{a,l(a)}$, $a \in \mathcal{L}$) do not induce additional m-degrees via the functional Ψ_A. The stages $3e+1$ and $3e+2$ are devoted to this, and Lemma 3.12 ensures the correctness.

Each of c.e. sets $R_{a,s}$ consists of smaller c.e. blocks $P_{b,s}$, where $b \leq_{L_s} a$. While satisfying requirements, we can move some elements of a block $P_{a,s}$ only to the left blocks (i.e. the blocks $P_{b,s}$ with $b <_{\mathcal{L}} a$) meanwhile keeping $P_{a,s}$ infinite. Note that if we keep each $P_{a,s}$ infinite, then it helps to distinguish $R_{a,s}$ from $R_{b,s}$, where $b <_{L_s} a$. The equivalence classes $[x]_{E_s}$ constitute a tool for working with m-reductions between $\Psi_A(W_e)$ and $\Psi_A(R_{a,s})$. Thus, moving an equivalence class to the left block can be considered as a correction of the reduction.

The role of stages $3e+3$ is ensuring immunity of A; also at these stages we split a block $P_{b,s}$ into two blocks $P_{a,s}$ and $P_{b,s}$, where a is a fresh element in L_s.

We note that the construction is non-uniform in the following sense. At a stage s, we don't have an effective procedure to determine which elements a belong to L_s. Thus, one can consider our actions as working with all possible approximations (or variants) of L_s, moreover, each of these approximations works with its own $P_{a,s+1}$. Then for a given approximation, we can uniformly construct $P_{a,s+1}$, and this allows us to make $P_{a,s+1}$ as c.e. and E_{s+1} as computable. One can imagine this process as working on a tree, where at a given level we have the results of the work at the corresponding stage with different approximations. Using the oracle $D(\mathcal{L}) \oplus \emptyset^{(2)}$, we can recognize the true path, where the real construction happens, and thus, the sets A and B will be c.e. relative to the approximations along this path. Now we proceed to the formal construction.

Construction. At stage 0, we set $A_0 = B_0 = \emptyset$, $P_{0,0} = 2\mathbb{N}$, $P_{1,0} = 2\mathbb{N}+1$, and $E_0 = id_\omega$.

Stage $3e+1$. If the set W_e is finite, then we proceed to the next stage. Assume that W_e is infinite. Let $s = 3e$. We find the $\leq_{\mathcal{L}}$-greatest element $a \in L_s$ such that $W_e \cap P_{a,s}$ is infinite. For $b >_{L_s} a$, set $P_{b,s+1} = P_{b,s}$.

The sets $P_{b,s+1}$, $b \leq_{L_s} a$, and the relation E_{s+1} are constructed as follows. At step 0, set $P_b^0 = P_{b,s}$, $b \leq_{L_s} a$, and $\tilde{E}^0 = E_s$. At an odd step $(t+1)$, find (some) elements $u \in W_e \cap P_a^t$ and $v \in P_a^t$ such that $[u]_{\tilde{E}^t} = [u]_{E_s}$, $[v]_{\tilde{E}^t} = [v]_{E_s}$, and $[u]_{E_s} \neq [v]_{E_s}$. Set $\tilde{E}^{t+1} = \tilde{E}^t \cup ([u]_{E_s} \cup [v]_{E_s})^2$ and $P_b^{t+1} = P_b^t$ for $b \leq_{L_s} a$.

At an even step $(t+1)$, find (some) elements $u \in W_e \cap P_a^t$ and $v \in \bigcup_{b \leq_{L_s} a} P_b^t$ such that $[u]_{\tilde{E}^t} = [u]_{E_s}$, $[v]_{\tilde{E}^t} = [v]_{E_s}$, $[u]_{E_s} \neq [v]_{E_s}$, and $[v]_{E_s} \cap W_{e,t} = \emptyset$. Suppose that $v \in P_c^t$. For $b \leq_{L_s} a$, define

$$\tilde{E}^{t+1} = \tilde{E}^t \cup ([u]_{E_s} \cup [v]_{E_s})^2,$$

$$P_b^{t+1} = \begin{cases} P_a^t \setminus [u]_{E_s}, & \text{if } b = a \neq c \\ P_c^t \cup [u]_{E_s}, & \text{if } b = c \neq a, \\ P_b^t, & \text{otherwise.} \end{cases}$$

We set $E_{s+1} = \bigcup_{t \in \omega} \tilde{E}^t$ and $P_{b,s+1} = \lim_t P_b^t$.

It is not hard to show that the sets $P_{b,s+1}$, $b \neq a$, are c.e. Note that $P_{a,s+1} \subseteq P_{a,s}$. Furthermore, we may assume that for any $x \in P_{a,s}$, there is a step t such that $[x]_{\tilde{E}^t} = [x]_{E_s}$, $[x]_{\tilde{E}^{t+1}} = [x]_{E_{s+1}} \supsetneq [x]_{E_s}$, and one of the following holds:

- $[x]_{E_{s+1}} \subseteq P_a^{t+1} \cap P_{a,s+1}$, or
- $[x]_{E_{s+1}} \subseteq P_b^{t+1} \cap P_{b,s+1}$ for some $b <_{L_s} a$.

Therefore, the set $P_{a,s+1}$ is c.e. and E_{s+1} is computable.

The stage $3e+1$ ensures the following: If W_e is infinite, then for any $u \in R_{a,s+1}$, the intersection $[u]_{E_{s+1}} \cap W_e$ is not empty.

Stage $3e+2$. Assume that $e = \langle a, b, k\rangle$ and $s = 3e+1$. Suppose that a, b, k satisfy the following conditions:

(i) $a, b \in L_s$ and $a <_{\mathcal{L}} b$,
(ii) $dom(\varphi_k) \supseteq R_{b,l(b)}$
(iii) $range(\varphi_k) \subseteq R_{a,l(a)}$.

Choose a fresh witness $w \in P_{b,l(b)} \cap P_{b,s}$ (here the freshness means that no $v \in [w]_{E_s}$ has been a witness at earlier stages). If $\varphi_k(w) \in A_s$, then enumerate the class $[w]_{E_s}$ into B. If $\varphi_k(w) \in B_s$, then enumerate $[w]_{E_s}$

into A. If $\varphi_k(w) \notin A_s \cup B_s$, then enumerate $[w]_{E_s}$ into A and $[\varphi_k(w)]_{E_s}$ into B, respectively. These actions ensure the following condition: $(w \in A) \Leftrightarrow (\varphi_k(w) \notin A)$. For $a \in L_s$, set $P_{a,s+1} = P_{a,s} \setminus ([w]_{E_s} \cup [\varphi_k(w)]_{E_s})$.

If a, b, k do not satisfy the conditions (i)–(iii) above, then proceed to the next stage.

Stage $3e+3$. First, let $s = 3e+2$ and let a be the (unique) element from $L_{s+1} \setminus L_s$. Suppose that b is the least element from L_s such that $a <_{\mathcal{L}} b$. Choose a computable enumeration $\{x_m\}_{m\in\omega}$ such that $\bigcup_{m\in\omega}[x_m]_{E_s} = P_{b,s}$ and $[x_m]_{E_s} \neq [x_k]_{E_s}$ for $m \neq k$. For $c \in L_{s+1}$, define

$$P^0_{c,s+1} = \begin{cases} \bigcup_{k\in\omega}[x_{2k}]_{E_s}, & \text{if } c = a,\\ \bigcup_{k\in\omega}[x_{2k+1}]_{E_s}, & \text{if } c = b,\\ P_{c,s}, & \text{otherwise.} \end{cases}$$

If the set W_e is finite, then set $P_{c,s+1} = P^0_{c,s+1}$ and proceed to the next stage. If W_e is infinite, then choose a fresh witness $w \in W_e \setminus A_s$. Enumerate $[w]_{E_s}$ into B and define $P_{c,s+1} = P^0_{c,s+1} \setminus [w]_{E_s}$ for $c \in L_{s+1}$.

Verification. Let $A = \bigcup_{s\in\omega} A_s$ and $B = \bigcup_{s\in\omega} B_s$. We prove a series of lemmas.

Lemma 3.9: *Suppose that $s \in \omega$.*

(1) For any $a \in L_s$, the set $P_{a,s}$ is infinite c.e.
(2) Given $x \in \omega$, one can effectively find the canonical index of the finite set $[x]_{E_s}$.

In particular, these properties imply that the sets $R_{a,s}$, $a \in L_s$, are c.e., and E_s is computable.

Proof: The proof is by induction on s. Assume that $L_s = \{a_0 < a_1 < \dots < a_n\}$, $P_{a_i,s} = W_{e_i}$ for $i \leq n$, and φ_j is a total computable function such that for any x, $\varphi_j(x)$ is the canonical index of $[x]_{E_s}$. Consider the case $s = 3e$. Suppose that W_e is an infinite set. The stage $(3e+1)$ describes an effective procedure such that given indices $e, e_0, e_1, \dots, e_n, j$ and a number $a \in L_s$ with infinite $W_e \cap P_{a,s}$, it produces c.e. indices for the sets $P_{a_i,s+1}$ and a computable index for the function which calculates canonical indices of $[x]_{E_{s+1}}$. The stages $(3e+2)$ and $(3e+3)$ describe similar effective procedures. □

Lemma 3.10: *The sets A and B are c.e. in $(D(\mathcal{L}) \oplus \emptyset^{(2)})$.*

Proof: First, note that each of the following conditions on $e, k \in \omega$ is Π^0_2:

- W_e is infinite,
- $W_e \subseteq dom(\varphi_k)$,
- $range(\varphi_k) \subseteq W_e$.

Thus, in order to construct the enumerations $\{A_s\}_{s\in\omega}$ and $\{B_s\}_{s\in\omega}$, it is sufficient to use the oracle $(D(\mathcal{L}) \oplus \emptyset^{(2)})$. □

Lemma 3.11: *The set A is immune.*

Proof: For an infinite c.e. set W_e, the step $(3e+3)$ ensures that $W_e \cap B \neq \emptyset$. Since $A \cap B = \emptyset$, we have $W_e \not\subseteq A$. □

Lemma 3.12:

(1) For any $a \in \mathcal{L}$ and $s \geq l(a)$, we have $\Psi_A(R_{a,s}) \equiv_m \Psi_A(R_{a,l(a)})$.
(2) Assume that W_e is an infinite c.e. set, and a is the $\leq_{\mathcal{L}}$-greatest element from L_{3e} such that the set $W_e \cap P_{a,3e}$ is infinite. Then $\Psi_A(W_e) \equiv_m \Psi_A(R_{a,3e+1})$.
(3) If $a, b \in \mathcal{L}$ and $a \neq b$, then $\Psi_A(R_{a,l(a)}) \not\equiv_m \Psi_A(R_{b,l(b)})$.

Proof: The proof is essentially the same as in [36, Proposition VI.2.2]. Here we only note that the stage $3e+1$ ensures the second condition, and stages $3\langle a,b,e\rangle + 2$ ensure the third condition. In addition, we have $R_{a,3e+3} =^* R_{a,3e+2}$ and $\Psi_A(R_{a,3e+3}) \equiv_m \Psi_A(R_{a,3e+2})$. □

Lemma 3.12 shows that the function

$$F\colon a \in \mathcal{L} \mapsto \Psi_A(R_{a,l(a)})$$

is an isomorphism from $\mathcal{L}$ onto $\mathbf{D}_m(\leq A)$. This concludes the proof of Proposition 3.4.

References

1. C. J. Ash, C. G. Jockusch Jr., and J. F. Knight, Jumps of orderings, *Trans. Am. Math. Soc.* **319** (1990), 573–599. doi: 10.1090/S0002-9947-1990-0955487-0.
2. S. Badaev and S. Goncharov. Computability and numberings. In eds. S. B. Cooper, B. Löwe, and A. Sorbi, *New Computational Paradigms*, pp. 19–34. Springer, New York, 2008. doi: 10.1007/978-0-387-68546-5_2.

3. S. Badaev, S. Goncharov, S. Podzorov, and A. Sorbi. Algebraic properties of Rogers semilattices of arithmetical numberings. In eds. S. S. Goncharov and S. B. Cooper, *Computability and Models*, pp. 45–78. Springer, New York, 2003. doi: 10.1007/978-1-4615-0755-0_3.
4. S. Badaev, S. Goncharov, and A. Sorbi. Isomorphism types and theories of Rogers semilattices of arithmetical numberings. In eds. S. S. Goncharov and S. B. Cooper, *Computability and Models*, pp. 79–92. Springer, New York, 2003. doi: 10.1007/978-1-4615-0755-0_4.
5. S. A. Badaev, Computable enumerations of families of general recursive functions, *Algebra Logic.* **16** (1977), 83–98. doi: 10.1007/BF01668593.
6. S. A. Badaev, Minimal numerations of positively computable families, *Algebra Logic.* **33** (1994), 131–141. doi: 10.1007/BF00750228.
7. S. A. Badaev and S. S. Goncharov. Theory of numberings: open problems. In eds. P. Cholak, S. Lempp, M. Lerman, and R. Shore, *Computability Theory and Its Applications*, vol. 257, *Contemp. Math.*, pp. 23–38. American Mathematical Society, Providence, 2000. doi: 10.1090/conm/257/04025.
8. S. A. Badaev, S. S. Goncharov, and A. Sorbi, Elementary theories for Rogers semilattices, *Algebra Logic.* **44** (2005), 143–147. doi: 10.1007/s10469-005-0016-x.
9. S. A. Badaev, S. S. Goncharov, and A. Sorbi, Isomorphism types of Rogers semilattices for families from different levels of the arithmetical hierarchy, *Algebra Logic.* **45** (2006), 361–370. doi: 10.1007/s10469-006-0034-3.
10. N. Bazhenov, M. Mustafa, and M. Yamaleev, Elementary theories and hereditary undecidability for semilattices of numberings, *Arch. Math. Logic.* **58** (2019), 485–500. doi: 10.1007/s00153-018-0647-y.
11. M. V. Dorzhieva, Elimination of metarecursive in Owing's theorem, *Vestn. Novosib. Gos. Univ., Ser. Mat. Mekh. Inform.* **14** (2014), 35–43. In Russian.
12. M. V. Dorzhieva, Undecidability of elementary theory of Rogers semilattices in analytical hierarchy, *Sib. Elektron Mat. Izv.* **13** (2016), 148–153. doi: 10.17377/semi.2016.13.013. In Russian.
13. R. Downey and J. F. Knight, Orderings with αth jump degree $\mathbf{0}^{(\alpha)}$, *Proc. Am. Math. Soc.* **114** (1992), 545–552. doi: 10.2307/2159679.
14. Y. L. Ershov, On computable enumerations, *Algebra Logic.* **7** (1968), 330–346. doi: 10.1007/BF02219286.
15. Y. L. Ershov, The upper semilattice of numerations of a finite set, *Algebra Logic.* **14** (1975), 159–175. doi: 10.1007/BF01668550.
16. Y. L. Ershov, *Theory of numberings.* Nauka, Moscow, 1977. In Russian.
17. Y. L. Ershov, Theory of numberings. In ed. E. R. Griffor, *Handbook of Computability Theory*, vol. 140, *Stud. Logic Found. Math.*, pp. 473–503. North-Holland, Amsterdam, 1999. doi: 10.1016/S0049-237X(99)80030-5.
18. Y. L. Ershov, Necessary isomorphism conditions for Rogers semilattices of finite partially ordered sets, *Algebra Logic.* **42** (2003), 232–236. doi: 10.1023/A:1025005309632.
19. Y. L. Ershov and I. A. Lavrov, The upper semilattice $L(\gamma)$, *Algebra Logic.* **12** (1973), 93–106. doi: 10.1007/BF02219292.
20. R. M. Friedberg, Three theorems on recursive enumeration. I. Decomposition.

II. Maximal set. III. Enumeration without duplication, *J. Symb. Logic.* **23** (1958), 309–316. doi: 10.2307/2964290.
21. K. Gödel, Über formal unentscheidbare Sätze der Principia Mathematica und verwandter Systeme. I., *Monatsh. Math. Phys.* **38** (1931), 173–198. doi: 10.1007/BF01700692.
22. S. S. Goncharov, Computable single-valued numerations, *Algebra Logic.* **19** (1980), 325–356. doi: 10.1007/BF01669607.
23. S. S. Goncharov, Positive numerations of families with one-valued numerations, *Algebra Logic.* **22** (1983), 345–350. doi: 10.1007/BF01982111.
24. S. S. Goncharov and A. Sorbi, Generalized computable numerations and nontrivial Rogers semilattices, *Algebra Logic.* **36** (1997), 359–369. doi: 10.1007/BF02671553.
25. I. Herbert, S. Jain, S. Lempp, M. Mustafa, and F. Stephan, Reductions between types of numberings, *Ann. Pure Appl. Logic.* **170** (2019), 102716. doi: 10.1016/j.apal.2019.102716.
26. I. S. Kalimullin, V. G. Puzarenko, and M. K. Faizrakhmanov, Positive presentations of families in relation to reducibility with respect to enumerability, *Algebra Logic.* **57** (2018), 320–323. doi: 10.1007/s10469-018-9503-8.
27. A. B. Khutoretskii, On the cardinality of the upper semilattice of computable enumerations, *Algebra Logic.* **10** (1971), 348–352. doi: 10.1007/BF02219842.
28. S. C. Kleene, *Introduction to metamathematics.* Van Nostrand, New York, 1952.
29. J. F. Knight, Degrees coded in jumps of orderings, *J. Symb. Logic.* **51** (1986), 1034–1042. doi: 10.2307/2273915.
30. A. N. Kolmogorov and V. A. Uspenskii, On the definition of an algorithm, *Uspehi Mat. Nauk.* **13** (1958), 3–28. In Russian.
31. A. H. Lachlan, Standard classes of recursively enumerable sets, *Z. Math. Logik Grundlagen Math.* **10** (1964), 23–42. doi: 10.1002/malq.19640100203.
32. A. H. Lachlan, On recursive enumeration without repetition, *Z. Math. Logik Grundlagen Math.* **11** (1965), 209–220. doi: 10.1002/malq.19650110305.
33. A. H. Lachlan, Initial segments of many-one degrees, *Can. J. Math.* **22** (1970), 75–85. doi: 10.4153/CJM-1970-010-6.
34. A. I. Mal'cev, Positive and negative numerations, *Sov. Math. Dokl.* **6** (1965), 75–77.
35. A. A. Mal'tsev, On the structure of families of immune, hyperimmune, and hyperhyperimmune sets, *Math. USSR Sb.* **52** (1985), 301–313. doi: 10.1070/SM1985v052n02ABEH002892.
36. P. Odifreddi, *Classical recursion theory.* vol. 125, *Stud. Logic Found. Math.*, Elsevier, Amsterdam, 1992.
37. E. A. Palyutin, Supplement to Yu. L. Ershov's article "The upper semilattice of numerations of a finite set", *Algebra Logic.* **14** (1975), 176–178. doi: 10.1007/BF01668551.
38. S. Y. Podzorov, Initial segments in Rogers semilattices of Σ^0_n-computable numberings, *Algebra Logic.* **42** (2003), 121–129. doi: 10.1023/A:1023354407888.

39. S. Y. Podzorov, Arithmetical D-degrees, *Sib. Math. J.* **49** (2008), 1109–1123. doi: 10.1007/s11202-008-0107-8.
40. M. B. Pour-El, Gödel numberings versus Friedberg numberings, *Proc. Am. Math. Soc.* **15** (1964), 252–256. doi: 10.2307/2034045.
41. H. Rogers, Gödel numberings of partial recursive functions, *J. Symb. Logic.* **23** (1958), 331–341. doi: 10.2307/2964292.
42. V. L. Selivanov, Two theorems on computable numberings, *Algebra Logic.* **15** (1976), 297–306. doi: 10.1007/BF01875946.
43. V. A. Uspenskii, Systems of denumerable sets and their enumeration, *Dokl. Akad. Nauk SSSR.* **105** (1958), 1155–1158. In Russian.
44. V. V. V'yugin, On some examples of upper semilattices of computable enumerations, *Algebra Logic.* **12** (1973), 287–296. doi: 10.1007/BF02218587.

https://doi.org/10.1142/9789811278631_0005

Immunity, Diagonalization and the Complexity of Mass Problems

Achilles A. Beros
Department of Mathematics
Miami University
Oxford, OH 45056, USA
berosaa@MiamiOH.edu

Mushfeq Khan
Department of Mathematics
University of Hawaiʻi at Mānoa
Honolulu, HI 96822, USA
mushfeq.khan@gmail.com

Bjørn Kjos-Hanssen
Department of Mathematics
University of Hawaiʻi at Mānoa
Honolulu, HI 96822, USA
bjoernkh@hawaii.edu

André Nies
Department of Computer Science
University of Auckland
Private Bag 92019, Auckland, New Zealand
andre@cs.auckland.ac.nz

Greenberg and Miller (Bull. London Math. Soc. 43, 2011) have shown that for any oracle X, there are arbitrarily slow-growing diagonally non-recursive functions relative to X that fail to compute a Martin-Löf random set. We show that the same holds when Martin-Löf randomness is replaced with effective bi-immunity. It follows that there are sequences of effective Hausdorff dimension 1 that compute no effectively bi-immune set.

A function is strongly non-recursive (SNR) if it is eventually different from each recursive function. We obtain hierarchy results for the mass problems given by classes of such functions with varying growth bounds. In particular, there is no least and no greatest Muchnik degree among mass problems of the form SNR_g, the SNR functions dominated by a recursive function g.

1. Introduction

Computability theorists study the complexity of incomputable objects with algorithmic tools. From the early days of computability, researchers have introduced notions of high complexity for objects by saying that they differ from all algorithmically given objects of a certain type. For instance, an infinite set A of natural numbers is called *immune* [22] if it contains no infinite r.e. subset. A function f on the natural numbers is strongly non-partial recursive [10] if f differs from each partial recursive function ψ on almost all inputs at which ψ is defined.

Sometimes it suffices to differ from only one algorithmically given object that is universal: a function g is diagonally non-recursive (DNR) if $g(n) \neq J(n)$ for each n; here $J(n) = \phi_n(n)$ is a universal partial recursive function.

Our purpose is to study variants of these properties with respect to their *relative* complexity. The relevant notions are explications of the meta-concepts of immunity and diagonalization. If the algorithmic tools are recursively enumerable, these are effective immunity and diagonally non-recursive functions. If they are computable, these are canonical immunity and strongly non-recursive functions, both defined below.

To study the relative complexity of the notions, we use the language of mass problems and Muchnik reducibility. A mass problem is a subset of Baire space ω^ω. Each notion discussed above can be viewed as a mass problem. Identifying a set with its characteristic function, we can also see a subset of $\mathcal{P}(\omega)$ as a mass problem.

Definition 1.1: A mass problem $\mathcal{A}$ is *Muchnik reducible* to a mass problem $\mathcal{B}$, written $\mathcal{A} \leq_w \mathcal{B}$ (and also read "weakly reducible"), if for each $B \in \mathcal{B}$, there is an $A \in \mathcal{A}$ such that $A \leq_T B$, where $\leq_T$ is Turing reducibility. If there is a single Turing reduction Φ such that $\Phi^A \in \mathcal{A}$ for all $B \in \mathcal{B}$, then $\mathcal{A}$ is said to be *Medvedev reducible* to $\mathcal{B}$, written $\mathcal{A} \leq_s \mathcal{B}$ (and also read "strongly reducible").

1.1. *Classes defined via r.e. tools*

In the first part of this chapter (up to the end of Sec. 5), the algorithmic tools to define high complexity are of the "r.e." type, i.e. defined via Σ_1^0 conditions. We focus on strengthenings of immunity: a set A is called *effectively immune* if there is a recursive function f such that for all e, if W_e is a subset of A, then $|W_e| \leq f(e)$. Here as usual $W_0, W_1, W_2, \ldots$ denotes an effective enumeration of the r.e. sets of natural numbers. The interest in sets, whose immunity is effectively witnessed in this manner, originally arose in the search for a solution to Post's problem posed in [22]: whether there is an incomputable r.e. set of incomplete Turing degree [27, Exercise V.4.2].

For some background, recall that an r.e. set is called *effectively simple* if its complement is effectively immune. Smullyan [26] appears to be the first to have explicitly isolated this notion. He observed that Post's construction [22] of a simple set actually produces an effectively simple set. Sacks [24] established the existence of a simple set that is not effectively simple. Subsequently, Martin [18] showed that every effectively simple set is Turing complete, and hence cannot constitute a solution to Post's problem. It is well-known that every nonzero r.e. Turing degree contains a simple set, so the two notions are actually very far apart.

Jockusch [10] established a key result showing the significance of the notion of effective immunity outside the context of the co-r.e. sets. He proved that the Turing degrees of the effectively immune sets coincide with those of the DNR functions.

One says that a set A is *bi-immune* if both A and its complement $\overline{A}$ are immune, and similarly for other set notions. Jockusch and Lewis [11] proved that every DNR function computes a *bi-immune* set. They left open the question whether the result can be extended to state that every DNR function computes an *effectively bi-immune set* [11, Question 1].

Definition 1.2: ([11], Def. 6.1) A set X is *effectively bi-immune* (EBI) if X and its complement, $\bar{X}$, are effectively immune. If f is a recursive function that witnesses the effective immunity of both X and $\bar{X}$, we say that X is *effectively bi-immune via* f, or f-EBI.

The first author [5] has provided a negative answer to Jockusch and Lewis's question. (In Sec. 2.1 we will provide an alternative proof.) To summarize, every DNR function computes an effectively immune set (in fact, of the same Turing degree) and a bi-immune set, but not every DNR function computes an EBI set.

Among randomness notions, the main notion, due to Martin-Löf, is based on an r.e. test concept. The following is well-known.
Every Martin-Löf random set is EBI via a function of type $e \mapsto e + O(1)$.

Proof: Given c, let $S^c_e = \{Z \colon W_e \subseteq Z \vee W_e \subseteq \overline{Z}\}$ in case $|W_e| \geq e+c+3$, and $S^c_e = \emptyset$ otherwise. Let $U_c = \bigcup_e S^c_e$. Clearly (U_c) is a ML-test. If Z is ML-random there is c such that $Z \notin U_c$. Then c is a constant as required. □

How close are these properties in terms of weak reducibility? Greenberg and Miller [9] have shown that there is a set of effective Hausdorff dimension 1 that computes no Martin-Löf random set. The main result of Sec. 3 will show that there is a set of effective Hausdorff dimension 1 that computes no EBI set. This is a potential strengthening of the Greenberg-Miller result.

It is not known whether every EBI set computes a Martin-Löf random set. However, existing results imply that the Turing degrees of these two classes do not coincide. Barmpalias, Lewis, and Ng [2] have shown that every PA degree is the join of two Martin-Löf random degrees. The effective join of two EBI sets is easily seen to be EBI, and so EBI sets are present in every PA degree, in particular, the incomplete ones. Such a degree cannot contain a Martin-Löf random set, by a theorem of Stephan [28].

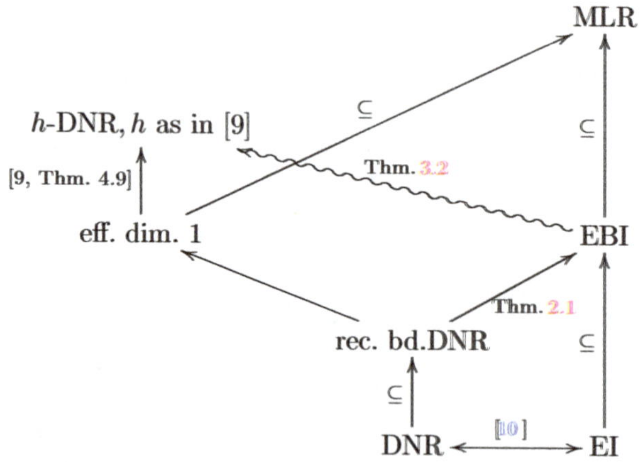

Fig. 1. A straight arrow indicates a known Muchnik reduction. A wavy arrow indicates a known non-reduction. Some arrows are trivially given by reverse inclusion of the mass problems.

1.2. *Classes defined via computable tools*

In the second part of this chapter, from Sec. 6 onwards, the algorithmic tools to define high complexity are computable. We consider strongly non-recursive (SNR) functions, functions that are eventually different from each recursive function. The corresponding randomness notion is the one due to Schnorr, which can be defined in terms of computable betting strategies (e.g. [21, Sec. 7.2]). Recursive traceability (e.g. [21, Sec. 8.2]) is equivalent to the lowness notion associated with Schnorr randomness. The concept of a recursive trace inspired a new notion of immunity of an infinite set R, which we call canonical immunity: for a fixed recursive function h, for each canonical listing (D_e) of the finite sets, $D_e \subseteq R$ implies $|D_e| \leq h(e)$ for almost all e. In analogy with Jockusch's result [10] that effective immunity is Muchnik equivalent to being DNR, we will show that canonical immunity is Muchnik equivalent to being SNR.

For an unbounded nondecreasing computable function g, let SNR_g denote the class of SNR functions bounded by g. We obtain a hierarchy result for these classes: if the function grows significantly slower, the relative Muchnik complexity goes up. We reduce the problem to a known hierarchy result for DNR functions due to Khan and Miller [15]. Monin, Miller and Nies, in the final section of [19], have obtained a similar hierarchy result for the dual classes IOE_g of g-bounded functions that agree with each computable function infinitely often.

Rupprecht [23] introduced an analogy of highness properties of oracles (i.e., natural Muchnik degrees) with cardinal characteristics. For a readable introduction to this recent area see Brendle *et al.* [6]. The analogy only exists for properties defined via computable tools, which correspond to a ground model in set theory. There appears to be no analogue in set theory of partiality and being r.e. More detail will be given in Sec. 8 below.

2. Locating the class EBI

Towards understanding the newly introduced class of effectively bi-immune sets, we compare it to known classes in the senses of Muchnik reducibility and inclusion. We also analyze its descriptive complexity.

2.1. *EBIs compute recursively bounded DNR functions*

In Corollary 2.2 we will provide a short proof of the main result from [5] that some DNR function does not compute an EBI set. It builds on previous work by Ambos-Spies, Kjos-Hanssen, Lempp, and Slaman [1].

In the language of weak reducibility, our first result states that the mass problem of recursively bounded DNR functions is weakly (or Muchnik) below the mass problem of EBI sets.

Theorem 2.1: *Every* EBI *set computes a recursively bounded* DNR *function. Moreover, a recursive bound for the* DNR *function can be obtained uniformly from a witnessing function for the* EBI *set.*

Proof: Let γ be any recursive bijection from ω to the collection of finite subsets of ω, and for an infinite set $Y \subseteq \omega$, let Y_n denote the set consisting of the first n elements of Y.

Suppose X is effectively bi-immune via f. Let h be a recursive function such that for all n,

$$W_{h(n)} = \begin{cases} \gamma(\varphi_n(n)), & \text{if } \varphi_n(n)\downarrow \\ \emptyset, & \text{otherwise.} \end{cases}$$

Now, let $g(n) = \gamma^{-1}(X_{f(h(n))+1})$ and let $\bar{g}(n) = \gamma^{-1}(\bar{X}_{f(h(n))+1})$. We claim that both g and $\bar{g}$ are DNR. Suppose that for some e, $\varphi_e(e) = g(e)$. Then $X_{f(h(e))+1} = W_{h(e)}$. But then $W_{h(e)} \subset X$ and $|W_{h(e)}| > f(h(e))$, a contradiction. The argument for $\bar{g}$ is identical.

Finally, let $\tilde{g} = \min(g, \bar{g})$. Clearly, $\tilde{g}$ is DNR and recursive in X. Given any $n \in \omega$, the largest elements in $X_{f(h(n))+1}$ and $\bar{X}_{f(h(n))+1}$ cannot both be larger than $2f(h(n)) + 1$. Thus, letting

$$\pi(n) = \max\{\gamma^{-1}(D) : \max(D) \leq 2f(h(n)) + 1\},$$

we have $\tilde{g}(n) \leq \pi(n)$. □

Ambos-Spies *et al.* [1] have shown that there is a DNR function that computes no recursively bounded DNR function, and so we reprove the main result of [5]:

Corollary 2.2: *There is a* DNR *function that computes no* EBI *set.*

It is worth noting that the construction in [5] achieves significantly more than was claimed in that paper. It partially relativizes, and Turing reduction can be replaced with recursive enumeration:

Theorem 2.3: (Beros) *For any set A, there is a function f that is* DNR *relative to A, such that no* EBI *set is r.e. in f.*

2.2. *A class between EBI and EI in terms of inclusion*

Theorem 2.4: *Some bi-immune set A is effectively immune while its complement is not.*

Proof: We build a set A in stages by describing its characteristic function, g.

Stage 0: Define g_0 to be the function with empty domain.
Stage $2e+1$*:* Define $g_{2e+1} \upharpoonright \mathrm{dom}(g_{2e}) = g_{2e}$. Let $m = \min(\mathbb{N} \setminus \mathrm{dom}(g_{2e}))$ and set $g_{2e+1}(m) = 1$. If $|W_e| > 2e+1$ and there is no $a \in W_e \cap \mathrm{dom}(g_{2e})$ such that $g_{2e}(a) = 0$, pick $x \in W_e \setminus \mathrm{dom}(g_{2e})$ and set $g_{2e+1}(x) = 0$. If W_e is infinite, select a $y \neq x$ such that $y \in W_e \setminus \mathrm{dom}(g_{2e})$ and set $g_{2e+1}(y) = 1$.
Stage $2e+2$*:* Define $g_{2e+2} \upharpoonright \mathrm{dom}(g_{2e+1}) = g_{2e+1}$. If ϕ_e is total, pick an r.e. set W_a such that $\phi_e(a) < |W_a| < \infty$ and $|W_a| \cap \mathrm{dom}(g_{2e+1}) = \emptyset$. Set $g_{2e+2}(x) = 0$ for all $x \in W_a$.

Notice that there are no more than $2s$ elements x such that $g_s(x) = 1$. So, either it is possible to pick an x as in the odd stages, $2e+1$, whenever $|W_e| > 2e+1$, or there is already an element of the domain of g_{2e}, which is in W_e, on which g_{2e} takes the value 0. Let $g = \bigcup_{s\in\mathbb{N}} g_s$. Observe that g is total and $\{0,1\}$-valued.

Let A be the set whose characteristic function is g. The effective immunity of A is witnessed by $f(x) = 2x+1$ and $\overline{A}$ is clearly immune, however, for any total function h there is an r.e. set W_a, such that $h(a) < |W_a|$ and $W_a \subseteq \overline{A}$. Thus, $\overline{A}$ is not effectively immune. □

2.3. *Descriptive complexity of the class of EBI sets*

Recall that a class of sets is F_σ if it is a countable union of closed sets.

Lemma 2.5: *Let f be a recursive function. The class of sets that are effectively immune via f is closed.*

Proof: Suppose that A is not effectively immune via f. Then there is a e, such that W_e is a subset of A and $|W_e| > f(e)$. If W_e is a finite set, then there is an initial segment σ of A, such that W_e is contained in any set, whose characteristic function extends σ, and so no extension of σ is effectively immune via f. So suppose that W_e is infinite. By the recursion theorem, there exists an e', such that $W_{e'}$ consists of the first $f(e')+1$ elements of W_e. Thus, in this case, there is also an initial segment σ of A

such that any set whose characteristic function extends σ contains $W_{e'}$, and is therefore not effectively immune via f. $\square$

Theorem 2.6: *The class of* EBI *sets is Wadge complete for* F_σ.

Proof: By Lemma 2.5, the class of EBI sets is F_σ.

For the hardness part, let $A \subset 2^\omega$ be the class of infinite strings that are eventually zero. It is well-known that A is Wadge complete for the F_σ sets (see, for example, [13, Exercise 21.17]). We construct a continuous $h : 2^\omega \to 2^\omega$ such that $X \in A$ iff $h(X)$ is EBI, showing that A is Wadge reducible to the class of EBI reals.

We first define a function $f : 2^{<\omega} \to 2^{<\omega}$ recursively. Let $e_0, e_1, \ldots$ be an increasing list of all codes for total computable functions. Also, for each $\tau \in 2^{<\omega}$, let e_τ be the least number e such that some ϕ_e-EBI real extending τ exists. Then let g_τ be such a ϕ_{e_τ}-EBI real.

Let α^i_τ be an extension of τ such that no real extending α^i_τ is ϕ_{e_i}-EBI. Note that α^i_τ exists by the argument in the proof of Theorem 2.5.

Let $f(\langle\rangle) = \langle\rangle$, where $\langle\rangle$ is the empty string. Suppose $\sigma \in 2^{<\omega}$, let $n = |\sigma|$ and k be the number of bits of σ which are 1. Given $f(\sigma) = \tau$, we define

$$f(\sigma\hat{\ }0) = g_\tau \restriction (|\tau| + 1),$$
$$f(\sigma\hat{\ }1) = \alpha^k_\tau.$$

Here we are using the fact that the set of ϕ_{e_τ}-EBI reals is closed. Finally, define h so that the nth bit of $h(x)$ is the nth bit of $f(x \restriction (n + 1))$. $\square$

3. Slowly-growing DNR functions

Theorem 2.1 says that the recursively bounded DNR functions are weakly reducible to the EBI sets. We show that there is no reduction in the converse direction.

Given that, one might still hope that sufficiently slow-growing DNR functions suffice: perhaps there is a slow enough recursive bound g such that each g-bounded DNR function computes an EBI set? This may seem plausible because Khan and Miller [15] have shown that by making g grow slower, one can get higher up among mass problems. Nonetheless, we show that this hope is in vain. See Fig. 1.1.

Definition 3.1: An *order function* is a recursive, unbounded and non-decreasing function from ω to $\omega \setminus \{0, 1\}$.

Theorem 3.2: *For each order function g, there is a g-bounded* DNR *function f that computes no* EBI *set.*

In other words, there is an arbitrarily slow-growing DNR function that computes no effectively bi-immune set. On the other hand, a sufficiently slow-growing DNR function is known to compute a set of effective Hausdorff dimension 1:

Theorem 3.3: (Greenberg and Miller [9]) *There is an order function h, such that every h-bounded* DNR *function computes a set of effective Hausdorff dimension 1.*

Together, these theorems imply the following (see Fig. 1.1):

Corollary 3.4: *There is a set of effective Hausdorff dimension 1 that computes no* EBI *set.*

In order to prove Theorem 3.2, we force with bushy trees, first used in recursion theory in [17].

3.1. *Definitions and combinatorial lemmas*

The following definitions can also be found in [9] and [15].

Definition 3.5: Given $\sigma \in \omega^{<\omega}$, we say that a tree $T \subseteq \omega^{<\omega}$ is *n-bushy above σ* if every element of T is comparable with σ, and for every $\tau \in T$ that extends σ and is not a leaf of T, τ has at least n immediate extensions in T. We refer to σ as the *stem* of T.

Definition 3.6: Given $\sigma \in \omega^{<\omega}$, we say that a set $B \subseteq \omega^{<\omega}$ is *n-big above σ* if there is a finite n-bushy tree T above σ such that all its leaves are in B. If B is not n-big above σ then we say that B is *n-small* above σ.

Proofs of the following lemmas can be found in [9, Lemmas 5.4 and 5.5] and [15, Lemmas 2.7 and 2.8].

Lemma 3.7: (Smallness preservation property) *Suppose that B and C are subsets of $\omega^{<\omega}$ and that $\sigma \in \omega^{<\omega}$. If B and C are respectively n- and m-small above σ, then $B \cup C$ is $(n+m-1)$-small above σ.*

Lemma 3.8: (Small set closure property) *Suppose that $B \subset \omega^{<\omega}$ is n-small above σ. Let $C = \{\tau \in \omega^{<\omega} : B$ is n-big above $\tau\}$. Then C is n-small above σ. Moreover C is n-*closed*, meaning that if C is n-big above a string ρ, then $\rho \in C$.*

Definition 3.9: Given an order function g, let $g^{<\omega}$ denote the set of finite strings σ in $\omega^{<\omega}$, whose entries are bounded by g in the sense that $\sigma(i) < g(i)$ for each $i < |\sigma|$. Let g^n denote the set of such strings of length n. Furthermore, let g^ω be those infinite sequences whose entries are bounded by g.

3.2. *Proof of Theorem 3.2*

We work entirely in $g^{<\omega}$, forcing with conditions of the form (σ, B), where $\sigma \in g^{<\omega}$ and $B \subset g^{<\omega}$ and B is $g(|\sigma|)$-small above σ. A condition (σ, B) *extends* another condition (τ, C) if $\sigma \succeq \tau$ and $C \subseteq B$. Let $\mathbb{P}$ denote this partial order of conditions. Let $[\sigma]$ denote the elements of g^ω that extend σ, and let $[B]^{\prec}$ denote the set of elements of g^ω that extend an element of B.

For a functional Γ and a recursive function q, let $\mathcal{H}_{\Gamma,q}$ be the set of all conditions (σ, B) such that if $f \in [\sigma] \setminus [B]^{\prec}$, then Γ^f is not effectively bi-immune as witnessed by the function q.

We assume that for all $f \in g^\omega$, for any functional Γ, the domain of Γ^f is an initial segment of ω.

Lemma 3.10: *$\mathcal{H}_{\Gamma,q}$ is dense in $\mathbb{P}$.*

Proof: Let (σ, B) be any condition and suppose B is k-small (and k-closed) above σ. By suitably extending σ, we may assume that $g(|\sigma|) \geq 8k$.

Suppose first that there is a $\tau \notin B$ extending σ such that for some m,

$$C_m = \{\rho : \Gamma^\rho(m)\downarrow\}$$

is $7k$-small above τ. Then $(\tau, B \cup C_m)$ is a condition extending (σ, B), and for every $f \in [\tau] \setminus [B \cup C_m]^{\prec}$, Γ^f is not total. So we assume from now on that for every $\tau \notin B$ extending σ and every $m \in \omega$, C_m is $7k$-big above τ.

It now follows that for every $\tau \notin B$ extending σ, there is an infinite exactly $6k$-bushy tree T_τ without leaves above τ such that for every $f \in [T_\tau]$, Γ^f is total: Let S_0 consist of τ and its initial segments. Next, suppose we have already constructed a finite tree S_n that is exactly $6k$-bushy above τ and such that for each leaf ρ of this tree, $\rho \notin B$ and Γ^ρ is defined up to $n-1$. By our assumption above, C_n is $7k$-big above each leaf, so $C_n \setminus B$ is $6k$-big above each leaf by Lemma 3.7. For a leaf ρ of S_n, let A_ρ be a finite exactly $6k$-bushy tree above τ with leaves in $C_n \setminus B$. We construct S_{n+1} by appending A_ρ to each leaf ρ of S_n. Finally, let $T_\tau = \bigcup_{n\in\omega} S_n$.

Definition 3.11: Let τ be any extension of σ that is not in B. We say τ *admits fusion* if there are infinitely many $m \in \omega$ such that for some

$i \in \{0, 1\}$,

$$\Delta_{\tau,m,i} = \{\rho \in T_\tau : \Gamma^\rho(m) = i\}$$

is $4k$-big above τ.

In Definition 3.11, by the pigeonhole principle it is equivalent to require that there is an i, such that there are infinitely many m, for which $\Delta_{\tau,m,i}$ is $4k$-big above τ.

Claim 3.12: If τ admits fusion, then there is a subtree T' of T_τ, which is $2k$-bushy above τ and such that for infinitely many $(m, i) \in \omega \times \{0, 1\}$, $\Gamma^f(m) = i$ for all $f \in [T']$.

Proof: Let $I_0 \subseteq \omega$ be such that for all $l \in I_0$, either $\Delta_{\tau,l,0}$ or $\Delta_{\tau,l,1}$ is $4k$-big above τ, and let Δ_l denote whichever one is. Let S_0 consist of τ and its initial segments, and note that S_0 is $2k$-bushy above τ.

Next, suppose that we have constructed a finite tree $S_k \subseteq T_\tau$, $2k$-bushy above τ, and a subset I_k of ω such that:

(1) There are $n_0 < n_1 < \cdots < n_{k-1}$, such that for each $i < k$, $\Gamma^\rho(n_i)$ is constant as ρ ranges over the leaves of S_k.
(2) For all $l \in I_k$, there is a tree which is $4k$-bushy above τ and contains S_k, whose leaves are in Δ_l.

Let n be the least element in I_k greater than n_{k-1}, and let C be a finite $4k$-bushy tree above τ containing S_k whose leaves are in Δ_n. Now for any $l > n$ in I_k, if F_l is any $4k$-bushy tree above τ containing S_k with leaves in Δ_l, then $F_l \cap C$ is a $2k$-bushy tree above τ that contains S_k. To see this, let $\rho \in F_l \cap C$. If ρ has an immediate extension in $F_l \cap C$, then ρ has $4k$ many extensions in each of F_l and C. But T_τ is exactly $6k$-bushy above τ, so at least $2k$ of these must be in $F_l \cap C$.

It follows from the pigeonhole principle (note that C is finite) that there is an infinite subset I_{k+1} of I_k, such that for all $l \in I_{k+1}$, there are $4k$-bushy subtrees above τ with leaves in Δ_l that intersect C in the *same* $2k$-bushy subtree S_{k+1} above τ that contains S_k.

This completes the definition of the sequence $\langle S_k \rangle_{k \in \omega}$. Let $T' = \bigcup_{k \in \omega} S_k$. Then T' is as desired. □

Case 1: Some $\tau \succeq \sigma$ admits fusion. We begin by extending σ to τ obtaining the condition (τ, B) (note that τ is by definition not in B). Let

Leaves(T) denote the set of leaves of a tree T. Claim 3.12 implies that, uniformly in k, we can find a finite $2k$-bushy tree R_k above τ such that

$$(\exists^{\geq k} n)(\exists i \in \{0,1\})(\forall \rho \in \text{Leaves}(R_k))(\Gamma^\rho(n) = i).$$

In particular, for any m by the pigeonhole principle it follows that

$$(\exists i \in \{0,1\})(\exists^{>m} n)(\forall \rho \in \text{Leaves}(R_{2m+1}))(\Gamma^\rho(n) = i). \tag{3.1}$$

For $i \in \{0,1\}$, let W_{e_i} be the r.e. set defined as follows: Using the Recursion Theorem, let $m = \max(q(e_0), q(e_1))$. Then let

$$W_{e_i} = \{n : \Gamma^\rho(n) = i \text{ for each leaf } \rho \text{ of } R_{2m+1}\}.$$

By (3.1) we know that for some $i \in \{0,1\}$, $|W_{e_i}| > m$. Suppose $i = 0$ (the argument for the other case is symmetric). Let $\rho \succeq \tau$ be a string in $R_{2m+1} \setminus B$ (note that B is k-small above τ). Then (ρ, B) is a condition, and for all $f \in [\rho] \setminus [B]^{\prec}$, if Γ^f is total, then W_{e_0} is contained in its complement.

Case 2: No extension of σ admits fusion. This means that for every extension τ of σ, such that $\tau \notin B$, there is an $m_\tau \in \omega$, such that for all $l \geq m_\tau$, both $\Delta_{\tau,l,0}$ and $\Delta_{\tau,l,1}$ are $4k$-small above τ. Recall that T_τ is exactly $6k$-bushy above τ. Therefore, $\Delta_{\tau,l,0} \cup \Delta_{\tau,l,1}$ is $6k$-big above τ. By Lemma 3.7, if one of these sets is $2k$-small above τ, the other is $4k$-big, so both must be $2k$-big above τ.

Let S_0 consist of σ and its initial segments. Note that no leaf of S_0 is in B.

Proceeding by induction, suppose we have constructed a finite k-bushy tree $S_k \subseteq T_\sigma$ above σ with the following properties:

(1) There are $n_0 < n_1 < \cdots < n_{k-1}$ such that for every leaf ρ of S_k, $\Gamma^\rho(n_i) = 0$ for each $i < k$.
(2) None of the leaves of S_k is in B.

Let $n_k = \max\{m_\tau : \tau \text{ a leaf of } S_k\} + 1$. By the observation above, for each leaf τ of S_k, $\Delta_{\tau,n_k,0}$ is $2k$-big above τ, so $\Delta_{\tau,n_k,0} \setminus B$ is k-big above τ. Let F_τ be a finite k-bushy tree above τ with leaves in $\Delta_{\tau,n_k,0} \setminus B$, and let S_{k+1} be obtained from S_k by extending each leaf τ of S_k by F_τ.

Finally, let $T' = \bigcup_{k\in\omega} S_k$. Then for all k, for every $f \in [T']$, $\Gamma^f(n_k) = 0$. A strategy similar to the one employed in case 1 now diagonalizes against the pair (Γ, q). This concludes the proof of the lemma.

To conclude the proof of Theorem 3.2, let B_{DNR^X} be the set of finite strings that cannot be extended to a DNR function relative to X, and let

$\mathcal{G}$ be any filter containing $(\langle\rangle, B_{\mathrm{DNR}^X})$ that meets $\mathcal{H}_{\Gamma,q}$ for each functional Γ and recursive function q. Then $f_{\mathcal{G}}$ is a g-bounded DNR function relative to X and does not compute an effectively bi-immune set.

4. Jump traceability implies a double lowness property

We focus on a partial relativization of effective bi-immunity. The idea of partial relativization goes back to Simpson. Suppose we are looking at a notion whose definition includes several occurrences of words like "computable", "recursive", "c.e.". To partially relativize the notion to the oracle A, we attach the clause "relative to A" to some (but not all) occurrences of "computable" etc.

Recursive bounds, for instance, are typically left unrelativized. The preposition "by", rather than "in", indicates that the relativization is partial. For background see [3, Sec. 2.1].

Definition 4.1: An infinite set R is *effectively immune by* G if there is a recursive function h such that for all e, if $W_e^G \subseteq R$ then $|W_e^G| \leq h(e)$. R is *effectively bi-immune by* G if both R and its complement are effectively immune by G.

Note that this is *more* restrictive than the standard relativization EBI^G. This partial relativization is motivated by the following fact that follows from the proof of Theorem 4.4 below. A set that is Martin-Löf random relative to an oracle G is effectively immune by G via the function $h(e) = e + c$ for some $c \in \omega$.

Definition 4.2: A set G is Low(MLR, *by* EBI) if each MLR set R is EBI by G.

For background on the following lowness notion see [21, Sec. 8.4].

Definition 4.3: A *recursively enumerable trace* T is a sequence of sets $T^{[e]} = W_{g(e)}$, $e \in \omega$ such that $|W_{g(e)}| \leq h(e)$ for all e, where g and h are recursive functions. For a function f, we say that T *traces* f on input n if $f(n) \in T^{[n]}$. A set G is *jump traceable* if there is an r.e. trace T such that for all e, if $J^G(e)\downarrow$ then $J^G(e) \in T^{[e]}$.

Theorem 4.4: *Each jump traceable set G is* Low(MLR, *by* EBI).

Note that this contrasts MLR with EBI. There is a jump traceable set G that is not low for ML-randomness. For such a G, MLR is not contained in MLR^G, while it is contained in EBI^G.

Proof: Let G be jump traceable via h, and let J^G denote a universal partial recursive function relative to G.

We define a recursive function f knowing its index in advance by the recursion theorem. Let φ be the function partial recursive in G that on input e, waits for W_e^G to enumerate at least $f(e)+1$ elements, and then outputs the natural number that encodes the finite set B_e consisting of the first $f(e)+1$ of these. Next, let p be a recursive function such that $J^G \circ p = \varphi$. Note that p can be obtained uniformly from an index for f. Now define f so that

$$h(p(e))\, 2^{-(f(e)+1)} \leq 2^{-e}.$$

We have an r.e. trace $S^{[p(e)]}$ for (the code for) B_e, and there are at most $h(p(e))$ many elements in it. Let $B_e^{(i)}$ denote the ith candidate for B_e if it exists, for $i < h(p(e))$. In other words, $B_e^{(i)}$ is the ith element of $S^{[p(e)]}$. Now let

$$U_c = \{A : (\exists e > c)(\exists i < h(p(e)))(B_e^{(i)} \downarrow\subseteq A)\}.$$

Then

$$\mu(U_c) \leq \sum_{e>c} h(p(e))\, 2^{-(f(e)+1)} \leq \sum_{e>c} 2^{-e} = 2^{-c}.$$

If A is MLR then there exists c such that for all $e > c$ and i, it is not the case that $B_e^{(i)} \downarrow\subseteq A$. Thus, for all $e \geq c$, if $W_e^G \subseteq A$ then W_e^G has size at most $f(e)$. Thus, A is EBI by G. □

5. DNR and SNPR functions

It is known [1,25] that bounding diagonally non-recursive functions by varying computable functions leads to a hierarchy of computational strength. This hierarchy interacts with Martin-Löf random reals and completions of Peano Arithmetic [9, 14].

Definition 5.1: (Kjos-Hanssen, Merkle, and Stephan [16]) A function $f : \omega \to \omega$ is *strongly non-recursive* (or SNR) if for every recursive function g, for all but finitely many $n \in \omega$, $f(n) \neq g(n)$.

f is *strongly non-partial-recursive* (or SNPR) if for every partial recursive function g, for all but finitely many n, if $g(n)$ is defined, $f(n) \neq g(n)$.

The strongly non-recursive functions also form a natural class. Here we start developing analogous hierarchy results for it.

Note that every SNPR function f is trivially SNR. Furthermore, a finite variant of it is almost DNR because for all but finitely many n, if $J(n)$ is defined, then $f(n) \neq J(n)$.

Definition 5.2: For a class $\mathcal{C}$ of functions from ω to ω, let $\mathcal{C}_h$ denote the subclass consisting of those members of $\mathcal{C}$ that are bounded by h.

Theorem 5.3: *For each order function h, there exists an order function g such that every* DNR_g *function computes an* SNPR_h *function.*

In order to prove this theorem, we will need a result due to Cenzer and Hinman [7], in a form presented in Greenberg and Miller [9].

Definition 5.4: (Greenberg and Miller [9]) Let $a \geq 2$ and let $0 < c < a$. Let $\mathcal{P}^c_a$ denote the class of functions $f \in a^\omega$ such that for all e and for all $x < c$, if $\varphi_e(x)\downarrow$, then $f(e) \neq \varphi_e(x)$.

Theorem 5.5: (Cenzer and Hinman [7]) *Let $a \geq 2$ and $c > 0$. Then any* DNR_a *function computes a function in $\mathcal{P}^c_{ca}$. Moreover, the reduction is uniform in a and c.*

Proof: [Proof of Theorem 5.3] Let r be the recursive function such that $\varphi_{r(x)}(e) = \varphi_e(x)$. For each $n \geq 2$, let $x_n \in \omega$ be the least such that $h(x_n) \geq n^2$.

We construct g to ensure that any DNR_g function computes a function f that is bounded by h and such that for all $x > x_n$ and for all $e < n$, if $\varphi_e(x)\downarrow$, then $f(x) \neq \varphi_e(x)$.

In order to compute f on the interval $[x_n, x_{n+1})$, a function in $\mathcal{P}^n_{n^2}$ suffices. Such a function can be uniformly obtained from a DNR_n function, by Theorem 5.5. However, we only need a finite part of this function, and we can recursively determine how much. Note that the reductions in Theorem 5.5 can be assumed to be total. For $n \geq 2$, let γ_n denote the use of reduction that, given a DNR_n function, computes a $\mathcal{P}^n_{n^2}$ function.

Then, letting

$$m_n = \max(\{\gamma_n(r(x)) : x \in [x_n, x_{n+1})\}),$$

it suffices for g to be any recursive function such that for all $x \leq m_n$, $g(x) \leq n$. □

We also have the following counterpart to Theorem 5.3:

Theorem 5.6: *For each order function g, there exists an order function h such that every* DNR_g *function computes an* SNPR_h *function.*

Proof: For $n \geq 1$, let $\tau_n : \omega \to \omega^n$ be a uniformly recursive sequence of bijections, and for $i < n$, $\pi_i^n : \omega^n \to \omega$ denote the projection function onto the i-th coordinate.

Let r be the recursive function such that for $n \geq 1$ and $e < n$, if $\varphi_e(n)$ converges, then $\varphi_{r(e,n)}$, on any input, outputs $\pi_e^n(\tau_n(\varphi_e(n)))$.

Now, given a DNR_g function f, let $j(0) = 0$ and for $n \geq 1$, let

$$j(n) = \tau_n^{-1}(\langle f(r(0,n)), \ldots, f(r(n-1,n))\rangle).$$

Then $j(n) \neq \varphi_e(n)$ for any $e < n$: If it were, then we would have

$$f(r(e,n)) = \pi_e^n(\tau_n(j(n))) = \pi_e^n(\tau_n(\varphi_e(n))) = \varphi_{r(e,n)}(r(e,n)),$$

which contradicts the fact that f is DNR. Thus, j is SNPR_h where for $n \geq 1$,

$$h(n) = \max(\{\tau_n^{-1}(\langle i_1, \ldots, i_{n-1}\rangle) : i_k < g(r(k,n)) \text{ for all } k < n\}). \qquad \square$$

6. Canonical immunity

Schnorr randomness is a weakening of ML-randomness where one requires that the measure of the test components is uniformly computable. For background see e.g. [21, Sec. 3.5]. Every high degree contains a Schnorr random [20]. This implies that not every Schnorr random is EBI: for instance, an incomplete high r.e. set does not even compute a DNR function.

A *canonical numbering of the finite sets* is a surjective function $D : \omega \to \{A : A \subseteq \omega \text{ and } A \text{ is finite}\}$ such that $\{(e,x) : x \in D(e)\}$ is recursive and the cardinality function $e \mapsto |D(e)|$, or equivalently, $e \mapsto \max D(e)$, is also recursive. We write $D_e = D(e)$.

Definition 6.1: R is *canonically immune* if R is infinite and there is a recursive function h, such that for each canonical numbering of the finite sets D_e, $e \in \omega$, for all but finitely many e, if $D_e \subseteq R$ then $|D_e| \leq h(e)$.

Proposition 6.2: *Each Schnorr random set is canonically immune.*

Proof: Fix a canonical numbering of the finite sets, $\{D_e\}_{e\in\omega}$. Define $U_c = \{X : (\exists e > c)(|D_e| \geq 2e \wedge D_e \subset X)\}$. Since $e \mapsto |D_e|$ is recursive, $\mu(U_c)$ is recursive and bounded by 2^{-c}. Thus, the sequence $\{U_c\}_{c\in\omega}$ is a Schnorr test. If A is a Schnorr random, then $A \in U_c$ for only finitely many $c \in \omega$. We conclude that A is canonically immune. $\square$

Proposition 6.3: *Each canonically immune set is immune.*

Proof: Suppose A has an infinite recursive subset R. Let h be any recursive function. Let R_n denote the set of the first n elements of R, and let $\{D_e : e \in \omega\}$ be a canonical numbering of the finite sets, such that $D_{2n} = R_{h(2n)+1}$ for all $n \in \omega$. For all n, $D_{2n} \subseteq R \subseteq A$ and $|D_{2n}| = h(2n) + 1 > h(2n)$, and so h does not witness the canonical immunity of A. □

Strongly non-recursive (SNR) and strongly non-partial recursive (SNPR) functions were defined in Definition 5.1. We now show how canonical immunity is an analogue of effective immunity.

Proposition 6.4: *Each canonically immune set computes a strongly non-recursive function.*

Proof: Let R be canonically immune as witnessed by the recursive function h. Define $f(e)$ to be (a code for) the first $h(2e)+1$ many elements of R, and note that f is recursive in R. We claim that f is strongly non-recursive.

Suppose that the recursive function g is infinitely often equal to f. Let $\{D_e : e \in \omega\}$ be any canonical numbering of finite sets, such that for all e, D_{2e} is the finite set coded by $g(e)$. We now have that for infinitely many e, D_{2e} is the set consisting of the first $h(2e) + 1$ many elements of R, a contradiction. □

Interestingly, Proposition 6.4 shows that we can strengthen "$D_e \subseteq R$" to "D_e is an initial segment of R".

Corollary 6.5: *The following are equivalent for an oracle A:*

(1) A computes a canonically immune set,
(2) A computes an SNR function,
(3) A computes an infinite subset of a Schnorr random.

Proof: (1) implies (2) is proved in Proposition 6.4. (2) implies (3) follows from older results: each SNR is high or computes a DNR [16], hence computes a Schnorr random [20] or computes an infinite subset of an MLR [8], hence either way computes an infinite subset of a Schnorr random. (3) implies (1) is proved in Proposition 6.2 because canonical immunity is closed under taking infinite subsets. □

Theorem 6.6: *If A is non-high and canonically immune, then A is effectively immune.*

Proof: Let us introduce the notation

$$W_{e,s}\#u$$

to mean the set of $k \in W_{e,s}$ such that when k enters $W_{e,s}$, at most u other numbers have already entered $W_{e,s}$. Thus, $D\#u$ consists of the first (up to) $u+1$ elements of D.

Let A be non-high and not effectively immune. We need to show that A is not canonically immune.

Since A is not effectively immune, for each computable $h : \omega \to \omega$ there are infinitely many e, for which there is an s with $\langle e, s\rangle \in \mathcal{F}$, where

$$\mathcal{F} = \{\langle e, s\rangle : |W_{e,s}| > h(e), \text{and } W_{e,s}\#h(e) \subseteq A\}.$$

Here we assume h is nondecreasing. So

$$\forall d \quad \exists s \quad \exists e_d > d \quad \langle e_d, s\rangle \in \mathcal{F}.$$

Let $f(d) = s$. Then there is a recursive function $g(d)$ which is not dominated by f.

We may assume $s > e_d$ since increasing s will keep $\langle e, s\rangle \in \mathcal{F}$. Let

$$D_{2\langle e,d\rangle} = W_{e,g(d)}\#h(d).$$

Let $\tilde{h}(\langle e, d\rangle) = h(d)$ for $d \leq e \leq g(d)$. Using a recursive bijection between the domain of $\tilde{h}$,

$$\{\langle e, d\rangle : d \leq e \leq g(d)\}$$

and ω, we may assume the domain of $\tilde{h}$ is ω.

Then for infinitely many e (namely, there are infinitely many d with $f(d) \leq g(d)$, and for each such d there is an e with $d \leq e \leq g(d)$ that works) we have

(1) $D_{2\langle e,d\rangle} = W_{e,g(d)}\#h(d) \subseteq W_{e,g(d)}\#h(e) \subseteq A$, and
(2) $|D_{2\langle e,d\rangle}| > \tilde{h}(\langle e, d\rangle) = h(d)$.

We have $\lim_{n\to\infty} \tilde{h}(n) = \infty$ since for each d we only include e up to the e_d above.

Let the sets D_{2k+1} be a canonical list of all the finite sets, just in case we missed some of them using the sets D_{2k}. □

For completeness' sake we include a proof of the following.

Theorem 6.7: (Kjos-Hanssen, Merkle, and Stephan [16]) *Every non-high* SNR *function is* SNPR.

Proof: Suppose that $f : \omega \to \omega$ is not high, and that ψ is a partial recursive function that is infinitely often equal to it. For each $n \in \omega$, let $g(n)$ be the least stage such that $|\{x \in \omega : \psi(x)[g(n)]\downarrow = f(x)\}| \geq n+1$. Then g is recursive in f.

Since f is not high, there is a recursive function h that escapes g infinitely often. We define a recursive function j that is infinitely often equal to f. Let $j_0 = \emptyset$. Given j_n, let

$$A = \{\langle x, \psi(x)\rangle : x \notin \mathrm{dom}(j_n), \psi(x)[h(n)]\downarrow\}.$$

Let y be the least number not in the domain of $j_n \cup A$, and let $j_{n+1} = j_n \cup A \cup \langle y, 0\rangle$. It is easily seen that $j = \bigcup_n j_n$ is recursive and infinitely often equal to f. □

Since an SNPR function has a finite variant that is DNR, we obtain:

Corollary 6.8: *For each order function h, every non-high* SNR_h *function computes a* DNR_h *function.*

7. The SNR hierarchy

Let $\mathfrak{O}$ denote the set of all order functions. In Corollaries 7.5 and 7.6 below we will obtain the following relationships of mass problems:

$$\forall h \in \mathfrak{O} \quad \exists g \in \mathfrak{O} \qquad \mathrm{SNR}_g >_w \mathrm{SNR}_h;$$
$$\forall g \in \mathfrak{O} \quad \exists h \in \mathfrak{O} \qquad \mathrm{SNR}_g >_w \mathrm{SNR}_h\,.$$

In particular, the Muchnik degrees of mass problems SNR_f for varying order function f have no least or greatest element.

We first elaborate on the DNR hierarchy, and then use the transfer results between SNR and DNR from the preceding section.

Theorem 7.1: *Uniformly in an order function h, we can find a recursive function π with the following property: If g is any order function such that $h(n)/g(\pi(n))$ is unbounded, then there is a low $f \in \mathrm{DNR}_h$ that computes no* DNR_g *function.*

Proof: In Sec. 3.1 we defined what it means for a tree $B \subseteq \omega^{<\omega}$ to be n-bushy, n-big and n-small over a string σ. We also stated the smallness preservation property and the small set closure property.

Given $\sigma \in h^{<\omega}$, let $q(\sigma, e, k)$ be an index for the partial recursive function that searches for a k-big set $A \subset h^{<\omega}$ above σ, such that $\Phi_e^\tau(q(\sigma, e, k))$

converges and is constant as τ ranges over A, and which then outputs this constant value. Let

$$\pi(n) = \max\{q(\sigma, e, k) : \sigma \in h^n, e, k \leq n\}.$$

Next, we describe a $0'$-recursive construction of f. We define a sequence

$$f_0 \preceq f_1 \preceq f_2 \preceq \dots$$

of finite strings in $h^{<\omega}$, and

$$B_0 \subseteq B_1 \subseteq B_2 \subseteq \dots$$

of r.e. subsets of $h^{<\omega}$, such that for each $s \in \omega$, B_s is $h(|f_s|)$-small and $h(|f_s|)$-closed above f_s.

Let $f_0 = \langle\rangle$, and let B_0 be the set of non-DNR strings. Next, we describe how to construct f_{s+1} and B_{s+1} given f_s and B_s.

If $s = 2e$ is even: We ensure that Φ_e^f is not DNR_g. Let $k = h(|f_s|)$ and let $n \geq k, e$ be the least such that $h(n) \geq k(g(\pi(n)) + 1)$. We begin by extending f_s to a string $\sigma \notin B_s$ of length n. Note that B_s is k-small and k-closed above σ. Let $x = q(\sigma, e, k)$, and note that $x \leq \pi(n)$.

Now, if $\varphi_x(x)\downarrow$ to some value i less than $g(x)$, then the set

$$A_i = \{\tau \succeq \sigma : \Phi_e^\tau(x)\downarrow = i\}$$

is k-big above σ, so there is an extension τ of σ, such that $\tau \in A_i \setminus B_s$. Let $f_{s+1} = \tau$ and $B_{s+1} = B_s$. This forces Φ_e^f to not be DNR.

Otherwise, for each $i < g(x)$, A_i is k-small above σ, and so

$$C = \bigcup_{i<g(x)} A_i$$

is $kg(x)$-small above σ, and $C \cup B_s$ is $k(g(x)+1)$-small above σ. Since $h(n) \geq k(g(\pi(n)) + 1) \geq k(g(x) + 1)$, we can let $B_{s+1} = B_s \cup C$ and $f_{s+1} = \sigma$. This forces $\Phi_e^f(x)$ to either diverge, or to converge to value greater than or equal to $g(x)$.

If $s = 2e + 1$ is odd: We ensure that f is low. We begin by extending f_s to a string $\sigma \notin B_s$, such that $h(|\sigma|) \geq 2h(|f_s|)$. If the set

$$F_e = \{\tau \succeq \sigma : \varphi_e^\tau(e)\downarrow\}$$

is $h(|f_s|)$-big above σ, then there is a $\tau \in F_e \setminus B_s$. Let $f_{s+1} = \tau$ and $B_{s+1} = B_s$. This forces e into the jump of f. Otherwise, let $f_{s+1} = \sigma$ and let $B_{s+1} = B_s \cup F_e$, which is $h(|\sigma|)$-small above σ. This forces e out of the jump of f. $\square$

By making g grow slowly enough, we get:

Corollary 7.2: *For every order function h there is an order function g, such that there is a low DNR_h that computes no DNR_g.*

Additionally, we have:

Corollary 7.3: *For every order function g there is an order function h, such that there is a low DNR_h that computes no DNR_g.*

Proof: Using the uniformity in Theorem 7.1 along with the recursion theorem, we construct h knowing its index in advance, thereby obtaining π, and ensuring that $h(n)/g(\pi(n))$ is unbounded. □

By combining the strategies for the two corollaries above, we get:

Corollary 7.4: *For every order function h there is an order function g, such that there is a low $f_1 \in \mathrm{DNR}_g$ that computes no DNR_h, as well as a low $f_2 \in \mathrm{DNR}_h$ that computes no DNR_g.*

Corollary 7.5: *Given any order function h there is an order function g, such that there is a function in SNR_h that computes no function in SNR_g.*

We may assume that $h \geq g$ by replacing h with $\max(g, h)$ if necessary. Thus, $\mathrm{SNR}_g \subset \mathrm{SNR}_h$ and hence $\mathrm{SNR}_g >_w \mathrm{SNR}_h$ as stated above.

Proof: By Theorem 5.3, there is an order function h', such that any $\mathrm{DNR}_{h'}$ computes an SNR_h. By Corollary 7.2, there is an order function g, such that there is a low $\mathrm{DNR}_{h'}$ function f' that computes no DNR_g function. Then f' computes an SNR_h function f that computes no SNR_g function: if j is recursive in f and is an SNR_g function and since it is low, it is itself DNR_g by Corollary 6.8, a contradiction. □

Corollary 7.6: *Given any order function g there is an order function h, such that there is a function in SNR_h that computes no function in SNR_g.*

Proof: By Corollary 7.3, there is an h' and a low $\mathrm{DNR}_{h'}$ function f' that computes no DNR_g function. By Theorem 5.6 there is an h, such that f' computes an SNR_h function f. Then f cannot compute an SNR_g function since the latter would be DNR_g by Corollary 6.8. □

8. Cardinal characteristics and prevalent numberings

As already mentioned in the introduction, Rupprecht [23] explored an analogy between the theory of cardinal characteristics in set theory and highness properties in computability theory. Some of the concepts behind this thesis work was made accessible to a wider audience in Brendle *et al.* [6]. The main results concern a version of Cichoń's diagram [4] in computability theory. Cardinal characteristics in Cichoń's diagram can be expressed as dual pairs:

$$\begin{aligned}\mathfrak{d}(R) &= \min\{|F| : F \subseteq Y \text{ and } \forall x \in X\, \exists y \in F\, (x\, R\, y)\},\\ \mathfrak{b}(R) &= \min\{|G| : G \subseteq X \text{ and } \forall y \in Y\, \exists x \in G\, \neg(x\, R\, y)\},\end{aligned}$$

where X and Y are two spaces and R is a relation on $X \times Y$. As the spaces considered admit a notion of relative computability, it is natural to say that an element $x \in X$ is computable (in $A \subset \omega$).

The computability-theoretic notions corresponding to $\mathfrak{d}(R)$ and $\mathfrak{b}(R)$ are as follows:

$$\begin{aligned}\mathcal{B}(R) &= \{A : \exists y \leq_T A\, \forall x\, (x \text{ is computable } \to x\, R\, y)\},\\ \mathcal{D}(R) &= \{A : \exists x \leq_T A\, \forall y\, (y \text{ is computable } \to \neg(x\, R\, y))\}.\end{aligned}$$

It turns out that $\mathcal{B}(R)$ and $\mathcal{D}(R)$ tend to be highness properties in computability theory, often equivalent to well-known notions.

Cardinal characteristics $\mathfrak{d}(R)$ or $\mathfrak{b}(R)$ are mapped to $\mathcal{D}(R)$ or $\mathcal{B}(R)$ respectively; if $\mathfrak{b}(R) \leq \mathfrak{d}(S)$, then $\mathcal{B}(R) \subseteq \mathcal{D}(S)$ and so on. The analog of SNR is a cardinal characteristic called $\mathfrak{b}(\neq^*)$. There is a result in set theory of Kamo and Osuga [12] roughly corresponding to our SNR hierarchy result.

The computability theoretic analog of Cichoń's diagram is not isomorphic to the original diagram, in that some strict inequalities of cardinal characteristics are consistent with ZFC but the corresponding computability-theoretic notions coincide. While in set theory there are 10 nodes, in computability there are only 7.

Using the cardinal characteristic point of view, we find a notion dual to canonical immunity: a numbering such that no recursive set is large with respect to it.

Definition 8.1: Let $D = (e \mapsto D_e)$ be a numbering of the finite subsets of ω and let $R \subseteq \omega$. We say that D is *h-endemic to R* if there are infinitely many e with $|D_e| \geq h(e)$ and $D_e \subseteq R$. D is a *prevalent numbering* if there is an order function h, such that for all infinite recursive sets R, D is h-endemic to R.

Recall also that an *escaping function* is a function $f : \omega \to \omega$, such that f is not dominated by any recursive function.

Theorem 8.2: *The Muchnik degrees of prevalent numberings and of hyperimmune sets coincide.*

Proof: In one direction, if $e \mapsto \max D_e$ is recursively bounded then we show that D is not a prevalent numbering. Namely, we construct a recursive set R which waits for h to get large ($h(e) > d$) and only then lets its dth element r_d enter R, and lets r_d be large enough (larger than $\max D_k$, $k \leq e$) to prevent $|D_e| \geq h(e)$, $D_e \subseteq R$.

In the other direction, given a hyperimmune set, we (as is well-known) also have an escaping function f. At those inputs e, where f is greater than a function associated with a potential infinite recursive set R_k (given as the graph of a partial recursive $\{0, 1\}$-valued function), we define the next D_e so as to ensure $D_e \subseteq R$ and $|D_e| \geq h(e)$. Namely, the function to escape is the time $e \mapsto t_k(e) = t(\langle e, k\rangle)$ it takes for R_k to get $h(e)$ many elements. If R_k is really infinite then t_k is total and so $f(e) \geq t_k(e)$ for some (infinitely many) e. So we define $D_{\langle e,k\rangle}$ to consist of the first $h(\langle e, k\rangle)$ elements of R_k, if any, as found during a search time of $f(\langle e, k\rangle)$ or even just $f(e)$. □

By Corollary 6.5 and Theorem 8.2, the duality between immunity and prevalent numberings is the same, Muchnik-degree-wise, as that between, in the notation of Brendle *et al.* [6],

- eventually different functions, $\mathcal{B}(\neq^*)$, and
- functions that are infinitely often equal to each recursive function, $\mathcal{D}(\neq^*)$.

Remark 8.3: As the recursive sets are closed under complement, a notion of *bi-prevalent* would be the same as *prevalent.* Thus, while we do not know whether canonical bi-immunity is Muchnik equivalent to canonical immunity, there is no corresponding open problem on the dual side.

References

1. Klaus Ambos-Spies, Bjørn Kjos-Hanssen, Steffen Lempp, and Theodore A. Slaman. Comparing DNR and WWKL. *J. Symbolic Logic*, 69(4):1089–1104, 2004.
2. George Barmpalias, Andrew E. M. Lewis, and Keng Meng Ng. The importance of Π^0_1 classes in effective randomness. *J. Symbolic Logic*, 75(1):387–400, 2010.

3. George Barmpalias, Joseph S. Miller, and André Nies. Randomness notions and partial relativization. *Israel J. Math.*, 191(2):791–816, 2012.
4. Tomek Bartoszyński, Haim Judah, and Saharon Shelah. The Cichoń diagram. *J. Symbolic Logic*, 58(2):401–423, 1993.
5. Achilles A. Beros. A DNC function that computes no effectively bi-immune set. *Arch. Math. Logic*, 54(5-6):521–530, 2015.
6. Jörg Brendle, Andrew Brooke-Taylor, Keng Meng Ng, and André Nies. An analogy between cardinal characteristics and highness properties of oracles. In *Proceedings of the 13th Asian Logic Conference*, pp. 1–28. World Sci. Publ., Hackensack, NJ, 2015.
7. Douglas Cenzer and Peter G. Hinman. Degrees of difficulty of generalized r.e. separating classes. *Arch. Math. Logic*, 46(7-8):629–647, 2008.
8. Noam Greenberg and Joseph S. Miller. Lowness for Kurtz randomness. *J. Symbolic Logic*, 74(2):665–678, 2009.
9. Noam Greenberg and Joseph S. Miller. Diagonally non-recursive functions and effective Hausdorff dimension. *Bull. Lond. Math. Soc.*, 43(4):636–654, 2011.
10. Carl G. Jockusch, Jr. Degrees of functions with no fixed points. In *Logic, Methodology and Philosophy of Science, VIII (Moscow, 1987)*, vol. 126 of *Stud. Logic Found. Math.*, pp. 191–201. North-Holland, Amsterdam, 1989.
11. Carl G. Jockusch, Jr. and Andrew E. M. Lewis. Diagonally non-computable functions and bi-immunity. *J. Symbolic Logic*, 78(3):977–988, 2013.
12. Shizuo Kamo and Noboru Osuga. Many different covering numbers of Yorioka's ideals. *Archive for Mathematical Logic*, 53(1-2):43–56, 2014.
13. Alexander S. Kechris. *Classical descriptive set theory*, vol. 156 of *Graduate Texts in Mathematics*. Springer-Verlag, New York, 1995.
14. Mushfeq Khan. *Some results on algorithmic randomness and computability-theoretic strength*. ProQuest LLC, Ann Arbor, MI, 2014. Thesis (Ph.D.)–The University of Wisconsin - Madison.
15. Mushfeq Khan and Joseph S. Miller. Forcing with bushy trees. *Bull. Symb. Log.*, 23(2):160–180, 2017.
16. Bjørn Kjos-Hanssen, Wolfgang Merkle, and Frank Stephan. Kolmogorov complexity and the recursion theorem. *Trans. Amer. Math. Soc.*, 363(10):5465–5480, 2011.
17. Masahiro Kumabe and Andrew E. M. Lewis. A fixed-point-free minimal degree. *J. Lond. Math. Soc. (2)*, 80(3):785–797, 2009.
18. Donald A. Martin. Completeness, the recursion theorem, and effectively simple sets. *Proceedings of the American Mathematical Society*, 17(4):838–842, 1966.
19. Benoit Monin and André Nies. Muchnik degrees and cardinal characteristics. *The Journal of Symbolic Logic* 86.2 (2021): 471–498.
20. André Nies, Frank Stephan, and Sebastiaan A. Terwijn. Randomness, relativization and Turing degrees. *J. Symbolic Logic*, 70(2):515–535, 2005.
21. André (ditto Otti) Nies. *Computability and Randomness*, vol. 51 of *Oxford Logic Guides*. Oxford University Press, Oxford, 2009. 444 pages. Paperback version 2011.

22. Emil L. Post. Recursively enumerable sets of positive integers and their decision problems. *Bull. Amer. Math. Soc.*, 50:284–316, 1944.
23. Nicholas Rupprecht. Relativized Schnorr tests with universal behavior. *Arch. Math. Logic*, 49(5):555–570, 2010.
24. Gerald E. Sacks. A simple set which is not effectively simple. *Proceedings of the American Mathematical Society*, 15(1):51–55, 1964.
25. Stephen G. Simpson. Turing degrees and Muchnik degrees of recursively bounded DNR functions. In *Computability and Complexity*, vol. 10010 of *Lecture Notes in Comput. Sci.*, pp. 660–668. Springer, Cham, 2017.
26. Raymond M. Smullyan. Effectively simple sets. *Proceedings of the American Mathematical Society*, 15(6):893–895, 1964.
27. Robert I. Soare. *Recursively enumerable sets and degrees.* Perspectives in Mathematical Logic. Springer-Verlag, Berlin, 1987. A study of computable functions and computably generated sets.
28. Frank Stephan. Martin-Löf random and PA-complete sets. In *Logic Colloquium '02*, vol. 27 of *Lect. Notes Log.*, pp. 342–348. Assoc. Symbol. Logic, La Jolla, CA, 2006.

https://doi.org/10.1142/9789811278631_0006

Computability and Categoricity of Weakly Homogeneous Boolean Algebras and p-Groups

Douglas Cenzer

Department of Mathematics, University of Florida
P.O. Box 118105, Gainesville, Florida 32611
cenzer@math.ufl.edu

Francis Adams

Oxford College at Emory University, Atlanta, Georgia
fsadams@emory.edu

Keng Meng Ng

School of Physical and Mathematical Sciences
Nanyang Technical University, Singapore
kmng@ntu.edu.sg

This chapter continues the study of weakly homogeneous structures. It is shown that a countable Boolean algebra is weakly homogeneous if and only if it has finitely many atoms. Hence every countable weakly homogeneous Boolean algebra has a computable copy, and a computable Boolean algebra is weakly homogeneous if and only if it is computably categorical. We also characterize countable weakly homogeneous Boolean algebras in various signatures. The countable weakly homogeneous abelian p-groups are characterized, and it is shown that every such group has a computable copy.

Contents

1. Introduction

A structure $\mathcal{A}$ is said to be *ultrahomogeneous*, or simply *homogeneous*, if any isomorphism between finitely generated substructures extends to an automorphism of $\mathcal{A}$. Homogeneous structures were first studied by Fraïssé [7], who defined the *age* of a structure to be the family of finitely generated substructures of $\mathcal{A}$ and gave properties which characterized the age of a homogeneous structure. In [1], we defined the notion of *weakly homogeneous* structures, where $\mathcal{A}$ is weakly homogeneous if there is a finite (*exceptional*) set of elements $a_1, \ldots, a_n$ of $\mathcal{A}$, such that any isomorphism between finitely generated substructures $\mathcal{A}$, which maps each a_i to itself may be extended to an automorphism of the structure.

Here are some well-known examples of countable homogeneous structures. See [14, 18] for more details. The linear ordering $(\mathbb{Q}, <)$ of the rationals is the unique ultrahomogeneous countable linear ordering. The age here is just the set of all finite linear orderings. An equivalence structure (A, E) is ultrahomogeneous if and only if all equivalence classes have the same size k, $1 \leq k \leq \aleph_0$; then the age is the set of all finite equivalence structures with all classes of size $\leq k$.

This notion turns out to have many connections with computability. For example, Csima, Harizanov, R. Miller and A. Montalban [5] studied computable ages and the computability of the canonical homogeneous structures, called *Fraïssé limits*.

Here we are concerned with countably infinite structures. We say a countable structure (model) $\mathcal{A}$ is computable if its universe A is computable and all of its functions and relations are uniformly computable. Given two computable structures, we will say they are computably isomorphic if there exists an isomorphism between them that is computable. For a single computable structure $\mathcal{A}$, we will say $\mathcal{A}$ is computably categorical if every computable structure isomorphic to $\mathcal{A}$ is in fact computably isomorphic to $\mathcal{A}$. More generally, we will say two computable structures are Δ^0_α isomorphic if there exists an isomorphism between them that is Δ^0_α, and we will say a computable structure $\mathcal{A}$ is Δ^0_α categorical if every computable structure isomorphic to $\mathcal{A}$ is Δ^0_α isomorphic to $\mathcal{A}$, and will say that a computable structure $\mathcal{A}$ is *relatively* Δ^0_α categorical if, for every structure $\mathcal{B}$ (which is not necessarily computable) isomorphic to $\mathcal{A}$, there is an isomorphism which is Δ^0_α relative to the diagram of $\mathcal{B}$. More generally, an arbitrary structure $\mathcal{A}$ is relatively Δ^0_α categorical, if for any structure $\mathcal{B}$ isomorphic to $\mathcal{A}$, there is an isomorphism which is Δ^0_α relative to the diagrams of $\mathcal{A}$ and $\mathcal{B}$. For computable structures, the last two notions agree.

The computability and categoricity of homogeneous structures was explored by the authors in a recent paper [1]. We introduced the notion of weakly ultrahomogeneous (weakly homogeneous) structures and examined several types of countable homogeneous and weakly homogeneous structures. A structure $\mathcal{A}$ is said to be *relatively computably categorical* if, for any structure $\mathcal{B}$ isomorphic to $\mathcal{A}$, there is an isomorphism which is computable from the diagrams of $\mathcal{A}$ and $\mathcal{B}$. Similarly, $\mathcal{A}$ is said to be *relatively* Δ^0_n *categorical* if,for any structure $\mathcal{B}$ isomorphic to $\mathcal{A}$, there is an isomorphism which is Δ^0_n-computable from the diagrams of $\mathcal{A}$ and $\mathcal{B}$. A structure is said to be *locally finite* if every finite set generates a finite substructure (for example, in a pure relational language). Here are the key results from [1], which we will need for this chapter.

Theorem 1.1:

(1) If a structure $\mathcal{A}$ is weakly homogeneous, then it is relatively Δ^0_2 categorical.
(2) If a structure $\mathcal{A}$ is locally finite and weakly homogeneous, then it is relatively computably categorical.

We then looked at specific classes of structures. Every countable homogeneous linear ordering and also every countable homogeneous equivalence structure has a computable copy. For computable linear orders and computable equivalence structures, the weakly homogeneous structures are exactly the computably categorical structures. For computable injection structures, we observed that there are continuum many homogeneous structures with no computable copy, and there are homogeneous structures which are not computably categorical, although every computably categorical structure is weakly homogeneous. There are continuum many homogeneous trees (T, f) under the predecessor function and they are all are relatively computably categorical. We made a connection between trees under the predecessor function and nested equivalence structures. For these structures, index sets were used to determine the complexity of the property of being homogeneous, and the complexity of the property of being weakly homogeneous.

Characterizations of exceptional sets were given for homogeneous linear orders, equivalence structures, injection structures, and trees as partial orders.

In this chapter, we study weakly homogeneous abelian p-groups and Boolean algebras.

Section 2 contains our results on abelian p-groups. It was shown by Cherlin and Felgner [4] that an abelian p-group is homogeneous if and only if is either divisible or homocylic. Here is the main theorem of this chapter.

Theorem 1.2: *An infinite countable abelian p-group $\mathcal{G}$ is weakly homogeneous if and only if it has one of the following forms:*

(1) $\mathcal{G} = \bigoplus_\alpha \mathbb{Z}(p^\infty)$ *for some* $\alpha \leq \omega$.
(2) $\mathcal{G} = \bigoplus_{i<\omega} \mathbb{Z}(p^n) \oplus \mathcal{F}$, *where n is finite and $\mathcal{F}$ is a finite product of finite cyclic groups each having order* $\geq n$.

It follows that every countable weakly homogenous abelian p-group has a computable copy. It also follows that every countable weakly homogeneous abelian p-group is computably categorical. Obviously, the converse does not hold.

Section 3 contains our results on Boolean algebras. We consider the standard language $\{\wedge, \vee, \neg, 0, 1\}$ for Boolean algebras as well as subsets of this language. It is well-known that, in the standard language, a Boolean algebra is homogeneous if and only if it is atomless. Here is the second main result.

Theorem 1.3: *A countable Boolean algebra $\mathcal{B}$ is weakly homogeneous if and only if it has finitely many atoms.*

It follows that every countable weakly homogeneous Boolean algebra has a computable copy. It also follows that from the results of Goncharov [9] and LaRoche [15], that a computable Boolean algebra is weakly homogeneous if and only if it is computably categorical. If we omit the complement and just use the language $\{\wedge, \vee, 0, 1\}$ of lattices, then $\mathcal{B}$ is homogeneous if and only if it is either atomless or has at most 4 elements, and $\mathcal{B}$ is weakly homogeneous if and only if it has finitely many atoms. If we view a Boolean algebra as a partial order with the language $\{\leq, 0, 1\}$, then $\mathcal{B}$ is homogeneous if and only if it is finite with at most 4 elements, and $\mathcal{B}$ is weakly homogeneous if and only if it is finite.

2. Abelian p-groups

In this section, we will characterize the weakly homogeneous abelian p-groups, and more generally the torsion abelian groups. The computability and categoricity of countable abelian p-groups have been studied for many years.

The computably categorical abelian p-groups were characterized by Goncharov [10] and Smith [22]. We present a slightly stronger formulation of the results as follows. Let us say that $\mathcal{G}$ is *strongly relatively computably categorical* if, for any group $\mathcal{H}$ isomorphic to $\mathcal{G}$, there is an isomorphism which is computable from the diagrams of $\mathcal{G}$ and $\mathcal{H}$.

Theorem 2.1: *(Goncharov, Smith) Let $\mathcal{G}$ be an abelian p-group $\mathcal{G}$ is strongly relatively computably categorical if and only if either*

(1) $\mathcal{G} \cong \bigoplus_\alpha \mathbb{Z}(p^\infty) \oplus \mathcal{F}$, where $\alpha \leq \omega$, or
(2) $\mathcal{G} \cong \bigoplus_r \mathbb{Z}(p^\infty) \oplus \bigoplus_\omega \mathbb{Z}(p^m) \oplus \mathcal{F}$, where $\mathcal{F}$ is a finite abelian p-group and $r, m \in \omega$.

Note that the decomposition of $\mathcal{G}$ as a product need not be effective (see Khisamiev [13]). We will refer to $\mathbb{Z}(p^\infty)$ as the quasicyclic p-group, as opposed to the cyclic p-groups $\mathbb{Z}(p^n)$.

For any group of the form

$$\mathcal{G} = \bigoplus_{i<\omega} \mathbb{Z}(p^{n_i}) \oplus \bigoplus_\alpha \mathbb{Z}(p^\infty),$$

we define the *character* of $\mathcal{G}$ to be

$$\chi(\mathcal{A}) = \{(n,k) : card(\{i : n_i = n\}) \geq k\}.$$

We say that $\mathcal{G}$ has *bounded character* if for some finite b and all $(n,k) \in \chi(\mathcal{G})$, $n \leq b$, and is said to have *unbounded character* otherwise.

The relatively Δ^0_2 categorical abelian p-groups were characterized by Calvert, Cenzer, Harizanov and Morozov [2] as follows.

Theorem 2.2: *(CCHM) A computable abelian p-group $\mathcal{G} = \bigoplus_{i<\omega} \mathbb{Z}(p^{n_i}) \oplus \bigoplus_\alpha \mathbb{Z}(p^\infty)$ for some $\alpha \leq \omega$ is relatively Δ^0_2 categorical if and only if either*

(1) $\alpha = 0$, that is, every element of $\mathcal{G}$ has finite height, or
(2) $\mathcal{G}$ has bounded character.

The first step in our characterization is to show that every weakly homogeneous abelian p-group is of the form given by Theorem 2.1.

Theorem 2.3: *Let $\mathcal{G}$ be any weakly homogeneous abelian p-group. Then either*

(1) $\mathcal{G} \cong \bigoplus_\alpha \mathbb{Z}(p^\infty) \oplus \mathcal{F}$, where $\alpha \leq \omega$, or
(2) $\mathcal{G} \cong \bigoplus_r \mathbb{Z}(p^\infty) \oplus \bigoplus_\omega \mathbb{Z}(p^m) \oplus \mathcal{F}$, where $\mathcal{F}$ is a finite abelian p-group and $r, m \in \omega$.

Proof: It follows from Theorem 1.1 that every weakly homogeneous abelian p-group $\mathcal{G}$ satisfies the hypothesis of Theorem 2.1 and hence must have one of the forms described above in Theorem 2.1. $\square$

Corollary 2.4: *For any countable weakly homogeneous abelian p-group $\mathcal{G}$, there is a computable group isomorphic to $\mathcal{G}$.*

We will show that, in fact, the weakly homogeneous abelian p-groups are a proper subset of the relatively computably categorical abelian p-groups.

It is easy to see that a countable homogeneous abelian p-group must either be divisible, that is, of the form $\oplus_\alpha \mathbb{Z}(p^\infty)$ for some $\alpha \leq \omega$, or it must be *homocyclic*, that is, of the form $\bigoplus_\omega \mathbb{Z}(p^n)$, for some fixed n. (See Cherlin and Felgner [4].) Thus every countable homogeneous abelian p-group has a computable copy, and any such copy is computably categorical.

Note, in particular, that the finite group $\mathbb{Z}(p^m) \oplus \mathbb{Z}(p^n)$ is not homogeneous for any $m \neq n$. To see this, fix $m < n$, and observe that both $(1, 0)$ and $(0, p^{n-m})$ have order p^m, so there is an isomorphism between the corresponding cyclic subgroups mapping $(1, 0)$ to $(0, p^{n-m})$. But this mapping cannot be extended to an automorphism since $(0, p^{n-m})$ is divisible by p but $(1, 0)$ is not.

We want to give another way to do this, which will prove to be more useful. That is, the elements $a = (0, p)$ and $b = (1, p)$ both have order p^{n-1}, so there is an isomorphism between the corresponding cyclic subgroups $\langle a \rangle$ and $\langle b \rangle$ mapping $(0, jp)$ to (j, jp). This cannot be extended to an automorphism since $(0, p)$ is divisible by p but $(1, p)$ is not. However, it is important to note that it can be extended to an automorphism of $\mathbb{Z}(p^m) \oplus \langle p \rangle$, by mapping $(1, 0)$ to $(1, 0)$ and hence mapping (i, jp) to $(i + j, jp)$. It can be checked that this mapping is well-defined and one-to-one, using the fact that, for any j, $jp = 0(mod p^n)$ implies $j = 0(mod p^{n-1})$, which implies that $j = 0(mod p^m)$. Thus for $m < n$, there is a subgroup $\mathcal{A}$ of $\mathbb{Z}(p^m) \oplus \mathbb{Z}(p^n)$, including $\mathbb{Z}(p^m) \oplus 0$ and an isomorphism $f : \mathcal{A} \to \mathcal{A}$, which fixes $\mathbb{Z}(p^m) \oplus 0$ but which cannot be extended to an automorphism. This idea will be used below to show that certain infinite products of cyclic p-groups are not weakly homogeneous.

On the other hand, it is important to note that the roles of $\mathbb{Z}(p^m)$ and $\mathbb{Z}(p^n)$ cannot be reversed here.

We need a bit of terminology. Any element a of a product $\mathcal{G} = \bigoplus_n \mathcal{G}_n$ is an infinite sequence $a = (x_0, x_1, \dots)$, where all but finitely many values x_i are zero. The *support* $Supp(a)$ of a is the finite set $\{n : x_n \neq 0\}$. The support of a finite subset S of $\mathcal{G}$ is $Supp(S) = \cup_{a \in S} Supp(a)$.

Proposition 2.5: *For any $m < n$, $\mathbb{Z}(p^m) \oplus \bigoplus_\omega \mathbb{Z}(p^n)$ is not weakly homogeneous.*

Proof: Let $\mathcal{G} = \mathbb{Z}(p^m) \oplus \bigoplus_\omega \mathbb{Z}(p^n)$. Elements of $\mathcal{G}$ will be infinite sequences $(x_0, x_1, \dots)$, where $x_0 \in \mathbb{Z}(p^m)$ and each x_{i+1} comes from the ith copy of $\mathbb{Z}(p^n)$. Suppose that S is a finite exceptional set. By expanding to a larger (hence still exceptional) set, we may assume that the element $(1, 0, \dots)$ from the $\mathbb{Z}(p^m)$ component is in S and that S is a subgroup of $\mathcal{G}$. Since S is finite, there is some i such that $i+1 \notin \mathit{Supp}(S)$; without loss of generality, we may assume that $1 \notin \mathit{Supp}(S)$. Thus, for any $(x_0, x_1, \dots) \in S$, $x_1 = 0$. As in the discussion above, we want to map $a = (0, p, 0, 0, \dots)$ to $b = (1, p, 0, 0, \dots)$, since both have order p^{n-1} and a is divisible by p whereas b is not, so the map would have no extension to an automorphism of $\mathcal{G}$. This map needs to be extended to include S. Let $k = \max \mathit{Supp}(S)$, so that for any $a = (x_0, x_1, \dots) \in S$, and any $i > k$, we have $x_i = 0$. Now define the finite subgroup $\mathcal{A} = \{(x_0, x_1, \dots) \in \mathcal{G} : p \mid x_1 \ \& \ (\forall i > k) x_i = 0\}$ and define the automorphism f of $\mathcal{A}$ by $f(i, jp, x_2, x_3, \dots)$ to $(i + j, jp, x_2, x_3, \dots)$. Since any element $a = (x_0, x_1, \dots)$ of S has $x_1 = 0$, it follows that $f(a) = a$ for any element of S, as desired. Since S was arbitrary, it follows that $\mathcal{G}$ is not weakly homogeneous. □

Next we want to consider products of the form $\mathbb{Z}(p^\infty) \oplus \mathbb{Z}(p^m)$. This is more difficult because we cannot factor out components as we could in the cases above. The following lemma refines the mappings used in the previous arguments, and will be needed.

Lemma 2.6: *For any prime p and any m, n, r with $n = 2m + r + 1$, there is an automorphism ϕ of a subgroup $\mathcal{A}$ of $\mathcal{G} = \mathbb{Z}(p^m) \oplus \mathbb{Z}(p^n)$, which fixes every element of order $\leq p^{m+r}$ (hence fixes all elements of $\mathbb{Z}(p^m) \oplus 0$ and also $0 \oplus \langle p^{m+1} \rangle$), but which cannot be extended to an automorphism*

Proof: Fix p, m, n, r as above and let $\mathcal{A} = \mathbb{Z}(p^m) \oplus p\mathbb{Z}(p^n)$. Define the isomorphism ϕ from $\mathcal{A}$ to itself, as discussed above, by $\phi(i, jp) = (i + j, jp)$. Now suppose that (i, jp) has order $\leq p^{m+r}$. Then $p^{m+r} \cdot jp = 0(mod p^{2m+r+1})$, so that $j = 0(mod p^m)$ and hence $i + j = i(mod p^m)$. This shows that elements of order $\leq p^{m+r}$ are fixed by f. As we have seen, ϕ cannot be extended to an automorphism of $\mathcal{G}$ since it maps the divisible element $(0, p)$ to the element $(1, p)$, which is not divisible. □

This leads to the following somewhat surprising result.

Theorem 2.7: *For any finite m, and any s and t with $1 \leq s, t \leq \omega$, the group $\mathcal{G} = \bigoplus_s \mathbb{Z}(p^\infty) \oplus \bigoplus_t \mathbb{Z}(p^m)$ is not weakly homogeneous.*

Proof: Suppose, by way of contradiction, that $\bigoplus_s \mathbb{Z}(p^\infty) \oplus \bigoplus_t \mathbb{Z}(p^m)$ is weakly homogeneous and let S be an exceptional set. Choose r large enough so that every element of S has order $\leq p^{m+r}$ and let $n = 2m + r + 1$. Now let a be an element of any $\mathbb{Z}(p^\infty)$ component with order p^n, and let b be an element of any $\mathbb{Z}(p^m)$ component with order p^m. The subgroup $\mathcal{B} = \langle a \rangle \oplus \langle b \rangle$ is clearly isomorphic to $\mathbb{Z}(p^n) \oplus \mathbb{Z}(p^m)$. It follows from Lemma 2.6 that there is an automorphism ϕ of $\mathcal{A} = \mathbb{Z}(p^m) \oplus \langle pa \rangle$, which cannot be extended to an automorphism of $\mathcal{B}$. Then ϕ also cannot be extended to an automorphism of $\mathcal{G}$. The simple argument here is that ϕ maps an element $(0, pa)$, which is divisible by p to an element $(1, pa)$, which is not divisible by p. We also observe that S is contained in $\mathcal{B}$ as all elements of S have order $\leq p^{m+r} < p^n$. Thus, ϕ must fix all elements of S. □

This can be extended using the following lemma.

Lemma 2.8: *If the abelian p-group $\mathcal{G} = \mathcal{G}_0 \oplus \mathcal{G}_1$ is weakly homogeneous, and $\mathcal{G}_1$ is finite, then $\mathcal{G}_0$ is weakly homogeneous.*

Proof: Let $C = \{(a_1, b_1), \ldots, (a_n, b_n)\}$ be an exceptional set for $\mathcal{G}$; we may assume that $0 \oplus \mathcal{G}_1$ is a subgroup of C and that C is a subgroup of $\mathcal{G}$. Let $A = \{a_1, \ldots, a_n\}$, which will be a subgroup of $\mathcal{G}_0$. Now suppose that ϕ_0 is an isomorphism from a subgroup $\mathcal{F}_0$ of $\mathcal{G}_0$ to a subgroup $\mathcal{H}_0$ of $\mathcal{G}_0$, where $A \subseteq \mathcal{F}_0 \cap \mathcal{H}_0$ and $\phi_0(a_i) = a_i$ for $i = 1, \ldots, n$. We can extend this to an isomorphism ϕ from $\mathcal{F}_0 \oplus \mathcal{G}_1$ to $\mathcal{H}_0 \oplus \mathcal{G}_1$ by letting $\phi(x, y) = (\phi_0(x), y)$. Then $\phi(a_i, b_i) = (a_i, b_i)$ for $i = 1, \ldots, n$. Since C is an exceptional set for $\mathcal{G}$, it follows that there is an automorphism ψ of $\mathcal{G}_0 \oplus \mathcal{G}_1$ extending ϕ. Then the mapping $\psi_0 : \mathcal{G}_0 \to \mathcal{G}_0$ defined by $\psi_0(x) = \psi(x, 0)$ is an automorphism of $\mathcal{G}_0$ extending ϕ_0. □

Corollary 2.9: *For any nonzero finite group $\mathcal{F}$ and any α with $0 < \alpha \leq \omega$, $\bigoplus_\alpha \mathbb{Z}(p^\infty) \oplus \mathcal{F}$ is not weakly homogeneous.*

To complete our characterization of the weakly homogeneous abelian p-groups, we need the following extension lemma.

Lemma 2.10: *Suppose that $\mathcal{A}, \mathcal{B}, C$ are abelian p-groups, that $\mathcal{B} \subseteq C$ and that f is an automorphism of $\mathcal{A} \oplus \mathcal{B}$ with $f(0, b) = (0, b)$ for all $b \in \mathcal{B}$.*

Then the map f may be extended to an automorphism g of $\mathcal{A} \oplus \mathcal{C}$, such that $g(0,c) = (0,c)$ for all $c \in \mathcal{C}$.

Proof: We just let $g(a,c) = f(a,0) + (0,c)$. This is clearly a homomorphism from $\mathcal{A} \oplus \mathcal{C}$ to itself. It agrees with f on $\mathcal{A} \oplus \mathcal{B}$, since $g(a,b) = f(a,0) + (0,b) = f(a,0) + f(0,b) = f(a,b)$. To see that g is one-to-one, suppose that $g(a,c) = (0,0)$ and let $f(a,0) = (a_1, b_1)$. Then $(a_1, b_1 + c) = (0,0)$. It follows that $a_1 = 0$ and $b_1 + c = 0$. But this means that $c \in \mathcal{B}$ and therefore $(0,0) = g(a,c) = f(a,0) + (0,c) = f(a,c)$. Since f is one-to-one, it follows that $a = c = 0$. To check that g is onto, let $(a_1, c_1) \in \mathcal{A} \oplus \mathcal{C}$. Since f is onto, there exist $a \in \mathcal{A}$ and $b \in \mathcal{B}$ such that $f(a,b) = (a_1, 0)$. Then $g(a, b + c_1) = g(a,b) + g(0,c_1) = f(a,b) + g(0,c_1) = (a_1, 0) + (0, c_1) = (a_1, c_1)$. □

Theorem 2.11: *For any n and any finite product $\mathcal{K}$ of cyclic p-groups each having order $> p^n$, $\mathcal{G} = \bigoplus_{i<\omega} \mathbb{Z}(p^n) \oplus \mathcal{K}$ is weakly homogeneous.*

Proof: We let $S = 0 \oplus \mathcal{K}$ be our exceptional set. Let $\mathcal{G}_1$ and $\mathcal{G}_2$ be finite subgroups of $\mathcal{G}$ each including S and let $h : \mathcal{G}_1 \to \mathcal{G}_2$ be an isomorphism with $h(0,k) = (0,k)$ for all $k \in \mathcal{K}$. Our goal is to show that h can be extended to an automorphism F of $\mathcal{G}$ which fixes S.

We can write $\mathcal{G}_i = \mathcal{A}_i \oplus \mathcal{K}$, where $\mathcal{A}_i = \{a \in \bigoplus_{i<\omega} \mathbb{Z}(p^n) : (a,0) \in \mathcal{G}_i\}$, for $i = 1,2$. Let $\mathcal{K} = \mathbb{Z}(p^{n_1}) \oplus \cdots \oplus \mathbb{Z}(p^{n_k})$, where each $n_j > n$. Consider the subgroup $\mathcal{K}^- = p^{n_1 - n}\mathbb{Z}(p^{n_1}) \oplus p^{n_2 - n}\mathbb{Z}(p^{n_2}) \oplus \cdots \oplus p^{n_k - n}\mathbb{Z}(p^{n_k})$ of $\mathcal{K}$. $\mathcal{K}^-$ is isomorphic to k copies of $\mathbb{Z}(p^n)$ and it consists exactly of the elements of $\mathcal{K}$ of order $\leq p^n$. Thus $\mathcal{A}_1 \oplus \mathcal{K}^-$ is the set of elements of $\mathcal{G}_1$ of order $\leq p^n$ and must be mapped by h to $\mathcal{A}_2 \oplus \mathcal{K}^-$, the elements of $\mathcal{A}_2 \oplus \mathcal{K}$ of order $\leq p^n$. Now $\mathcal{G}^- = \bigoplus_{i<\omega} \mathbb{Z}(p^n) \oplus \mathcal{K}^-$ is isomorphic to $\bigoplus_{i<\omega} \mathbb{Z}(p^n)$ and is therefore homogeneous. It follows that there is an automorphism f of $\mathcal{G}^-$ which extends the restriction of h to $\mathcal{A}_1 \oplus \mathcal{K}^-$, and thus $f(0,k) = (0,k)$ for all $k \in \mathcal{K}^-$. Finally, by Lemma 2.10, f may be extended to an automorphism of $\mathcal{G}$, say F, such that $F(0,k) = (0,k)$ for all $k \in \mathcal{K}$. In that case, it is easy to see that F extends h. □

We are now ready to state the characterization of weakly homogeneous abelian p-groups:

Theorem 2.12: *An infinite countable abelian p-group $\mathcal{G}$ is weakly homogeneous if and only if it has one of the following forms:*

(1) $\mathcal{G} = \bigoplus_{\alpha} \mathbb{Z}(p^\infty)$ for some $\alpha \leq \omega$.

(2) $\mathcal{G} = \bigoplus_{i<\omega} \mathbb{Z}(p^n) \oplus \mathcal{F}$*, where* n *is finite and* $\mathcal{F}$ *is a finite product of cyclic p-groups each having order* $> p^n$*.*

In the second case, an exceptional set may be given to contain a generator for each factor of $\mathcal{F}$*.*

Proof: First, suppose that $\mathcal{G}$ is weakly homogeneous. Then, it has one of the two forms specified by Theorem 2.3. If $\mathcal{G}$ has the first such form, then by Corollary 3.3 it must have form (1). If $\mathcal{G}$ has the second form from Theorem 2.3, then by Theorem 2.7 and Lemma 2.8 it must have no quasicyclic component, and then it follows from Proposition 2.5 and Lemma 2.8 that $\mathcal{G}$ must have form (2).

Next suppose that $\mathcal{G}$ has the prescribed form. If $\mathcal{G}$ is of form (1), then $\mathcal{G}$ is in fact homogeneous. If $\mathcal{G}$ is of form (2), then $\mathcal{G}$ is weakly homogeneous by Theorem 2.11. □

On the other hand, it is clear from Theorem 2.1 of Goncharov and Smith that not every computably categorical abelian p-group is weakly homogeneous.

We point out a similar, and similarly named, concept which nevertheless differs from our definition of weakly homogeneous. Melnikov and Ng [17] defined an abelian p-group to have the weak homogeneity property (WHP) if it is either divisible, or for each non-zero element a of order p and finite height, there exist at most finitely many elements with order p and height greater than that of a. They prove that this is equivalent to an abelian p-group being of the form $U \oplus H$, where U is any finite direct sum of cyclic and quasi-cyclic p-groups, H is a direct power of some fixed cyclic or quasi-cyclic p-group $\mathbb{Z}(p^\lambda)$, and the least α such that $\mathbb{Z}(p^\alpha)$ occurs in U (if there are any) is at least λ. So every abelian p-group with the WHP is (relatively) computably categorical. It follows from Theorem 2.12 that any weakly homogeneous p-group has the weak homogeneity property. On the other hand, the group $\mathbb{Z}(p^\infty) \oplus \mathbb{Z}(p)$ has the WHP but is not weakly homogeneous. Melnikov and Ng use the WHP property to show that the notion of computable categoricity for computable torsion abelian groups is Π^0_4 complete.

2.1. *Torsion abelian groups*

Finally, we want to briefly consider arbitrary torsion abelian groups. Any such group $\mathcal{G}$ may be represented as a direct sum $\bigoplus_p \mathcal{G}_p$, where $\mathcal{G}_p$ is the subgroup of $\mathcal{G}$ of elements having order p^n for some n.

Theorem 2.13: *For any torsion abelian group $\mathcal{G}$:*

(1) $\mathcal{G}$ is homogeneous if and only if $\mathcal{G}_p$ is homogeneous for each prime p;

(2) $\mathcal{G}$ is weakly homogeneous if and only if $\mathcal{G}_p$ is weakly homogeneous for each prime p and $\mathcal{G}_p$ is homogeneous for all but finitely many primes.

Proof: (1) Assume that $\mathcal{G}$ is homogeneous. Fix p and let ϕ be an isomorphism between two finitely generated subgroups $\mathcal{A}$ and $\mathcal{B}$ of $\mathcal{G}_p$. Then $\mathcal{A}$ and $\mathcal{B}$ are also subgroups of $\mathcal{G}$ and therefore ϕ may be extended to an automorphism ψ of $\mathcal{G}$. Then the restriction of ψ to $\mathcal{G}_p$ is clearly an automorphism of $\mathcal{G}_p$ which extends ϕ.

Next, assume that each $\mathcal{G}_p$ is homogeneous. Suppose that $\mathcal{A}$ and $\mathcal{B}$ are finitely generated subgroups of $\mathcal{G}$ and that $\phi : \mathcal{A} \to \mathcal{B}$ is an isomorphism. Then $\mathcal{A}$ and $\mathcal{B}$ are finite and there is some finite set P of primes such that $\mathcal{A} = \bigoplus_{p\in P} \mathcal{A}_p$ and $\mathcal{B} = \bigoplus_{p\in P} \mathcal{B}_p$. It follows that ϕ induces isomorphisms $\phi_p : \mathcal{A}_p \to \mathcal{B}_p$ for each $p \in P$. Now, for each $p \in P$, $\mathcal{A}_p$ and $\mathcal{B}_p$ are finitely generated subgroups of G_p. Since each $\mathcal{G}_p$ is homogeneous, it follows that ϕ_p may be extended to an automorphism ψ_p of $\mathcal{G}_p$. Now we can define an automorphism ψ of $\mathcal{G}$ extending ϕ by letting ψ be the identity on each $\mathcal{G}_p$ for $p \notin P$.

(2) Assume that $\mathcal{G}$ is weakly homogeneous with exceptional finite subgroup $\mathcal{H} = \bigoplus_p \mathcal{H}_p$. Fix p and let ϕ be an isomorphism between two finitely generated subgroups $\mathcal{A}$ and $\mathcal{B}$ of $\mathcal{G}_p$, which fixes each element of $\mathcal{H}_p$. Now extend ϕ to an isomorphism of $\mathcal{A} \oplus \bigoplus_{q\neq p} \mathcal{H}_q \mapsto \mathcal{B} \oplus \bigoplus_{q\neq p} \mathcal{H}_q$ by fixing each element of $\mathcal{H}_q$ for all $q \neq p$. This may now be extended to an automorphism ψ of $\mathcal{G}$ and the restriction of ψ to $\mathcal{G}_p$ will be an automorphism which extends ϕ and fixes $\mathcal{H}_p$. For all but finitely many primes p, $\mathcal{H}_p = \emptyset$, and it follows that for those p, $\mathcal{G}_p$ is homogeneous.

For the other direction, suppose that for some finite set P of primes, $\mathcal{G}_p$ is weakly homogeneous with exceptional subgroup $\mathcal{H}_p$ and that $\mathcal{G}_p$ is homogeneous for all other primes p. It is easy to see that $\mathcal{G}$ will be weakly homogeneous with exceptional subgroup $\bigoplus_{p\in P} \mathcal{H}_p$. □

The condition that $\mathcal{G}_p$ is homogeneous for almost every component p cannot be dropped in part (2) of Theorem 2.13, as the following example shows.

Example. Let $\mathcal{G} = \bigoplus_p (\mathbb{Z}(p) \oplus \mathbb{Z}(p^2))$. Then $\mathcal{G}_p = (\mathbb{Z}(p) \oplus \mathbb{Z}(p^2))$ is a finite group and is weakly homogeneous but not homogeneous for every p. We check that $\mathcal{G}$ is not weakly homogeneous. Let $\mathcal{H}$ be any potential finite exceptional subgroup of $\mathcal{G}$. Then for some finite set P of primes, $\mathcal{H} \subseteq \bigoplus_{p\in P} (\mathbb{Z}(p) \oplus \mathbb{Z}(p^2))$. Now let q be any prime not in P and consider the subgroups $\mathcal{A} = \mathcal{H} \oplus \mathbb{Z}(q) \oplus 0$ and $\mathcal{B} = \mathcal{H} \oplus 0 \oplus q\mathbb{Z}(q^2)$ of $\bigoplus_{p\in P\cup\{q\}} (\mathbb{Z}(p) \oplus \mathbb{Z}(p^2))$ and consider the isomorphism from $\mathcal{A} \mapsto \mathcal{B}$ fixing each element of $\mathcal{H}$ and mapping $(x, 0)$ to $(0, xq)$ for the components corresponding to $\mathbb{Z}(q) \oplus \mathbb{Z}(q^2)$. This cannot be extended to an automorphism of $\mathcal{G}$ since $(0, q)$ is divisible by q but $(1, 0)$ is not.

3. Boolean Algebras

The standard signature for the study of Boolean algebras is $\{\wedge, \vee, \neg, 0, 1\}$. Here $a \wedge b$ is the *meet* of elements a and b, $a \vee b$ is the *join*, $\neg a$ is the *complement* of a, and 0 and 1 are the identity elements. A Boolean algebra $\mathcal{B} = (B, \wedge, \vee, \neg, 0, 1)$ may be viewed as a 2-group under the operation $a \oplus b = (a \wedge \neg b) \vee (b \wedge \neg a)$. $\mathcal{B}$ also has a natural partial ordering, defined by $a \leq b \iff a \vee b = b$.

The computability-theoretic properties of Boolean algebras have been well studied. The computably categorical Boolean algebras were determined independently by LaRoche [15] and Goncharov [9].

Theorem 3.1: *(LaRoche, Goncharov) A computable Boolean algebra is computably categorical iff it is relatively computably categorical iff it has finitely many atoms.*

Further investigations on the existence of effective isomorphisms appear in [20] and [16]. It is well-known that a countable Boolean algebra is homogeneous if it is atomless or has size at most 4. Details can be found in [11]. So we proceed to the weakly homogeneous case.

Theorem 3.2: *Assume the language is $\{\wedge, \vee, \neg, 0, 1\}$. Then a countable Boolean algebra $\mathcal{B}$ is weakly homogeneous if and only if $\mathcal{B}$ has finitely many atoms.*

Proof: First, suppose that $\mathcal{B}$ is weakly homogeneous. It follows from Theorem 1.1 that $\mathcal{B}$ is relatively computably categorical. Then by Theorem 3.1, $\mathcal{B}$ has finitely many atoms.

Next, suppose that the Boolean algebra $\mathcal{C}$ has only finitely many atoms. We can clearly take $\mathcal{C}$ to be infinite (otherwise it is trivially weakly homogeneous). Then we may assume that $\mathcal{C}$ has the form $\mathcal{B} \otimes \mathcal{A}$, where $\mathcal{B}$ is atomless and $\mathcal{A}$ is finite. Let $\{0\} \otimes \mathcal{A}$ be the exceptional set. Now suppose that we are given an isomorphism $F : \mathcal{C}_1 \to \mathcal{C}_2$, where $\mathcal{C}_1$ and $\mathcal{C}_2$ are finite subalgebras of $\mathcal{C}$, which both include $\{0\} \otimes \mathcal{A}$ and such that $F(0,a) = (0,a)$ for all $a \in \mathcal{A}$. Let $\mathcal{B}_1 = \pi_0(\mathcal{C}_1)$ and $\mathcal{B}_2 = \pi_0(\mathcal{C}_2)$, where $\pi_0(x,y) = x$. Note that $\mathcal{C}_1 = \mathcal{B}_1 \otimes \mathcal{A}$ and $\mathcal{C}_2 = \mathcal{B}_2 \otimes \mathcal{A}$ and that (b,a) is an atom of $\mathcal{C}_i$ if and only if either $b = 0$ and a is an atom of $\mathcal{A}$ or $a = 0$ and b is an atom of $\mathcal{B}_i$.

Now define $G(x) = \pi_0(F(x,0))$ for $x \in \mathcal{B}_1$. G is a composition of homomorphisms and therefore is a homomorphism. To see that G is one-to-one, suppose that $G(x) = G(y)$ for some $x, y \in \mathcal{B}_1$, and thus $F(x,0) = (b,i)$ and $F(y,0) = (b,j)$ for some $i, j \in \mathcal{A}$. Then $F(x \oplus y, 0) = (0, i \oplus j) = F(0, i \oplus j)$, so that $x \oplus y = 0$ and hence $x = y$ and also $i = j$. This shows that G is one-to-one.

We claim that, in addition, $F(x,0) = (G(x),0)$ for all $x \in \mathcal{B}_1$, and thus $F(x,i) = (G(x),i)$ for all x, i. In particular, since F is onto, this claim will show that $G : \mathcal{B}_1 \to \mathcal{B}_2$ is also onto. Since $\mathcal{B}_1$ is finite, it suffices to show this for all atoms b of $\mathcal{B}_1$. Let $F(b,0) = (y,i)$. Since $(b,0)$ is an atom of $\mathcal{C}_1$, (y,i) must be an atom of $\mathcal{C}_2$. If $i \neq 0$, then y must equal 0, otherwise (y,i) cannot be an atom of $\mathcal{C}_2$. But then $F(b,0) = (0,i) = F(0,i)$, so that $b = i = 0$. Therefore, $i = 0$ and hence $F(b,0) = (y,0) = (G(b),0)$.

Since $\mathcal{B}$ is homogeneous, G may be extended to an automorphism of $\mathcal{B}$. Finally, let $H : \mathcal{C} \to \mathcal{C}$ be given by $H(x,i) = (G(x),i))$. This is an automorphism of $\mathcal{C}$ which extends F. □

Corollary 3.3: *A computable Boolean algebra $\mathcal{B}$ is weakly homogeneous iff $\mathcal{B}$ is computably categorical iff $\mathcal{B}$ is relatively computably categorical.*

Corollary 3.4: *Every countable weakly homogeneous Boolean algebra has a computable copy.*

The previous proof shows that a finite subalgebra of a weakly homogeneous Boolean algebra is exceptional when it includes every atom. It is natural to conjecture that including all the atoms is not only sufficient, but also necessary. However, the following proposition shows this is not the case. This suggests a classification of the exceptional sets for weakly homogeneous Boolean algebras would require a finer combinatorial analysis like for linear orders and trees as found in [1].

Proposition 3.5: *Suppose that $\mathcal{C} = \mathcal{A} \otimes \mathcal{B}$, where both $\mathcal{A}$ and $\mathcal{B}$ are homogeneous. (That is, each is either atomless or has size ≤ 4.) Then $S = \{(0,1)\}$ is an exceptional set for $\mathcal{C}$.*

Proof: Let $\mathcal{C}_1$ and $\mathcal{C}_2$ be finite subalgebras of $\mathcal{C}$ each containing $(0,1)$, and hence also containing the complement $(1,0)$. Let $F : \mathcal{C}_1 \to \mathcal{C}_2$ be an isomorphism with $F(0,1) = (0,1)$ and $F(1,0) = (1,0)$. Let $\mathcal{A}_i$ be the projection $\pi_0(\mathcal{C}_i)$ of $\mathcal{C}_i$ onto $\mathcal{A}$ and let $\mathcal{B}_i = \pi_1(\mathcal{C}_i)$, for $i = 1, 2$.

Claim 3.6:

(1) For any $a \in \mathcal{A}$, any $b \in \mathcal{B}$, and for $i = 1, 2$, $a \in \mathcal{A}_i \iff (a, 0) \in \mathcal{C}_i$, and similarly $b \in \mathcal{B}_i \iff (0, b) \in \mathcal{C}_i$.
(2) $\mathcal{C}_i = \mathcal{A}_i \otimes \mathcal{B}_i$ for $i = 1, 2$.

Proof: For part (1), certainly $(a, 0) \in \mathcal{C}_i$ implies that $a \in \mathcal{A}_i$. Now suppose that $a \in \mathcal{A}_i$. Then for some $b \in B$, $(a, b) \in \mathcal{C}_i$. But then $(a, 0) = (a, b) \wedge (1, 0) \in \mathcal{C}_i$ as well, since $(1, 0) \in \mathcal{C}_i$. Similarly, $b \in \mathcal{B}_i$ implies that some $(a, b) \in \mathcal{C}_i$ and hence $(0, b) = (a, b) \wedge (0, 1) \in \mathcal{C}_i$.

For part (2), it is clear that $\mathcal{C}_i \subseteq \mathcal{A}_i \otimes \mathcal{B}_i$. Now suppose that $(x, y) \in \mathcal{A}_i \otimes \mathcal{B}_i$. Then by part (1), $(x, 0) \in \mathcal{C}_i$ and $(0, y) \in \mathcal{C}_i$. Thus $(x, y) = (x, 0) \vee (0, y) \in \mathcal{C}_i$. □

Define $G : \mathcal{A}_1 \to \mathcal{A}_2$ by $G(a) = \pi_0(F(a, 0))$ and define $H : \mathcal{B}_1 \to \mathcal{B}_2$ by $H(b) = \pi_1(F(0, b))$.

Claim 3.7:

(3) For any $(a, b) \in \mathcal{C}$, $F(a, b) = (G(a), H(b))$.
(4) G and H are isomorphisms.

Proof: For part (3), let $F(a, b) = (c, d)$. Since $F(1, 0) = (1, 0)$ and $(a, 0) = (a, b) \wedge (1, 0)$, it follows that $F(a, 0) = (c, d) \wedge (1, 0) = (c, 0)$, so that $c = G(a)$. Similarly, $d = H(b)$.

Next, we prove part (4). To see that G is a homomorphism, let $G(a) = b$ and $G(c) = d$. Then, by part (3), $F(a, 0) = (b, 0)$ and $F(c, 0) = (d, 0)$. Thus, $F(a \vee c, 0) = (b \vee d, 0)$ and hence $G(a \vee c) = b \vee d$. Similarly, $F(a \wedge c, 0) = (b \wedge d, 0)$ so $G(a \wedge c) = b \wedge d$. A similar argument shows that H is a homomorphism. To see that G is one-to-one, suppose that $G(a) = G(c)$. Then $F(a, 0) = F(c, 0)$, so that $a = c$. In the same way, we can see that H is one-to-one. To see that G is onto, let $c \in \mathcal{A}_2$. Then $(c, 0) \in \mathcal{C}_2$, so that $F(a, b) = (c, 0)$ for some $(a, b) \in \mathcal{C}_1$. It follows from part (3) that $c = G(a)$. Similarly, H is onto. □

Since $\mathcal{A}$ and $\mathcal{B}$ are homogeneous, it follows that there are automorphisms $\hat{G} : \mathcal{A} \to \mathcal{A}$ extending G and $\hat{H} : \mathcal{B} \to \mathcal{B}$ extending H. Thus for $a \in \mathcal{A}_1$ and $b \in \mathcal{B}_1$, we have $\hat{G}(a) = G(a)$ and $\hat{H}(b) = H(b)$. Now let $\hat{F}(a, b) =$

$(\hat{G}(a), \hat{H}(b))$. Then $\hat{H}$ is an automorphism of $\mathcal{C}$ and furthermore, for $(a,b) \in \mathcal{C}_1$, we have $\hat{F}(a,b) = (G(a), H(b)) = F(a,b)$.

It should be mentioned that there is a natural correspondence between Boolean algebras and linear orders. Given a (computable) linear order $\mathcal{L}$, the left-closed, right-open intervals of $\mathcal{L}$ generate a (computable) Boolean algebra, called the interval algebra of $\mathcal{L}$. That is, for each $a < b$ in $\mathcal{L}$, let $[a, b) = \{x : a \leq x < b\}$ and let $\mathcal{B}(\mathcal{L})$ be the family of finite unions of such intervals and their complements. It is easy to see that a linear order has finitely many successivities iff its interval algebra has finitely many atoms. So from [1] it immediately follows that a countable linear order is weakly homogeneous iff its interval algebra is weakly homogeneous. However, it does not seem possible to use this correspondence to directly obtain a proof of Theorem 3.2. More discussion about the interval algebra of a linear ordering in the context of computability can be found in [6] and [19].

Next, we consider other signatures for Boolean algebras.

By omitting the complement operation, Boolean algebras with the meet and join operations may be viewed as bounded, distributive lattices in the signature $\{\vee, \wedge, 0, 1\}$. Note that the complement $\neg a$ is definable in the lattice as the unique element x, such that $x \vee a = 1$ and $x \wedge a = 0$.

Boolean algebras come with a natural ordering definable from the algebraic operations, where $a \leq b$ iff $a \wedge b = a$ iff $a \vee b = b$, so may be considered as partial orderings in the signature $\{\leq, 0, 1\}$. Note that the join and meet operations are definable in the partial ordering. For example, $a \vee b = c$ means that $a \leq c$ and $b \leq c$ and that for any d such that $a \leq d$ and $b \leq d$, $c \leq d$.

We will classify the homogeneous and weakly homogeneous Boolean algebras as lattices and as partial orderings.

Proposition 3.8:

(1) If $\mathcal{B}$ is homogeneous as a lattice or as a partial order, then it is also homogeneous as a Boolean algebra.

(2) If $\mathcal{B}$ is weakly homogeneous as a lattice or as a partial order, then it is also weakly homogeneous as a Boolean algebra.

Proof: Suppose that $\mathcal{B}$ is homogeneous in one of the two other signatures and let $\mathcal{B}_1$ and $\mathcal{B}_2$ be finitely generated subalgebras and $F : \mathcal{B}_1 \to \mathcal{B}_2$ an isomorphism. Then $\mathcal{B}_1$ and $\mathcal{B}_2$ are in fact finite, and hence are finitely generated in the smaller signature as well. Hence F may be extended to an automorphism H of $\mathcal{B}$ as a partial order, or as a lattice. But this means that H is in fact an automorphism of $\mathcal{B}$ as a Boolean algebra. Next, suppose that $\mathcal{B}$ is weakly homogeneous in a smaller signature with some exceptional set S and suppose that $F : \mathcal{B}_1 \to \mathcal{B}_2$ is an isomorphism of finitely generated substructures each including S and fixing every element of S. Then F is also an isomorphism in the smaller signature, and hence may be extended to an automorphism H of $\mathcal{B}$ in the smaller signature. As above, H is in fact an automorphism of $\mathcal{B}$ as a Boolean algebra. □

We can now classify the homogeneous and weakly homogeneous Boolean algebras as lattices. We will need a fact about the lattices and Boolean algebras they generate.

Lemma 3.9: *([12] Sec. II.4.1) Let $\mathcal{B}$ be a Boolean algebra in the language $\{\wedge, \vee, 0, 1\}$ and let $\varphi : L_1 \to L_2$ be an isomorphism of sublattices. Then φ extends to an isomorphism of subalgebras of $\mathcal{B}$ generated by L_1, L_2.*

Proposition 3.10: *For any countable Boolean algebra $\mathcal{B}$, $\mathcal{B}$ is homogeneous in the language of Boolean algebras if and only if it is homogeneous in the language of lattices, and similarly for weakly homogeneous.*

Proof: In each case, one direction is immediate from Proposition 3.8. Now suppose that $\mathcal{B}$ is homogeneous as a Boolean algebra and let $\phi : L_1 \to L_2$ be an isomorphism of two finitely generated sublattices. By Lemma 3.9, ϕ extends to an isomorphism F of the subalgebras $\mathcal{B}_1$ and $\mathcal{B}_2$ generated by L_1 and L_2. Since $\mathcal{B}$ is homogeneous as a Boolean algebra, F extends to an automorphism H of $\mathcal{B}$ as a Boolean algebra. Then H is also a lattice automorphism and extends ϕ. If L_1 and L_2 are required to include some exceptional set S and ϕ is required to have $\phi(x) = x$ for $x \in S$, then $\mathcal{B}_1$ and $\mathcal{B}_2$ above will also include S and the extension F above will still have $F(x) = x$ for $x \in S$. So F will again have an extension to an automorphism H of $\mathcal{B}$ as a Boolean algebra. □

We next consider Boolean algebras in the language of partial orders $\{\leq, 0, 1\}$; the classification here might be a bit surprising.

Proposition 3.11: *Assume the language is the language of partial orders $\{\leq, 0, 1\}$. Then given a countable Boolean algebra $\mathcal{B}$,*

(1) $\mathcal{B}$ is homogeneous if and only if $\mathcal{B}$ is finite with at most 4 elements.
(2) $\mathcal{B}$ is weakly homogeneous if and only if $\mathcal{B}$ is finite.

Proof: (1): If $\mathcal{B}$ has $1, 2$ or 4 elements then it is clearly homogeneous. Now suppose $\mathcal{B}$ has at least 8 elements. Then $\mathcal{B}$ contains elements a and b such that $0_{\mathcal{B}} < a < b < 1_{\mathcal{B}}$. Let $c = a \vee \neg b$. We now claim that $c \not\geq b$. Suppose not, then $c \wedge (b \wedge \neg a) = b \wedge \neg a$. Since $c \wedge (b \wedge \neg a) = 0_{\mathcal{B}}$, this means that $b \wedge \neg a = 0_{\mathcal{B}}$ and hence $a \geq b$, a contradiction. So b and c are incomparable. Now we can consider $\phi(b) = b$ and $\phi(c) = \neg b$, which is an isomorphism of the suborderings $\{0_{\mathcal{B}}, 1_{\mathcal{B}}, b, c\}$ and $\{0_{\mathcal{B}}, 1_{\mathcal{B}}, b, \neg b\}$. This cannot be extended to an automorphism of $\mathcal{B}$ as a partial order because a is a non-zero element below both b and c, while no such element exists below b and $\neg b$.

(2): Every finite set is trivially weakly homogeneous. Now assume that $\mathcal{B}$ is an infinite Boolean algebra. Fix a finite subalgebra $\mathcal{A}$ of $\mathcal{B}$. We can find an infinite set C so that for any $x, y \in C$ and any $a \in \mathcal{A}$, $x < a \iff y < a$ and $a < x \iff a < y$, since there are only finitely many such configurations. Let m be the meet of all elements of $\mathcal{A}$ above every element of C and let j be the join of all elements of $\mathcal{A}$ below C. Then all elements of C come from the interval $[j, m]$. Using elements of C we can generate from C, relativized to $[j, m]$, a subalgebra of $\mathcal{B}$ with 8 elements; here the complement $\neg x = m \wedge \neg x$. So this algebra has m as its largest element and j as its smallest element. Now repeat the argument from the homogeneous case. This means we obtain elements $x, y, z \in C$ so x is incomparable with both y and z, but while x, y have an element below them in $\mathcal{B}$ and strictly above j, the pair x, z does not. Then the partial isomorphism fixing all of $\mathcal{A}$, fixing x, and sending y to z cannot extend to an automorphism of $\mathcal{B}$. So $\mathcal{A}$ is not exceptional, hence $\mathcal{B}$ is not weakly homogeneous. $\square$

Acknowledgements

This chapter is based on a lecture given at the IMS of National University of Singapore during the "Aspects of Computation" program in 2017. The authors thank the IMS and the conference organizers for the opportunity. This chapter is dedicated to longtime colleague and friend Rod Downey.

This research was partially supported by the National Science Foundation SEALS grant NSF DMS-1362273. The third author was also partially supported by MOE2015-T2-2-055 and RG131/17.

References

1. F. Adams and D. Cenzer, Computability and categoricity of weakly ultrahomogeneous structures, *Computability* **6** (2017), 365–389.
2. W. Calvert, D. Cenzer, V. Harizanov, and A. Morozov, Effective categoricity

of equivalence structures, *Annals of Pure and Applied Logic* **159** (2009), pp. 87–97.

3. D. Cenzer and J. B. Remmel, Polynomial-Time Abelian Groups, *Ann. Pure and Applied Logic* **56** (1992), pp. 313–363.
4. G. Cherlin and U. Felgner, Homeogeneous solvable groups, *J. London Math. Soc.* **89** (1991), pp. 102–120.
5. B. Csima, V. Harizanov, R. Miller and A. Montalban, Computability of Fraisse limits, *J. Symbolic Logic* **76** (2011), pp. 66–93.
6. R. Downey, Computability theory and linear orderings, In *Studies in Logic and the Foundations of Mathematics*, vol. 139, pp. 823–976, Elsevier, 1998.
7. R. Fraisse, Theory of Relations, North-Holland, Amsterdam, 1986.
8. L. Fuchs, Infinite Abelian Groups. Vol. I. *Pure and Applied Mathematics*, Vol. 36, Academic Press (1970) xi + 290 pp.
9. S. S. Goncharov, Constructivizability of superatomic Boolean algebras. *Algebra and Logic* **12** (1973), pp. 31–40 (Russian), pp. 17–22 (English translation).
10. S. S. Goncharov, Autostability of models and abelian groups, *Algebra and Logic* **19** (1980), pp. 23–44 (Russian), pp. 13–27 (English translation).
11. S. Givant and P. Halmos, Introduction to Boolean algebras. Springer Science & Business Media, 2008.
12. G. Gratzer, Lattice theory: foundation. Springer Science & Business Media, 2011.
13. N. G. Khisamiev, Constructive abelian groups. Handbook of Recursive Mathematics, Vol. 2. *Studies in Logic and Foundations of Mathematics*, Vol. 139. Elsevier (1998), pp. 1177–1231.
14. A. H. Lachlan, Homogeneous structures, in: *Proceedings of the International Congress of Mathematicians (Berkeley, Calif., 1986), Vol. 1*, Amer. Math. Soc., Providence, RI, 1987, pp. 314–321.
15. P. LaRoche, Recursively presented Boolean algebras. *Notices AMS*, **24** (1977), pp. 552–553.
16. C. F. McCoy, Δ_2^0-categoricity in Boolean algebras and linear orderings. *Annals of Pure and Applied Logic* **119** (2003), pp. 85–120.
17. A. G. Melnikov and K. M. Ng, Computable torsion abelian groups, *Advances in Mathematics*, **325** (2018) pp. 864–907.
18. D. Marker, Model Theory: An Introduction, Springer, New York, 2002.
19. J. B. Remmel, Recursively Categorical Linear Orders, *Proceedings of the AMS* **83** (1981) pp. 387–391.
20. J. B. Remmel, Recursive isomorphism types of recursive Boolean algebras. *The Journal of Symbolic Logic* **46** (1981) pp. 572–594.
21. R. Soare, Recursive Enumerable Sets and Degrees, Springer-Verlag, 1987.
22. R. L. Smith, Two theorems on autostability in *p*-groups, in *Logic Year* 1979–80, University of Connecticut, Storrs, Lecture Notes in Mathematics 859, Springer, Berlin (1981), pp. 302–311.

https://doi.org/10.1142/9789811278631_0007

From Quasi-Dominions to Progress Measures

Massimo Benerecetti[1,†], Daniele Dell'Erba[2],
Marco Faella[1,‡] and Fabio Mogavero[1,♯]
[1] *Università degli Studi di Napoli Federico II*
[†] *massimo.benerecetti@unina.it*, [‡] *marco.faella@unina.it*, [♯] *fabio.mogavero@unina.it*
[2] *University of Liverpool*
daniele.dell-erba@liverpool.ac.uk

We revisit the approaches to the solution of *parity games* based on *progress measures* and show how the notion of *quasi dominions* can be integrated with those approaches. The idea is that, while progress measure based techniques typically focus on one of the two players, little information is gathered on the other player during the solution process. Adding quasi dominions provides additional information on this player that can be leveraged to greatly accelerate convergence to a progress measure. To accommodate quasi dominions, however, non trivial refinements of the approach are necessary. In particular, we need to introduce a novel notion of measure and a new method to prove correctness of the resulting solution technique.

1. Introduction

Parity games are two-player infinite-duration games on graphs, which play a crucial role in various fields of theoretical computer science. These are games played on graphs, whose nodes, called *positions*, are labelled with natural numbers, called *priorities*, and controlled by one of two players: player 0 and player 1. Each player can choose edges, called *moves*, when the game is at one of its positions. The goal of player 1 is to force a play π, namely an infinite path in the underlying graph, such that the maximal priority occurring infinitely often along π is of odd parity. If such a play cannot be enforced by player 1, player 0 wins the game. In this case, player 0 can indeed force a play whose maximal recurring priority is even.

Finding efficient algorithms to solve these games in practice is a core problem in formal verification and reactive synthesis, as it leads to efficient solutions of the model-checking and satisfiability problems of expressive temporal logics. These algorithms can, indeed, be used as back-ends of satisfiability and model-checking procedures [1–3]. In particular, the solution problem for these games has been proved linear-time interreducible with the model-checking problem for the *modal* μ*Calculus* [2] and it is closely related to other games of infinite duration, such as *mean payoff* [4,5], *discounted payoff* [6], *simple stochastic* [7], and *energy* [8] games. Parity games are also central to several techniques employed in automata theory [9–12]. Besides the practical importance, parity games are also interesting from a computational complexity point of view, since their solution problem is one of the few inhabitants of the UPTime $\cap$ CoUPTime class [13]. That result improves the NPTime $\cap$ CoNPTime membership [2], which easily follows from the property of *memoryless determinacy* [10,14].

A number of quite different approaches to solve parity games have been proposed in the literature that exhibit quite different characteristics. Typically, the most efficient ones in practice are those based on game decomposition, such as the Recursive Algorithm [15,16], Priority Promotion [17–19], and Tangle Learning [20], which, however, usually suffer from poor worst-case bounds. On the other hand, the approaches based on progress measures [21,22] often lead to good worst-case behaviours [23,24], but typically perform very poorly in practice. The main reason for this inefficiency resides in the fact that those algorithms iteratively explore a space of functions assigning some value, called *measure*, to each position in the game. At each iteration, the measures of some of the positions may increase and when they become stationary for all the positions, a fixpoint is reached and a solution can be extracted from the resulting measures. In order to guarantee correctness, measures are allowed to increase very slowly, which often leads to slow convergence to a solution and makes these approaches less appealing in practical contexts. The slow growth is the result of a uniform measure update policy for one of the two players, specifically player 1, which only allows for a minimal measure increase for each of its positions that must be updated.

In this chapter, we show that the update policy can be considerably improved upon, without sacrificing correctness. Instead of relying on the minimal increase policy to ensure soundness of measure updates, we propose an approach that brings *quasi dominions* into the equation. Note that the same idea has been implemented for Mean Payoff Games in [25] where the resulting algorithm proved to be very efficient. Informally, a quasi dominion is a set of positions from which one of the players, say player 0, can win the game as long as the opponent, player 1, chooses not to escape from that set. As such, the notion is not new and is at the very heart of the Priority Promotion algorithms [26]. The idea is to leverage quasi dominions to justify a larger, but still sound, increase in the measure for positions controlled by player 1. The crucial observation is that player 1 surely loses from any position of a quasi dominion for player 0, unless it can escape that set by taking some exiting move. Therefore, player 1 can safely increase the measure of the escaping position according to the exiting moves chosen, regardless of the fact that the increase may not be minimal. In this way, we are able to skip lower measures and jump directly to measures that would be reached anyway, albeit with a number of iterations that is usually much higher.

The integration of progress measures and quasi dominions, however, requires (i) a richer form of measure, able to encode additional information that allows us to identify quasi dominions in the game, and (ii) a new update algorithm that takes this information into account when increasing the measures.

The main contributions of this chapter can be summarised as follows: (a) a novel solution algorithm for parity games based on the integration of progress measures and quasi dominions; (b) the experimental results showing an improvement on the performance of orders of magnitude *w.r.t.* the classic and quasi-polynomial small progress measure algorithm; (c) the present approach can pave the way to practically efficient quasi-polynomial algorithms based on the integration of succinct progress measures, such as those in [23, 24].

2. Preliminaries

A two-player turn-based *arena* is a tuple $\mathcal{A} = \langle \mathrm{Ps}_0, \mathrm{Ps}_1, Mv \rangle$, with $\mathrm{Ps}_0 \cap \mathrm{Ps}_1 = \emptyset$ and $\mathrm{Ps} \triangleq \mathrm{Ps}_0 \cup \mathrm{Ps}_1$, such that $\langle \mathrm{Ps}, Mv \rangle$ is a finite directed graph without sinks. Ps_0 (*resp.*, Ps_1) is the set of positions of player 0 (*resp.*, 1) and $Mv \subseteq \mathrm{Ps} \times \mathrm{Ps}$ is a left-total relation describing all possible moves. A *path* in $\mathrm{V} \subseteq \mathrm{Ps}$ is a finite (possible empty) or infinite sequence $\pi \in \mathrm{Pth}(\mathrm{V})$ of positions in V compatible with the move relation, *i.e.*, $(\pi_i, \pi_{i+1}) \in Mv$, for all $0 \leq i < |\pi| - 1$. The set $\mathrm{FPth}(v)$ contains all the finite (possible empty) paths originating at the position v. For a finite path π, with $\mathsf{lst}(\pi)$ we denote the last position of π. Finally, $\mathrm{SPth}(v, \mathrm{V})$ is the set of simple paths in $\mathrm{FPth}(v)$ that are completely composed of positions in V. Notice that, if v does not belong to V, then we have that $\mathrm{SPth}(v, \mathrm{V}) = \{\varepsilon\}$. A positional *strategy* for player $\alpha \in \{0,1\}$ on $\mathrm{V} \subseteq \mathrm{Ps}$ is a function $\sigma_\alpha \in \mathrm{Str}_\alpha(\mathrm{V}) \triangleq (\mathrm{V} \cap \mathrm{Ps}_\alpha) \to \mathrm{Ps}$, mapping each α-position v in V to a position $\sigma_\alpha(v)$ compatible with the move relation, *i.e.*, $(v, \sigma_\alpha(v)) \in Mv$. With $\mathrm{Str}_\alpha(\mathrm{V})$ we denote the set of all α-strategies on V. Given an α-strategy $\sigma_\alpha \in \mathrm{Str}_\alpha(\mathrm{V})$ and a set of positions $\mathrm{U} \subseteq \mathrm{Ps}$, the operator $\sigma_\alpha \downarrow \mathrm{U}$ restricts σ_α to the positions in $\mathrm{V} \cap \mathrm{U}$. A *play* in $\mathrm{V} \subseteq \mathrm{Ps}$ from a position $v \in \mathrm{V}$ *w.r.t.* a pair of strategies $(\sigma_0, \sigma_1) \in \mathrm{Str}_0(\mathrm{V}) \times \mathrm{Str}_1(\mathrm{V})$, called $((\sigma_0, \sigma_1), v)$-*play*, is a path $\pi \in \mathrm{Pth}(\mathrm{V})$ such that $(\pi)_0 = v$ and, for all $0 \leq i < |\pi| - 1$, if $(\pi)_i \in \mathrm{Ps}_0$ then $(\pi)_{i+1} = \sigma_0((\pi)_i)$ else $(\pi)_{i+1} = \sigma_1((\pi)_i)$. The *play function* $\mathsf{play} : (\mathrm{Str}_0(\mathrm{V}) \times \mathrm{Str}_1(\mathrm{V})) \times \mathrm{V} \to \mathrm{Pth}(\mathrm{V})$ returns, for each position $v \in \mathrm{V}$ and pair of strategies $(\sigma_0, \sigma_1) \in \mathrm{Str}_0(\mathrm{V}) \times \mathrm{Str}_1(\mathrm{V})$, the maximal $((\sigma_0, \sigma_1), v)$-play $\mathsf{play}((\sigma_0, \sigma_1), v)$. A path $\pi \in \mathrm{Pth}(v)$ is called a (σ_α, v)-*play in* V, if $\pi = \mathsf{play}((\sigma_0, \sigma_1), v)$, for some $\sigma_{\overline{\alpha}} \in \mathrm{Str}_{\overline{\alpha}}(\mathrm{V})$.

A *parity game* is a tuple $\partial = \langle \mathcal{A}, \mathrm{Pr}, \mathsf{pr} \rangle \in \mathrm{PG}$, where $\mathcal{A}$ is an arena, $\mathrm{Pr} \subset \mathbb{N}$ is a finite set of priorities, and $\mathsf{pr} : \mathrm{Ps} \to \mathrm{Pr}$ is a *priority function* assigning a priority to each position. The priority function can be naturally extended to games and paths as follows: $\mathsf{pr}(\partial) \triangleq \mathsf{max}_{v \in \mathrm{Ps}} \mathsf{pr}(v)$; for a path $\pi \in \mathrm{Pth}$, we set $\mathsf{pr}(\pi) \triangleq \mathsf{max}_{0 \leq i < |\pi|} \mathsf{pr}((\pi)_i)$, if π is finite, and $\mathsf{pr}(\pi) \triangleq \limsup_{i \in \mathbb{N}} \mathsf{pr}((\pi)_i)$, otherwise. A set of positions $\mathrm{V} \subseteq \mathrm{Ps}$ is an α-*dominion*, with $\alpha \in \{0,1\}$, if there exists an α-strategy $\sigma_\alpha \in \mathrm{Str}_\alpha(\mathrm{V})$ such that, for all $\overline{\alpha}$-strategies $\sigma_{\overline{\alpha}} \in \mathrm{Str}_{\overline{\alpha}}(\mathrm{V})$ and positions $v \in \mathrm{V}$, the induced play $\pi = \mathsf{play}((\sigma_0, \sigma_1), v)$ is infinite and $\mathsf{pr}(\pi) \equiv_2 \alpha$. In other words, σ_α only induces on V infinite plays whose maximal priority visited infinitely often has parity α. By $\partial \setminus \mathrm{V}$ we denote the maximal subgame of ∂ with set of positions Ps' contained in $\mathrm{Ps} \setminus \mathrm{V}$ and move relation Mv' equal to the restriction of Mv to Ps'.

3. Solving Parity Games via Progress Measures

The abstract notion of *progress measure* [27] has been introduced as a way to encode *global properties* on paths of a graph by means of simpler *local properties* of adjacent vertexes, *i.e.*, of edges. In particular, this notion has been successfully employed in the literature, *e.g.*, for the solution of automata theory [28–32] and program verification [33, 34] problems.

In the context of parity games [21], the graph property of interest, called *parity property*, asserts that, along every path in the graph, the maximal priority occurring infinitely often is of odd parity. More precisely, in game theoretic terms, a *parity progress measure* witnesses the existence of a strategy σ for one of the two players, from now on player 1, such that each path in the graph induced by fixing σ satisfies the desired property, where the graph induced by that strategy is obtained from the game arena by removing all the moves exiting from position owned by player 1, except those specified by σ itself. A parity progress measure associates with each vertex of the underlying graph a value, called *parity measure* (or simply *measure*, for short), taken from some totally-ordered set. Measures are thus associated with positions in the game and the measure η of a position v can intuitively be interpreted as a local assessment of *how far* v is from satisfying the parity property, with the maximal value $\eta = \top$ denoting failure in the satisfaction of the property for v. More precisely, a progress measure implicitly identifies a strategy σ with the following characteristic: in the graph induced by σ, along every path, measures cannot increase and they strictly decrease when passing through an even-priority position. This ensures that every path eventually gets trapped into a cycle whose maximal priority is odd.

In order to obtain a progress measure, we start from some well-behaved assignment of measures to positions of the game. The local information encoded by these measures is, then, propagated back along the edges of the underlying graph so as to associate with each position the information on the priorities occurring along the plays of some finite length starting at that position. The propagation process is performed by means of a low-level measure-update operator, called *stretch* operator $+$. The operator computes the contribution that a given position v would provide to a given measure η. Consider, for instance, a position v that has an adjacent position u with measure η. Then $\eta + v$ is the measure that v would obtain by choosing the move leading to u, namely, the one obtained by augmenting η with the contribution of (the priority of) position v. When v is an even-priority position, the augmented measure $\eta + v$ strictly increases, moving further away from the parity property.

The process described above terminates when no position can be pushed further away from the property. More specifically, each even position has to strictly dominate the measures obtainable through all, respectively one of, its adjacent positions, depending on whether that position belongs to player 0 or to player 1, respectively. Similarly, each odd position must have measures no lower than those obtainable through all, respectively one of, its adjacent moves, again depending on the owner of the position. When this happens, the positions with measure $\top$ are the ones from which player 0 wins the parity game, while the remaining ones are those from which the opponent can win, by simply forcing plays of non-increasing measures. The measures currently associated with this second set of positions form a progress measure for the game.

Different notions of parity measures have been proposed in the literature, see, *e.g.*, [21–24]. In this section we introduce an abstract concept of measure space and progress measure. All the progress measure based approaches in the literature, including the one presented in this chapter, can be viewed as instantiations of this abstract schema.

3.1. *Measure-Function Spaces*

As mentioned above, techniques based on progress measures rely on attaching, at each step of the computation, suitable information to all positions in the game and updating it until a fixpoint is reached. The piece of information associated with every single position is called the current *measure* of that position, whereas the set of all possible measures is called a *measure space*. Such a space is a totally ordered set, with minimum and maximum elements, and provides the two special operations of *truncation* and *stretch* that evaluate and update the measure of a given position *w.r.t.* another position. Intuitively, the truncation operator $\restriction_v$ disregards the contribution to a given measure that is due to positions with priority lower than that of v along the explored finite plays. The stretch operator $+$, introduced in the previous paragraph, propagates the contribution that the position would provide to a given measure.

These two operators essentially embed the semantics of the parity property into the propagation operation that sits at the basis of the computation of a progress measure. At the abstract level, canonical instances of these operators can be any functions that preserve the maximum element and the order, except for possibly mapping different measures onto the same one. All these requirements are formalised by the following definition.

Definition 3.1: (Measure Space) A *measure space* is a mathematical structure $\mathcal{M} \triangleq \langle \mathrm{Ms}, <, \bot, \top, \restriction, + \rangle$, whose components enjoy the following properties:

(1) $\langle \mathrm{Ms}, <, \bot, \top \rangle$ is a *strict total order* with *minimum* and *maximum* on elements referred to as *measures*;
(2) the function $\restriction\colon \mathrm{Ms} \times \mathrm{Ps} \to \mathrm{Ms}$, called *truncation operator*, maps a measure $\eta \in \mathrm{Ms}$ and a position $v \in \mathrm{Ps}$ to a measure $\eta\restriction_v \in \mathrm{Ms}$; this operator is *canonical* whenever (i) $\eta = \top$ *iff* $\eta\restriction_v = \top$ and (ii) if $\eta \leq \eta^\star$ then $\eta\restriction_v \leq \eta^\star\restriction_v$, for all $\eta^\star \in \mathrm{Ms}$;
(3) the function $+\colon \mathrm{Ms} \times \mathrm{Ps} \to \mathrm{Ms}$, called *stretch operator*, maps a measure $\eta \in \mathrm{Ms}$ and a position $v \in \mathrm{Ps}$ to a measure $\eta + v \in \mathrm{Ms}$;

this operator is *canonical* whenever (i) $\eta = \top$ *iff* $\eta + v = \top$ and (ii) if $\eta \leq \eta^\star$ then $\eta + v \leq \eta^\star + v$, for all $\eta^\star \in \text{Ms}$.

Notice that both operators are canonical if they are *monotone* in their first argument and preserve the distinction between the measure $\top$ and the other measures, in the sense that $\top$ cannot be obtained by truncating or stretching a non-top measure and, *vice versa*, no non-$\top$ measure is derivable by truncating or stretching $\top$.[a]

Given a measure space $\mathcal{M}$, a *measure function* μ formalises the intuitive association discussed above by mapping each position v in the game to a measure $\mu(v)$ in $\mathcal{M}$. In addition, the order relation $<$ between measures declared in $\mathcal{M}$ induces a pointwise partial order $\sqsubseteq$ on the set measure functions MF defined in the usual way. This set together with its induced order form what we call a *measure-function space.*

Definition 3.2: (Measure-Function Space) The *measure-function space* induced by a given measure space $\mathcal{M}$ is the *partial order* $\mathcal{F} \triangleq \langle \text{MF}, \sqsubseteq \rangle$, whose components are defined as prescribed in the following:

(1) $\text{MF} \triangleq \text{Ps} \to \text{Ms}$ is the set of all functions $\mu \in \text{MF}$, named *measure functions*, mapping each position $v \in \text{Ps}$ to a measure $\mu(v) \in \text{Ms}$;
(2) for all $\mu_1, \mu_2 \in \text{MF}$, it holds that $\mu_1 \sqsubseteq \mu_2$ *iff* $\mu_1(v) \leq \mu_2(v)$, for all positions $v \in \text{Ps}$.

By taking $\mu_\bot$ as the measure function associating measure $\bot$ with every position, the following property of measure-function spaces immediately follows.

Proposition 3.3: *Every measure-function space $\mathcal{F}$ contains a unique minimal element $\mu_\bot \in \text{MF}$.*

---

[a]Readers familiar with the research published in [21, 23] might find it interesting to observe that both the small progress measure [21] and the succinct progress measure [23] algorithms make implicit use of a measure space with a canonical truncation operator, but a non-canonical stretch operator. In more detail, the *d/2-tuples* associated with the positions during an execution form a totally ordered set with minimum and maximum, once extended with the value $\top$ and where the value $\bot$ is identified with the unique all-zero $d/2$-tuple. Moreover, the truncation operator is represented by the function that zeros all components of a $d/2$-tuple with indexes smaller than the priority of the position given as second argument. Finally, the maps over $d/2$-tuples induced by the ternary functions $\mathsf{Prog}(\cdot,\cdot,\cdot)$ [21] and $\mathsf{lift}(\cdot,\cdot,\cdot)$ [23], used in the definition of the function $\mathsf{Lift}(\cdot,\cdot)$ at the core of the algorithms, implement the corresponding stretch operators. Such operators are, however, not canonical, since they can map some of the non-$\top$ measures to $\top$, failing so to satisfy the *if* direction of Condition 3i of Definition 3.1.

The 0-*denotation* (*resp.*, 1-*denotation*) of a measure function $\mu \in \mathrm{MF}$ is the set $\|\mu\|_0 \triangleq \{v \in \mathrm{Ps} : \mu(v)\restriction_v = \top\}$ (*resp.*, $\|\mu\|_1 \triangleq \overline{\|\mu\|_0}$) of all positions having maximal (*resp.*, non-maximal) measures associated with them in μ, once truncated. Similarly, the $\bot$-*denotation* (*resp.*, $+$-*denotation*) of μ is the set $\|\mu\|_\bot \triangleq \{v \in \mathrm{Ps} : \mu(v)\restriction_v = \bot\}$ (*resp.*, $\|\mu\|_+ \triangleq \overline{\|\mu\|_\bot}$) of all positions having minimal (*resp.*, non-minimal) measures.

According to the intuition reported at the beginning of this section, the measure associated with a given position v is meant to encode information about the priorities encountered along the plays starting at that positions. More specifically, each measure contains the information gathered along some finite path and can be obtained by repeatedly applying the stretch operator backwards from the last position of that path. To formalise this intuition, we introduce the notion of measure $\eta(\pi)$ of a finite path π, including the empty one ε, that can be recursively computed via the function η: FPth $\to$ MF as follows:

$$\eta(\pi) \triangleq \begin{cases} \bot, & \text{if } \pi = \varepsilon; \\ \eta(\pi') + v, & \text{otherwise.} \end{cases}$$

Where $\pi = v \cdot \pi'$, for some unique $v \in \mathrm{Ps}$ and $\pi' \in \mathrm{FPth}$.

At this point, we can constrain a measure function μ, by requiring the measure $\mu(v)$ of a position v to be witnessed by some finite path π starting at v, *i.e.*, $\mu(v) = \eta(\pi)$. By doing this, we obtain a *ground measure function*.

Definition 3.4: (Ground Measure-Function Space) The *ground measure-function space* induced by a given measure space $\mathcal{M}$ is the *subspace* $\langle \mathrm{GM}, \sqsubseteq \rangle$ of the measure-function space $\mathcal{F}$, where $\mathrm{GM} \triangleq \{\mu \in \mathrm{MF} : \forall v \in \mathrm{Ps}. \mu(v) \in \mathrm{GMs}(v)\}$ with $\mathrm{GMs}(v) \triangleq \{\eta(\pi) : \pi \in \mathrm{FPth}(v)\}$.

All progress-measure approaches proposed in the literature implicitly work by updating ground measure functions only. The same holds true for the algorithm proposed in the current work, which actually runs on the even more restricted set of *simple measure functions* introduced in Sec. 4.

3.2. *Progress-Measure Functions*

The following definition, which tightly connects the truncation and stretch operators, formalises the essential semantic features of a measure space that are required to provide a meaningful notion of progress measure, as proven in Theorem 3.7.

Definition 3.5: (Progress Measure Space) A measure space $\mathcal{M}$ is a *progress measure space* if the following properties hold true, for each measure $\eta \in \text{Ms}$ and position $v \in \text{Ps}$:

(1) $\eta\restriction_v < (\eta + v)\restriction_v$, if $\eta\restriction_v < \top$ and $\mathsf{pr}(v)$ is even;
(2) $\eta\restriction_u \leq (\eta + v)\restriction_u$, for all even-priority positions u with $\mathsf{pr}(v) \leq \mathsf{pr}(u)$.

The first condition requires that the contribution to a measure due to an even position v cannot be cancelled out by truncating at v itself and that such contribution is always meaningful, namely strictly increasing. This matches the intuition that even priorities tend to move away from the parity property, therefore increasing the associated measure. The second property, instead, ensures that no lower-priority position v can overcome a higher even priority position u, in the sense that the contribution of v to the measure cannot move closer to the parity property, once the stretched measure is analysed *w.r.t.* u. Technically, this means that the stretch forced by a lower priority position is always viewed as a non-strict improvement by any even position with higher priority, regardless of the parity of the first one.

We can now turn our attention to the notion of *progress measure*. Intuitively, a measure function over a progress measure space is a progress measure if it guarantees that every position with a non-$\top$ measure can progress toward the parity property, namely toward lower measures. In other words, this establishes a type of stability property on the positions of a game according to the following intuition, which takes into account the opposite attitude of the two players. While player 0 tries to push toward higher measures, the opponent will try to keep the measure as low as possible. If player 0 cannot increase the measures of its positions and the opponent is not forced to increase the measures of its own positions, then player 0 cannot prevent player 1 from winning the game from all those positions whose measure did not reach value $\top$. This corresponds to requiring that player 0 cannot increase the measure of its positions by stretching the measure of any adjacent move, while the opponent can always choose a move whose corresponding stretch does not force the increment of the measure. The following definition precisely formalises this intuitive explanation.

Definition 3.6: (Progress Measure) A measure function $\mu \in \text{MF}$ is a *progress measure* if the following conditions hold, for all positions $v \in \text{Ps}$:

(1) $\mu(w) + v \leq \mu(v)$, for all adjacents $w \in Mv(v)$ of v, if $v \in \text{Ps}_0$;
(2) $\mu(w) + v \leq \mu(v)$, for some adjacent $w \in Mv(v)$ of v, if $v \in \text{Ps}_1$.

A 1-strategy $\sigma \in \mathrm{Str}_1$ is *μ-coherent* if $\mu(\sigma(v)) + v \leq \mu(v)$, for all 1-positions $v \in \mathrm{Ps}_1$.

Assuming a progress measure space with a canonical truncation operator, the notion of progress measure actually ensures that any play satisfying this progress condition will eventually be trapped in a cycle in which the maximal priority is odd, thereby witnessing a win for player 1. This is established by the following theorem.

Theorem 3.7: *(Progress Measure) Let $\mu \in \mathrm{MF}$ be a progress measure w.r.t. a progress measure space $\mathcal{M}$ with a canonical truncation operator. Then, $\|\mu\|_1$ is a 1-dominion for which all μ-coherent 1-strategies are 1-winning.*

Proof: Let us consider an arbitrary μ-coherent 1-strategy $\sigma_1 \in \mathrm{Str}_1$. All measures $\mu(v)$ of positions $v \in \|\mu\|_1 \cap \mathrm{Ps}_1$ are a progress for v *w.r.t.* the measures $\mu(\sigma_1(v))$ of their adjacents $\sigma_1(v)$, *i.e.*, formally, $\mu(\sigma_1(v))+v \leq \mu(v)$. The existence of such a coherent strategy is ensured by the fact that μ is a progress measure. Indeed, by Condition 2 of Definition 3.6, there necessarily exists an adjacent $w^\star \in Mv(v)$ of v such that $\mu(w^\star) + v \leq \mu(v)$.

It can be shown that σ_1 is a winning strategy for player 1 from all the positions in $\|\mu\|_1$, which implies that $\|\mu\|_1 \subseteq \mathrm{Win}_1$. To do this, consider a 0-strategy $\sigma_0 \in \mathrm{Str}_0$ and the associated play $\pi = \mathsf{play}((\sigma_0, \sigma_1), v)$ starting at a position $v \in \|\mu\|_1$. Assume, by contradiction, that π is won by player 0. Since the game ∂ is finite and the strategies are memoryless, π must contain a finite simple cycle, and so a finite simple path, and the maximal priority seen infinitely often along it needs to be even. In other words, there exist two natural numbers $h \in \mathbb{N}$ and $k \in \mathbb{N}_+$, such that $(\pi)_h = (\pi)_{h+k}$ and $\mathsf{pr}(\rho) \equiv_2 0$, where $\rho \triangleq ((\pi)_{\geq h})_{<h+k}$ is the simple path named above. Moreover, one can choose the value of the index h in such a way that $\mathsf{pr}((\pi)_h) \geq \mathsf{pr}((\pi)_i)$, for all $i \in \mathbb{N}$ with $h < i < h+k$. Recall that $((\pi)_i, (\pi)_{i+1}) \in Mv$, for all indexes $i \in \mathbb{N}$. Thanks to the two conditions of Definition 3.6 and the notion of play, it holds that

$$\mu((\pi)_{i+1}) + (\pi)_i \leq \mu((\pi)_i).$$

By Item 2ii of Definition 3.1, for all indexes $h \leq i < h+k$, it immediately follows that

$$(\mu((\pi)_{i+1}) + (\pi)_i)\restriction_{(\pi)_h} \leq \mu((\pi)_i)\restriction_{(\pi)_h}. \qquad (*)$$

At this point, it is important to observe that μ cannot associate the maximum value $\top$ with any position in the play, in particular when restricted to $(\pi)_h$, which means that the entire path is contained into $\|\mu\|_1$. The first element of the play trivially satisfies such a constraint, as $\mu((\pi)_0) = \mu(v) \neq \top$, since $v \in \|\mu\|_1$. Hence, $\mu((\pi)_0)\!\upharpoonright_{(\pi)_h} \neq \top$, by Item 2i of Definition 3.1. Now, suppose by contradiction that $\mu((\pi)_{i+1})\!\upharpoonright_{(\pi)_h} = \top$, for some index $i \in \mathbb{N}_+$. By Inequality $(*)$ and Item 2 of Definition 3.5, it follows that

$$\top = \mu((\pi)_{i+1})\!\upharpoonright_{(\pi)_h} \leq (\mu((\pi)_{i+1}) + (\pi)_i)\!\upharpoonright_{(\pi)_h} \leq \mu((\pi)_i)\!\upharpoonright_{(\pi)_h},$$

being $\mathsf{pr}((\pi)_h)$ the maximal priority along the path ρ that is also even. Due to the maximality of $\top$ ensured by Item 1 of Definition 3.1, it obviously follows that $\mu((\pi)_i)\!\upharpoonright_{(\pi)_h} = \top$ as well. Therefore, by iterating this process until index 0 is reached, one would obtain $\mu((\pi)_0)\!\upharpoonright_{(\pi)_h} = \top$, which is impossible, as previously observed.

To complete the proof, we can exploit the properties of the progress measure space. Recall that $\mu((\pi)_{h+1})\!\upharpoonright_{(\pi)_h} < \top$, thanks to the above observation. Thus, by Item 1 of Definition 3.5 and the fact that $\mathsf{pr}((\pi)_h)$ is an even priority, one can derive that

$$\mu((\pi)_{h+1})\!\upharpoonright_{(\pi)_h} < (\mu((\pi)_{h+1}) + (\pi)_h)\!\upharpoonright_{(\pi)_h}. \qquad (<)$$

Moreover, by applying again Item 2 of Definition 3.5 and due to the fact that $\mathsf{pr}((\pi)_h)$ is the maximal priority in the cycle, for all indexes $h < i < h + k$, it holds that

$$\mu((\pi)_{i+1})\!\upharpoonright_{(\pi)_h} \leq (\mu((\pi)_{i+1}) + (\pi)_i)\!\upharpoonright_{(\pi)_h}. \qquad (\leq)$$

As a consequence of the transitivity of the order relation between measures, by putting together Inequalities $(*)$, $(<)$, and $(\leq)$, one would therefore obtain

$$\mu((\pi)_{h+k})\!\upharpoonright_{(\pi)_h} < \mu((\pi)_h)\!\upharpoonright_{(\pi)_h}.$$

However, $(\pi)_{h+k} = (\pi)_h$, leading to $\mu((\pi)_h)\!\upharpoonright_{(\pi)_h} < \mu((\pi)_h)\!\upharpoonright_{(\pi)_h}$, which is clearly impossible, being $<$ an irreflexive relation. □

4. Solving Parity Games via Quasi-Dominion Measures

The framework set forth in the previous section is already sufficient to define a sound and complete approach for the solution of parity games, as shown in [21]. Here, however, we shall further refine the measure space in order to accommodate the notion of quasi dominion into the measure functions.

4.1. *Quasi-Dominion-Measure Functions*

The notion of *quasi dominion* was originally introduced for parity games in [17, 26] and, in a slightly different form, in [35]. Here we provide a somewhat more general version that can be easily integrated with measure functions.

Definition 4.1: (Quasi Dominion) A set of positions $\mathrm{Q} \subseteq \mathrm{Ps}$ is a *weak quasi* 0*-dominion* if there exists a 0-strategy $\sigma_0 \in \mathrm{Str}_0(\mathrm{Q})$, called 0*-witness for* Q, such that, for all 1-strategies $\sigma_1 \in \mathrm{Str}_1(\mathrm{Q})$ and positions $v \in \mathrm{Q}$, if the induced play $\pi = \mathsf{play}((\sigma_0, \sigma_1), v)$ is infinite then $\mathsf{pr}(\pi)$ is even. If the even-parity property holds also for finite plays, then Q is called *quasi* 0*-dominion.*

Essentially, a quasi 0-dominion consists in a set Q of positions, starting from which player 0 can force plays whose maximal prefixes contained in Q have even maximal priority. Observe that, in case the maximal prefix contained in Q is infinite, then the play is actually winning for player 0. When the condition holds only for infinite plays, the set is called weak quasi 0-dominion. Clearly, any quasi 0-dominion is also a weak quasi 0-dominion. Moreover, the latter are closed under subsets, while the former are not. It is an immediate consequence of the definition above that all infinite plays induced by the 0-witness, if any, are winning for player 0. This also entails that any subset $\mathrm{Q}^\star$ of a weak quasi 0-dominion Q, in which the player 0 can remain and from where the opponent cannot escape, is a 0-dominion. Indeed, in such a set of positions, player 0 always has moves that remain in $\mathrm{Q}^\star$, while the opponent can only choose moves remaining in $\mathrm{Q}^\star$. Hence, any play compatible with the 0-witness for Q that starts in $\mathrm{Q}^\star$ is infinite and entirely contained in $\mathrm{Q}^\star$. We have, so, the following result.

Corollary 4.2: *(Quasi Dominion) Let* $\mathrm{Q} \subseteq \mathrm{Ps}$ *be a* weak quasi 0-dominion, $\sigma_0 \in \mathrm{Str}_0(\mathrm{Q})$ *one of its* 0*-witnesses, and* $\mathrm{Q}^\star \subseteq \mathrm{Q}$ *a subset such that for all positions* $v \in \mathrm{Q}^\star \cap \mathrm{Ps}_0$ *it holds* $\sigma_0(v) \in \mathrm{Q}^\star$ *and for all positions* $v \in \mathrm{Q}^\star \cap \mathrm{Ps}_1$ *it holds* $Mv(v) \subseteq \mathrm{Q}^\star$*. Then,* $\mathrm{Q}^\star$ *is a* 0*-dominion.*

The notion of progress measure introduced in the previous section basically gives us positions that do satisfy the parity property and, thus, are winning for player 1, namely the non-$\top$ positions. This is done by enforcing on the measure space the conditions of Definition 3.5 that formally captures the idea that even priority positions push further away from the parity property. However, the progress measure is an asymmetric notion, centred around one of the two players, specifically player 1, and does not provide any meaningful information on the other player. More specifically, no estimation on how far player 0 is from winning the game starting in a given position, *i.e.*, from satisfying the dual/even parity property, can be extracted from it.

Quasi dominions, however, are precisely intended to encode the dual information for player 0. In this case, the odd priority positions are those that push further away from satisfying the dual/even parity property. A natural way to embed information about quasi dominions into the measures is, thus, to enforce the dual conditions of Definition 3.5, which leads us to the notion of *regress measure*. Here we constrain the behaviour of the truncation and stretch operators *w.r.t.* the odd positions, instead of the even ones.

Definition 4.3: (Regress Measure Space) A measure space $\mathcal{M}$ is a *regress space* if the following properties hold true, for each measure $\eta \in \mathrm{Ms}$ and position $v \in \mathrm{Ps}$:

(1) $(\eta + v)\restriction_v < \eta\restriction_v$, if $\bot < \eta\restriction_v < \top$ and $\mathsf{pr}(v)$ is odd;
(2) $(\eta + v)\restriction_u \leq \eta\restriction_u$, for all odd-priority positions u with $\mathsf{pr}(v) \leq \mathsf{pr}(u)$.

We can now define the notion of *regress measure* as the dual of the progress measure.

Definition 4.4: (Regress Measure) A measure function $\mu \in \mathrm{MF}$ is a *regress measure* if the following conditions hold, for all positions $v \in \|\mu\|_+ \setminus \|\mu\|_0$:

(1) $\mu(v) \leq \mu(w) + v$, for some adjacent $w \in Mv(v)$ of v, if $v \in \mathrm{Ps}_0$;
(2) $\mu(v) \leq \mu(w) + v$, for all adjacents $w \in Mv(v)$ of v, if $v \in \mathrm{Ps}_1$.

A 0-strategy $\sigma \in \mathrm{Str}_0$ is *μ-coherent* if $\mu(v) \leq \mu(\sigma(v)) + v$, for all 0-positions $v \in \mathrm{Ps}_0$.

Regress measures are meant to ensure that all the positions whose truncation is neither $\bot$ nor $\top$ form a weak quasi dominion for player 0, as established by the following theorem.

Theorem 4.5: *(Regress Measure) Let $\mu \in \mathrm{MF}$ be a regress measure w.r.t. a regress measure space $\mathcal{M}$ with a canonical truncation operator. Then, $\|\mu\|_+ \setminus \|\mu\|_0$ is a weak quasi 0-dominion for which all μ-coherent 0-strategies are 0-witnesses, once restricted to $\|\mu\|_+ \setminus \|\mu\|_0$.*

Proof: *Mutatis mutandis*, the proof proceeds similarly to the one previously presented for Theorem 3.7.

First of all, let us consider an arbitrary μ-coherent 0-strategy $\sigma_0 \in \mathrm{Str}_0$. All measures $\mu(v)$ of positions $v \in \|\mu\|_0 \cap \mathrm{Ps}_0$ are a regress for v *w.r.t.* the measures $\mu(\sigma_0(v))$ of their adjacents $\sigma_0(v)$, *i.e.*, formally, $\mu(v) \leq \mu(\sigma_0(v)) + v$. The existence of such a coherent strategy is ensured by the fact that μ is a regress measure. Indeed, by Condition 1 of Definition 4.4, there necessarily exists an adjacent $w^\star \in Mv(v)$ of v such that $\mu(v) \leq \mu(w^\star) + v$.

To prove that $\mathrm{Q} \triangleq \|\mu\|_+ \setminus \|\mu\|_0$ is a weak quasi 0-dominion with $\sigma_0 \downarrow \mathrm{Q} \in \mathrm{Str}_0(\mathrm{Q})$ as 0-witness, consider a 1-strategy $\sigma_1 \in \mathrm{Str}_1(\mathrm{Q})$ such that the associated play $\pi = \mathsf{play}((\sigma_0 \downarrow \mathrm{Q}, \sigma_1), v)$ starting at a position $v \in \mathrm{Q}$ is infinite. Now, one needs to show that $\mathsf{pr}(\pi)$ is even. Assume by contradiction that this condition on the parity of the priority does not hold. Since the game $\eth$ is finite and the strategies are memoryless, π must contain a finite simple cycle, and so a finite simple path, and the maximal priority seen infinitely often along it needs to be odd. In other words, there exist two natural numbers $h \in \mathbb{N}$ and $k \in \mathbb{N}_+$, such that $(\pi)_h = (\pi)_{h+k}$ and $\mathsf{pr}(\rho) \equiv_2 1$, where $\rho \triangleq ((\pi)_{\geq h})_{<h+k}$ is the simple path named above. Moreover, one can choose the value of the index h in such a way that $\mathsf{pr}((\pi)_h) \geq \mathsf{pr}((\pi)_i)$, for all $i \in \mathbb{N}$ with $h < i < h+k$. Recall that $((\pi)_i, (\pi)_{i+1}) \in Mv$, for all indexes $i \in \mathbb{N}$. Thanks to the two conditions of Definition 4.4 and the notion of play, it holds that

$$\mu((\pi)_i) \leq \mu((\pi)_{i+1}) + (\pi)_i.$$

By Item 2ii of Definition 3.1, for all indexes $h \leq i < h+k$, it immediately follows that

$$\mu((\pi)_i)\restriction_{(\pi)_h} \leq (\mu((\pi)_{i+1}) + (\pi)_i)\restriction_{(\pi)_h}. \qquad (*)$$

At this point, we can exploit the properties of the regress measure space. By construction, $\pi \in \mathrm{Pth}(\mathrm{Q})$, where we recall that $\mathrm{Q} \triangleq \|\mu\|_+ \setminus \|\mu\|_0$. Thus, observe that both $\mu((\pi)_i)\restriction_{(\pi)_i} \neq \bot$ and $\mu((\pi)_i)\restriction_{(\pi)_i} \neq \top$ hold, for all $i \in \mathbb{N}$. This implies that $\bot < \mu((\pi)_i)\restriction_{(\pi)_i}$ and $\mu((\pi)_i)\restriction_{(\pi)_h} < \top$, thanks to Item 2i of Definition 3.1. By Item 2 of Definition 4.3 and the fact that $\mathsf{pr}((\pi)_h)$ is an odd priority, one can derive that

$$(\mu((\pi)_{h+1}) + (\pi)_h)\restriction_{(\pi)_h} \leq \mu((\pi)_{h+1})\restriction_{(\pi)_h}.$$

Hence, by Inequality $(*)$ and the above observations, it follows that

$$\bot < \mu((\pi)_h)\restriction_{(\pi)_h} \leq (\mu((\pi)_{h+1}) + (\pi)_h)\restriction_{(\pi)_h} \leq \mu((\pi)_{h+1})\restriction_{(\pi)_h} < \top,$$

which in turn implies

$$\bot < \mu((\pi)_{h+1})\restriction_{(\pi)_h} < \top.$$

Now, by Item 1 of Definition 4.3, one can obtain that

$$(\mu((\pi)_{h+1}) + (\pi)_h)\restriction_{(\pi)_h} < \mu((\pi)_{h+1})\restriction_{(\pi)_h}. \qquad (<)$$

Moreover, by applying again Item 2 of Definition 4.3 and due to the fact that $\mathsf{pr}((\pi)_h)$ is the maximal priority in the cycle, for all indexes $h < i < h + k$, it holds that

$$(\mu((\pi)_{i+1}) + (\pi)_i)\restriction_{(\pi)_h} \leq \mu((\pi)_{i+1})\restriction_{(\pi)_h}. \qquad (\leq)$$

As a consequence of the transitivity of the order relation between measures, by putting together Inequalities (∗), (<), and (≤), one would derive

$$\mu((\pi)_h)\restriction_{(\pi)_h} < \mu((\pi)_{h+k})\restriction_{(\pi)_h}.$$

However, $(\pi)_{h+k} = (\pi)_h$, leading to $\mu((\pi)_h)\restriction_{(\pi)_h} < \mu((\pi)_h)\restriction_{(\pi)_h}$, which is obviously impossible, being $<$ an irreflexive relation. □

A *quasi-dominion measure* is, then, defined as a regress measure with the additional property that all the positions with measure $\top$ form a 0-dominion, *i.e.*, are indeed winning for player 0.

Definition 4.6: (Quasi-Dominion Measure) A measure function $\mu \in \mathrm{MF}$ is a *quasi-dominion measure* (QDM, for short) if it is a regress measure for which $\|\mu\|_0$ is a 0-dominion. QDM denotes the set of all QDMs.

The following theorem establishes the main property of quasi-dominion measures, namely that the set of non-$\bot$ positions always forms a weak quasi 0-dominion.

Theorem 4.7: *(Quasi-Dominion Measure I) Let $\mu \in \mathrm{MF}$ be a quasi-dominion measure w.r.t. a regress measure space $\mathcal{M}$ with a canonical truncation operator. Then, $\|\mu\|_+$ is a weak quasi 0-dominion for which all μ-coherent 0-strategies that are winning on $\|\mu\|_0$ are 0-witnesses, once restricted to $\|\mu\|_+$.*

Proof: Let $\mathrm{Q} \triangleq \|\mu\|_+$ and $\sigma_0 \in \mathrm{Str}_0$ be an arbitrary μ-coherent 0-strategy that is winning on $\|\mu\|_0$. To prove that Q is a weak quasi 0-dominion with $\sigma_0 \downarrow \mathrm{Q} \in \mathrm{Str}_0(\mathrm{Q})$ as 0-witness, consider any 1-strategy $\sigma_1 \in \mathrm{Str}_1(\mathrm{Q})$ such that the play $\pi = \mathsf{play}((\sigma_0 \downarrow \mathrm{Q}, \sigma_1), v)$ from position $v \in \mathrm{Q}$ is infinite. We need to show that $\mathsf{pr}(\pi)$ is even. The following two different cases may arise, where $\mathrm{H} \triangleq \|\mu\|_+ \setminus \|\mu\|_0 \subseteq \mathrm{Q}$:

- $[\pi \in \mathrm{Pth}(\mathrm{H})]$. By Theorem 4.5, H is a weak quasi 0-dominion with $\sigma_0 \downarrow \mathrm{H} = (\sigma_0 \downarrow \mathrm{Q}) \downarrow \mathrm{H}$ as 0-witness. Moreover, π is a $(\sigma_0 \downarrow \mathrm{H}, v)$-play in H. Hence, the thesis immediately follows from the definition of weak quasi 0-dominion.
- $[\pi \notin \mathrm{Pth}(\mathrm{H})]$. Since $\pi \notin \mathrm{Pth}(\mathrm{H})$, there clearly exists and index $i \in \mathbb{N}$ such that $(\pi)_{\geq i} \in \mathrm{Pth}(\|\mu\|_0)$. This follows from the fact that $\|\mu\|_0$ is a 0-dominion with σ_0 as a 0-winning strategy, since every play compatible with σ_0 gets necessarily trapped in $\|\mu\|_0$. Moreover, $(\pi)_{\geq i}$ is a $(\sigma_0 \downarrow \|\mu\|_0, (\pi)_i)$-play in $\|\mu\|_0$. Hence, we immediately obtain the thesis from the definition of 0-dominion, since $\mathsf{pr}(\pi) = \mathsf{pr}((\pi)_{\geq i})$. □

4.2. *Simple-Measure Functions*

While the solution of a parity game involves checking the parity property along infinite plays, as the solution algorithm proceeds, the measure of a position v encodes a finite horizon approximation of that condition for v. This approximation is progressively refined during the computation, by exploring longer and longer finite prefixes of the possible plays starting from v. Clearly, any play that contains a cycle is either winning for player 0 or for the opponent. In a sense, the shortest prefix of the play that ends with a repetition of some position already provides all the necessary information to assess the winning player of the entire infinite play. This observation suggests that measures need only encode information of finite simple paths in the game, since those are the only prefixes that need to be extended to obtain finer approximations.

From now on, we shall fix a measure space $\mathcal{M}$ and, thus, the induced measure-function space $\mathcal{F}$. Given a position $v \in \mathrm{Ps}$ and a set of positions $\mathrm{X} \subseteq \mathrm{Ps}$, we introduce the set

$$\mathrm{SMs}(v, \mathrm{X}) \triangleq \{\eta(\pi) \in \mathrm{Ms} : \pi \in \mathrm{SPth}(v, \mathrm{X})\} \cup \{\top\}$$

of the *simple measures* of v *w.r.t.* X. This set contains, besides the measure $\top$, only those measures induced by the finite simple paths originating at v and composed only of positions in X. It is immediate to observe that $\mathrm{SMs}(v, \mathrm{X}) = \{\top, \bot\}$, whenever $v \notin \mathrm{Ps}$, since $\mathrm{SPth}(v, \mathrm{V}) = \{\varepsilon\}$.

Proposition 4.8: *$\langle \mathrm{SMs}(v, \mathrm{X}), <, \bot, \top \rangle$ is a finite strict total order with minimum and maximum, for all positions $v \in \mathrm{Ps}$ and sets of positions $\mathrm{X} \subseteq \mathrm{Ps}$.*

Following the observations above, *simple measure functions* restrict the possible measures of each position to those induced by finite simple paths contained in the quasi 0-dominion $\|\mu\|_+$.

Definition 4.9: (Simple Measure Function) A measure function $\mu \in \mathrm{MF}$ is a *simple measure* (SM, for short) if $\mu(v) \in \mathrm{SMs}(v, \|\mu\|_+)$, for all positions $v \in \mathrm{Ps}$. SM denotes the set of all SMs.

The next proposition shows that, for simple measures μ, if the truncation of the measure $\mu(v)$ *w.r.t.* the position v itself is $\bot$, then the original measure $\mu(v)$ needs to be $\bot$ as well, since (i) $\mathrm{SMs}(v, \|\mu\|_+) = \{\top, \bot\}$ and (ii) $\top\restriction_v = \top$, due to Item 2i of Definition 3.1.

Proposition 4.10: *For every simple measure* $\mu \in \mathrm{SM}$ *and position* $v \in \|\mu\|_\bot$, *it holds that* $\mu(v) = \bot$.

The measure function $\mu_\bot$ is clearly the minimal element *w.r.t.* $\sqsubseteq$ in the set SM of simple measures.

Proposition 4.11: $\langle \mathrm{SM}, \sqsubseteq \rangle$ *is a finite partial order with* $\mu_\bot \in \mathrm{SM}$ *as unique minimal element.*

Putting together Definitions 4.6 and 4.9, we obtain *simple quasi-dominion measures*, which enjoy a stronger property than the one stated in Theorem 4.7.

Theorem 4.12: *(Quasi-Dominion Measure II) Let* $\mu \in \mathrm{MF}$ *be a simple quasi-dominion measure w.r.t. a regress measure space* $\mathcal{M}$ *with a canonical truncation operator satisfying the equality* $\bot\restriction_v = \bot$, *for all odd priority positions* v. *Then,* $\|\mu\|_+$ *is a quasi 0-dominion for which all* μ*-coherent 0-strategies that are winning on* $\|\mu\|_0$ *are 0-witnesses, once restricted to* $\|\mu\|_+$.

Proof: By Theorem 4.7, $\mathrm{Q} \triangleq \|\mu\|_+$ is a weak quasi 0-dominion. Therefore, to prove that it is a quasi 0-dominion with $\sigma_0 \downarrow \mathrm{Q} \in \mathrm{Str}_0(\mathrm{Q})$ as 0-witness, for some arbitrary μ-coherent 0-strategy $\sigma_0 \in \mathrm{Str}_0$ that is winning on $\|\mu\|_0$, consider a 1-strategy $\sigma_1 \in \mathrm{Str}_1(\mathrm{Q})$ for which the associated play $\pi = \mathsf{play}((\sigma_0 \downarrow \mathrm{Q}, \sigma_1), v)$ starting at a position $v \in \mathrm{Q}$ is finite. Now, one needs to show that $\mathsf{pr}(\pi)$ is even.

Suppose by contradiction that $\mathsf{pr}(\pi)$ is odd. Then, there exists an index $h \in \mathbb{N}$ with $0 \leq h \leq n \triangleq |\pi| - 1$, such that $\mathsf{pr}((\pi)_h) \equiv_2 1$ and $\mathsf{pr}((\pi)_h) \geq \mathsf{pr}((\pi)_i)$, for all $i \in \mathbb{N}$ with $h \leq i \leq n$. Thanks to Definition 4.4,

Items 2i and 2ii of Definition 3.1, Item 2 of Definition 4.3, and the fact that $\mu((\pi)_h)\lceil_{(\pi)_h} \neq \bot$ and $\mathsf{pr}((\pi)_h)$ is the maximal priority along the finite path $(\pi)_{\geq h}$ that is also odd, one can obtain the following inequality, by applying the same inductive reasoning employed in the second half of the proof of Theorem 4.5:

$$\bot < \mu((\pi)_h)\lceil_{(\pi)_h} \leq \mu((\pi)_n)\lceil_{(\pi)_h}. \qquad (*)$$

Since π if finite, either one of the two strategies σ_0 and σ_1 have to terminate in $(\pi)_n$, *i.e.*, we necessarily have that $w^\star \in \|\mu\|_\bot$, where $w^\star \triangleq \sigma_0((\pi)_n)$, if $(\pi)_n \in \mathrm{Ps}_0$, and $w^\star \triangleq \sigma_1((\pi)_n)$, otherwise. Hence, by Proposition 4.10, it holds that $\mu(w^\star) = \bot$, due to the fact that μ is a simple measure. Moreover, again by Definition 4.4, it holds that

$$\mu((\pi)_n) \leq \mu(w^\star) + (\pi)_n = \bot + (\pi)_n,$$

from which, by Item 2ii of Definition 3.1, it follows that

$$\mu((\pi)_n)\lceil_{(\pi)_h} \leq (\bot + (\pi)_n)\lceil_{(\pi)_h}. \qquad (\diamond)$$

At this point, by Item 2 of Definition 4.3 and the equality $\bot\lceil_v = \bot$, for the odd-priority position v, one can obtain that

$$(\bot + (\pi)_n)\lceil_{(\pi)_h} \leq \bot\lceil_{(\pi)_h} = \bot. \qquad (\bot)$$

Thus, as an immediate consequence of Inequalities $(*)$, $(\diamond)$, and $(\bot)$, one would derive $\bot < \bot$, which is obviously impossible, being $<$ an irreflexive relation. □

Since $\mu_\bot$ is the measure function induced by the empty path on each position, the following property is immediate.

Proposition 4.13: *The minimal measure function $\mu_\bot \in \mathrm{MF}$ is a simple quasi-dominion measure.*

5. A Concrete Algorithm

This section describes an algorithm that solves any parity game by maintaining and updating a simple quasi-dominion measure function, until it reaches a fixpoint that is both a progress and a quasi-dominion measure. At that point, the results in the previous sections ensure that the winning positions for both players are determined and easily recovered from the final measure by computing the 0- and 1-denotations.

5.1. *A Concrete Measure Space*

The measures used by the concrete algorithm associate a non-negative integer with each priority in the game, in other words, they are sequences of naturals, one for each priority. For technical reasons, however, we introduce a more general class of concrete measures $\mathrm{CMs} \triangleq \mathrm{Pr} \to \mathbb{Z}$, whose range also includes the negative integers. This allows us to provide algebraic operations on measures, such as addition and subtraction, which will prove instrumental in the implementation of the basic operators used by the algorithm. More specifically: (i) the null element $\mathbf{0} \in \mathrm{CMs}$ is the distinguished measure such that $(\mathbf{0})(k) \triangleq 0$, for all indexes $k \in \mathrm{Pr}$; (ii) the opposite of a measure and the sum of two measures are defined point-wise: $(-\phi)(k) \triangleq -\phi(k)$ and $(\phi_1 + \phi_2)(k) \triangleq \phi_1(k) + \phi_2(k)$, for all $k \in \mathrm{Pr}$. Recall that, according to Definition 3.1, measures need to be totally ordered. For our concrete measures, we employ an *alternate lexicographic order*, that is, if we interpret a measure as a sequence of integers, with decreasing indexes from left to right, values that are later in the sequence are less important than those that come earlier: as in the standard numeric representation, the left-most integer is the most-significant digit, while the right-most integer is the least-significant one. Moreover, values with even indexes are ordered in the natural way, namely by increasing magnitude, whereas those with odd indexes are ordered in the opposite fashion, *i.e.*, by decreasing magnitude. Formally, $< \subseteq \mathrm{CMs} \times \mathrm{CMs}$ is the strict total order defined as follows: $\phi_1 < \phi_2$ if there exists an index $k \in \mathrm{Pr}$ such that (i) k is the greatest index for which $\phi_1(k) \neq \phi_2(k)$ and (ii) $\phi_1(k) < \phi_2(k)$, if k is even, and $\phi_2(k) < \phi_1(k)$, otherwise. A special family of measures is given by the Kronecker delta $\delta \colon \mathrm{Pr} \to (\mathrm{Pr} \to \{0, 1\})$, where $\delta_i(i) \triangleq 1$, and $\delta_i(j) \triangleq 0$, for all $j \neq i$. Obviously, $\delta_i \in \mathrm{CMs}$, for every index $i \in \mathrm{Pr}$. It is quite immediate to show the following property.

Proposition 5.1: *The structure* $\mathcal{C} \triangleq \langle \mathrm{CMs}, <, \mathbf{0}, -, + \rangle$ *is a totally-ordered Abelian group.*

We can now define the measure space used by the concrete algorithm. It suffices to restrict the measures to only assign non-negative values to the existing priorities in the game, and then define the appropriate canonical truncation and stretch operators. Let CMs^+ denote the set of all the concrete measures $\phi \in \mathrm{CMs}$ with the following two properties: (i) the highest priority k for which $\phi(k) > 0$ is even; (ii) $\phi(k) \geq 0$, for all $k \in \mathrm{Pr}$.

Definition 5.2: (Concrete Measure Structure) The *concrete measure structure* is the tuple $\mathcal{M} \triangleq \langle \mathrm{Ms}, <, \bot, \top, \upharpoonright, + \rangle$, whose components are defined as follows:

(1) $\mathrm{Ms} \triangleq \mathrm{CMs}^{+} \cup \{\top\}$, where $\top$ is a distinguished fresh element and $\bot \triangleq \mathbf{0}$;
(2) $< \subseteq \mathrm{Ms} \times \mathrm{Ms}$ is the order on CMs^{+} extended with $\eta < \top$, for every measure $\eta \in \mathrm{CMs}^{+}$;
(3) $\restriction : \mathrm{Ms} \times \mathrm{Ps} \to \mathrm{Ms}$ is the operator such that, for all positions $v \in \mathrm{Ps}$, the following holds: (i) $\top\restriction_v \triangleq \top$; (ii) $(\eta\restriction_v)(p) \triangleq \eta(p)$, if $p \geq \mathsf{pr}(v)$, and $(\eta\restriction_v)(p) \triangleq 0$, otherwise, for all $\eta \in \mathrm{CMs}^{+}$ and $p \in \mathrm{Pr}$;
(4) $+ : \mathrm{Ms} \times \mathrm{Ps} \to \mathrm{Ms}$ is the operator such that $\eta + v \triangleq \max\{0, \eta + \delta_{\mathsf{pr}(v)}\}$, for all positions $v \in \mathrm{Ps}$.

Truncating a measure *w.r.t.* a position v consists in setting to zero the value associated with all priorities smaller than the priority of v. Stretching a measure *w.r.t.* a position v, instead, means incrementing the value associated with the priority of v, unless the result is lower than 0, in which case the stretch is set to 0, so as to enforce it to belong to CMs^{+}. For example, consider a game with priorities from 0 to 4 and the three measures $\eta_1 \triangleq (0,0,1,0,1)$, $\eta_2 \triangleq (0,0,1,1,0)$, and $\eta_3 \triangleq (0,0,1,1,1)$. Then, we have $\eta_2 < \eta_3 < \eta_1$. Indeed, 0 is the greatest priority in which η_2 and η_3 differ, it is even, and $\eta_2(0) = 0 < 1 = \eta_3(0)$. Moreover, 1 is the greatest priority in which both η_2 and η_3 differ from η_1, it is odd, and $\eta_2(1) = \eta_3(1) = 1 > 0 = \eta_1(1)$. By truncating the three measures at a position v with priority 1, we obtain $\eta_2\restriction_v = \eta_3\restriction_v = (0,0,1,1,0) < \eta_1\restriction_v = (0,0,1,0,0)$. We also have $\eta_1 + u = \eta_2 + u = \eta_3 + u = \bot$, if u is a position with priority 3 and $(\eta_2 + w) + u = (1,1,1,1,0) < (1,0,1,2,0) = (\eta_2 + w) + v$, if w is a position with priority 4.

The following straightforward result states that the concrete measure structure satisfies all the desired properties introduced in the previous sections, plus some additional ones. Specifically, the truncation operator preserves the $\bot$ measure, the truncation of a non-$\bot$ stretch cannot lead to the $\bot$ measure, and the stretch operator commutes with the addition on measures.

Proposition 5.3: *The concrete measure structure is a progress and regress measure space, whose restriction and stretch operators are canonical. Moreover, the following properties hold true, for all positions $v \in \mathrm{Ps}$, measures $\eta \in \mathrm{Ms}$, and evaluations $\phi \in \mathrm{CMs}$: (i) $\bot\restriction_v = \bot$; (ii) if $\eta + v \neq \bot$ then $(\eta + v)\restriction_v \neq \bot$; (iii) if $\eta + v \neq \bot$ then $(\eta + v) + \phi = (\eta + \phi) + v$.*

5.2. *The Solution Algorithm*

The algorithm we now propose, which makes implicit use of the concrete measure structure just introduced, is based on repeatedly applying two *progress operators*, prg_+ and $\mathsf{prg}_\bot$, to an initial quasi-dominion measure function μ, until a fixpoint $\mu^\star$ is reached. At that point, the positions whose measure is $\top$ in $\mu^\star$ are winning for player 0, while all other positions are winning for player 1.

Recall that a function f on an ordered set is *inflationary* if for all elements x of its domain it holds $x \leq f(x)$. We show in the following that both progress operators are inflationary, so that the fixpoint $\mu^\star$ is the limit of the ascending sequence of measures obtained by repeated application of those progress operators (*a.k.a.* the *inflationary fixpoint*). We denote by $\mathsf{sol}(\mu)$ such limit, as a partial mapping from MF to MF, when starting from measure μ:

$$\mathsf{sol} \triangleq \mathsf{ifp}\, \mu \,.\, \mathsf{prg}_+(\mathsf{prg}_\bot(\mu)) \colon \mathrm{MF} \rightharpoonup \mathrm{MF}.$$

Intuitively, all positions with a non-$\bot$ measure in μ, *i.e.*, $\mathrm{Q} \triangleq \|\mu\|_+$, form a quasi 0-dominion and the prg_+ (*resp.*, $\mathsf{prg}_\bot$) operator is responsible for enforcing the progress condition on the positions inside (*resp.*, outside) Q that do not satisfy the proper inequalities between the measures along the moves. This is done in such a way to preserve the properties of quasi-dominion measure function and represents the main point where the classic progress-measure approaches and the proposed technique diverge.

Both operators prg_+ and $\mathsf{prg}_\bot$ internally employ a *lift operator* $\mathsf{lift} \colon \mathrm{MF} \times 2^{\mathrm{Ps}} \times 2^{\mathrm{Ps}} \rightarrow \mathrm{MF}$, that adjusts a measure function so that it locally satisfies the conditions of both the progress and regress measures (Definitions 3.6 and 4.4). The two operators need to selectively adjust the measure of specific sets of positions. For this reason, besides the current measure function, the lift operator carries two additional arguments: (i) the set of positions S whose measure we want to update, and (ii) the set of successor positions T that the update must be based on. Formally, we obtain the following definition:

$$\mathsf{lift}(\mu, \mathrm{S}, \mathrm{T})(v) \triangleq \begin{cases} \mathsf{max}\{\mu(w) + v : w \in \mathit{Mv}(v) \cap \mathrm{T}\}, & \text{if } v \in \mathrm{S} \cap \mathrm{Ps}_0; \\ \mathsf{min}\{\mu(w) + v : w \in \mathit{Mv}(v) \cap \mathrm{T}\}, & \text{if } v \in \mathrm{S} \cap \mathrm{Ps}_1; \\ \mu(v), & \text{otherwise.} \end{cases}$$

The $\mathsf{prg}_\bot$ operator is tasked with adjusting the measure of the positions that currently have the minimal measure $\bot$. After the update, some of them will acquire a positive measure, thus entering into the current quasi

0-dominion. From an operational viewpoint, $\mathsf{prg}_\bot$ consists of a single call to lift:

$$\mathsf{prg}_\bot(\mu) \triangleq \mathsf{lift}(\mu, \|\mu\|_\bot, \mathrm{Ps}) : \mathrm{MF} \to \mathrm{MF}.$$

Applying the definitions, one can easily see that $\mathsf{prg}_\bot$ raises the measure of those positions such that: (i) they have the minimal measure $\bot$, (ii) they either belong to player 0 and have an adjacent with positive measure or belong to player 1 and have all adjacents with positive measures, and (iii) the stretch of the adjacent measures is greater than $\bot$.

The following lemma states the main properties of interest for the $\mathsf{prg}_\bot$ operator, where we assume $\mathrm{MF}_\bot \triangleq \{\mu \in \mathrm{MF} : \forall v \in \|\mu\|_\bot \,.\, \mu(v) = \bot\}$ and $\mathrm{QDM}_\bot \triangleq \mathrm{QDM} \cap \mathrm{MF}_\bot$. Observe that, by Proposition 4.10, it holds that $\mathrm{SM} \subseteq \mathrm{MF}_\bot$.

Lemma 5.4: *The progress operator* $\mathsf{prg}_\bot$ *enjoys the following properties: (i) it is an inflationary function from* $\mathrm{MF}_\bot$ *to* $\mathrm{MF}_\bot$*; (ii) it maps* $\mathrm{QDM}_\bot$ *into* $\mathrm{QDM}_\bot$*; (iii) it maps* SM *into* SM*; (iv) every fixpoint* $\mu \in \mathrm{MF}$ *of* $\mathsf{prg}_\bot$ *is a progress measure over* $\|\mu\|_\bot$.

Proof: We analyse the four properties separately, where we recall that $\mathsf{prg}_\bot(\mu) = \mathsf{lift}(\mu, \|\mu\|_\bot, \mathrm{Ps})$.

- **[i].** Let v be a position such that $\mu^\star(v) \neq \mu(v)$, where $\mu^\star \triangleq \mathsf{prg}_\bot(\mu)$. By definition of the lift operator, it holds that $v \in \|\mu\|_\bot$. Obviously, $\mu(v) = \bot$, since $\mu \in \mathrm{MF}_\bot$. Thus, by Item 1 of Definition 3.1, it follows that $\mu(v) = \bot < \mu^\star(v)$, being $\bot$ the minimal measure. Hence, $\mu \sqsubseteq \mu^\star$, due to the arbitrary choice of v, which means that $\mathsf{prg}_\bot$ is inflationary on $\mathrm{MF}_\bot$, as required by the lemma statement. In addition, it holds that $\mu^\star \in \mathrm{MF}_\bot$. Indeed, again by definition of the lift operator, there exists an adjacent $w \in Mv(v)$ of v, such that $\mu^\star(v) = \mu(w) + v$. If v has even priority, then either $\mu(w){\restriction_v} = \top$ and, so, $\mu^\star(v){\restriction_v} = \top$, by Item 1 of Definition 3.1 and Item 2 of Definition 3.5, or $\mu(w){\restriction_v} < \top$ and $\mu^\star(v){\restriction_v} = (\mu(w)+v){\restriction_v} > \mu(w){\restriction_v}$, by Item 1 of Definition 3.5. If v has odd priority, instead, by Item ii of Proposition 5.3, it holds that if $\mu(w)+v \neq \bot$ then $(\mu(w)+v){\restriction_v} \neq \bot$, from which it follows that $\mu^\star(v){\restriction_v} = (\mu(w) + v){\restriction_v} \neq \bot$, since $\mu(w) + v = \mu^\star(v) \neq \mu(v) = \bot$. Thus, in all cases, $\mu^\star(v){\restriction_v} \neq \bot$, *i.e.*, $v \notin \|\mu^\star\|_\bot$, which vacuously satisfies the definitional requirement of $\mathrm{MF}_\bot$.

- **[ii].** Let μ be a QDM in $\mathrm{MF}_{\perp}$. Thanks to the above item, it suffices to prove that $\mu^{\star} \triangleq \mathsf{prg}_{\perp}(\mu)$ is a QDM as well. To do this, one first needs to show that $\mu^{\star}$ is a regress measure. Consider an arbitrary position $v \in \|\mu^{\star}\|_{+} \setminus \|\mu^{\star}\|_{0}$. Then, two cases may arise.
  - $[v \in \|\mu\|_{+}]$. $\mu^{\star}(v) = \mu(v)$, so, $v \in \|\mu\|_{+} \setminus \|\mu\|_{0}$. Since μ is a quasi-dominion measure and, therefore, also a regress measure, it satisfies both conditions of Definition 4.4 at v. It is quite immediate to prove that the same holds for $\mu^{\star}$ at the same position v as well, thanks to Item 3ii of Definition 3.1, since $\mu \sqsubseteq \mu^{\star}$ as proved in Item i above.
  - $[v \notin \|\mu\|_{+}]$. In this case, one needs to analyse the following two subcases. If $v \in \mathrm{Ps}_0$, it holds that

$$\begin{aligned}\mu^{\star}(v) &= \mathsf{max}\{\mu(w) + v : w \in Mv(v)\}\\ &\leq \mathsf{max}\{\mu^{\star}(w) + v : w \in Mv(v)\} = \mu^{\star}(w) + v\end{aligned}$$

    for some $w \in Mv(v)$ adjacent to v. Thus, Condition 1 of Definition 4.4 is satisfied for $\mu^{\star}$. If $v \in \mathrm{Ps}_1$, instead, it holds that $\mu^{\star}(v) = \mathsf{min}\{\mu(w) + v : w \in Mv(v)\} \leq \mathsf{min}\{\mu^{\star}(w) + v : w \in Mv(v)\} \leq \mu^{\star}(w) + v$, for all $w \in Mv(v)$ adjacent to v. Hence, also Condition 2 of Definition 4.4 is satisfied for $\mu^{\star}$. Observe that, to prove both conditions, we applied again Item 3ii of Definition 3.1 and the fact that $\mu \sqsubseteq \mu^{\star}$, as proved in Item i above.

At this point, it only remains to prove that $\|\mu^{\star}\|_{0}$ is a 0-dominion. By hypothesis, it is known that $\|\mu\|_{0}$ is a 0-dominion, being μ a QDM. Therefore, let us consider a position $v \in \|\mu^{\star}\|_{0} \setminus \|\mu\|_{0}$. We can show that, again by definition of the lift operator, there exists an adjacent $w \in Mv(v)$ of v, such that $w \in \|\mu\|_{0}$, if $v \in \mathrm{Ps}_0$, and all adjacents $w \in Mv(v)$ of v satisfy $w \in \|\mu\|_{0}$, otherwise. Indeed, if $v \in \mathrm{Ps}_0$, there exists an adjacent $w \in Mv(v)$ of v, such that $\mu(w) + v = \mu^{\star}(v) = \top$. This implies that $\mu(w) = \top$, due to Item 3i of Definition 3.1. Hence, $w \in \|\mu\|_{0}$. Similarly, if $v \in \mathrm{Ps}_1$, all adjacents $w \in Mv(v)$ of v satisfy the equality $\mu(w) + v = \mu^{\star}(v) = \top$. Thus, again by Item 3i of Definition 3.1, it holds that $\mu(w) = \top$, which means that $w \in \|\mu\|_{0}$. As an immediate consequence, every play starting at v and compatible with the 0-winning strategy on $\|\mu\|_{0}$, suitably extended to $\|\mu^{\star}\|_{0}$, is won by player 0. Therefore, $\|\mu^{\star}\|_{0}$ is necessarily a 0-dominion, as required by the definition of QDM.

- [iii]. Let μ be a SM and $\mu^\star \triangleq \mathsf{prg}_\bot(\mu)$ the result of the $\mathsf{prg}_\bot$ operator. One needs to prove that the latter is a SM too. To do this, let us focus on a position v, such that $\mu^\star(v) \neq \mu(v)$. If $\mu^\star(v) = \top$, there is nothing more to show, as $\top \in \mathrm{SMs}(v, \|\mu^\star\|_+)$, as required by Definition 4.9. Therefore, assume $\mu^\star(v) \neq \top$. By definition of the lift operator, there exists an adjacent $w \in Mv(v)$ of v, such that $\mu^\star(v) = \mu(w) + v$. Now, by Item 3i of Definition 3.1, it follows that $\mu(w) \neq \top$. Thus, thanks to the fact that μ is a SM, it holds that $\mu(w) \in \mathrm{SMs}(w, \|\mu\|_+)$, which means that there exists a simple path $\pi \in \mathrm{SPth}(w, \|\mu\|_+)$, such that $\mu(w) = \eta(\pi)$. Obviously, $\mu^\star(v) = \mu(w) + v = \eta(\pi) + v = \eta(v \cdot \pi)$. Moreover, $v \cdot \pi$ is a simple path passing through positions in $\{v\} \cup \|\mu\|_+$, *i.e.*, $v \cdot \pi \in \mathrm{SPth}(v, \{v\} \cup \|\mu\|_+)$, since $v \notin \|\mu\|_+$. As shown at the end of the proof of the first item of this lemma, $\mu^\star(v)\restriction_v \neq \bot$, so, $v \in \|\mu^\star\|_+$. Thus, as an immediate consequence, one obtains that $v \cdot \pi \in \mathrm{SPth}(v, \|\mu^\star\|_+)$, being $\{v\} \cup \|\mu\|_+ \subseteq \|\mu^\star\|_+$, which implies that $\mu^\star(v) \in \mathrm{SMs}(v, \|\mu^\star\|_+)$. Hence, $\mu^\star$ is a SM.
- [iv]. Let μ be a fixpoint of $\mathsf{prg}_\bot$, *i.e.*, $\mu = \mathsf{lift}(\mu, \|\mu\|_\bot, \mathrm{Ps})$, and $v \in \|\mu\|_\bot$ an arbitrary position. If $v \in \mathrm{Ps}_0$, it holds that $\mu(w) + v \leq \max\{\mu(w) + v : w \in Mv(v)\} = \mu(v)$, for all adjacents $w \in Mv(v)$ of v. Thus, Condition 1 of Definition 3.6 is satisfied on $\|\mu\|_\bot$. If $v \in \mathrm{Ps}_1$, instead, it holds that $\mu(w) + v = \min\{\mu(w) + v : w \in Mv(v)\} = \mu(v)$, for some adjacent $w \in Mv(v)$ of v. Hence, Condition 2 of Definition 3.6 is satisfied on $\|\mu\|_\bot$ as well. □

Algorithm 1: Operator prg_+

signature $\mathsf{prg}_+ : \mathrm{MF} \to \mathrm{MF}$
function $\mathsf{prg}_+(\mu)$

$\quad \mathrm{Q} \leftarrow \|\mu\|_+$
$\quad$ **while** $\mathsf{esc}(\mu, \mathrm{Q}) \neq \emptyset$ **do**
$\quad\quad \mathrm{E} \leftarrow \mathsf{bep}(\mu, \mathrm{Q})$
$\quad\quad \mu \leftarrow \mathsf{lift}(\mu, \mathrm{E}, \overline{\mathrm{Q}})$
$\quad\quad \mathrm{Q} \leftarrow \mathrm{Q} \setminus \mathrm{E}$
$\quad \mu \leftarrow \mu[\mathrm{Q} \mapsto \top]$
$\quad$ **return** μ

We now turn our attention to the progress operator prg_+, whose pseudo-code is reported in Algorithm 1. Besides the lift function, this operator employs two other functions, $\mathsf{esc} \colon \mathrm{MF} \times 2^{\mathrm{Ps}} \to 2^{\mathrm{Ps}}$ and $\mathsf{bef} \colon \mathrm{MF} \times 2^{\mathrm{Ps}} \times \mathrm{Ps} \to$

CMs, called, respectively, *escape function* and *best-escape position function*. Given a set of positions Q, the escape function collects the subset of positions in Q from which their owner wants or is forced to exit from Q, according to their objective. Specifically, those are (i) the 1-positions having a successor outside Q and (ii) the 0-positions v, such that none of their successors w belonging to Q support the measure of v in the current measure function. Formally:

$$\mathsf{esc}(\mu, \mathrm{Q}) \triangleq \{v \in \mathrm{Q} \cap \mathrm{Ps}_1 : Mv(v) \setminus \mathrm{Q} \neq \emptyset\} \\ \cup \\ \{v \in \mathrm{Q} \cap \mathrm{Ps}_0 : \forall w \in Mv(v) \cap \mathrm{Q} \,.\, \mu(w) + v < \mu(v)\}.$$

All positions belonging to Q must be lifted during the execution of prg_+. However, they must be lifted in the appropriate order: namely, the first positions to be lifted are those whose measure will rise the least. This is the role of the bep function and its supporting *best escape forfeit function* bef. The bep simply collects the positions v having minimal *forfeit* $\mathsf{bef}(\mu, \mathrm{Q}, v)$ (defined below).

$$\mathsf{bep}(\mu, \mathrm{Q}) \triangleq \operatorname*{argmin}_{v \in \mathrm{Q}} \mathsf{bef}(\mu, \mathrm{Q}, v) : \mathrm{MF} \times 2^{\mathrm{Ps}} \to 2^{\mathrm{Ps}}.$$

Assuming that v is a position in the escape set $\mathsf{esc}(\mu, \mathrm{Q})$, its forfeit is the difference between the measure that v would acquire if lifted and its current measure. In the following, we use the difference operation $\phi_1 - \phi_2 \triangleq \phi_1 + (-\phi_2)$ defined as usual.

$$\mathsf{bef}(\mu, \mathrm{Q}, v) \triangleq \begin{cases} \max\{(\mu(w) + v) - \mu(v) : w \in Mv(v) \setminus \mathrm{Q}\}, \\ \qquad\qquad\qquad\qquad \text{if } v \in \mathsf{esc}(\mu, \mathrm{Q}) \cap \mathrm{Ps}_0; \\ \min\{(\mu(w) + v) - \mu(v) : w \in Mv(v) \setminus \mathrm{Q}\}, \\ \qquad\qquad\qquad\qquad \text{if } v \in \mathsf{esc}(\mu, \mathrm{Q}) \cap \mathrm{Ps}_1; \\ \top, \qquad\qquad\qquad\qquad \text{otherwise.} \end{cases}$$

Notice that the following inclusions are an immediate consequence of the definitions: $\mathsf{bep}(\mu, \mathrm{Q}) \subseteq \mathsf{esc}(\mu, \mathrm{Q}) \subseteq \mathrm{Q}$.

We can now describe Algorithm 1, which computes operator prg_+. The purpose of this operator is to enforce the progress condition on the positions in Q, by lifting their value. Line 1 identifies Q as the set of positions with non-minimal measures. If Q is closed (condition at Line 2), namely the escape set of Q is empty, then every position in Q is lifted to value $\top$ (Line 6). Indeed, by Theorem 4.7, Q is a weak quasi 0-dominion and any strategy compatible with μ is a possible 0-witness. The emptiness of $\mathsf{esc}(\mu, \mathrm{Q})$ implies that player 1 cannot exit from Q and player 0 has a strategy compatible with μ to remain in the set. Hence, by Corollary 4.2, Q is a dominion for player 0. Otherwise, if Q is not closed, Line 3 collects the best escape forfeit set of positions in E. This step ensures that the positions are lifted in increasing order of forfeit. Once the positions in E have been lifted, Lines 4 and 5 remove them from Q, and the function iterates Steps 2–5 until Q is either empty or closed. The following lemma is the core result of this section and proves the key properties of the progress operator prg_+.

Lemma 5.5: *The progress operator* prg_+ *enjoys the following properties: (i) it maps* $\mathrm{MF}_\bot$ *into* $\mathrm{MF}_\bot$*; (ii) it is an inflationary function from* QDM *to* QDM*; (iii) it maps* SM *into* SM*; (iv) every fixpoint* $\mu \in \mathrm{MF}$ *of* prg_+ *is a progress measure over* $\|\mu\|_+$*.*

Proof: Let us assume $\|\mu\|_+ \neq \emptyset$, since there is nothing to prove, otherwise, being $\mu^\star \triangleq \mathsf{prg}_+(\mu) = \mu$, and consider the three (potentially) infinite sequences $\mathrm{Q}_0, \mathrm{Q}_1, \ldots$, $\mathrm{E}_0, \mathrm{E}_1, \ldots$, and $\mu_0, \mu_1, \ldots$ generated by Algorithm 1, which explicitly implements the progress operator prg_+. These sequences are defined as follows: (i) $\mathrm{Q}_0 \triangleq \|\mu\|_+$ and $\mu_0 \triangleq \mu$; (ii) $\mathrm{Q}_{i+1} \triangleq \mathrm{Q}_i \setminus \mathrm{E}_i$ and $\mu_{i+1} = \mathsf{lift}(\mu_i, \mathrm{E}_i, \overline{\mathrm{Q}_i})$, where $\mathrm{E}_i \triangleq \mathsf{bep}(\mu_i, \mathrm{Q}_i) \subseteq \mathsf{esc}(\mu_i, \mathrm{Q}_i)$, for all $i \in \mathbb{N}$. Since $|\mathrm{Q}_0| < \infty$ and $\mathrm{Q}_{i+1} \subseteq \mathrm{Q}_i$, there necessarily exists an index $k \in \mathbb{N}$ such that $\mathrm{Q}_{k+1} = \mathrm{Q}_k$, $\mu_{k+1} = \mu_k$, $\mathrm{E}_k = \emptyset$, and $\mathrm{E}_j \neq \emptyset$, for all $j < k$. Moreover, observe that $\mu^\star = \mu_k[\mathrm{Q}_k \mapsto \top]$. At this point, we analyse the four properties separately.

- **[i].** To prove that $\mu^\star \in \mathrm{MF}_\bot$, whenever $\mu \in \mathrm{MF}_\bot$, one can focus on those positions v that changed their measure from μ to $\mu^\star$, *i.e.*, such that $\mu^\star(v) \neq \mu(v)$. If $v \in \|\mu^\star\|_+$, there is nothing to prove, as v vacuously satisfies the definitional requirement of $\mathrm{MF}_\bot$. If $v \notin \|\mu^\star\|_+$, instead, it holds that $\mu^\star(v){\restriction_v} = \bot$. Due to the fact that the position changed its measure, there is an index $i \in [0, k]$, such that $v \in \mathrm{E}_i$ and, so, $\mu^\star(v) = \mu_{i+1}(v)$. Therefore, by definition of the lift operator, there exists an adjacent $w \in Mv(v) \setminus \mathrm{Q}_i$, such that $\mu_{i+1}(v) = \mu_i(w) + v$. At this point, we can observe that v has odd priority. Indeed, if, by contradiction, v has even priority, by Item 1 of Definition 3.1 and Item 2 of Definition 3.5, one would have $\mu_i(w){\restriction_v} = \bot \neq \top$, since $\mu_i(w){\restriction_v} \leq (\mu_i(w) + v){\restriction_v} = \mu^\star(v){\restriction_v} = \bot$, which would in turn imply $\bot = \mu_i(w){\restriction_v} < (\mu_i(w) + v){\restriction_v} = \mu^\star(v){\restriction_v} = \bot$, due to Item 1 of the same definition, which is obviously impossible, being $<$ an irreflexive relation. Thus, as a consequence of Item ii of Proposition 5.3, $\mu_i(w) + v = \bot$, since $(\mu_i(w) + v)\restriction_v = \bot$, which means that $\mu^\star(v) = \bot$, as required by the definition of the set $\mathrm{MF}_\bot$. Hence, $\mu^\star \in \mathrm{MF}_\bot$.

- **[ii].** We first prove the inflationary property of the progress operator, assuming $\mu = \mu_0 \in \mathrm{QDM}$ is a quasi-dominion measure, hence also a regress measure. To do this, consider the sequence of forfeit values $\phi_0, \ldots, \phi_{k-1} \in \mathrm{CMs}$ defined as follows: $\phi_i \triangleq \mathsf{min}_{v \in \mathrm{Q}_i} \mathsf{bef}(\mu_i, \mathrm{Q}_i, v)$, for all indexes $i \in [0, k)$. In addition, let $\iota \colon \|\mu\|_+ \setminus \mathrm{Q}_k \to [0, k)$ be the function associating each position $v \in \|\mu\|_+ \setminus \mathrm{Q}_k$ with the index $\iota(v) \in [0, k)$, such that $v \in \mathrm{E}_{\iota(v)}$. Due to the way the sequence of measure functions $\mu_0, \mu_1, \ldots$ is constructed, it is immediate to observe that, for all positions $v \in \mathrm{Ps}$ and indexes $i \in [0, k]$, it holds that

$$\text{if } v \in \|\mu\|_+ \setminus \mathrm{Q}_k \text{ and } \iota(v) < i \text{ then } \mu_i(v) = \mu_{\iota(v)}(v) + \phi_{\iota(v)} \text{ else } \mu_i(v) = \mu(v). \qquad (*)$$

At this point, by induction on the index $i \in [0, k)$ and in that specific order, we can prove the following four auxiliary properties: (a) $\mu \sqsubseteq \mu_i$; (b) for each position $v \in \|\mu\|_+ \setminus \mathrm{Q}_{i+1}$, there exists an adjacent $w \in Mv(v) \setminus \mathrm{Q}_i$ such that $\mu_{i+1}(v) = \mu_i(w) + v \neq \top$; (c) if $i > 0$ then $\phi_{i-1} \leq \phi_i$; (d) $\mathbf{0} \leq \phi_i$.

  - **[a].** If $i = 0$ then $\mu \sqsubseteq \mu_i$, since $\mu_0 = \mu$. If $i > 0$, instead, by the Inductive Hypotheses a and d, it holds that $\mu \sqsubseteq \mu_{i-1}$ and $\mathbf{0} \leq \phi_{i-1}$. Moreover, by the previous Observation $(*)$, it follows that $\mu_i(v) \neq \mu_{i-1}(v)$ only if $v \in \mathrm{E}_{i-1}$ and, in this case, $\mu_i(v) = \mu_{i-1}(v) + \phi_{i-1}$, since $\iota(v) = i - 1$. As a consequence, if $\mu_i(v) \neq \mu_{i-1}(v)$ then $\mu_i(v) > \mu_{i-1}(v)$, thanks to Proposition 5.1, which implies, in general, that $\mu_i(v) \geq \mu_{i-1}(v) \geq \mu(v)$. Hence, $\mu \sqsubseteq \mu_i$ holds true.
  - **[b].** Let $v \in \|\mu\|_+ \setminus \mathrm{Q}_{i+1}$. If $v \in \|\mu\|_+ \setminus \mathrm{Q}_i$ then $i > 0$, since $\mathrm{Q}_0 = \|\mu\|_+$. By the Inductive Hypothesis b, there exists an adjacent $w \in Mv(v) \setminus \mathrm{Q}_{i-1}$ such that $\mu_i(v) = \mu_{i-1}(w) + v \neq \top$. Thanks to Observation $(*)$, we have that $\mu_{i+1}(v) = \mu_i(v)$ and $\mu_i(w) = \mu_{i-1}(w)$, since $\iota(v) < i$ and $\iota(w) < i - 1$. Thus, $\mu_{i+1}(v) = \mu_i(w) + v \neq \top$, as required. If $v \notin \|\mu\|_+ \setminus \mathrm{Q}_i$, instead, it holds that $v \in \mathrm{E}_i$. As a consequence, by definition of the lift operator, there exists an adjacent $w \in Mv(v) \setminus \mathrm{Q}_i$ such that $\mu_{i+1}(v) = \mu_i(w) + v$. Notice now that $\mu_i(w) \neq \top$.

Indeed, if $w \in \|\mu\|_{+}$ then $\mu_i(w) \neq \top$ directly follows from the Inductive Hypothesis b. Otherwise, $w \in \|\mu\|_{\perp}$, which implies that $\mu(w) \neq \top$, thanks to Item 2i of Definition 3.1. Moreover, $\mu_i(w) = \mu(w)$, by Observation $(*)$. To conclude, $\mu_{i+1}(v) = \mu_i(w) + v \neq \top$, due to Item 3i of Definition 3.1.

- **[c]**. Let $i > 0$ and $v \in \mathrm{E}_i$. Thanks to Observation $(*)$, we have that $\mu_{i+1}(v) = \mu_i(v) + \phi_i$ and, so, $\phi_i = \mu_{i+1}(v) - \mu_i(v)$, due to Proposition 5.1. To continue, we need to consider the following case analysis in three parts, which is partially based on ownership of the positions v.

  * **[$v \in \mathrm{Ps}_0 \cap \mathsf{esc}(\mu_{i-1}, \mathrm{Q}_{i-1})$]**. Since $v \in \mathrm{Ps}_0 \cap \mathsf{esc}(\mu_{i-1}, \mathrm{Q}_{i-1})$, it holds that $\mu_{i-1}(u) + v < \mu_{i-1}(v)$, for all adjacents $u \in Mv(v) \cap \mathrm{Q}_{i-1}$. However, by Condition 1 of Definition 4.4, there exists an adjacent $w \in Mv(v)$ such that $\mu(v) \leq \mu(w) + v$. By Observation $(*)$, $\mu_{i-1}(v) = \mu(v)$, as $\iota(v) = i$. Thus, $\mu_{i-1}(v) = \mu(v) \leq \mu(w) + v \leq \mu_{i-1}(w) + v$, thanks to the Inductive Hypothesis a and Item 3ii of Definition 3.1. As a consequence, $w \notin \mathrm{Q}_{i-1}$ and, so, $w \notin \mathrm{Q}_i$, since all adjacents of v inside Q_{i-1} falsify the inequality, as shown before. Moreover, $v \notin \mathrm{E}_{i-1}$, as $v \in \mathrm{E}_i$. Hence, $\phi_{i-1} < \mathsf{bef}(\mu_{i-1}, \mathrm{Q}_{i-1}, v)$. At this point, the following inequalities hold:

$$
\begin{aligned}
\phi_{i-1} &< \mathsf{bef}(\mu_{i-1}, \mathrm{Q}_{i-1}, v)\\
&= \min\{(\mu_{i-1}(u) + v) - \mu_{i-1}(v) : u \in Mv(v) \setminus \mathrm{Q}_{i-1}\}\\
&\leq (\mu_{i-1}(w) + v) - \mu_{i-1}(v)\\
&= (\mu_i(w) + v) - \mu_i(v)\\
&\leq \max\{\mu_i(u) + v : u \in Mv(v) \setminus \mathrm{Q}_i\} - \mu_i(v)\\
&= \mathsf{lift}(\mu_i, \mathrm{E}_i, \overline{\mathrm{Q}_i})(v) - \mu_i(v)\\
&= \mu_{i+1}(v) - \mu_i(v)\\
&= \phi_i.
\end{aligned}
$$

  Notice that the first equality follows from the definition of the best escape forfeit function, while the second one is due to Observation $(*)$. Indeed, $\mu_{i-1}(v) = \mu_i(v)$ and $\mu_{i-1}(w) = \mu_i(w)$, since $\iota(v) = i$ and $\iota(w) < i - 1$. Finally, from the last inequality onward, we applied the definition of the lift operator.

* $[v \in \mathrm{Ps}_0 \setminus \mathsf{esc}(\mu_{i-1}, \mathrm{Q}_{i-1})]$. Since $v \in \mathrm{Ps}_0 \setminus \mathsf{esc}(\mu_{i-1}, \mathrm{Q}_{i-1})$, there exists an adjacent $w \in Mv(v) \cap \mathrm{Q}_{i-1}$, such that $\mu_{i-1}(v) \leq \mu_{i-1}(w) + v$. Moreover, by Observation $(*)$, $\mu_{i-1}(v) = \mu_i(v)$, as $\iota(v) = i$, from which we derive $\mu_i(v) \leq \mu_{i-1}(w) + v$ and, so, $(\mu_{i-1}(w) + v) - \mu_i(v) \geq \mathbf{0}$, due to Proposition 5.1. Obviously, $\iota(w) \geq i - 1$, due to the definition of the function ι, since the membership of w in Q_{i-1} implies $w \notin \mathrm{E}_j$, for all indexes $j < i - 1$. Actually, it holds true that the value $\iota(w)$ is precisely $i - 1$. Indeed, suppose, by contradiction, that $\iota(w) > i - 1$. Then, we would have $\mu_{i-1}(w) = \mu_i(w)$ and, thus, $\mu_i(v) \leq \mu_{i-1}(w) + v = \mu_i(w) + v$, contradicting the fact that $v \in \mathrm{E}_i \subseteq \mathsf{esc}(\mu_i, \mathrm{Q}_i)$. Now, the equality $\iota(w) = i - 1$ implies $w \in \mathrm{E}_{i-1}$ and, so, $w \notin \mathrm{Q}_i$. Moreover, $\mu_i(w) = \mu_{i-1}(w) + \phi_{i-1}$, again due to Observation $(*)$. At this point, the following holds:

$$
\begin{aligned}
\phi_{i-1} &\leq ((\mu_{i-1}(w) + v) - \mu_i(v)) + \phi_{i-1} \\
&= ((\mu_{i-1}(w) + \phi_{i-1}) + v) - \mu_i(v) \\
&= (\mu_i(w) + v) - \mu_i(v) \\
&\leq \mathsf{max}\{\mu_i(u) + v : u \in Mv(v) \setminus \mathrm{Q}_i\} - \mu_i(v) \\
&= \mathsf{lift}(\mu_i, \mathrm{E}_i, \overline{\mathrm{Q}_i})(v) - \mu_i(v) \\
&= \mu_{i+1}(v) - \mu_i(v) \\
&= \phi_i.
\end{aligned}
$$

Notice that the first two derivation steps follow from the Abelian group properties of the evaluation structure stated in Proposition 5.1 and from Item iii of Proposition 5.3. Moreover, from the last inequality onward, we applied the definitions of the lift operator and forfeit values.

* $[v \in \mathrm{Ps}_1]$. Since $v \in \mathrm{E}_i$, it holds that $v \in \|\mu\|_+ \setminus \mathrm{Q}_{i+1}$. By the Inductive Hypothesis b, there exists an adjacent $w \in Mv(v) \setminus \mathrm{Q}_i$, such that $\mu_{i+1}(v) = \mu_i(w) + v$, which implies $\phi_i = (\mu_i(w) + v) - \mu_i(v)$. Now, consider the nested case analysis on the membership of the position w *w.r.t.* E_{i-1}.

  · $[w \notin \mathrm{E}_{i-1}]$. Since $w \notin \mathrm{E}_{i-1}$ and $w \notin \mathrm{Q}_i$, it holds that $w \notin \mathrm{Q}_{i-1}$. This means that $w \in Mv(v) \setminus \mathrm{Q}_{i-1} \neq$

$\emptyset$, which in turn implies that $v \in \mathsf{esc}(\mu_{i-1}, \mathrm{Q}_{i-1})$, since $v \in \mathrm{Ps}_1$. Moreover, $v \notin \mathrm{E}_{i-1}$, as $v \in \mathrm{E}_i$. Thus, $\phi_{i-1} < \mathsf{bef}(\mu_{i-1}, \mathrm{Q}_{i-1}, v)$. At this point, the following inequalities hold:

$$\begin{aligned}\phi_{i-1} &< \mathsf{bef}(\mu_{i-1}, \mathrm{Q}_{i-1}, v)\\ &= \min\{(\mu_{i-1}(u) + v) - \mu_{i-1}(v) : u \in Mv(v) \setminus \mathrm{Q}_{i-1}\}\\ &\leq (\mu_{i-1}(w) + v) - \mu_{i-1}(v)\\ &= (\mu_i(w) + v) - \mu_i(v)\\ &= \phi_i.\end{aligned}$$

Notice that the first equality follows from the definition of the best escape forfeit function, while the second one is due to Observation $(*)$. Indeed, $\mu_{i-1}(v) = \mu_i(v)$ and $\mu_{i-1}(w) = \mu_i(w)$, since $\iota(v) = i$ and $\iota(w) < i - 1$.

· $[w \in \mathrm{E}_{i-1}]$. Since $w \in \mathrm{E}_{i-1}$, by Observation $(*)$, it follows that $\mu_i(w) = \mu_{i-1}(w) + \phi_{i-1}$, as $\iota(w) = i-1$. Similarly, $\mu_i(v) = \mu(v)$, as $\iota(v) = i$. Now, by Condition 2 of Definition 4.4, $\mu_i(v) = \mu(v) \leq \mu(u) + v \leq \mu_{i-1}(u) + v$, for all adjacents $u \in Mv(v)$, thanks to the Inductive Hypothesis a and Item 3ii of Definition 3.1. As an immediate consequence, $\mu_i(v) \leq \mu_{i-1}(w) + v$, *i.e.*, $(\mu_{i-1}(w) + v) - \mu_i(v) \geq \mathbf{0}$. At this point, the following holds:

$$\begin{aligned}\phi_{i-1} &\leq ((\mu_{i-1}(w) + v) - \mu_i(v)) + \phi_{i-1}\\ &= ((\mu_{i-1}(w) + \phi_{i-1}) + v) - \mu_i(v)\\ &= (\mu_i(w) + v) - \mu_i(v)\\ &= \phi_i.\end{aligned}$$

Notice that the first two derivation steps follow from the Abelian group properties of the evaluation structure stated in Proposition 5.1 and from Item iii of Proposition 5.3.

Summing up, in all cases we have $\phi_{i-1} \leq \phi_i$.

– **[d].** If $i > 0$, by the Inductive Hypotheses d and c, it holds that $\mathbf{0} \leq \phi_{i-1}$ and $\phi_{i-1} \leq \phi_i$. Hence, $\mathbf{0} \leq \phi_i$. If $i = 0$, instead, let $v \in \mathrm{E}_0$. Thanks to Observation $(*)$, one has that $\mu_1(v) = \mu_0(v) + \phi_0$ and, so, $\phi_0 = \mu_1(v) - \mu_0(v)$, due to

Proposition 5.1. To continue, we need to consider the following case analysis on the ownership of the position v.

* $[v \in \mathrm{Ps}_0]$. By Condition 1 of Definition 4.4, there exists an adjacent $w \in Mv(v)$, such that $\mu_0(v) \leq \mu_0(w) + v$, as $\mu_0 = \mu$. Since $v \in \mathrm{E}_0$, thanks to the definitions of both the best escape forfeit and escape functions, it holds that $\mu_0(u) + v < \mu_0(v)$, for all adjacents $u \in Mv(v) \cap \mathrm{Q}_0$. Hence, as an immediate consequence, $w \notin \mathrm{Q}_0$. Now, by definition of the lift operator, $\mu_1(v) = \mathsf{max}\{\mu_0(u) + v : u \in Mv(v) \setminus \mathrm{Q}_0\} \geq \mu_0(w) + v \geq \mu_0(v)$, from which it follows that $\phi_0 = \mu_1(v) - \mu_0(v) \geq \mathbf{0}$, due to Proposition 5.1.
* $[v \in \mathrm{Ps}_1]$. By the Inductive Hypothesis b, there exists an adjacent $w \in Mv(v) \setminus \mathrm{Q}_0$, such that $\mu_1(v) = \mu_0(w) + v$, which implies $\phi_0 = (\mu_0(w) + v) - \mu_0(v)$. Moreover, by Condition 2 of Definition 4.4, $\mu_0(v) \leq \mu_0(u) + v$, for all adjacents $u \in Mv(v)$, since $\mu_0 = \mu$. Thus, as an obvious consequence, $\mu_0(w) + v \geq \mu_0(v)$, which immediately implies $\phi_0 = (\mu_0(w) + v) - \mu_0(v) \geq \mathbf{0}$, as required, again due to Proposition 5.1.

Summing up, in both cases we have $\mathbf{0} \leq \phi_0$.

We now have the necessary tool to show that the progress operator is inflationary. Indeed, by Property a, $\mu \sqsubseteq \mu_k$. Moreover, $\mu^\star$ and μ_k differ only on positions $v \in \|\mu^\star\|_+$ such that $\mu^\star(v) = \top$. Hence, it easily follows that $\mu \sqsubseteq \mu^\star$, thanks to Item 1 of Definition 3.1. Observe also that $\|\mu\|_+ \subseteq \|\mu_k\|_+ \subseteq \|\mu^\star\|_+$ and $\|\mu\|_0 \subseteq \|\mu_k\|_0 \subseteq \|\mu^\star\|_0$, due to Item 2ii of the same definition.

It remains to prove that $\mu^\star$ is a QDM. First notice that, for any position $v \in \mathrm{Ps}$, if $\mu_k(v) \neq \mu(v)$ then $v \in \|\mu\|_+ \setminus \mathrm{Q}_k$, due to Observation $(*)$. Thus, by Property b, for such a position v, there exists an adjacent $w \in Mv(v) \setminus \mathrm{Q}_{k-1} \subseteq Mv(v) \setminus \mathrm{Q}_k$, such that $\mu_k(v) = \mu_{k-1}(w) + v \neq \top$. Since $w \notin \mathrm{Q}_{k-1}$, it holds that $\iota(w) < k - 1$, so, $\mu_k(w) = \mu_{k-1}(w)$, again due to Observation $(*)$. Hence, for all positions $v \in \mathrm{Ps}$, either one of the following two possibilities holds:

$$\left.\begin{array}{l} v \notin \|\mu\|_+ \setminus \mathrm{Q}_k \text{ and } \mu_k(v) = \mu(v); \\ v \in \|\mu\|_+ \setminus \mathrm{Q}_k \text{ and there exists an adjacent } w \in Mv(v) \setminus \mathrm{Q}_k \\ \qquad\qquad \text{such that } \mu_k(v) = \mu_k(w) + v \neq \top. \end{array}\right\} \quad (\sharp)$$

From this, we easily derive that $\|\mu\|_{+} = \|\mu_k\|_{+} = \|\mu^\star\|_{+}$ and $\|\mu\|_{0} = \|\mu_k\|_{0} \subseteq \|\mu^\star\|_{0} = \mathrm{Q}_k$. Indeed, the only positions that can change their measure are those in $\|\mu\|_{+}$ and they are not set to $\top$ in μ_k. Moreover, $\|\mu\|_{0} \subseteq \|\mu\|_{+}$, so, $\|\mu_k\|_{0} \subseteq \|\mu\|_{+}$. Now, $(\|\mu\|_{+} \setminus \mathrm{Q}_k) \cap \|\mu_k\|_{0} = \emptyset$, hence, $\|\mu_k\|_{0} \subseteq \mathrm{Q}_k$. Finally, $\|\mu^\star\|_{0} = \mathrm{Q}_k$, since, by construction, $\|\mu^\star\|_{0} = \|\mu_k\|_{0} \cup \mathrm{Q}_k$.

We can now show that $\mu^\star$ is a regress measure, *i.e.*, that every position $v \in \|\mu^\star\|_{+} \setminus \|\mu^\star\|_{0}$ satisfies the suitable condition of Definition 4.4. We do this, via a case analysis on the ownership of v.

- $[v \in \mathrm{Ps}_0]$. Since $\|\mu^\star\|_{+} \setminus \|\mu^\star\|_{0} = \|\mu\|_{+} \setminus \mathrm{Q}_k$, by Observation ($\sharp$), there exists an adjacent $w \in Mv(v) \setminus \mathrm{Q}_k$, such that $\mu_k(v) = \mu_k(w) + v$. Obviously, $\mu^\star(v) = \mu_k(v)$ and $\mu^\star(w) = \mu_k(w)$, since $v, w \notin \mathrm{Q}_k$. Thus, $\mu^\star(v) = \mu^\star(w) + v$, as required by Condition 1.
- $[v \in \mathrm{Ps}_1]$. Suppose by contradiction that $\mu^\star$ does not satisfy Condition 2 on v. Then, there exists one of its adjacents $w \in Mv(v)$, such that $\mu^\star(w)+v < \mu^\star(v)$. Obviously, $\mu^\star(w)+v \neq \top$, so, $\mu^\star(w) \neq \top$, due to Item 3i of Definition 3.1, which in turn implies that $w \notin \mathrm{Q}_k$. As a consequence, $v \notin \mathrm{Q}_k$ too, since we would have had, otherwise, $Mv(v) \subseteq \mathrm{Q}_k$, as $\mathsf{esc}(\mu_k, \mathrm{Q}_k) = \emptyset$. Thus, $v \in \|\mu\|_{+} \setminus \mathrm{Q}_k$, from which it follows that $\mu^\star(v) = \mu_k(v) = \mu_{\iota(v)+1}(v) = \mu_{\iota(v)}(v) + \phi_{\iota(v)} = \mu(v) + \phi_{\iota(v)}$, due to Observation ($*$). To proceed, we now need the following nested case analysis, which allows to prove that $w \notin \mathrm{Q}_{\iota(v)}$ and $\mu^\star(w) = \mu_{\iota(v)}(w)$.
  * $[w \in \|\mu^\star\|_{\perp}]$. Notice that $w \notin \mathrm{Q}_o = \|\mu\|_{+} = \|\mu^\star\|_{+}$ and, so, $w \notin \mathrm{Q}_{\iota(v)}$, as $\mathrm{Q}_{\iota(v)} \subseteq \mathrm{Q}_o$. Moreover, $\mu^\star(w) = \mu_k(w) = \mu_{\iota(v)}(w)$, due to Observations ($*$).
  * $[w \in \|\mu^\star\|_{+}]$. By Observations ($*$), it holds that $\mu^\star(w) = \mu_k(w) = \mu_{\iota(w)}(w) + \phi_{\iota(w)} = \mu(w) + \phi_{\iota(w)}$. Now, by substituting in $\mu^\star(w) + v < \mu^\star(v)$ both $\mu^\star(w)$ and $\mu^\star(v)$ with $\mu(w) + \phi_{\iota(w)}$ and $\mu(v) + \phi_{\iota(v)}$, respectively, and exploiting Proposition 5.1 and Item iii of Proposition 5.3, one can obtain $(\mu(w) + v) - \mu(v) < \phi_{\iota(v)} - \phi_{\iota(w)}$. Since μ is a QDM and, so, a regress measure, we have that it satisfies Condition 2 on v, *i.e.*, $\mu(v) \leq \mu(w) + v$, which implies $(\mu(w) + v) - \mu(v) \geq \mathbf{0}$. Hence, $\phi_{\iota(v)} - \phi_{\iota(w)} > \mathbf{0}$ and, so, $\iota(v) > \iota(w)$, thanks to Property c. From this we can derive

that $w \notin \mathrm{Q}_{\iota(v)}$ and $\mu^\star(w) = \mu_k(w) = \mu_{\iota(v)}(w)$, due to Observations $(*)$.

At this point, the following impossible inequality chain should hold:

$$\begin{aligned}\mu^\star(v) &= \mu_{\iota(v)+1}(v)\\ &= \mathsf{lift}(\mu_{\iota(v)}, \mathrm{E}_{\iota(v)}, \overline{\mathrm{Q}_{\iota(v)}})(v)\\ &= \min\{\mu_{\iota(v)}(u) + v : u \in Mv(v) \setminus \mathrm{Q}_{\iota(v)}\}\\ &\leq \mu_{\iota(v)}(w) + v\\ &= \mu^\star(w) + v\\ &< \mu^\star(v).\end{aligned}$$

Finally, we can conclude the proof of this item by showing that $\|\mu^\star\|_0$, which we now know to be equal to Q_k, is a 0-dominion. Let $\sigma_0 \in \mathrm{Str}_0$ be a μ-coherent 0-strategy winning on $\|\mu\|_0$ such that $\sigma_0(v) \in \mathrm{Q}_k$, for all 0-positions $v \in \mathrm{Q}_k \cap \mathrm{Ps}_0$. Such a strategy surely exists, since μ is, by hypothesis, a QDM and Q_k is closed, *i.e.*, $\mathsf{esc}(\mu, \mathrm{Q}_k) = \mathsf{esc}(\mu_k, \mathrm{Q}_k) = \emptyset$. To state the first equality we exploited the fact that $\mu_k(v) = \mu(v)$, for all positions $v \in \mathrm{Q}_k$, due to Observation ($\sharp$). Now, by Theorem 4.7, $\|\mu\|_+$ is a weak quasi 0-dominion, for which $\sigma_0 \downarrow \|\mu\|_+ \in \mathrm{Str}_0(\|\mu\|_+)$ is a 0-witness. Hence, Q_k and, so, $\|\mu^\star\|_0$, is a 0-dominion, thanks to Corollary 4.2.

- **[iii]**. To show that $\mu^\star$ is a SM, whenever μ is a SM, we first prove the following statement: $\mu_i(v) \in \mathrm{SMs}(v, \|\mu\|_+ \setminus \mathrm{Q}_i) \setminus \{\top\}$, for all indexes $i \in [0, k]$ and positions $v \in \mathrm{Ps} \setminus \mathrm{Q}_i$.

  The proof proceeds by induction on i. The base case $i = 0$ trivially follows from the hypothesis, since $\mu_0 = \mu$. Indeed, it holds that $\mathrm{Ps} \setminus \mathrm{Q}_0 = \|\mu\|_\perp$ and $\mu_i(v) = \perp$, for all positions $v \in \|\mu\|_\perp$, due to Proposition 4.10. For the inductive case $i > 0$, let $v \in \mathrm{Ps} \setminus \mathrm{Q}_i$ and assume that $\mu_{i-1}(w) \in \mathrm{SMs}(w, \|\mu\|_+ \setminus \mathrm{Q}_{i-1}) \setminus \{\top\}$, for all positions $w \in \mathrm{Ps} \setminus \mathrm{Q}_{i-1}$. If $v \in \mathrm{Ps} \setminus \mathrm{Q}_{i-1}$, there is nothing more to prove, since $\mu_i(v) = \mu_{i-1}(v)$ and, by inductive hypothesis, $\mu_{i-1}(v) \in \mathrm{SMs}(v, \|\mu\|_+ \setminus \mathrm{Q}_{i-1}) \setminus \{\top\} \subseteq \mathrm{SMs}(v, \|\mu\|_+ \setminus \mathrm{Q}_i) \setminus \{\top\}$, being $\|\mu\|_+ \setminus \mathrm{Q}_{i-1} \subseteq \|\mu\|_+ \setminus \mathrm{Q}_i$. Therefore, consider the case $v \notin \mathrm{Ps} \setminus \mathrm{Q}_{i-1}$, which implies that $v \in \mathrm{E}_{i-1} \subseteq \mathrm{Q}_{i-1}$. By definition of the lift operator, there exists an adjacent $w \in Mv(v) \setminus \mathrm{Q}_{i-1} \subseteq \mathrm{Ps} \setminus \mathrm{Q}_{i-1}$, such that $\mu_i(v) = \mu_{i-1}(w) + v$. By the inductive hypothesis, $\mu_{i-1}(w) \neq \top$. Thus, by Item 3i of Definition 3.1, it holds that $\mu_i(v) \neq \top$. Moreover, there exists a simple path $\pi \in \mathrm{SPth}(w, \|\mu\|_+ \setminus \mathrm{Q}_{i-1})$, such that

$\mu_{i-1}(w) = \eta(\pi)$. Obviously, $\mu_i(v) = \mu_{i-1}(w) + v = \eta(\pi) + v = \eta(v \cdot \pi) \neq \top$. Now, it is quite clear that $v \cdot \pi$ is a simple path passing through positions in $\{v\} \cup \|\mu\|_+ \setminus \mathrm{Q}_{i-1}$, *i.e.*, $v \cdot \pi \in \mathrm{SPth}(v, \|\mu\|_+ \setminus \mathrm{Q}_i)$, since $v \in \mathrm{Q}_{i-1}$. Hence, $\mu_i(v) \in \mathrm{SMs}(v, \|\mu\|_+ \setminus \mathrm{Q}_i) \setminus \{\top\}$. This concludes the inductive proof.

At this point, it immediately follows, from what we have just proved, that μ_k is SM. In addition, $\mu^\star$ potentially differs from μ_k only on positions $v \in \|\mu^\star\|_+$ such that $\mu^\star(v) = \top$. Consequently, $\mu^\star$ is a SM as well.

- **[iv]**. By hypothesis, $\mu^\star = \mu$, which implies that $\mu_i = \mu$, for all $i \in \mathbb{N}$, due to the way the sequence $\mu_0, \mu_1, \ldots$ is constructed. Now, let us consider an arbitrary position $v \in \|\mu\|_+$. Due to the definition of the sequence $\mathrm{Q}_0, \mathrm{Q}_1, \ldots$, it obviously holds that either $v \in \mathrm{Q}_k$ or there is a unique index $i \in [0, k)$, such that $v \in \mathrm{Q}_i \setminus \mathrm{Q}_{i+1}$, *i.e.*, $v \in \mathrm{E}_i$. In the first case, we have $\mu(v) = \top$, due to the assignment $\mu^\star = \mu_k[\mathrm{Q}_k \mapsto \top]$. Therefore, v is a progress position, *i.e.*, it satisfies both conditions of Definition 3.6. In the other case, the proof proceeds by a case analysis on the ownership of the position v itself.
  - $[v \in \mathrm{Ps}_0]$. First recall that $\mathrm{E}_i = \mathsf{bep}(\mu, \mathrm{Q}_i) \subseteq \mathsf{esc}(\mu, \mathrm{Q}_i)$. Thus, due to the definition of the escape function, we have that $\mu(w) + v < \mu(v)$, for all positions $w \in Mv(v) \cap \mathrm{Q}_i$. Now, by definition of the lift operator, we have that $\mu(w) + v \leq \mathsf{max}\{\mu(w) + v : w \in Mv(v) \cap \overline{\mathrm{Q}_i}\} = \mu(v)$, for all adjacents $w \in Mv(v) \cap \overline{\mathrm{Q}_i}$ of v. Thus, $\mu(w) + v \leq \mu(v)$, for all positions $w \in Mv(v)$, as required by Condition 1 of Definition 3.6 on $\|\mu\|_+$.
  - $[v \in \mathrm{Ps}_1]$. Again by definition of the lift operator, we have that $\mu(w) + v \leq \mathsf{min}\{\mu(w) + v : w \in Mv(v) \cap \overline{\mathrm{Q}_i}\} = \mu(v)$, for some adjacent $w \in Mv(v) \cap \overline{\mathrm{Q}_i} \subseteq Mv(v)$ of v. Hence, Condition 2 of Definition 3.6 is satisfied on $\|\mu\|_+$ as well. □

As an example, consider the simple game $\eth$ in Fig. 1. Starting from the minimal measure function $\mu_\perp$, the solution algorithm first computes $\mu_1 = \mathsf{prg}_\perp(\mu_\perp)$, by lifting the four even-priority positions a, b, c, and h to their respective measures $\eta_{\mathsf{a}} = \eta_{\mathsf{h}} \triangleq (0,0,0,0,1,0,0)$, $\eta_{\mathsf{b}} \triangleq (1,0,0,0,0,0,0)$, and $\eta_{\mathsf{c}} \triangleq (0,0,1,0,0,0,0)$. Figure 1(1) reports the situation after this initial phase, where the blue (*resp.*, dashed red) edges indicate the moves satisfying (*resp.*, not satisfying) the progress condition. Since inside the quasi

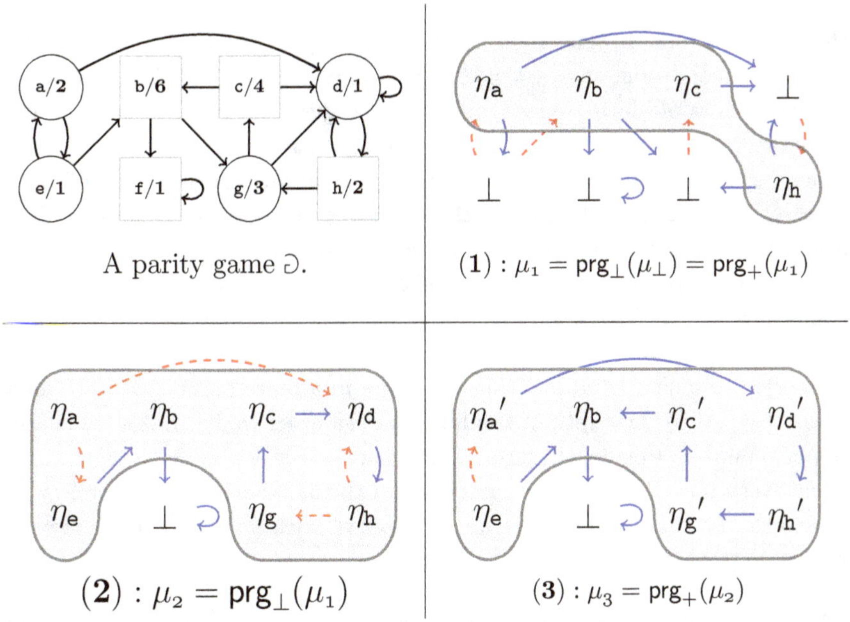

Fig. 1. Simulating the first steps of the concrete algorithm on a simple game. Positions of player 0 are circles and positions of player 1 are squares. The label inside each position indicates its name and priority.

0-dominion $\|\mu_1\|_+ = \{\mathsf{a}, \mathsf{b}, \mathsf{c}, \mathsf{h}\}$, identified by the grey area, all positions are in progress, the prg_+ operator does not change their measures, *i.e.*, $\mu_1 = \mathsf{prg}_+(\mu_1)$. The three odd-priority positions d, e, and g outside $\|\mu_1\|_+$ do not satisfy the progress condition, so the $\mathsf{prg}_\bot$ operator lifts their measure to $\eta_\mathsf{d} \triangleq (0,0,0,0,1,1,0)$, $\eta_\mathsf{e} \triangleq (1,0,0,0,0,1,0)$, and $\eta_\mathsf{g} \triangleq (0,0,1,1,0,0,0)$, as reported in Fig. 1(2). Now, positions a and h inside the quasi 0-dominion do not satisfy the progress condition anymore. Therefore, the prg_+ operator tries to recover the condition as follows. It starts by identifying the escape positions $\mathsf{esc}(\mu_2, \mathrm{Q}_0) = \{\mathsf{b}\}$ of $\mathrm{Q}_0 \triangleq \|\mu_2\|_+ = \{\mathsf{a}, \mathsf{b}, \mathsf{c}, \mathsf{d}, \mathsf{e}, \mathsf{g}, \mathsf{h}\}$. Since b has a progress move exiting from Q_0, its measure remains unchanged. Now, $\mathsf{esc}(\mu_2, \mathrm{Q}_1) = \{\mathsf{c}, \mathsf{e}\}$, where $\mathrm{Q}_1 \triangleq \mathrm{Q}_0 \setminus \{\mathsf{b}\}$. Since e has a progress move exiting from Q_1, while c can escape from Q_1 only by increasing its measure, we have $\mathsf{bep}(\mu_2, \mathrm{Q}_1) = \{\mathsf{e}\}$ and $\mathrm{Q}_2 \triangleq \mathrm{Q}_1 \setminus \{\mathsf{e}\}$. Also in this case, e does not change its measure. The process continues by extracting and lifting the measures of all the remaining positions in Q_2 in the following order: (i) c with $\eta_\mathsf{c}' \triangleq (1,0,1,0,0,0,0)$; (ii) g with $\eta_\mathsf{g}' \triangleq (1,0,1,1,0,0,0)$; (iii) h with $\eta_\mathsf{h}' \triangleq (1,0,1,1,1,0,0)$; (iv) d with $\eta_\mathsf{d}' \triangleq (1,0,1,1,1,1,0)$; (v) a with $\eta_\mathsf{a}' \triangleq (1,0,1,1,2,1,0)$. Note that player 0 positions g and d are forced to exit the quasi 0-dominion, since their internal moves (g, d) and (d, d) do not satisfy the regress condition, as $\eta_{\mathsf{d}} + \mathsf{g} = \bot < \eta_{\mathsf{g}}$ and $\eta_{\mathsf{d}} + \mathsf{d} = (0,0,0,0,1,2,0) < \eta_{\mathsf{d}}$; these moves would form, indeed, odd cycles. Figure 1(3) reports the situation after the complete execution of prg_{+}, where position e has the non-progress move (e, a). Another application of prg_{+} modifies the measure of e to $\eta_{\mathsf{e}}' \triangleq (1,0,1,1,2,2,0)$, triggering the non-progress move (a, e). After a final application of prg_{+}, positions a and e are lifted to $\top$ and the algorithm reaches its fixpoint. All positions, except a and e, satisfy the progress conditions and are, thus, winning for player 1; a and e are won by player $\mathsf{0}$.

We can prove that the solver operator is well-defined and that, when it is applied to a simple measure, it converges in a finite number of iterations, at most equal to the depth of the finite partial order $\langle \mathrm{SM}, \sqsubseteq \rangle$. A very coarse upper bound on this depth, for a game with n positions, is given by $(n+1)!$, since every non-$\top$ position is associated with the measure of a simple path of length less than n and there are at most $n!$ such paths. In the following, we use $\mathrm{SQDM} \triangleq \mathrm{QDM} \cap \mathrm{SM}$.

Theorem 5.6: *(Termination) The solver operator* $\mathsf{sol}\colon \mathrm{SQDM} \to \mathrm{SQDM}$ *is a well-defined function. Moreover, for every* $\mu \in \mathrm{SQDM}$, *there exists an index* $k \leq d$, *such that* $\mathsf{sol}(\mu) = (\mathsf{ifp}_k\, \nu \,.\, \mathsf{prg}_{+}(\mathsf{prg}_{\bot}(\nu)))(\mu)$, *where* $d \in \mathbb{N}$ *is the depth of the finite partial order* $\langle \mathrm{SM}, \sqsubseteq \rangle$.

Proof: By Items i–iii of Lemma 5.4 and Items i–iii of Lemma 5.5, we have that $\mathsf{prg}_{\bot}$ and prg_{+} are inflationary total functions on SQDM, which implies that their composition $\mathsf{prg}_{+} \circ \mathsf{prg}_{\bot}$ is both inflationary and total on SQDM as well. Consider now the infinite sequence $\mu_0, \mu_1, \ldots$ of measure functions recursively derived from an arbitrary input element $\mu \in \mathrm{SQDM}$ as follows: $\mu_0 \triangleq (\mathsf{ifp}_0\, \nu \,.\, \mathsf{prg}_{+}(\mathsf{prg}_{\bot}(\nu)))(\mu) = \mu$ and $\mu_{i+1} \triangleq (\mathsf{ifp}_{i+1}\, \nu \,.\, \mathsf{prg}_{+}(\mathsf{prg}_{\bot}(\nu)))(\mu) = \mathsf{prg}_{+}(\mathsf{prg}_{\bot}(\mu_i))$, for all $i \in \mathbb{N}$. Obviously, $\mu_i \sqsubseteq \mu_{i+1}$. Moreover, each element μ_i is a SM. Since every strict chain in $\langle \mathrm{SM}, \sqsubseteq \rangle$ can be composed of at most d elements, there necessarily exists an index $k \leq d$ such that $\mu_{k+1} = \mu_k$, as required by the theorem statement. □

The next theorem stating the soundness and completeness of the solution algorithm is a simple consequence of the properties of the $\mathsf{prg}_{\bot}$ and prg_{+} operators, combined with the general results about the measure-function spaces discussed in the previous sections.

Theorem 5.7: *(Solution)* $\mathrm{Win}_0 = \|\mathsf{sol}(\mu_{\bot})\|_0$ *and* $\mathrm{Win}_1 = \|\mathsf{sol}(\mu_{\bot})\|_1$.

Proof: Let $\mu^\star \triangleq \mathsf{sol}(\mu_\bot)$ be the result of the application of the solver operator to the minimal measure function $\mu_\bot \in \text{SQDM}$. By the notion of inflationary fixpoint, $\mu^\star$ is a fixpoint of the composition of the two progress operators, *i.e.*, $\mu^\star = \mathsf{prg}_+(\mathsf{prg}_\bot(\mu^\star))$, which are inflationary functions on SQDM, due to Items i and ii of Lemma 5.4 and Items i and ii of Lemma 5.5. Therefore, it holds that $\mu^\star \sqsubseteq \mathsf{prg}_\bot(\mu^\star) \sqsubseteq \mathsf{prg}_+(\mathsf{prg}_\bot(\mu^\star)) = \mu^\star$, which implies that $\mathsf{prg}_\bot(\mu^\star) = \mu^\star$ and, so, $\mathsf{prg}_+(\mu^\star) = \mu^\star$. As a consequence of Item iv of Lemma 5.4 and Item iv of Lemma 5.5, it holds that $\mu^\star$ is a progress measure. Hence, $\|\mu^\star\|_1 \subseteq \text{Win}_1$ follows from Theorem 3.7. By Theorem 5.6, it holds that $\mu^\star \in \text{SQDM}$, which implies that $\|\mu^\star\|_0 \subseteq \text{Win}_0$, due to Definition 4.6. Hence, the thesis follows, since $\|\mu^\star\|_0$ and $\|\mu^\star\|_1$ partition the set of positions. □

6. Experimental Evaluation

The algorithm proposed in this chapter has been implemented in OINK [36], a C++ framework supporting different parity game solvers[b] and providing tools to compare their performance on various worst-case families. The solvers considered in the experiments include the original priority promotion algorithm *PP* [26] and the progress measure version presented in this chapter *QDPM*, the optimised version [37] of the Recursive algorithm *Rec* [16], the optimised version of the Small Progress Measure algorithm *SPM* [21] and its quasi-polynomial version *SSPM* [23], the quasi-polynomial algorithm *QPT* [24], the Tangle Learning algorithm *TL* [20], and the Distraction-based Fixpoint Iteration algorithm with justifications *FPJ* [38]. The benchmarks include worst-case games for the considered solvers and clustered random games generated with the PGSolver framework [39]. The latter are games that exhibit a complex structure *w.r.t.* the class of randomly generated games. Indeed, while most of the random games consist in a single *strongly connected component (SCC)* and are easily solved by any attractor-based approach, clustered games rely, instead, on a tree-like structure with multiple SCCs.

Table 1 displays the results on the following worst-case families[c]: the

---

[b]Experiments were carried out on a 64-bit 1.6GHz INTEL quad-core machine, with i5-8250U processor and 8GB of RAM, running UBUNTU 18.04.5 with LINUX kernel version 3.28.2. OINK was compiled with gcc version 7.4.

[c]The instances were generated by issuing the following OINK commands: tc+ n; counter_qpt n; counter_m n; counter_dp n. The Robust Worst Case has been implemented according to [42].

Table 1. Biggest (index of the) instance of the worst-case families solved within 30s. If the 1000th instance is solved, its approximated solution time is reported in brackets.

	Exponential						Quasi Polynomial	
Benchmarks	Rec	PP	TL	FPJ	SPM	**QDPM**	SSPM	QPT
Two Counters [40]	20	107	11	26	7	85	4	5
QPT [24]	[0*s*]	[0*s*]	[0*s*]	[0*s*]	[0*s*]	[0*s*]	[0.41*s*]	30
Gazda's wc [41]	35	35	[0.01*s*]	37	380	[0.67*s*]	22	29
DP [18]	36	23	[0.19*s*]	38	20	[0*s*]	[0.17*s*]	28
Divide&Impera [42]	17	91	[27.75*s*]	16	7	[7.45*s*]	173	4

family for TL "*Two Counters*" [40], the one for QPT algorithm "*QPT*" [24], the family for Zielonka's Optimised algorithm "*Gazda's wc*" [41], the family for the Delayed Priority Promotion algorithm "*DP*" [18], and the Robust Worst Case for Divide-et-Impera Algorithms "*Divide&Impera*" [42]. Each row reports the biggest instance each solver could solve within the time limit of 30 seconds. The "Two Counters" family proved to be very demanding for all the solvers, as none of them could solve the 108th instance within the time limit. On the contrary, the "QPT" family can easily be solved by all the solvers except QPT. The proposed solver QDPM performs extremely well on all the families, being able to solve the 1000th instance faster than the competitors, except for the "Two Counters", on which it is outperformed only by PP, and Gazda's family, where TL is slightly better.

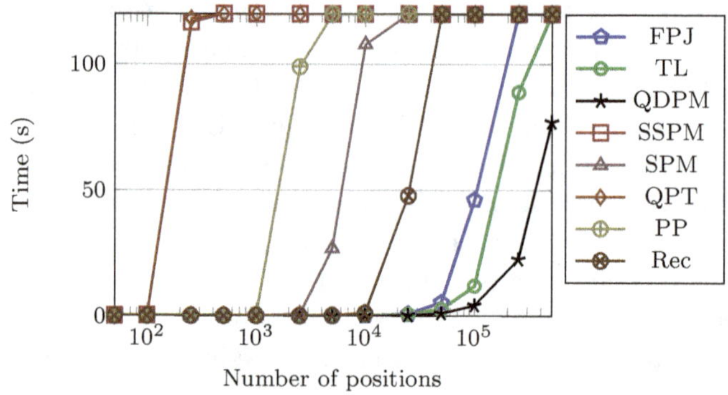

Fig. 2. Time on clustered random games with 2 moves per position.

Figure 2 compares the running times on 1300 random clustered games of size ranging from 50 to $5 \cdot 10^5$ positions and 2 outgoing moves per position.[d]

[d]The instances were generated by issuing the following PGSolver command: clusteredrandomgame n $n/10$ 2 2 5 3 7 3 7.

We set the time-out at 120 seconds. Each point in the graph shows the average time over a cluster of 100 different games of the same size shown on a logarithmic scale. For a game of size n, we set the number of priorities to $k = n/10$. The performance of the quasi-polynomial solvers (QPT and SSPM) reaches the time-out already for the smaller instances: for games with 250 positions they could solve less than 20% of the instances. Almost all of the quasi-dominion-based algorithms, namely TL, FPJ, and QDPM, instead, scale quite well. Their behaviour start to differentiate for games with at least 10^5 positions. On the biggest instances ($5 \cdot 10^5$), QDPM is the only algorithm to terminate within 2 minutes, with an average solution time of 76 seconds and only 14% of time-outs.

7. Discussion

We propose a revisited progress measures-based algorithm for parity games that integrates progress measures and quasi-dominions. This integration requires a novel notion of measure to encode the additional information needed to identify quasi-dominions and a new update policy that takes advantage of quasi-dominions and often allows to skip intermediate measures and reach a progress measure much more quickly than the classic progress measure algorithms. This motivates the conjecture that the integration significantly accelerates the convergence to a progress measure. The experiments show that the proposed approach scales better than any known algorithm on games with a complex structure, such as clustered random games and the worst-case families. In particular, the speed-up can be of several orders of magnitude when compared to other algorithms based on progress measures. We believe that this integration approach may also lead to practically efficient quasi-polynomial algorithms based on succinct progress measures.

Acknowledgments

F. Mogavero acknowledges a partial support by GNCS 2020 project "Ragionamento Strategico e Sintesi Automatica di Sistemi Multi-Agente".

This project has received funding from the European Union's Horizon 2020 research and innovation programme under the Marie Sklodowska-Curie grant agreement No 101032464. The work was supported by EPSRC grant EP/P020909/1.

References

1. E. Emerson and C. Jutla, The Complexity of Tree Automata and Logics of Programs (Extended Abstract). In *FOCS'88*, pp. 328–337, IEEECS, 1988.
2. E. Emerson, C. Jutla, and A. Sistla, On Model-Checking for Fragments of muCalculus. In *CAV'93*, LNCS 697, pp. 385–396, Springer, 1993.
3. O. Kupferman, M. Vardi, and P. Wolper, An Automata Theoretic Approach to Branching-Time Model Checking. *JACM.* **47** (2000), 312–360.
4. A. Ehrenfeucht and J. Mycielski, Positional Strategies for Mean Payoff Games. *IJGT.* **8** (1979), 109–113.
5. V. Gurvich, A. Karzanov, and L. Khachivan, Cyclic Games and an Algorithm to Find Minimax Cycle Means in Directed Graphs. *USSRCMMP.* **28** (1988), 85–91.
6. U. Zwick and M. Paterson, The Complexity of Mean Payoff Games on Graphs. *TCS.* **158** (1996), 343–359.
7. A. Condon, The Complexity of Stochastic Games. *IC.* **96** (1992), 203–224.
8. K. Chatterjee, L. Doyen, T. Henzinger, and J.-F. Raskin, Generalized Mean-Payoff and Energy Games. In *FSTTCS'10*, LIPIcs 8, pp. 505–516, Leibniz-Zentrum fuer Informatik, 2010.
9. A. Mostowski, Regular Expressions for Infinite Trees and a Standard Form of Automata. In *SCT'84*, LNCS 208, pp. 157–168, Springer, 1984.
10. E. Emerson and C. Jutla, Tree Automata, muCalculus, and Determinacy. In *FOCS'91*, pp. 368–377, IEEECS, 1991.
11. O. Kupferman and M. Vardi, Weak Alternating Automata and Tree Automata Emptiness. In *STOC'98*, pp. 224–233, ACM, 1998.
12. E. Grädel, W. Thomas, and T. Wilke, *Automata, Logics, and Infinite Games: A Guide to Current Research.* LNCS 2500, Springer, 2002.
13. M. Jurdziński, Deciding the Winner in Parity Games is in UP ∩ co-UP. *IPL.* **68** (1998), 119–124.
14. A. Mostowski, Games with Forbidden Positions. Technical report, University of Gdańsk, Gdańsk, Poland, 1991.
15. R. McNaughton, Infinite Games Played on Finite Graphs. *APAL.* **65** (1993), 149–184.
16. W. Zielonka, Infinite Games on Finitely Coloured Graphs with Applications to Automata on Infinite Trees. *TCS.* **200** (1998), 135–183.
17. M. Benerecetti, D. Dell'Erba, and F. Mogavero, Solving Parity Games via Priority Promotion. *FMSD.* **52** (2018), 193–226.
18. M. Benerecetti, D. Dell'Erba, and F. Mogavero, A Delayed Promotion Policy for Parity Games. *IC.* **262** (2018), 221–240.
19. M. Benerecetti, D. Dell'Erba, and F. Mogavero, Improving Priority Promotion for Parity Games. In *HVC'16*, LNCS 10028, pp. 1–17, Springer, 2016.
20. T. van Dijk. Attracting Tangles to Solve Parity Games. In *CAV'18*, LNCS 10982, pp. 198–215, Springer, 2018.
21. M. Jurdziński, Small Progress Measures for Solving Parity Games. In *STACS'00*, LNCS 1770, pp. 290–301, Springer, 2000.

22. M. Gazda and T. Willemse, Improvement in Small Progress Measures. In *GANDALF'15*, EPTCS 193, pp. 158–171, 2015.
23. M. Jurdziński and R. Lazic, Succinct Progress Measures for Solving Parity Games. In *LICS'17*, pp. 1–9, ACM, 2017.
24. J. Fearnley, S. Jain, S. Schewe, F. Stephan, and D. Wojtczak, An Ordered Approach to Solving Parity Games in Quasi Polynomial Time and Quasi Linear Space. In *SPIN'17*, pp. 112–121, ACM, 2017.
25. M. Benerecetti, D. Dell'Erba, and F. Mogavero, Solving Mean-Payoff Games via Quasi Dominions. In *TACAS'20*, LNCS 12079, pp. 289–306, 2020.
26. M. Benerecetti, D. Dell'Erba, and F. Mogavero, Solving Parity Games via Priority Promotion. In *CAV'16*, LNCS 9780 (Part II), pp. 270–290, Springer, 2016.
27. N. Klarlund, Liminf Progress Measures. In *MFPS'91*, LNCS 598, pp. 477–491, Springer, 1991.
28. R. Streett and E. Emerson, An Automata Theoretic Decision Procedure for the Propositional Mu-Calculus. *IC.* **81** (1989), 249–264.
29. N. Klarlund, Progress Measures for Complementation of omega-Automata with Applications to Temporal Logic. In *FOCS'91*, pp. 358–367, IEEECS, 1991.
30. N. Klarlund and D. Kozen, Rabin Measures and Their Applications to Fairness and Automata Theory. In *LICS'91*, pp. 256–265, IEEECS, 1991.
31. N. Klarlund, Progress Measures, Immediate Determinacy, and a Subset Construction for Tree Automata. In *LICS'92*, pp. 382–393, IEEECS, 1992.
32. N. Klarlund, Progress Measures, Immediate Determinacy, and a Subset Construction for Tree Automata. *APAL.* **69** (1994), 243–268.
33. N. Klarlund, Progress Measures and Stack Assertions for Fair Termination. In *PODC'92*, pp. 229–240, ACM, 1992.
34. M. Vardi, Rank Predicates vs. Progress Measures in Concurrent-Program Verification. *CJTCS.* **1996** (1996), 1–8.
35. J. Fearnley, Non-Oblivious Strategy Improvement. In *LPAR'10*, LNCS 6355, pp. 212–230, Springer, 2010.
36. T. van Dijk, Oink: an Implementation and Evaluation of Modern Parity Game Solvers. In *TACAS'18*, LNCS 10805, pp. 291–308, Springer, 2018.
37. Y. Liu, Z. D., and C. Tian, An Improved Recursive Algorithm for Parity Games. In *TASE'14*, pp. 154–161, IEEECS, 2014.
38. R. Lapauw, M. Bruynooghe, and M. Denecker, Improving Parity Game Solvers with Justifications. In *20*, LNCS 11990, pp. 449–470, Springer, 2020.
39. O. Friedmann and M. Lange, Solving Parity Games in Practice. In *ATVA'09*, LNCS 5799, pp. 182–196, Springer, 2009.
40. T. van Dijk, A Parity Game Tale of Two Counters. In *GANDALF'19*, EPTCS 305, pp. 107–122, 2019.
41. M. Gazda, *Fixpoint Logic, Games, and Relations of Consequence.* PhD thesis, Eindhoven University of Technology, Eindhoven, Netherlands, 2016.
42. M. Benerecetti, D. Dell'Erba, and F. Mogavero, Robust Worst Cases for Parity Games Algorithms. *IC.* **272** (2020), 104501:1–31.

https://doi.org/10.1142/9789811278631_0008

Effective Ultrapowers and Applications

Rumen D. Dimitrov[†] and Valentina Harizanov[‡]
[†]*Department of Mathematics and Philosophy*
Western Illinois University
Macomb, IL 61455, USA
rd-dimitrov@wiu.edu
[‡]*Department of Mathematics*
George Washington University
Washington, DC 20052, USA
harizanv@gwu.edu

We give a systematic account of the current state of knowledge of an effective analogue of the ultraproduct construction. We start with a product of a uniformly computable sequence of computable structures indexed by the set of natural numbers. The equality of elements and satisfaction of formulas are defined modulo a subset of the index set, which is cohesive, i.e., indecomposable with respect to computably enumerable sets. We present an analogue of Łoś's theorem for effective ultraproducts and a number of results on definability and isomorphism types of the effective ultrapowers of the field of rational numbers, when the complements of cohesive sets are computably enumerable. These effective ultraproducts arose naturally in the study of the automorphisms of the lattice of computably enumerable vector spaces. Previously, a number of authors considered related constructions in the context of nonstandard models of fragments of arithmetic.

1. Introduction

The idea of establishing an axiomatic system that captures the standard model of arithmetic, $\mathbb{N} = (N, +, \cdot, 0, 1, <)$, has been developed in the late 19th century by Dedekind and Peano [28]. Hilbert asked for proof that such axiomatizations are complete and categorical. Computable axiomatizations give incomplete theories according to Gödel. They are also not categorical. The existence of nonstandard models of *formal arithmetic* (the deductive closure of a fixed computable axiomatization) and of *true arithmetic*, $Th(\mathbb{N})$, can be established using ultrapower constructions. In particular, if F is a non-principal ultrafilter on ω, then the ultrapower $\mathbb{N}^\omega/F$ is not isomorphic to $\mathbb{N}$ and is hence a *nonstandard model of arithmetic.* The ultrafilter construction is very elegant but it uses the Axiom of Choice and also produces uncountable models: $|\mathbb{N}^\omega/F| = 2^{\aleph_0}$.

By the ultrapower construction, various first-order theories of arithmetic are not $2^{\aleph_0}$ categorical. It is natural to ask whether there are countable nonstandard models of arithmetic and if such models can be effective. The first question was answered positively by Skolem in 1934. In [32], Skolem constructed a countable nonstandard model of Peano arithmetic, PA (see also [33]). The second question was answered by Tennenbaum in 1959. He proved that if $\mathcal{M} = (M, +, \cdot, 0, 1, <)$ is a countable model of PA, which is not isomorphic to the standard model $\mathbb{N}$, then $\mathcal{M}$ is not computable.

In his construction of nonstandard countable models of PA, Skolem used a combinatorial lemma (Satz 1 in [32]) to define an equivalence relation $\sim$ on the countable set F_1 of arithmetically definable unary functions. The domain of his nonstandard model $\mathbb{N}^*$ is $F_1/\sim$. Skolem's construction has later been presented as an ultrapower construction (see [2], pp. 236–240) but it is Skolem's original proof that is of particular interest to us. The proof of his Satz 1 provides an idea for the construction of certain (more effective) nonstandard structures. Skolem constructed a total function the range of which is what we now call an *arithmetically indecomposable set* (see [20], p. 429, and Remark 2.3(iii)).

Following Skolem's idea, Feferman, Scott, and Tennenbaum in 1959 considered a quotient structure $\mathcal{R}/C$ as a possible model of arithmetic. Here, $\mathcal{R}$ is the set of all unary (total) computable functions, and C is an r-cohesive set of natural numbers. The domain of $\mathcal{R}/C$ consists of the equivalence classes of the elements in $\mathcal{R}$ modulo an equivalence relation $=_c$ defined using C (see Remark 2.3(i)) as follows:

$$f =_c g \Leftrightarrow C \subseteq^* \{n \in \omega \mid f(n) = g(n)\}.$$

It can be established that $\mathbb{N}$ and $\mathcal{R}/C$ are 2-elementary equivalent. Feferman, Scott, and Tennenbaum [12] proved that $\mathcal{R}/C$ is a model of only a fragment of arithmetic. They gave a specific Π^0_3 sentence σ such that $\mathbb{N} \models \sigma$ but $\mathcal{R}/C \not\models \sigma$. Hirshfeld, McLaughlin, and Wheeler later proved a general result that implies that the structure $\mathcal{R}/C$ has no nontrivial automorphisms (that is, it is rigid). The idea of the proof has been given by Hirshfeld and Wheeler in [18] and formally presented by McLaughlin in [22]. The idea for the construction of $\mathcal{R}/C$ can be generalized to higher levels of arithmetical hierarchy (see Definition 2.2). The structures $\mathcal{R}/C$ and the Skolem's $\mathbb{N}^*$ can be seen as the bottom and the top levels of the hierarchy of such structures, respectively. These structures have been studied as models of fragments of arithmetic by Hirshfeld, Lerman, McLaughlin and Wheeler in [17, 18, 20], and [21–23]. Recursive ultrafilter constructions have also been studied by Nelson in [26], where he provided a sufficient condition for recursive saturation of recursive ultrapowers.

More recently, the notion of *cohesive powers of computable structures* has been presented in [10]. The cohesive powers of fields appear in the characterization of certain filters in the lattice of computably enumerable (c.e.) vector spaces, which has been extensively studied in computable model theory. The vector space V_∞ has been introduced by Metakides and Nerode in [24] as a canonical infinite-dimensional, fully effective vector space over a computable field F. The lattice $\mathcal{L}(V_\infty)$ of c.e. subspaces of V_∞, and its factor lattice modulo finite dimension, $\mathcal{L}^*(V_\infty)$, are examples of modular counterparts to the classical distributive lattices of c.e. sets, $\mathcal{E}$, and c.e. sets modulo finite sets, $\mathcal{E}^*$, in computability theory.

In the characterization of certain principal filters in $\mathcal{L}^*(V_\infty)$ (see [9]), we encountered fields with elements that are the equivalence classes of partial computable functions from ω to F. We noticed that the structure of these fields has certain similarities with the classical ultrapower construction. In [10], we introduced the general notion of a cohesive power of a computable structure. It turns out that, in some instances, the cohesive powers of $\mathbb{N}$ are isomorphic to the structures $\mathcal{R}/C$ introduced by Feferman, Scott, and Tennenbaum. In this chapter, we present results from [3, 6, 10], together with certain generalizations and some new proofs.

In Sec. 2, we give some standard definitions from computability theory. In Sec. 3, we introduce the notion of cohesive products and prove the fundamental theorem of cohesive products, an effective analogue of Łoś's theorem for ultraproducts. We further explore the properties of cohesive products of computable structures and provide several examples. In Sec. 4, we present the following theorem from [3] for a field $\mathbb{Q}$ of rational numbers.

Theorem 4.4: For any two maximal sets M_1 and M_2 with complements $\overline{M_1}$ and $\overline{M_2}$, we have

$$\prod_{\overline{M_1}} \mathbb{Q} \cong \prod_{\overline{M_2}} \mathbb{Q} \text{ iff } M_1 \equiv_m M_2,$$

where $\equiv_m$ stands for m-equivalence.

In Sec. 5, we present a new proof of the following theorem from [3].

Theorem 5.1: If M is a maximal set, then $\prod_{\overline{M}} \mathbb{Q}$ has only trivial automorphisms.

Finally, in Sec. 6, we give an example of the role cohesive powers play in the study of the structure and automorphisms of the lattice $\mathcal{L}^*(V_\infty)$.

2. Computability-Theoretic Background

The following notions come from computability theory. The complement of a set $X \subseteq \omega$ is denoted by $\overline{X}$.

Definition 2.1: (i) An infinite set $C \subseteq \omega$ is *cohesive* iff for every c.e. set W, either $W \cap C$ or $\overline{W} \cap C$ is finite.

(ii) A set $M \subseteq \omega$ is *maximal* iff M is c.e. and $\overline{M}$ is cohesive.

(iii) An infinite set $C \subseteq \omega$ is *r-cohesive* iff for every computable set W, either $W \cap C$ or $\overline{W} \cap C$ is finite.

(iv) A set M is *r-maximal* iff M is c.e. and $\overline{M}$ is *r-cohesive.*

Equivalently, a set $M \subseteq \omega$ is maximal if M is c.e., $\overline{M}$ is infinite, and for every c.e. set E such that $M \subseteq E \subseteq \omega$, either $\omega - E$ or $E - M$ is finite. Maximal sets have been extensively studied within the lattice $\mathcal{E}$ of c.e. sets. For sets A and B, we use $A =^* B$ to denote that A and B differ on at most finitely many elements, and we use $A \subseteq^* B$ to denote that all but finitely many elements of A are also elements of B. For vector spaces, we use the same notation provided "modulo finitely many elements" is replaced by "modulo finite dimension." We will now introduce a more general notion than cohesiveness.

Definition 2.2: Let $\mathcal{F}$ be a countable family of subsets of ω. A set $C \subseteq \omega$ is called *$\mathcal{F}$-indecomposable* if C is infinite, and for every W in $\mathcal{F}$, either $W \cap C$ or $\overline{W} \cap C$ is finite.

Remark 2.3:

(i) If $\mathcal{E}$ is the collection of computably enumerable sets, then $\mathcal{E}$-indecomposable sets are cohesive.

(ii) If $\mathcal{R}$ is the collection of computable sets, then $\mathcal{R}$-indecomposable sets are r-cohesive.

(iii) If $\mathcal{A}$ is the collection of arithmetically definable sets, then $\mathcal{A}$-indecomposable sets will be called *arithmetically indecomposable.*

Definition 2.4: A set $B \subseteq \omega$ is called *quasimaximal* if it is the intersection of finitely many maximal sets. If $B = \bigcap_{i=1}^{n} M_i$ where M_i's are maximal, and for $i \neq j$, $M_i \neq^* M_j$, the number n is called the *rank* of B.

Next, we will introduce the notion of m-reducibility which is stronger than Turing reducibility.

Definition 2.5: Let $A, B \subseteq \omega$.

(i) The set A is m-reducible to B, denoted by $A \leq_m B$, if there is a computable function $f : \omega \to \omega$ such that

$$x \in A \Leftrightarrow f(x) \in B.$$

(ii) The sets A and B have the same m-*degree*, denoted by $A \equiv_m B$ (or $\deg_m(A) = \deg_m(B)$), iff $A \leq_m B$ and $B \leq_m A$.

We will now recall the definitions of a computable and a decidable structure.

Definition 2.6: Let $\mathcal{A}$ be a countable structure for a computable language L.

(i) $\mathcal{A}$ is *computable* if its domain and its atomic diagram are computable.

(ii) $\mathcal{A}$ is *decidable* if its domain and its elementary (complete) diagram are computable.

Clearly, every decidable structure is computable, and the theory of a structure is computable in its elementary diagram.

3. Cohesive Products

Cohesive powers of fields were used in [9] to characterize principal filters of quasimaximal spaces. They motivated the definition of cohesive powers of computable structures in [10]. We now give a more general definition of a cohesive product of computable structures. We will use $\simeq$ to denote the equality of partial functions.

Definition 3.1: Let $\mathcal{A}_i$, for $i \in \omega$, be a uniformly computable sequence of computable structures in a computable language L, and let $C \subseteq \omega$ be a cohesive set. The *cohesive product* $\mathcal{B}$ *of* $\mathcal{A}_i$ *over* C, in symbols $\mathcal{B} = \prod_C \mathcal{A}_i$, is a structure defined as follows.

(1) Let

$$D = \{\varphi \mid \varphi : \omega \to \bigcup_{i\in\omega} A_i \text{ is a partial computable function,}$$
$$C \subseteq {}^{*}dom(\varphi), \text{ and if } \varphi(i)\downarrow \text{ then } \varphi(i) \in A_i\}.$$

For $\varphi_1, \varphi_2 \in D$, let

$$\varphi_1 =_C \varphi_2 \quad \text{iff} \quad C \subseteq^* \{i : \varphi_1(i)\downarrow = \varphi_2(i)\downarrow\}.$$

The domain of $\prod\limits_C \mathcal{A}_i$ is the quotient set $D/_{=_C}$ and is denoted here by B.

(2) If $f \in L$ is an n-ary function symbol, then $f^{\mathcal{B}}$ is an n-ary function on B such that for every $[\varphi_1], \dots, [\varphi_n] \in B$, we have

$$f^{\mathcal{B}}([\varphi_1], \dots, [\varphi_n]) = [\varphi] \quad \text{iff} \quad (\forall i \in \omega)\left[\varphi(i) \simeq f^{\mathcal{A}_i}(\varphi_1(i), \dots, \varphi_n(i))\right].$$

(3) If $P \in L$ is an m-ary predicate symbol, then $P^{\mathcal{B}}$ is an m-ary relation on B such that for every $[\varphi_1], \dots, [\varphi_m] \in B$,

$$P^{\mathcal{B}}([\varphi_1], \dots, [\varphi_m]) \quad \text{iff} \quad C \subseteq^* \{i \in \omega \mid P^{\mathcal{A}_i}(\varphi_1(i), \dots, \varphi_m(i))\}.$$

(4) If $c \in L$ is a constant symbol, then $c^{\mathcal{B}}$ is the equivalence class (with respect to $=_C$) of the computable function $g : \omega \to \bigcup\limits_{i\in\omega} A_i$ such that $g(i) = c^{\mathcal{A}_i}$, for each $i \in \omega$.

Definition 3.2: (Dimitrov [10]) If $\mathcal{A}_i = \mathcal{A}$ for $i \in \omega$, then $\prod\limits_C \mathcal{A}_i$ is called the *cohesive power of* $\mathcal{A}$ *over* C and is denoted by $\prod\limits_C \mathcal{A}$.

Theorem 3.3: *(Fundamental theorem of cohesive products)*
(1) If $\tau(y_1, \dots, y_n)$ is a term in L and $[\varphi_1], \dots, [\varphi_n] \in B$, then

$$[\tau^{\mathcal{B}}([\varphi_1], \dots, [\varphi_n])]$$

is the equivalence class of a partial computable function such that

$$\tau^{\mathcal{B}}([\varphi_1], \dots, [\varphi_n])(i) \simeq \tau^{\mathcal{A}_i}(\varphi_1(i), \dots, \varphi_n(i)).$$

(2) If $\Phi(y_1, \dots, y_n)$ is a formula in L, which is a conjunction of Σ^0_1 and Π^0_1 formulas, and $[\varphi_1], \dots, [\varphi_n] \in B$, then

$$\mathcal{B} \models \Phi([\varphi_1], \dots, [\varphi_n]) \text{ iff } C \subseteq^* \{i : \mathcal{A}_i \models \Phi(\varphi_1(i), \dots, \varphi_n(i))\}.$$

(3) If Φ is a Π^0_2 sentence in L, then $C \subseteq^* \{i : \mathcal{A}_i \models \Phi\}$ implies $\mathcal{B} \models \Phi$.

Proof: (1) The proof is straightforward, but we note that we essentially use the fact that the operations in the structures $\mathcal{A}_i$ are uniformly computable.

(2) We proceed by induction.

(2.1) Let $\Phi(y_1, \ldots, y_n) = P(\tau_1(y_1, \ldots, y_n), \ldots, \tau_m(y_1, \ldots, y_n))$ be an atomic formula and suppose $[\psi_i] = \tau_i^{\mathcal{B}}([\varphi_1], \ldots, [\varphi_n])$. Then

$\mathcal{B} \models \Phi([\varphi_1], \ldots, [\varphi_n])$

iff

$\mathcal{B} \models P([\psi_1], \ldots, [\psi_m])$

iff

$C \subseteq^* \{i : \mathcal{A}_i \models P(\psi_1(i), \ldots, \psi_m(i))\}$

iff

$C \subseteq^* \{i : \mathcal{A}_i \models \Phi(\varphi_1(i), \ldots, \varphi_m(i))\}$.

(2.2) Suppose $\Phi(y_1, \ldots, y_n) = \Phi_1(y_1, \ldots, y_n) \wedge \Phi_2(y_1, \ldots, y_n)$ and the claim is true for $\Phi_i(y_1, \ldots, y_n)$, $i = 1, 2$. Then

$\mathcal{B} \models \Phi([\varphi_1], \ldots, [\varphi_n])$

iff

$\mathcal{B} \models \Phi_1([\varphi_1], \ldots, [\varphi_n])$ and $\mathcal{B} \models \Phi_2([\varphi_1], \ldots, [\varphi_n])$

iff

$C \subseteq^* \{i : \mathcal{A}_i \models \Phi_1(\varphi_1(i), \ldots, \varphi_n(i))\}$ and
$C \subseteq^* \{i : \mathcal{A}_i \models \Phi_2(\varphi_1(i), \ldots, \varphi_n(i))\}$

iff

$C \subseteq^* \{i : \mathcal{A}_i \models \Phi(\varphi_1(i), \ldots, \varphi_n(i))\}$.

(2.3) Suppose $\Phi(y_1, \ldots, y_n) = \exists y \Psi(y, y_1, \ldots, y_n)$ and $\Psi(y, y_1, \ldots, y_n)$ is a quantifier-free formula for which the claim is true.

(2.3a) Suppose $\mathcal{B} \models \exists y \Psi(y, [\varphi_1], \ldots, [\varphi_n])$ and suppose that the partial computable function φ is such that $\mathcal{B} \models \Psi([\varphi], [\varphi_1], \ldots, [\varphi_n])$. By the inductive hypothesis, $C \subseteq^* \{i : \mathcal{A}_i \models \Psi(\varphi(i), \varphi_1(i), \ldots, \varphi_n(i))\}$ and so

$$C \subseteq^* \{i : \mathcal{A}_i \models \exists y \Psi(y, \varphi_1(i), \ldots, \varphi_n(i))\}.$$

(2.3b) Suppose $C \subseteq^* \{i : \mathcal{A}_i \models \exists y \Psi(y, \varphi_1(i), \ldots, \varphi_n(i))\}$. Since the structures $\mathcal{A}_i$ are uniformly computable and $\Psi(y, y_1, \ldots, y_n)$ is quantifier-free, we can define a partial computable function

$$\varphi(i) = (\mu y \in A_i)\,[\mathcal{A}_i \models \Psi(y, \varphi_1(i), \ldots, \varphi_n(i))].$$

Then

$$\{i : \mathcal{A}_i \models \exists y \Psi(y, \varphi_1(i), \ldots, \varphi_n(i))\} = \{i : \mathcal{A}_i \models \Psi(\varphi(i), \varphi_1(i), \ldots, \varphi_n(i))\}$$

and so $C \subseteq^* \{i : \mathcal{A}_i \models \Psi(\varphi(i), \varphi_1(i), \ldots, \varphi_n(i))\}$. By the inductive hypothesis,

$$\mathcal{B} \models \Psi([\varphi], [\varphi_1], \ldots, [\varphi_n]), \text{ and so } \mathcal{B} \models \exists y \Psi(y, [\varphi_1], \ldots, [\varphi_n]).$$

(2.4) Suppose $\Phi(y_1, \ldots, y_n) = \neg\Psi(y_1, \ldots, y_n)$ where $\Psi(y_1, \ldots, y_n)$ is a Σ_1^0 formula for which the hypothesis is true.

(2.4a) Suppose $\mathcal{B} \models \Phi([\varphi_1], \ldots, [\varphi_n])$ and let

$$W = \{i : \mathcal{A}_i \models \Psi(\varphi_1(i), \ldots, \varphi_n(i))\}.$$

Since $\mathcal{B} \not\models \Psi([\varphi_1], \ldots, [\varphi_n])$, we have $C \not\subseteq^* W$. Since $\Psi(y_1, \ldots, y_n)$ is a Σ_1^0 formula, the structures $\mathcal{A}_i$ are uniformly computable, and φ_j, for $1 \leq j \leq n$, is partial computable, W is a c.e. set. Since C is cohesive, we have $C \cap W =^* \emptyset$. Also, since $C \subseteq^* \bigcap_{j=1}^{n} dom(\varphi_j)$, for almost all $i \in C$ we have

$$\mathcal{A}_i \not\models \Psi(\varphi_1(i), \ldots, \varphi_n(i)).$$

Therefore,

$$C \subseteq^* \{i : \mathcal{A}_i \models \neg\Psi(\varphi_1(i), \ldots, \varphi_n(i))\}.$$

(2.4b) Suppose $C \subseteq^* \{i : \mathcal{A}_i \models \neg\Psi(\varphi_1(i), \ldots, \varphi_n(i))\}$. Then

$$C \cap \{i : \mathcal{A}_i \models \Psi(\varphi_1(i), \ldots, \varphi_n(i))\} =^* \emptyset,$$

and, by the inductive hypothesis, $\mathcal{B} \not\models \Psi([\varphi_1], \ldots, [\varphi_n])$. Therefore,

$$\mathcal{B} \models \neg\Psi([\varphi_1], \ldots, [\varphi_n]).$$

(3) Assume that $C \subseteq^* \{i : \mathcal{A}_i \models \Phi\}$, where $\Phi = \forall y \exists z \Psi(y, z)$ for some quantifier-free formula $\Psi(y, z)$. Let $[\varphi] \in B$. Define a partial computable ψ as follows:

$$\psi(i) = (\mu u \in A_i)[\mathcal{A}_i \models \Psi(\varphi(i), u)].$$

Since $\mathcal{A}_i$'s are uniformly computable, we can deduce that ψ is partial computable. Note that

$$C \subseteq^* dom(\psi) \text{ and } C \subseteq^* \{i : \mathcal{A}_i \models \Psi(\varphi(i), \psi(i))\},$$

because $C \subseteq^* dom(\varphi)$ and $C \subseteq^* \{i : \mathcal{A}_i \models \forall y \exists z \Psi(y, z)\}$. Therefore, $\mathcal{B} \models \Psi([\varphi], [\psi])$, which means that $\mathcal{B} \models \exists z \Psi([\varphi], z)$, and so $\mathcal{B} \models \Phi$. □

Remark 3.4: The statement in Part (2) of Theorem 3.3 can be extended to Boolean combinations of Σ_1^0 and Π_1^0 formulas. (For further improvements see [5].)

Definition 3.5: (Dimitrov [10]) Let $\mathcal{B} = \prod_C \mathcal{A}$ be a cohesive power of $\mathcal{A}$. For $c \in A$, let $[\varphi_c] \in B$ be the equivalence class of the total function φ_c such that $\varphi_c(i) = c$ for every $i \in \omega$. The map $d : A \to B$ such that $d(c) = [\varphi_c]$ is called the *canonical embedding* of $\mathcal{A}$ into $\mathcal{B}$.

Theorem 3.6: *(Dimitrov [10]) Let $\mathcal{B} = \prod_C \mathcal{A}$ be a cohesive power of $\mathcal{A}$.*

(1) If Φ is a Π_3^0 sentence in L, then $\mathcal{B} \models \Phi$ implies $\mathcal{A} \models \Phi$.
(2) If Φ is a Σ_3^0 sentence in L, then $\mathcal{A} \models \Phi$ implies $\mathcal{B} \models \Phi$.
(3) If Φ is a Π_2^0 (or Σ_2^0) sentence in L, then $\mathcal{B} \models \Phi$ iff $\mathcal{A} \models \Phi$.
(4) Let $c_1, \dots, c_n \in A$ and let $\Phi(y_1, \dots, y_n)$ be a Π_2^0 or a Σ_2^0 formula in L. Then

$$\mathcal{A} \models \Phi(c_1, \dots, c_n) \text{ iff } \mathcal{B} \models \Phi([\varphi_{c_1}], \dots, [\varphi_{c_n}]).$$

(5) Let $c_1, \dots, c_n \in A$ and let $\Phi(y_1, \dots, y_n)$ be a Σ_3^0 formula in L. Then

$$\mathcal{A} \models \Phi(c_1, \dots, c_n) \text{ implies } \mathcal{B} \models \Phi([\varphi_{c_1}], \dots, [\varphi_{c_n}]).$$

Proof: (1) Let $\Phi = \forall y \exists z \forall t \Psi(y, z, t)$ where $\Psi(y, z, t)$ is a quantifier-free formula. Let $c \in A$ be arbitrary and let $\varphi_c(i) = c$ for every $i \in \omega$. Let $[\varphi] \in B$ be such that $\mathcal{B} \models \forall t \Psi([\varphi_c], [\varphi], t)$. By (2) of Theorem 3.3 above, we have $C \subseteq^* \{i : \mathcal{A} \models \forall t \Psi(\varphi_c(i), \varphi(i), t)\}$. Then $C \subseteq^* \{i : \mathcal{A} \models \exists z \forall t \Psi(c, z, t)\}$. The set C is infinite and so $\mathcal{A} \models \exists z \forall t \Psi(c, z, t)$. Therefore, $\mathcal{A} \models \Phi$.

(2) This is the contrapositive of (1).

(3) Let $\Phi = \forall y \exists z \Psi(y, z)$ where $\Psi(y, z)$ is a quantifier-free formula.

(3a) The fact that $\mathcal{A} \models \Phi$ whenever $\mathcal{B} \models \Phi$ follows from (1).

(3b) Suppose that $\mathcal{A} \models \Phi$ and let $[\varphi] \in B$ be arbitrary. We have that $C \subseteq^* dom(\varphi) = \{i : \mathcal{A} \models \exists z \Psi(\varphi(i), z)\}$. By (2) of Theorem 3.3, $\mathcal{B} \models \exists z \Psi([\varphi], z)$ and so $\mathcal{B} \models \Phi$.

(4) Let $\mathcal{A}_{\mathcal{K}}$ be the structure $\mathcal{A}$ in the language $L \cup \{\mathbf{c}_1, \dots, \mathbf{c}_n\}$ with the constant symbols $\mathbf{c}_1, \dots, \mathbf{c}_n$ interpreted as $c_1, \dots, c_n$, respectively. The result follows from part (3) of this theorem applied to $\mathcal{A}_{\mathcal{K}}$ and $\mathcal{B}_{\mathcal{K}} = \prod_C \mathcal{A}_{\mathcal{K}}$.

(5) It follows from part (2) of this theorem. □

Remark 3.7: If $\mathcal{A}$ is a (dense) linear order (without endpoints), a ring, (an algebraically closed) field (of characteristic p), lattice, (atomless) Boolean algebra, then so is $\prod_C \mathcal{A}$. Each of their theories is Π^0_2-axiomatizable, and so we can apply part (3) of the previous theorem.

Theorem 3.8: *(Dimitrov [10]) Let $\mathcal{A}$ be decidable and $\mathcal{B} = \prod_C \mathcal{A}$.*

(1) If $\Phi(y_1, \ldots, y_n)$ is a formula in L, and $[\varphi_1], \ldots, [\varphi_n] \in B$, then

$$\prod_C \mathcal{A} \models \Phi([\varphi_1], \ldots, [\varphi_n]) \text{ iff } C \subseteq^* \{i : \mathcal{A} \models \Phi(\varphi_1(i), \ldots, \varphi_n(i))\}.$$

(2) If Φ is a sentence, then $\prod_C \mathcal{A} \models \Phi$ iff $\mathcal{A} \models \Phi$.

(3) The canonical map d is an elementary embedding of $\mathcal{A}$ into $\mathcal{B}$.

Proof: (1) Note that for any formula $\Psi(y_1, \ldots, y_n)$ the set $\{(a_1, \ldots, a_n) \in A^n : \mathcal{A} \models \Psi(a_1, \ldots, a_n)\}$ is computable, and hence $\{i : \mathcal{A} \models \Psi(\varphi_1(i), \ldots, \varphi_n(i))\}$ is c.e. Then the steps 2.3 and 2.4 in the proof of Theorem 3.3 can be carried out for any formula Ψ.

(2) and (3) follow immediately from (1). □

Proposition 3.9: *(Dimitrov [10]) Let the structure $\mathcal{A}$ be finite and let $\mathcal{B} = \prod_C \mathcal{A}$. Then $\mathcal{B} \cong \mathcal{A}$.*

Proof: Let $[\varphi] \in B$ be arbitrary. For any $c \in A$, let $I_c = \{i : \varphi(i) = c\}$. Note that the set I_c is c.e. Since $dom(\varphi) = \bigcup_{c \in A} I_c$ and A is finite, for some $c_1 \in A$, the set $I_{c_1} \cap C$ is infinite. Since C is cohesive, we have that $C \subseteq^* I_{c_1}$, and so $[\varphi] = [\varphi_{c_1}]$. Thus, the canonical embedding of $\mathcal{A}$ into $\mathcal{B}$ is onto, and so $\mathcal{B} \cong \mathcal{A}$. □

Proposition 3.10: *(Dimitrov [10]) If there is a computable permutation σ of ω such that $\sigma(C_1) =^* C_2$, then $\prod_{C_1} \mathcal{A} \cong \prod_{C_2} \mathcal{A}$.*

Proof: Let the map $\Phi : \prod_{C_2} \mathcal{A} \to \prod_{C_1} \mathcal{A}$ be such that $\Phi([\psi]) = [\psi \circ \sigma]$.

The proof that Φ is an isomorphism is straightforward. □

Remark 3.11: (1) Let C be co-c.e. Then for every $[\varphi] \in \prod_C \mathcal{A}$ there is a computable function f such that $[f] = [\varphi]$.

(2) If M is a maximal set, then $\prod_{\overline{M}} \mathbb{N}$ and $\mathcal{R}/\sim_{\overline{M}}$ are isomorphic.

Proof: (1) Let $[\varphi] \in \prod_C \mathcal{A}$ be arbitrary, $a \in A$, and define

$$f(n) = \begin{cases} \varphi(n) & \text{if } \varphi(n)\downarrow \text{ first,} \\ a & \text{if } n \text{ is enumerated into } \overline{C} \text{ first.} \end{cases}$$

(2) It follows from (1). □

Remark 3.12: (1) The fields $\mathbb{Z}_{p_i}$ (p_i is the i-th prime number) are uniformly computable and, by part (3) of Theorem 3.3, $\prod_C \mathbb{Z}_{p_i}$ is a field, but of characteristic 0.

(2) Notice that, by Proposition 3.9, $\prod_C \mathbb{Z}_p \cong \mathbb{Z}_p$.

(3) No "nonstandard" element of the cohesive power $\prod_C \mathbb{Q}$ of the field $\mathbb{Q}$ is algebraic over $\mathbb{Q}$ (see [10]).

Let $\overline{\mathbb{Z}}_{p_i}$ be the algebraic closure of $\mathbb{Z}_{p_i}$. By a result of Rabin (Theorem 7 in [29]), each $\overline{\mathbb{Z}}_{p_i}$ has a computable presentation. If these presentations are uniformly computable, then we can form $\mathcal{B} = \prod_C \overline{\mathbb{Z}}_{p_i}$. By part (3) of Theorem 3.3, $\mathcal{B}$ is an algebraically closed field (ACF). The structure $\mathcal{B}$ is a countable model of the theory of ACF and we note that the characteristic of $\mathcal{B}$ is 0. Although the theory of ACF_0 is not $\aleph_0$-categorical, we know that all of its countable models are isomorphic to one of the following

$$\overline{\mathbb{Q}} \prec \overline{\mathbb{Q}(x_1)} \prec \overline{\mathbb{Q}(x_1, x_2)} \prec \cdots \prec \overline{\mathbb{Q}(x_1, x_2, \ldots)}.$$

Note that in the classical case the situation is quite different. If U is a nonprincipal ultrafilter on the set of prime numbers, then the classical ultraproduct $\prod_U \overline{\mathbb{Z}}_{p_i}$ is a model of ACF_0. Since $\left|\prod_U \overline{\mathbb{Z}}_{p_i}\right| = 2^{\aleph_0}$ and ACF_0 is categorical in every $\kappa \geq \aleph_1$, we have $\prod_U \overline{\mathbb{Z}}_{p_i} \cong \mathbb{C}$, where $\mathbb{C}$ is the field of complex numbers.

4. Isomorphism Types of Cohesive Powers of $\mathbb{Q}$

The results in this section were obtained by Dimitrov, Harizanov, Miller and Mourad and were originally presented in [3]. We assume that $\mathbb{N}$, $\mathbb{Z}$, and

$\mathbb{Q}$ are the standard computable presentations of the structures of natural numbers, integers, and rational numbers in the language $L = \{+, \cdot, 0, 1\}$. Before giving the characterizations of the isomorphism types of co-maximal powers of $\mathbb{Q}$, we will first prove that $\prod_C \mathbb{N}$ is definable both in $\prod_C \mathbb{Z}$ and $\prod_C \mathbb{Q}$. The definability of $\mathbb{Z}$ in $\mathbb{Q}$ (by a Π^0_3 formula) has been established by J. Robinson in [30]. More recently, Koenigsmann [19] gave a Π^0_1 definition of $\mathbb{Z}$ in $\mathbb{Q}$. He proved that there is a positive integer n and a polynomial $p \in \mathbb{Z}[y, z_1, \ldots, z_n]$ such that

$$y \in \mathbb{Z} \Leftrightarrow \mathbb{Q} \models \forall z_1 \cdots \forall z_n [p(y, z_1, \ldots, z_n) \neq 0].$$

The proof of Proposition 4.1 below makes essential use of Koenigsmann's definition of $\mathbb{Z}$ in $\mathbb{Q}$ and does not work with definitions of higher complexity.

The definability of $\mathbb{N}$ in $\mathbb{Z}$ (by various Σ^0_1 formulas) has been established by R. Robinson in [31]. We will use the formula that defines the natural numbers as sums of squares of four integers. Using these results, we obtain that $\mathbb{N}$ can be defined in $\mathbb{Q}$ as follows:

$$x \in \mathbb{N} \Leftrightarrow \exists y_1 \cdots \exists y_4 \left[\bigwedge_{1 \leq i \leq 4} y_i \in \mathbb{Z} \wedge x = y_1^2 + y_2^2 + y_3^2 + y_4^2 \right],$$

$$x \in \mathbb{N} \Leftrightarrow \mathbb{Q} \models \exists y_1 \cdots \exists y_4 \forall z_1 \cdots \forall z_n$$
$$\left[\bigwedge_{1 \leq i \leq 4} p(y_i, z_1, \ldots, z_n) \neq 0 \wedge x = y_1^2 + y_2^2 + y_3^2 + y_4^2 \right],$$

which we will abbreviate as

$$x \in \mathbb{N} \Leftrightarrow \mathbb{Q} \models \exists \overline{y} \forall \overline{z} \theta(x, \overline{y}, \overline{z}).$$

Hence $\theta(x, \overline{y}, \overline{z})$ is the quantifier-free formula above in the language of rings $L = \{+, \cdot, 0, 1\}$.

Proposition 4.1: *(Dimitrov, Harizanov, Miller, and Mourad [3]) The natural embedding of* $\prod_C \mathbb{N}$ *is definable in* $\prod_C \mathbb{Q}$ *by the same formula* $\exists \overline{y} \forall \overline{z} \theta(x, \overline{y}, \overline{z})$ *that defines* $\mathbb{N}$ *in* $\mathbb{Q}$.

Proof: First, assume that for some $[\varphi] \in \prod_C \mathbb{Q}$, we have

$$\prod_C \mathbb{Q} \models \exists \overline{y} \forall \overline{z} \theta([\varphi], \overline{y}, \overline{z}),$$

and that y_i's are such that $y_i = [\psi_i]$ and

$$\prod_C \mathbb{Q} \models \forall \overline{z} \theta([\varphi], \overline{[\psi_i]}, \overline{z}).$$

By Theorem 3.3, we have that

$$C \subseteq^* \{n : \mathbb{Q} \models \forall \overline{z} \theta(\varphi(n), \overline{\psi_i(n)}, \overline{z})\}.$$

Using the definition of $\theta(x, \overline{y}, \overline{z})$ we obtain that $C \subseteq^* \{n : \varphi(n) \in \mathbb{N}\}$, which means that $[\varphi] \in \prod_C \mathbb{N}$.

Now, assume that $[\varphi] \in \prod_C \mathbb{N}$. We will prove that

$$\prod_C \mathbb{Q} \models \exists \overline{y} \forall \overline{z} \theta([\varphi], \overline{y}, \overline{z}).$$

Define the partial computable functions $\xi_i : \omega \to \mathbb{Q}$, for $1 \leq i \leq 4$, as follows. If at stage s we have $\varphi^s(n) = m$ and $m \in \mathbb{N}$, then find the least $(b_1, \ldots, b_4) \in \mathbb{N}^4$ such that $m = \sum_{i=1}^{4} b_i^2$, and let $\xi_i(n) = b_i$.
By the definition of the functions ξ_i, we have that

$$C \subseteq^* \left\{ n : \mathbb{Q} \models \forall z_1 \cdots \forall z_n \right.$$

$$\left. \left[\bigwedge_{1 \leq i \leq 4} p(\xi_i(n), z_1, \ldots, z_n) \neq 0 \wedge \varphi(n) = \sum_{1 \leq i \leq 4} \xi_i(n)^2 \right] \right\},$$

which implies that

$$\prod_C \mathbb{Q} \models \exists \overline{y} \forall \overline{z} \theta([\varphi], \overline{y}, \overline{z}).$$

□

Notation 4.2: We will use $\phi^{\dagger}(\overline{x})$ to denote the formula $\phi(\overline{x})$ with the scope of its quantifiers restricted from $\mathbb{Q}$ to $\mathbb{N}$ (and from $\prod_C \mathbb{Q}$ to $\prod_C \mathbb{N}$ because of Proposition 4.1).

Proposition 4.3: *(Dimitrov, Harizanov, Miller, and Mourad [3]) The structures $\prod_C \mathbb{Q}$ and $\mathbb{Q}$ are not elementary equivalent.*

Proof: Let T be Kleene's predicate. Recall that $T(e, x, z)$ holds if the eth Turing machine program halts on input x in $\leq z$ steps. By a result by

Feferman, Scott, and Tennenbaum (see Theorem 2.1 in [20]), we know that for

$$\phi = \forall x \exists t \forall e \forall z[(e < x \wedge T(e,x,z)) \Rightarrow z < t],$$

$$\mathbb{N} \models \phi \quad \text{but} \quad \prod_C \mathbb{N} \not\models \phi.$$

We can assume that ϕ is a sentence in $L = \{+, \cdot, 0, 1\}$ since both Kleene's T predicate and $<$ are definable in $\mathbb{N}$. Then $\mathbb{Q} \models \phi^{\dagger}$ iff $\mathbb{N} \models \phi$, and $\prod_C \mathbb{Q} \not\models \phi^{\dagger}$ iff $\prod_C \mathbb{N} \not\models \phi$. This finally gives us

$$\mathbb{Q} \not\equiv \prod_C \mathbb{Q}.$$

By $\equiv$ we denoted elementary equivalence of structures. □

The following theorem is the main result in this section.

Theorem 4.4: *(Dimitrov, Harizanov, Miller, and Mourad [3]) Let $M_1 \subseteq \omega$ and $M_2 \subseteq \omega$ be maximal sets.*

(1) If $M_1 \equiv_m M_2$, then $\prod_{\overline{M_1}} \mathbb{Q} \cong \prod_{\overline{M_2}} \mathbb{Q}$.
(2) If $M_1 \not\equiv_m M_2$, then $\prod_{\overline{M_1}} \mathbb{Q} \not\equiv \prod_{\overline{M_2}} \mathbb{Q}$.

Proof: (1) It follows from a result by Hermann [16] that if $M_1 \equiv_m M_2$, then there is a computable permutation σ of ω such that $\sigma(M_1) =^* M_2$. Then, by Proposition 3.10, the map $\Phi : \prod_{\overline{M_2}} \mathbb{Q} \to \prod_{\overline{M_1}} \mathbb{Q}$ such that $\Phi([\psi]) = [\psi \circ \sigma]$ is an isomorphism.

(2) If $M_1 \not\equiv_m M_2$, then $\mathcal{R}/\overline{M_1} \not\equiv \mathcal{R}/\overline{M_2}$, by Theorem 2.3 from [20]. In fact, Lerman provided a specific sentence θ (originally in the language $L_< = \{+, \cdot, 0, 1, <\}$) for which $\mathcal{R}/\overline{M_1} \models \theta$ while $\mathcal{R}/\overline{M_2} \models \neg\theta$. By Remark 3.11, part (2), $\mathcal{R}/\overline{M_i} \cong \prod_{\overline{M_i}} \mathbb{N}$, for $i = 1, 2$, and so $\prod_{\overline{M_1}} \mathbb{N} \not\equiv \prod_{\overline{M_2}} \mathbb{N}$. Since $<$ is definable in $L = \{+, \cdot, 0, 1\}$, we can assume that the sentence θ is equivalent to a sentence in the language L. Thus, for the relativisation $\theta^{\dagger}$, we have $\prod_{\overline{M_1}} \mathbb{Q} \models \theta^{\dagger}$, while $\prod_{\overline{M_2}} \mathbb{Q} \models \neg\theta^{\dagger}$. □

5. Automorphisms of Cohesive Powers of $\mathbb{Q}$

The results in this section were obtained by Dimitrov, Harizanov, Miller and Mourad and were originally presented in [3]. Here, we give a new proof of Theorem 5.1.

Let M be a maximal set and let $C = \overline{M}$. We will prove that the field $\prod_C \mathbb{Q}$ is rigid. In this case, $\prod_C \mathbb{N}$ is a special case of the Δ_1^0 ultrapowers studied by Hirshfeld, Wheeler, and McLaughlin. Specifically, they studied the structures $\mathcal{F}_n/U$, where U is a non-principal ultrafilter in the Boolean algebra of Δ_n^0 sets, and $\mathcal{F}_n$ is the set of all total functions with Σ_n^0 graphs. In [22] (Theorem 2.11), McLaughlin proved a general result about rigidity of a class of ultrapowers. His result implies that the structure $\mathcal{F}_1/U$, which is the set $\mathcal{F}_1$ of all computable functions modulo a non-principal ultrafilter U in the Boolean algebra of Δ_1^0 sets, is rigid in the language $L_< = \{+, \cdot, 0, 1, <\}$. To apply McLaughlin's result, we first notice that his theorem also holds for the language $L = \{+, \cdot, 0, 1\}$ because of the definability of the relation "$<$". Next, the equivalence relation induced by the cohesive set C is equivalent to the one induced by a Δ_1^0 ultrafilter because

$$U_C =_{def} \{W \mid W \in \Delta_1^0 \text{ and } C \subseteq^* W\}$$

is an ultrafilter in the Boolean algebra Δ_1^0. Finally, the domain of $\prod_C \mathbb{N}$ consists of partial computable functions, while the functions in $\mathcal{F}_1$ are total. However, since the set C is co-c.e., by part (2) of Remark 3.11, $\prod_C \mathbb{N} \cong \mathcal{F}_1/U_C$ and, using McLaughlin's result, we have that $\prod_C \mathbb{N}$ is also rigid.

Theorem 5.1: *(Dimitrov, Harizanov, Miller, and Mourad [3])*
(1) The structure $\prod_C \mathbb{Z}$ *is rigid.*
(2) The structure $\prod_C \mathbb{Q}$ *is rigid.*

Proof: We will prove only (2). The idea for the proof of (1) is similar. Suppose that Γ is an automorphism of $\prod_C \mathbb{Q}$, and let $[\varphi] \in \prod_C \mathbb{Q}$ be arbitrary. We will prove that $\Gamma([\varphi]) = [\varphi]$ in the (nontrivial) case when $[\varphi] \neq 0$.

By Proposition 4.1, $\prod_C \mathbb{N}$ is first-order definable in $\prod_C \mathbb{Q}$, and so $\Gamma \restriction_{\prod_C \mathbb{N}}$ is an automorphism of $\prod_C \mathbb{N}$. Since $\prod_C \mathbb{N}$ is rigid, we have $\Gamma \restriction_{\prod_C \mathbb{N}} = id \restriction_{\prod_C \mathbb{N}}$.

Let $\overline{y}$ denote the sequence of variables

$$y_{11}, y_{12}, y_{13}, y_{14}, y_{21}, y_{22}, y_{23}, y_{24}, y_{31}, y_{32}, y_{33}, y_{34}.$$

Also, let

$$\theta^{\#}(x, x_1, x_2, x_3, \overline{y}) =_{def}$$

$$(\forall z_1 \cdots \forall z_n) \left[(x \cdot x_1 + x_2 = x_3) \wedge \bigwedge_{i=1}^{3} (x_i = \sum_{j=1}^{4} y_{ij}^2) \wedge \right.$$

$$\left. \bigwedge_{i=1}^{3} \bigwedge_{j=1}^{4} (p(y_{ij}, z_1, \ldots, z_n) \neq 0) \right].$$

Here, $p(y_{ij}, z_1, \ldots, z_n)$ is the polynomial from the Koenigsmann's formula. The intended interpretation of the Π_1^0 formula $\theta^{\#}$ is that

(i) y_{ij} (for $1 \leq i \leq 3, 1 \leq j \leq 4$) are elements of $\mathbb{Z}$,

(ii) x_i are elements of $\mathbb{N}$,

(iii) for $x \in \mathbb{Q}$ we have

$$x = \frac{x_3 - x_2}{x_1}.$$

We will now construct $[\varphi_i], [\varphi_{ij}] \in \prod_C \mathbb{N}$ (for $1 \leq i \leq 3, 1 \leq j \leq 4$) such that

$$\prod_C \mathbb{Q} \models \theta^{\#}([\varphi], [\varphi_1], [\varphi_2], [\varphi_3], \overline{[\varphi_{ij}]}).$$

For each $n \in \omega$ such that $\varphi(n) \downarrow = 0$, let:

$$m_i = 0 \text{ for } 1 \leq i \leq 3, \text{ and}$$
$$m_{ij} = 0 \text{ for } 1 \leq i \leq 3 \text{ and } 1 \leq j \leq 4.$$

For each $n \in \omega$ such that $\varphi(n) \downarrow \neq 0$, find the least

$$\langle m_1, m_2, m_3 \rangle \in \mathbb{N}^3$$

and

$$\langle m_{11}, m_{12}, m_{13}, m_{14}, m_{21}, m_{22}, m_{23}, m_{24}, m_{31}, m_{32}, m_{33}, m_{34} \rangle \in \mathbb{N}^{12}$$

such that

$$m_1 \neq 0$$
$$\varphi(n) \cdot m_1 + m_2 = m_3, \text{ and}$$
$$m_i = \sum_{j=1}^{4} m_{ij}^2 \text{ for } 1 \leq i \leq 3.$$

Then let $\varphi_i(n) = m_i$ (for $1 \leq i \leq 3$) and $\varphi_{ij}(n) = m_{ij}$ (for $1 \leq i \leq 3$ and $1 \leq j \leq 4$). Note that

$$dom(\varphi) = dom(\varphi_i) = dom(\varphi_{ij}),$$

for each $1 \leq i \leq 3$ and $1 \leq j \leq 4$, and moreover,

$$C \subseteq^* dom(\varphi) = \{n : \mathbb{Q} \models \theta^{\#}(\varphi(n), \varphi_1(n), \varphi_2(n), \varphi_3(n), \overline{\varphi_{ij}(n)})\}.$$

By Theorem 3.3,

$$\prod_C \mathbb{Q} \models \theta^{\#}([\varphi], [\varphi_1], [\varphi_2], [\varphi_3], \overline{[\varphi_{ij}]}).$$

In particular, this implies that $[\varphi]$ is a solution of the linear equation

$$x \cdot [\varphi_1] + [\varphi_2] = [\varphi_3] \text{ in } \prod_C \mathbb{Q}.$$

Since Γ is an automorphism of $\prod_C \mathbb{Q}$, we also have

$$\prod_C \mathbb{Q} \models \theta^{\#}(\Gamma([\varphi]), \Gamma([\varphi_1]), \Gamma([\varphi_2]), \Gamma([\varphi_3]), \overline{\Gamma([\varphi_{ij}])}).$$

Since $[\varphi_i]$ and $[\varphi_{ij}]$ are elements of $\prod_C \mathbb{N}$, we have

$$\Gamma([\varphi_i]) = [\varphi_i] \text{ and } \Gamma([\varphi_{ij}]) = [\varphi_{ij}],$$

for $1 \leq i \leq 3$ and $1 \leq j \leq 4$, and so

$$\prod_C \mathbb{Q} \models \theta^{\#}(\Gamma([\varphi]), [\varphi_1], [\varphi_2], [\varphi_3], \overline{[\varphi_{ij}]}).$$

This implies that $\Gamma([\varphi])$ is also a solution of the linear equation

$$x \cdot [\varphi_1] + [\varphi_2] = [\varphi_3] \text{ in } \prod_C \mathbb{Q}.$$

Therefore, $\Gamma([\varphi]) = [\varphi]$. □

6. An Application of Cohesive Powers to Computable Algebra

Assume that the canonical computable $\aleph_0$-dimensional vector space V_∞ is over the field $\mathbb{Q}$. As usual, we denote the lattice of c.e. subspaces of V_∞ by $\mathcal{L}(V_\infty)$, and the lattice $\mathcal{L}(V_\infty)$ modulo finite dimension by $\mathcal{L}^*(V_\infty)$. If $V \in \mathcal{L}(V_\infty)$, then by $\mathcal{L}^*(V^*, \uparrow)$ we will denote the principal filter of V^* in $\mathcal{L}^*(V_\infty)$. To simplify the notation, we will use $Q_{\mathbf{a}} =_{def} \prod_{\overline{M}} \mathbb{Q}$ to denote the cohesive power of the field $\mathbb{Q}$ over a co-maximal set $\overline{M}$ such that $deg_m(M) = \mathbf{a}$. This notation is justified by Theorem 5.1. In [6], Dimitrov and Harizanov characterized the orbits, in $\mathcal{L}^*(V_\infty)$, of the equivalence classes of the closures of quasimaximal subsets of computable bases. We introduced the notion of an *m-degree type* of a quasimaximal set $E = \bigcap_{i=1}^{n} M_i$ of rank n with respect to a fixed computable basis A of V_∞, denoted by $type_A(E)$. This notion captures the number and the m-degrees of the maximal sets M_i's. In the Main Theorem in [6], we proved that there is an automorphism Φ of $\mathcal{L}^*(V_\infty)$ such that $\Phi(cl(E_1)^*) = cl(E_2)^*$ if and only if $type_{A_1}(E_1) = type_{A_2}(E_2)$. We now give an example illustrating the notion of the type and apply the theorem to this case.

Example 6.1: Suppose A_1 and A_2 are computable bases of V_∞.

Suppose that, for $j = 1, 2$, the sets $M_1^j \equiv_m M_2^j \equiv_m M_3^j$ are different maximal subsets of A_j and $deg_m(M_i^j) = \mathbf{a}_j$. Let $E_j = \bigcap_{i=1}^{3} M_i^j$. In this case, $type_{A_j}(E_j) = (3; \mathbf{a}_j)$.

By the Main Theorem in [9], $\mathcal{L}^*(cl(E_j)^*, \uparrow) \cong \mathcal{L}(3, Q_{\mathbf{a}_j})$, where $\mathcal{L}(3, Q_{\mathbf{a}_j})$ denotes the lattice of all subspaces of a 3-dimensional vector space over the field $Q_{\mathbf{a}_j} =_{def} \prod_{\overline{M_j}} \mathbb{Q}$, for $j = 1, 2$.

By the Main Theorem in [6], there is an automorphism Φ of $\mathcal{L}^*(V_\infty)$ such that $\Phi(cl(E_1)^*) = cl(E_2)^*$ if and only if $\mathbf{a}_1 = \mathbf{a}_2$.

In the case of maximal sets, we have the following result.

Theorem 6.2: *(Dimitrov and Harizanov [6]) Let M_1 and M_2 be maximal subsets of computable bases A_1 and A_2 of V_∞, respectively. Then the following are equivalent:*

(1) There is an automorphism Φ of $\mathcal{L}^*(V_\infty)$ such that

$$\Phi(cl(M_1)^*) = cl(M_2)^*,$$

(2) $deg_m(M_1) = deg_m(M_2)$.

7. Additional Notes and Acknowledgements

Harizanov was partially supported by the Simons Foundation grant 429466 and by FRG NSF grant DMS-2152095. Harizanov thanks the organizers and IMS of NUS for inviting her and partially supporting her participation in the research workshop "Computable Structures and Reverse Mathematics" during September 11–15, 2018.

Since the original submission, there have been additional new results related to cohesive powers and linear orders (see [4, 5]).

References

1. Davis, M., Matiyasevich, Y., and Robinson, J., Hilbert's tenth problem: Diophantine equations: positive aspects of a negative solution, *Mathematical Developments Arising from Hilbert Problems* (*Proc. Sympos. Pure Math.*, vol. XXVIII, Northern Illinois Univ., De Kalb, Ill., 1974), Amer. Math. Soc., Providence, RI (1976), pp. 323–378.
2. Slomson, A.B., and Bell, J.L., *Models and Ultraproducts: An Introduction*, 3rd revised printing, North-Holland, Amsterdam, 1974.
3. Dimitrov, R., Harizanov, V., Miller, R., and Mourad, K.J., Isomorphisms of non-standard fields and Ash's conjecture, in: *Language, Life, Limits*, A. Beckmann, E. Csuhaj-Varjú, and K. Meer, editors, Lecture Notes in Computer Science 8493 (2014), Springer, pp. 143–152.
4. Dimitrov, R., Harizanov, V., Morozov, A., Shafer, P., Soskova, A., and Vatev, S., Cohesive powers of linear orders, in: *Computing with Foresight and Industry*, F. Manea, B. Martin, D. Paulusma and G. Primiero, editors, Lecture Notes in Computer Science 11558 (2019), Springer, pp. 168–180.
5. Dimitrov, R., Harizanov, V., Morozov, A., Shafer, P., Soskova, A., and Vatev, S., On cohesive powers of linear orders, *Journal of Symbolic Logic*, to appear.
6. Dimitrov, R.D., and Harizanov, V., Orbits of maximal vector spaces, *Algebra and Logic* 54 (2016), pp. 440–477.
7. Dimitrov, R., and Harizanov, V., The lattice of computably enumerable vector spaces, in: *Computability and Complexity*, A. Day, M. Fellows, N. Greenberg, B. Khoussainov, A. Melnikov, and F. Rosamond, eds., Springer (2017), pp. 366–393.
8. Dimitrov, R., Quasimaximality and principal filters isomorphism between $\mathcal{E}^*$ and $\mathcal{L}^*(V_\infty)$, *Arch. Math. Logic* 43 (2004), pp. 415–424.
9. Dimitrov, R., A class of Σ^0_3 modular lattices embeddable as principal filters in $\mathcal{L}^*(V_\infty)$, *Arch. Math. Logic* 47 (2008), pp. 111–132.
10. Dimitrov, R., Cohesive powers of computable structures, *Annuare de l'Université de Sofia "St. Kliment Ohridski"*, Fac. Math. and Inf., tome 99 (2009), pp. 193–201.
11. Dimitrov, R., Extensions of certain partial automorphisms of $\mathcal{L}^*(V_\infty)$, *Annuare de l'Université de Sofia "St. Kliment Ohridski"*, Fac. Math. and Inf., tome 99 (2009), pp. 183–191.

12. Feferman, S., Scott, D.S., and Tennenbaum, S., Models of arithmetic through function rings, *Notices Amer. Math. Soc.* 6, 173 (1959). Abstract #556-31.
13. Fokina, E.B., Harizanov, V., and Melnikov, A., Computable model theory, in: *Turing's Legacy: Developments from Turing Ideas in Logic*, R. Downey, editor, Cambridge University Press/ASL (2014), pp. 124–194.
14. Fröhlich, A. and Shepherdson, J.C., Effective procedures in field theory, *Philosophical Transactions of the Royal Society of London.* Ser. A., 248 (1956), pp. 407–432.
15. Guichard, D.R., Automorphisms of substructure lattices in recursive algebra, *Ann. Pure Appl. Logic* 25 (1983), pp. 47–58.
16. Hermann, E., 1-reducibility inside an m-degree with maximal set, *Journal of Symbolic Logic* 57 (1992), pp. 1046–1056.
17. Hirshfeld, Y., Models of arithmetic and recursive functions, *Israel Journal of Mathematics* 20 (1975), pp. 111–126.
18. Hirshfeld, Y. and Wheeler, W.H., Forcing, arithmetic, division rings, *Lecture Notes in Mathematics* 454, Springer, Berlin, 1975.
19. Koenigsmann, J., Defining $\mathbb{Z}$ in $\mathbb{Q}$, *Annals of Mathematics* 183 (2016), pp. 73–93.
20. Lerman, M., Recursive functions modulo co-r-maximal sets, *Transactions of the American Mathematical Society* 148 (1970), pp. 429–444.
21. McLaughlin, T.G., Some extension and rearrangement theorems for Nerode semirings, *Zeitschr. f. Math. Logic und Grundlagen d. Math.* 35 (1989), pp. 197–209.
22. McLaughlin, T.G., Sub-arithmetical ultrapowers: a survey, *Annals of Pure and Applied Logic* 49 (1990), pp. 143–191.
23. McLaughlin, T.G., Δ_1 ultrapowers are totally rigid, *Archive for Mathematical Logic* 46 (2007), pp. 379–384.
24. Metakides, G. and Nerode, A., Recursively enumerable vector spaces, *Annals of Mathematical Logic* 11 (1977), pp. 147–171.
25. Miller, R., **d**-computable categoricity for algebraic fields, *Journal of Symbolic Logic* 74 (2009), pp. 1325–1351.
26. Nelson, G.C., Constructive ultraproducts and isomorphisms of recursively saturated ultrapowers, *Notre Dame J. Formal Logic* 33 (1992), pp. 433–441.
27. Miller, R. and Shlapentokh, A., Computable categoricity for algebraic fields with splitting algorithms, *Transactions of the American Mathematical Society* 367 (2015), pp. 3955–3980.
28. Peano, Giuseppe, Arithmetices principia: nova methodo exposita (1889).
29. Rabin, M., Computable algebra, general theory and theory of computable fields, *Transactions of the American Mathematical Society* 95 (1960), pp. 341–360.
30. Robinson, J., Definability and decision problems in arithmetic, *Journal of Symbolic Logic* 14 (1949), pp. 98–114.
31. Robinson, R.M., Arithmetical definitions in the ring of integers, *Proceedings of the American Mathematical Society* 2 (1951), pp. 279–284.

32. Skolem, Th., Über die Nicht-charakterisierbarkheit der Zahlenreihe mittels endlich oder abzählbar unendlich vieler Aussagen mit ausschliesslich Zahlenvariablen, *Fund. Math.* 23 (1934), pp. 150–161.
33. Skolem, Th., Peano's axioms and models of arithmetic, *Mathematical Interpretation of Formal Systems*, North-Holland, Amsterdam (1955), pp. 1–14.
34. Soare, R.I., *Recursively Enumerable Sets and Degrees. A Study of Computable Functions and Computably Generated Sets*, Springer-Verlag, Berlin, 1987.
35. van der Waerden, B.L., Eine Bemerkung über die Unzerlegbarkeit von Polynomen, *Mathematische Annalen* 102 (1930), pp. 738–739.

https://doi.org/10.1142/9789811278631_0009

FPT-Inspired Approximations

Henning Fernau
Fachbereich IV, Abteilung Informatikwissenschaften,
Universität Trier
54296 Trier, Germany
fernau@uni-trier.de

Approximation algorithms and their performance analysis have a history of more than five decades, which turns them into one of the most traditional subjects in algorithmic sciences. By comparison, the field of fixed-parameter algorithms, pioneered by Rod Downey and Mike Fellows, can barely look at three decades of existence. Yet, the rapid development of techniques in fixed-parameter algorithms led us to the question if and possibly how this relatively young field can inspire new developments in the older field. In this chapter, we are proposing several such strategies, linking them with previous ones whenever possible.

The most promising strategy (for the discussion in this chapter) is to employ ideas typically used in kernelization algorithms in order to achieve polynomial-time approximation algorithms. In particular, the use of exact and approximative reduction rules offer the possibility to easily monitor the quality of the approximation, this way enabling mathematically sound ways to estimate the overall approximation ratio for a concrete instance, as opposed to the usual way of giving a bound on the approximation ratio for all instances (of a certain size). We hence propose to distinguish between the traditional *a priori* analysis of approximation algorithms, as opposed to the *a posteriori* analysis described here.

Contents

1. Introduction

The idea of using approximation algorithms for problems that are seemingly too hard to solve exactly can be traced back well into the 1960s. One of the first (if not the first) paper published in a journal is from R. L. Graham [1], who was probably also the first person obtaining a PhD on approximation algorithms. This is quite remarkable, as this clearly predates the theory of NP-completeness, showing up in the early 1970s.

In the 1970s and 1980s, the only accepted mathematical solutions[a] to the fact that some combinatorial problem had been proven to be NP-hard was either to look for approximative solutions or to study if certain structural restrictions could lead to feasible scenarios. In the first case, one has to deal with the fact that there are fundamental differences between optimization problems (which is the setting where approximation could be studied) and decision problems (which is the formal setting of notions like NP-completeness). It starts out with the fact that algorithms for solving optimization problems must return solutions (as otherwise there is no meaningful way to measure the quality of such a solution against an optimum solution), while (at least formally) algorithms that solve decision problems are only asked to return yes or no. This issue will become quite an interesting point later on in our discussions. For the second case, one should first look into the question what type of restrictions one is interested in. For instance, for problems dealing with graphs, it might be the case that the class of graphs that are actually to be solved turn out to be a collection of easy instances. Hence, there might be polynomial-time algorithms for such

[a]Clearly, in practice often heuristics were used in such cases. In actual fact, NP-hardness could then be taken as a kind of excuse to apply mathematically less satisfying approaches to solve instances of NP-hard problems, often admittedly to an extent that barely leaves any traces of intractability.

scenarios, although the general problem is NP-hard. The website `http://www.graphclasses.org/index.html` is a well-known collection of graph classes and the status of several basic decision problems on these classes. It is quite interesting to observe that the link `http://www.graphclasses.org/classes/problem_Independent_set.html` not only lists dozens of graph classes where INDEPENDENT SET (and hence VERTEX COVER) can be solved in polynomial time, and other dozens of classes where these problems are known to be NP-hard, but there are obviously also dozens of classes where the status of this problem is still unknown. For instance, VERTEX COVER can be solved in polynomial time on bipartite graphs via matching techniques.

Around 1990, the idea of finding structural properties that certify polynomial-time solvability has been expressed in a seemingly different way, as this was the time seeing the birth of what as of now has become a discipline of its own: fixed-parameter (in)tractability, an area mostly pioneered by Rod Downey and Mike Fellows in its early days. Quite logically, they also compiled the first textbook in this area, see [2]. In the setting of parameterized complexity, it is crucial to quantify the properties one likes to make use of for developing efficient algorithms. For instance, VERTEX COVER can be solved in time $O(f(k)p(n))$ on graphs of order n that have tree-width of at most k, or on graphs that have vertex cover number at most k, where f is some arbitrary (computable) function and p is some polynomial. The quantity k is known as the parameter, and we could also state that VERTEX COVER, parameterized by an upper-bound on the tree-width or on the vertex cover number of the given graph instance, is fixed-parameter tractable, or belongs to FPT for short.

One of the core techniques of FPT is kernelization, as kernelizability is equivalent to membership in FPT. Kernelization could be seen as a formal treatment of simple preprocessing rules, so-called reduction rules, that also play a crucial role in many successful (meta-)heuristic strategies to cope with intractability. Recently, a textbook dedicated to kernelization alone has been published [3]. We will show in this chapter that reduction rules can also lead to very good approximation algorithms.

This chapter is explaining a research direction rather than presenting new material. We also report on experimental evidence showing that the self-monitoring approach performs quite well. We will explain our approach with graph-theoretic examples, using standard graph-theoretic terminology, as $N(x)$ for the open neighborhood of vertex x, or $G - \{x, y\}$ for the graph obtained from G by deleting vertices x, y and all incident edges.

2. The general idea of using reduction rules for approximation

In this section, we are going to present basics of a framework that allows us to use reduction rules (very similar to kernelization strategies) for approximation purposes. A reader who is not that much interested in this framework can also skip this section and immediately move to the next two sections that provide concrete examples for this approach.

2.1. *Maximization problems*

We are first recalling some important notions and results from [4].

Specializing standard terminology from [5], we can express the following. A maximization problem $\mathcal{P}$ is specified by a triple $(I_\mathcal{P}, \mathrm{SOL}_\mathcal{P}, m_\mathcal{P})$, where

(1) $I_\mathcal{P}$ is the set of input instances of $\mathcal{P}$;
(2) $\mathrm{SOL}_\mathcal{P}$ is a function that associates to $x \in I_\mathcal{P}$ the set $\mathrm{SOL}_\mathcal{P}(x)$ of feasible solutions of x;
(3) $m_\mathcal{P}$ provides on (x, y), where $x \in I_\mathcal{P}$ and $y \in \mathrm{SOL}_\mathcal{P}(x)$, a positive integer which is the value of the solution y.

An optimum solution y^* to x satisfies: (i) $y^* \in \mathrm{SOL}_\mathcal{P}(x)$, and (ii) $m_\mathcal{P}(y^*) = \max\{m_\mathcal{P}(y) \mid y \in \mathrm{SOL}_\mathcal{P}(x)\}$. The value $m_\mathcal{P}(y^*)$ is also referred to as $m^*_\mathcal{P}(x)$ for brevity. It would be possible to allow for non-integer values of solutions, but this is not necessary in the context of this chapter, and this might create some problems in the context of parameterized algorithms; for a recent discussion on weighted problem variants in this context, we refer to [6].

Given a maximization problem $\mathcal{P}$, a *factor-α approximation*, $\alpha \geq 1$, associates to each $x \in I_\mathcal{P}$ some $y \in \mathrm{SOL}_\mathcal{P}(x)$, such that $\alpha \cdot m_\mathcal{P}(x, y) \geq m^*_\mathcal{P}(x)$. A solution $y \in \mathrm{SOL}_\mathcal{P}(x)$ satisfying $\alpha \cdot m_\mathcal{P}(x, y) \geq m^*_\mathcal{P}(x)$ is also called an *α-approximate solution* for x. As the reader will have noticed, we will only consider constant-factor approximations formally in this chapter, but it should be clear how to adapt these notions to more general circumstances. Also, we are mostly concerned about factor-α approximations that can be computed in polynomial time, a fact that we sometimes make explicit if there is some danger of confusion.

We are now going to present a first key notion for this chapter, plus some consequences of this, whose proofs can be found in [4].

Definition 2.1: Let $\mathcal{P} = (I_\mathcal{P}, \mathrm{SOL}_\mathcal{P}, m_\mathcal{P})$ be a maximization problem. An *α-preserving reduction*, with $\alpha \geq 1$, is a pair of mappings $\mathtt{inst}_\mathcal{P} : I_\mathcal{P} \to I_\mathcal{P}$

and $\mathtt{sol}_{\mathcal{P}}$ which, given $y' \in \mathrm{SOL}_{\mathcal{P}}(\mathtt{inst}_{\mathcal{P}}(x))$, produces some $y \in \mathrm{SOL}_{\mathcal{P}}(x)$ such that there are constants $a, b \geq 0$ satisfying the following inequalities:

(1) $a \leq \alpha \cdot b$,
(2) $m^*_{\mathcal{P}}(\mathtt{inst}_{\mathcal{P}}(x)) + a \geq m^*_{\mathcal{P}}(x)$, and
(3) $\forall y' \in \mathrm{SOL}_{\mathcal{P}}(\mathtt{inst}_{\mathcal{P}}(x)) : m_{\mathcal{P}}(\mathtt{inst}_{\mathcal{P}}(x), y') + b \leq m_{\mathcal{P}}(x, \mathtt{sol}_{\mathcal{P}}(y'))$.

We call the pair (a, b) a *factor proof* of the *local approximation factor* α.

Theorem 2.2: *Let $\mathcal{P} = (I_{\mathcal{P}}, \mathrm{SOL}_{\mathcal{P}}, m_{\mathcal{P}})$ be some maximization problem. If the pair $(\mathtt{inst}_{\mathcal{P}}, \mathtt{sol}_{\mathcal{P}})$ describes an α-preserving reduction and if, given some instance x, $y' \in \mathrm{SOL}_{\mathcal{P}}(\mathtt{inst}_{\mathcal{P}}(x))$ is an α-approximate solution for $\mathtt{inst}_{\mathcal{P}}(x)$, then $y = \mathtt{sol}_{\mathcal{P}}(y')$ is an α-approximate solution for x.*

We call an α-preserving reduction $(\mathtt{inst}_{\mathcal{P}}, \mathtt{sol}_{\mathcal{P}})$ *strict* if $|\mathtt{inst}_{\mathcal{P}}(x)| < |x|$ for all $x \in I_{\mathcal{P}}$, and it is called *polynomial-time computable* if the two mappings comprising the reduction can be computed in polynomial time.

Lemma 2.3: *If $(\mathtt{inst}_{\mathcal{P}}, \mathtt{sol}_{\mathcal{P}})$ and $(\mathtt{inst}'_{\mathcal{P}}, \mathtt{sol}'_{\mathcal{P}})$ are two α-preserving reductions, then the composition $(i, s) := (\mathtt{inst}_{\mathcal{P}} \circ \mathtt{inst}'_{\mathcal{P}}, \mathtt{sol}'_{\mathcal{P}} \circ \mathtt{sol}_{\mathcal{P}})$ is also an α-preserving reduction. More precisely, if (a, b) and (a', b') are factor proofs for $(\mathtt{inst}_{\mathcal{P}}, \mathtt{sol}_{\mathcal{P}})$ and $(\mathtt{inst}'_{\mathcal{P}}, \mathtt{sol}'_{\mathcal{P}})$, respectively, then $(a+a', b+b')$ is a factor proof for (i, s). If both $(\mathtt{inst}_{\mathcal{P}}, \mathtt{sol}_{\mathcal{P}})$ and $(\mathtt{inst}'_{\mathcal{P}}, \mathtt{sol}'_{\mathcal{P}})$ are strict (polynomial-time computable, resp.), then the composition (i, s) is strict (polynomial-time computable, resp.).*

In [4], it is also explained how *applicability conditions* can be integrated in α-preserving reductions, so that they will resemble even more what people more accustomed to kernelization algorithms will expect. Notice that such conditions are often given in an implicit form. For instance, if a rule (roughly) says that isolated vertices can be deleted from an instance without changing solutions, then such a rule can only be applied if the instance contains some isolated vertices. We will see concrete examples below. Then, we can paraphrase the first part of an approximation algorithm based on α-preserving reductions as follows:

As long as possible apply some of the α-preserving reductions.

Observe that if all α-preserving reductions are strict and polynomial-time computable, then this procedure will converge in the sense that after a polynomial number of steps, we reach a situation where none of the reduction rules applies. So far, the situation is completely analogous to kernelization.

However, while we can stop now in the kernelization setting, this is not the case for approximability, as we do not have any solution to return so far. As it is often the case with kernelization rules as well, the actual purpose of the reduction rules is to create a special structure in the graph, so that now some other algorithmic tools can be used to compute a good approximation to the reduced instance, which means that we apply Theorem 2.2 once more, so that finally we obtain the desired approximate solution to the original instance.

2.2. *Minimization problems*

We are now presenting a similar framework for minimization problems. A minimization problem $\mathcal{P}$ is specified by a triple $(I_\mathcal{P}, \mathrm{SOL}_\mathcal{P}, m_\mathcal{P})$, where

(1) $I_\mathcal{P}$ is the set of input instances of $\mathcal{P}$;
(2) $\mathrm{SOL}_\mathcal{P}$ is a function that associates to $x \in I_\mathcal{P}$ the set $\mathrm{SOL}_\mathcal{P}(x)$ of feasible solutions of x;
(3) $m_\mathcal{P}$ provides on (x, y), where $x \in I_\mathcal{P}$ and $y \in \mathrm{SOL}_\mathcal{P}(x)$, a positive integer which is the value of the solution y.

An optimum solution y^* to x satisfies: (i) $y^* \in \mathrm{SOL}_\mathcal{P}(x)$, and (ii) $m_\mathcal{P}(y^*) = \min\{m_\mathcal{P}(y) \mid y \in \mathrm{SOL}_\mathcal{P}(x)\}$. The value $m_\mathcal{P}(y^*)$ is also referred to as $m^*_\mathcal{P}(x)$ for brevity.

Given a minimization problem $\mathcal{P}$, a *factor-α approximation*, $\alpha \geq 1$, associates to each $x \in I_\mathcal{P}$ some $y \in \mathrm{SOL}_\mathcal{P}(x)$, such that $m_\mathcal{P}(x, y) \leq \alpha \cdot m^*_\mathcal{P}(x)$. A solution $y \in \mathrm{SOL}_\mathcal{P}(x)$ satisfying $m_\mathcal{P}(x, y) \leq \alpha \cdot m^*_\mathcal{P}(x)$ is also called an *α-approximate solution* for x.

Definition 2.4: Let $\mathcal{P} = (I_\mathcal{P}, \mathrm{SOL}_\mathcal{P}, m_\mathcal{P})$ be a minimization problem. An *α-preserving reduction*, with $\alpha \geq 1$, is a pair of mappings $\mathtt{inst}_\mathcal{P} : I_\mathcal{P} \to I_\mathcal{P}$ and $\mathtt{sol}_\mathcal{P}$ which, given $y' \in \mathrm{SOL}_\mathcal{P}(\mathtt{inst}_\mathcal{P}(x))$, produces some $y \in \mathrm{SOL}_\mathcal{P}(x)$ such that there are constants $a, b \geq 0$ satisfying the following inequalities:

(1) $\alpha \cdot a \geq b$,
(2) $m^*_\mathcal{P}(\mathtt{inst}_\mathcal{P}(x)) + a \leq m^*_\mathcal{P}(x)$, and
(3) $\forall y' \in \mathrm{SOL}_\mathcal{P}(\mathtt{inst}_\mathcal{P}(x)) : m_\mathcal{P}(\mathtt{inst}_\mathcal{P}(x), y') + b \geq m_\mathcal{P}(x, \mathtt{sol}_\mathcal{P}(y'))$.

We call the pair (a, b) a *factor proof* of the *local approximation factor α*.

Theorem 2.5: *Let $\mathcal{P} = (I_\mathcal{P}, \mathrm{SOL}_\mathcal{P}, m_\mathcal{P})$ be some minimization problem. If the pair $(\mathtt{inst}_\mathcal{P}, \mathtt{sol}_\mathcal{P})$ describes an α-preserving reduction and if, given some instance x, $y' \in \mathrm{SOL}_\mathcal{P}(\mathtt{inst}_\mathcal{P}(x))$ is an α-approximate solution for $\mathtt{inst}_\mathcal{P}(x)$, then $y = \mathtt{sol}_\mathcal{P}(y')$ is an α-approximate solution for x.*

Proof: We have to prove that $m_{\mathcal{P}}(x,y) \leq \alpha \cdot m^*_{\mathcal{P}}(x)$. Now,

$$\frac{m_{\mathcal{P}}(x,y)}{m^*_{\mathcal{P}}(x)} \leq \frac{m_{\mathcal{P}}(\mathtt{inst}_{\mathcal{P}}(x),y')+b}{m^*_{\mathcal{P}}(\mathtt{inst}_{\mathcal{P}}(x))+a} \leq \frac{\alpha m^*_{\mathcal{P}}(\mathtt{inst}_{\mathcal{P}}(x))+\alpha a}{m^*_{\mathcal{P}}(\mathtt{inst}_{\mathcal{P}}(x))+a} = \alpha$$

as required. □

Also, the notions of strictness and polynomial-time computable transfer in a literal sense to the minimization case, which allows us to conclude:

Lemma 2.6: *If* $(\mathtt{inst}_{\mathcal{P}}, \mathtt{sol}_{\mathcal{P}})$ *and* $(\mathtt{inst}'_{\mathcal{P}}, \mathtt{sol}'_{\mathcal{P}})$ *are two α-preserving reductions, then the composition* $(i,s) := (\mathtt{inst}_{\mathcal{P}} \circ \mathtt{inst}'_{\mathcal{P}}, \mathtt{sol}'_{\mathcal{P}} \circ \mathtt{sol}_{\mathcal{P}})$ *is also an α-preserving reduction. More precisely, if (a,b) and (a',b') are factor proofs for* $(\mathtt{inst}_{\mathcal{P}}, \mathtt{sol}_{\mathcal{P}})$ *and* $(\mathtt{inst}'_{\mathcal{P}}, \mathtt{sol}'_{\mathcal{P}})$*, respectively, then $(a+a', b+b')$ is a factor proof for (i,s). If both* $(\mathtt{inst}_{\mathcal{P}}, \mathtt{sol}_{\mathcal{P}})$ *and* $(\mathtt{inst}'_{\mathcal{P}}, \mathtt{sol}'_{\mathcal{P}})$ *are strict (polynomial-time computable, resp.), then the composition (i,s) is strict (polynomial-time computable, resp.).*

Proof: Consider a situation described as in the lemma, where the pair of numbers (a,b) shows that $(\mathtt{inst}_{\mathcal{P}}, \mathtt{sol}_{\mathcal{P}})$ is α-preserving and the pair of numbers (a',b') shows that $(\mathtt{inst}'_{\mathcal{P}}, \mathtt{sol}'_{\mathcal{P}})$ is α-preserving. Clearly, if $x \in I_{\mathcal{P}}$, then $\mathtt{inst}_{\mathcal{P}}(x) \in I_{\mathcal{P}}$ and hence $(\mathtt{inst}_{\mathcal{P}} \circ \mathtt{inst}'_{\mathcal{P}})(x) = \mathtt{inst}'_{\mathcal{P}}(\mathtt{inst}_{\mathcal{P}}(x)) \in I_{\mathcal{P}}$, as well. A similar observation applies to the solutions (in the reversed order). We now prove that the composition (i,s) is α-preserving, testified by the pair of numbers $(a+a', b+b')$ as a factor proof.

$$\begin{aligned} m^*_{\mathcal{P}}((\mathtt{inst}_{\mathcal{P}} \circ \mathtt{inst}'_{\mathcal{P}})(x)) + (a+a') &= (m^*_{\mathcal{P}}(\mathtt{inst}'_{\mathcal{P}}(\mathtt{inst}_{\mathcal{P}}(x))) + a') + a \\ &\leq m^*_{\mathcal{P}}(\mathtt{inst}_{\mathcal{P}}(x)) + a \\ &\leq m^*_{\mathcal{P}}(x). \end{aligned}$$

The computation for the bounds on the solution is similar. The claims on composability of the strictness and the polynomial-time computability are easy to see. □

Again, we can integrate applicability conditions and then use sets of strict, polynomial-time computable α-preserving reductions in order to produce a factor-α approximation, assuming that we can do something clever with an irreducible instance. However, with minimization problems, the obtained irreducible instance is often trivial to solve, so that this approach becomes somewhat easier in this situation, as we will also see with examples below.

2.3. *A priori versus a posteriori analysis*

Summarizing our observations, we can state:

Theorem 2.7: *Let $\mathcal{P}$ be some optimization problem. Let $\mathcal{R}$ be a finite collection of strict, polynomial-time computable factor-α-preserving reductions and let A be a polynomial-time computable factor-α approximation that is applicable whenever none of the rules in $\mathcal{R}$ applies. Then, there is a polynomial-time factor-α approximation for $\mathcal{P}$.*

Notice the idea of using approximation algorithms A that are not applicable to all possible instances (as it would be the case in the general setting of approximation), but only to those to which none of the rules in $\mathcal{R}$ applies. This can be decisive in the design and analysis of our algorithms, as we will see below.

Proof: Consider the following algorithm sketch:

(1) **As long as possible** apply some of the reductions from $\mathcal{R}$.
(2) Apply A.

This provides the claimed factor-α approximation for $\mathcal{P}$. Namely, consider some instance x_0 of $\mathcal{P}$. Step 1 of the algorithm will produce a sequence of instances $x_1, \ldots, x_n$, such that none of the rules from $\mathcal{R}$ applies to x_n. Hence, A is applicable to x_n, producing an α-approximate solution y_n to instance x_n. Let $r_i = (\mathtt{inst}^i_{\mathcal{P}}, \mathtt{sol}^i_{\mathcal{P}})$ for $1 \leq i \leq n$, so that $x_i = \mathtt{inst}^i_{\mathcal{P}}(x_{i-1})$. Now, defining recursively $y_{i-1} = \mathtt{sol}^i_{\mathcal{P}}(x_i)$, Theorems 2.2 and 2.5 show, by a simple inductive argument, that each y_i is a factor-α approximation of x_i, even for $i = 0$. This shows the claim, as the composition of polynomially many polynomial-time procedures is polynomial-time computable. The fact that the whole algorithm terminates that quickly is due to our assumption that all rules in $\mathcal{R}$ are strict (and polynomial-time computable). □

This is in the tradition of a classical worst-case analysis of approximation algorithms. It uses the knowledge available without looking at any concrete instance and will be hence called *a priori analysis* in the following. Alternatively, we can try to find (better) estimates for the concrete input instance x our algorithm has to deal with.

First, observe by looking at the definitions that whenever r is some α-preserving reduction, then r is also α'-preserving for any $\alpha' \geq \alpha$. Due to this monotonicity, it makes sense to associate to any reduction r the smallest $\alpha \geq 1$, such that r is α-preserving; denote this value by $\alpha(r)$. As a first, still quite rough idea, one could consider the scenario when (in the setting given by Theorem 2.7) A is in fact a factor-β approximation, with $\beta \leq \alpha$, and when processing x, only rules from $\mathcal{R}' \subseteq \mathcal{R}$ have been used. Then, the argument of the preceding proof shows that the solution y computed by the algorithm, given x, should produce a factor-$\max\{\max\{\alpha(r) \mid r \in \mathcal{R}'\}, \beta\}$ approximation, which could already be smaller than the *a priori* value of α.

However, more powerful is the following observation, inductively based on Lemmas 2.3 and 2.6.

Corollary 2.8: *Let $\mathcal{P}$ be some optimization problem. Let $\mathcal{R}$ be a finite collection of factor-α-preserving reductions, such that $(a,b)[r]$ denotes a factor proof for $r \in \mathcal{R}$. Let $r_1, \ldots, r_n \in \mathcal{R}$ and consider $\bar{r} = r_1 \circ \cdots \circ r_n$. Then, $\bar{r}$ is a factor-α-preserving reduction with factor proof $\sum_{i=1}^{n}(a,b)[r_i]$, where the sum is understood componentwise.*

This seemingly innocuous corollary is quite strong in practice, as it allows us to announce an *a posteriori* estimate of the approximation factor of a concrete solution to an instance x, basically by summing up the a- and b-values (of the known factor proofs) of the reduction rules that were actually applied for producing a solution. Even more, as the sum of the a-values gives a bound on any optimum solution, we can use this estimate to measure the quality of approximation of any other run of some approximation algorithm, a remark that is also helpful at times, as we prove by examples below.

Let us underline that a certain type of *a posteriori* analysis is well-known in the area of approximation algorithms, namely in the context of approximations based on integer linear programming, where dual programs can be used to estimate optimal solutions and hence improve on the *a priori* analysis. However, there are a couple of things that make a difference in the two approaches.

- The rule-based approach allows us to find an easy combinatorial interpretation of what is going on.
- Also, it is easy to locate the part of the instance where (relatively) bad approximation rules have been used. In such a (small) region of the instance, it might pay off to apply exact algorithmic techniques in order to improve the overall approximation.
- Rule-based techniques easily combine with FPT approximation algorithms, as it was shown in [7].

The last item is clearly also due to the fact that rule-based approximation techniques were inspired by kernelization ideas.

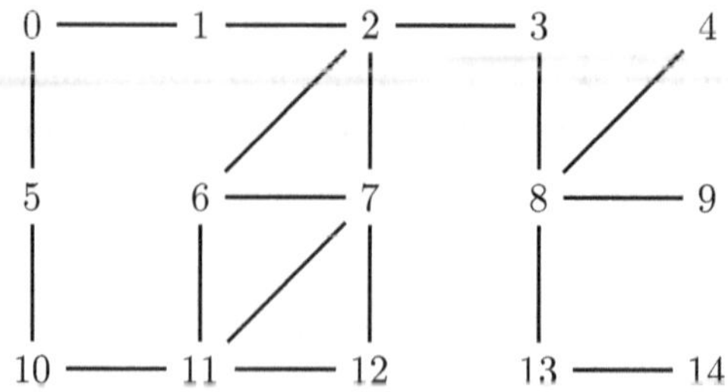

Fig. 1. Can we compute good total dominating sets in simple graphs?

3. Reduction rules for approximating Minimum Total Dominating Set

In this section, we are going to describe how reduction rules could be used to find approximate solutions for maximization problems, focussing on a maximization problem related to Minimum Total Dominating Set.

3.1. *Problem description*

Total Dominating Set is probably the most-studied variation of Dominating Set. This is also certified by the fact that a whole compendium book has been written on this topic [8]. We first recall the basic combinatorial notion. Let $G = (V, E)$ be a graph. $D \subseteq V$ is a *total dominating set* if

$$\forall v \in V \exists u \in D \cap N(v) .$$

Minimum Total Dominating Set asks, given a graph G, to determine a smallest total dominating set. This problem also shares the relatively bad approximability properties of Minimum Dominating Set, mostly due to the relation to Minimum Set Cover, also known as Minimum Hitting Set. In a nutshell, this means that a logarithmic approximation factor is possible, but nothing better is to be expected. Likewise, the standard parameterization of the associated parameterized problem, *i.e.*, the parameter is an upper bound on the number of vertices in the sought solution, would turn this problem into an intractable one from the angle of parameterized complexity, more precisely, it turns out to be W[2]-complete.

The negative picture of Total Dominating Set changes in many ways (both from the approximation point of view and from the point of view of parameterized complexity) if we do not try to minimize the size of a total dominating set, but rather look into maximizing the size of the complement of a total dominating set. Clearly, if we succeed in finding an optimum solution, this would be equally good. Otherwise, this type of optimization can be also seen as an alternative way to measure the quality of an approximative solution, confer also the notion of differential approximation [9–11]. If $D \subseteq V$ is a total dominating set in the graph $G = (V, E)$, then the complement $V \setminus D$ has been introduced as a *harmless set* or a *robust set* (with unaminity thresholds) in [12].

3.2. *Designing reduction rules*

We are going to first present a couple of reduction rules to simplify a given instance $G = (V, E)$ of Harmless Set, as we will call our maximization problem in the following.

It is often a good idea to start with looking at small-degree vertices and to see if one could develop simple reduction rules for these cases. We are following this approach here and we report on results from [4], starting with the *isolate rule.*

> If there is some isolated vertex, produce the instance $(\{x\}, \emptyset)$ that has trivially no solution. If there is some isolated edge xy, produce the instance $G[V \setminus \{x, y\}]$ from $G = (V, E)$.

Unfortunately, we could not find a general rule for vertices of degree one, rather a *twin leaf rule*, described as follows.

> If there are two leaf vertices u, v with common neighbor w, then delete u; vertex u would go into the harmless set.

Observe that the isolate rule (for edges) and the twin leaf rule are α-preserving for any $\alpha \geq 1$. Hence, they could be used both for exact and for approximate computation purposes. In our example in Fig. 1, only the twin leaf rule would apply and put, for instance, vertex 9 into the harmless set (and remove it from the graph).

We also employ some non-standard terminology on special paths that we introduce now. A *chain* is an induced path whose interior vertices are of degree two in G. A chain with one leaf endpoint is a *pendant chain.* A *floating chain* is a chain with two leaves. In other words, this is a path component. A *support vertex* is a non-leaf endpoint of a pendant chain. Support vertices may have more than one pendant chain. Based on this terminology, we can formulate some further reduction rules.

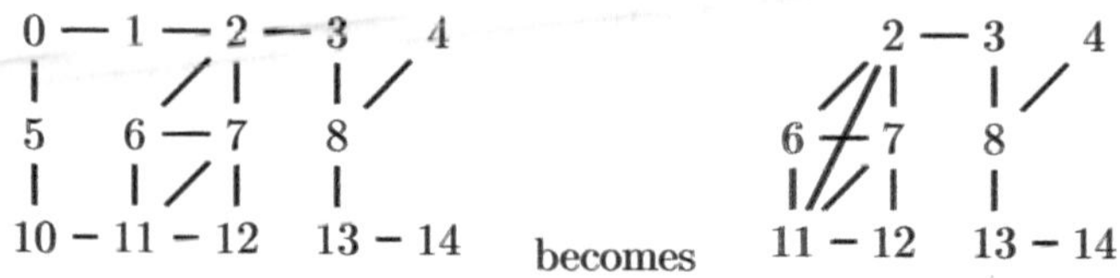

Fig. 2. The long chain rule at work.

Floating chain rule Delete all floating chains.

Long chain rule Assume that G is a graph that contains a path $x-u-v-w-y$, where u, v, w are three consecutive vertices of degree two, where $|N(y)| \geq 2$. Then, construct the graph G' by

- deleting x, u, v, w and
- making y adjacent to all vertices in $N(x) \setminus \{u\}$.

This corresponds to merging x and y and deleting u, v, w.

Cycle chain rule If G is a graph that contains a cycle $x-u-v-w-x$, where u, v, w are three consecutive vertices of degree two, then construct G' by deleting u.

Pendant chain rule Assume that $G = (V, E)$ is a graph that contains two pendant chains with common endpoint v, of which at least one path is of length two. Then, construct the graph $G' = (V', E')$ by deleting one of the two pendant chains, keeping one which is of length two.

As detailed in [4], these rules are α-preserving for any $\alpha \geq 1$. However, while the long chain rule has a factor proof of $(2, 2)$, the cycle chain rule and the pendant chain rule both have a factor proof of $(1, 1)$. Observe that the long chain rule is similar to the well-known folding rule for Vertex Cover, which we will encounter again in the next section.

Now, no further reduction rules apply. Notice that so far, we (only) applied exact reduction rules. In fact, we do not know of any useful genuinely approximative reduction rules for Harmless Set. However, we know of several ***structural insights*** that are helpful now, in fact both for obtaining small kernels (considering the standard parameterization of Harmless Set) and for approximation purposes.

3.3. *Useful combinatorial results*

A classical area of combinatorial graph theory deals with finding bounds on graph parameters assuming certain minimum degree conditions. What bounds are known for the total domination number $\gamma_t(G)$ of a graph G of order n, or for the harmless set number $hs(G)$, depending on the minimum degree $\delta(G)$?

$\delta(G)=0$	No solution exists.	No solution exists.	reference
$\delta(G)\geq 1$	$\gamma_t(G)\leq \frac{2}{3}n$	$hs(G)\geq \frac{1}{3}n$	[13]
$\delta(G)\geq 2$	$\gamma_t(G)\leq \frac{4}{7}n$	$hs(G)\geq \frac{3}{7}n$	[14]
$\delta(G)\geq 3$	$\gamma_t(G)\leq \frac{1}{2}n$	$hs(G)\geq \frac{1}{2}n$	[15]

Notice that this list of results already provides an immediate $3k$-kernel for the k-Harmless Set problem, after applying the isolate rules exhaustively. Namely, if G contains more than $3k$ vertices, we have identified a positive instance. This is not that simple for the approximation algorithm that we want to develop, as in that context, we actually must provide a good solution. Bazgan and Chopin [16] have found a nice exploit for the case when $\delta(G)\geq 1$. This is based on the next monotonicity lemma.

Lemma 3.1: *Let $G=(V,E)$ and $G'=(V,E')$ be graphs with $E'\subseteq E$. If $D'\subseteq V$ is a total dominating set of G', then D' is a total dominating set of G.*

Based on this lemma and the facts that we can compute approximations for connected components separately and that an optimum total dominating set of a tree can be computed in polynomial time, they proposed the following algorithm.

(1) Let $G=(V,E)$ be (w.l.o.g.) a connected graph of order n.
(2) Compute a spanning tree $T=(V,E_T)$ of G.
(3) Compute an optimum TDS D_T for T (by dynamic programming).
(4) $H:=V\setminus D_T$ is a harmless set for G satisfying

$$|H|=hs(T)\geq \frac{1}{3}n\geq \frac{1}{3}|H^*|,$$

where H^* is a maximum harmless set for G.

This reasoning shows:

Proposition 3.2: *Maximum Harmless Set can be approximated up to a factor of 3 in polynomial time.*

Can we improve on this with better use of combinatorial results? Unfortunately, as mentioned above, we do not know of general rules dealing with leaf vertices. This is even worse if it comes to deal with all degree-2 vertices. However, the colleagues working on graph combinatorics did find an interesting fact that we could make use of. Let $S_2(G)$ be the set all vertices of degree two within G. By courtesy to the authors, let us say that G *satisfies the Lam-Wei property* if it has minimum degree at least two and $G[S_2(G)]$ decomposes into K_1- and K_2-components. This property is equivalent to requiring that G has no isolates, no leaves and no three consecutive vertices of degree two.

Theorem 3.3: *(Lam and Wei [17]) Let G be a graph of order n_G that satisfies the Lam-Wei property. Then, $\gamma_t(G) \leq n_G/2$.*

At first glance, this may not appear to be that helpful for two reasons: (a) We still have to deal with all leaves somehow, (b) it is not clear if the theorem actually provides an algorithmic solution to produce a factor-2 approximation to MAXIMUM HARMLESS SET, given an instance that satisfies the Lam-Wei property. The latter point is in fact often a problem when trying to make exploits for approximation purposes. This is far less problematic when aiming for kernelization results, though, because there a simple yes-answer (or possibly no-answer) would suffice. Therefore, in [4] it is described how to find factor-2 approximations for graphs satisfying the Lam-Wei property. The first two simple observations are leading to:

- remove edges as long as the Lam-Wei property is not violated;
- process connected components separately.

Now, the algorithm distinguishes two main cases:

- either $S_2(G)$ is an independent set, then further case distinctions are necessary, depending on the length of the shortest cycle,
- or $S_2(G)$ contains a K_2-component. Lemma 6 in [17] shows how to construct a set Q of vertices such that the graph $G' = G[V \setminus Q]$ also satisfies the Lam-Wei property and, moreover, $\gamma_t(G) \leq \gamma_t(G') + \frac{|Q|}{2}$.

These considerations show that for Lam-Wei graphs, MAXIMUM HARMLESS SET is factor-2 approximable. More precisely, we know the following:

Theorem 3.4: [4] *For a given graph $G = (V, E)$ of order n_G that satisfies the Lam-Wei property, one can compute a TDS $D \subseteq V$ with $|D| \leq \frac{n_G}{2}$ in polynomial time.*

Observe that a quick analysis of the sketched algorithm indicates a bound of $\mathcal{O}(n_G^8)$ for the running time. We pose it as an open question to

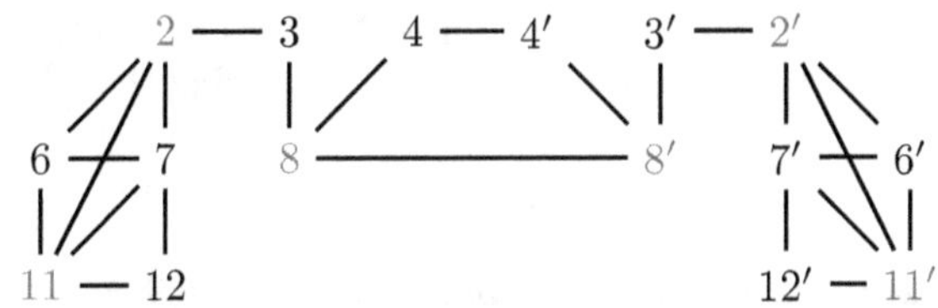

Fig. 3. How to build graph H' from the reduced graph from Fig. 2.

come up with an algorithm that shows a better polynomial-time behavior.

Now, the main question for us is how to make use of this algorithm as a subroutine of a factor-2 approximation algorithm for arbitrary graphs. Assume we face a graph instance $G = (V, E)$, to which none of our reduction rules apply. Hence, in particular, G satisfies the following properties.

- G contains no chain of three vertices of degree two.
- By the twin leaf rule, any vertex has at most one leaf neighbor.

We now construct a graph H' that satisfies the Lam-Wei property. Let G' be a graph isomorphic to G, so that each vertex v of G corresponds to a vertex v' of G', under the assumed isomorphism $f : V(G) \longrightarrow V(G')$. Let L be the set of leaves of G (and correspondingly L' the set of leaves of G'). Let N be the set of leaf neighbors of G and N' be the set of leaf neighbors of G'. By our properties of G and by construction, $|N'| = |N| = |L| = |L'|$. Let $\tilde{G} := G[V(G) \setminus L]$ and $\tilde{G'} := G'[V(G') \setminus L']$. We construct a graph H obtained from the graph union of $\tilde{G}$ and $\tilde{G'}$ simply by adding edges between each pair of corresponding leaf neighbors $v \in N$ of G and $v' = f(v) \in N'$.

H might contain chains with more than two vertices of degree two. This happens if G contains a pendant chain. Hence, we apply the long chain rule as long as possible. This results in a graph H' of order $n_{H'}$, with minimum degree at least two containing no chain of three vertices of degree two.

Let us return once more to our example graph from Fig. 2. The surgery described above would lead to the graph H' displayed in Fig. 3. In constructing H, still the vertices $13, 14, 14', 13'$ were present, all of degree two. So, the long chain rule applied to them, connecting 8 with $8'$.

Now, we apply a good approximation algorithm (like the one promised by Theorem 3.4) to obtain a solution for H'. The set of blue vertices $\bar{D} = \{2, 11, 8, 8', 11', 2'\}$ clearly forms a total dominating set that is good enough, because more than half of the vertices of H' belong to the complementary harmless set. By adding $13, 13'$ to $\bar{D}$, we obtain a total dominating set for H

that is still 2-approximative. In this (special) case, $D = \{1, 2, 8, 10, 11, 13\}$ is a total dominating set for G, obtained after undoing the effect of the long chain rule.

We finally mention the combinatorial reason why this approach works:

Theorem 3.5: *Let G be a graph of order n_G and let H be a graph of order n_H obtained from G by deleting Δ vertices and possibly adding some edges. Let D_G and D_H be TDS solutions of G and H, respectively, such that $a = |D_G| - |D_H| \leq \Delta$.*
If $|D_H| \leq c \cdot n_H$ and $\Delta \leq \gamma_t(G)$, then $V(G) \setminus D_G$ is a harmless set of G, whose size $n_G - |D_G|$ is within a factor of $(1-c)^{-1}$ from optimum.

In our concrete situation, G corresponds to the two copies (dubbed G and G' above), while H is the reduced graph H' in the situation above. Also, the bound $c = 0.5$ implies the claimed approximation factor of two. We omitted some details on how to produce a total dominating set for the original graph G from what has been processed for H' here, but the overall idea should have become clear.

3.4. *Discussion of our findings*

We have presented a strategy that uses reduction rules mainly in order to turn our instance into something that satisfies some conditions given by previous combinatorial insights.

When does this approach work?

- Look for algorithmic combinatorial results, based on conditions.
  This algorithmic aspect was irrelevant for kernelization.
- Design (approximation-preserving) reductions that work towards these conditions. Often, these rules will also work as kernelization rules.
- There is some transfer algorithm from special cases (like Wam-Lei property) to more general situations (as obtained by reduction rules).
- So far this approach works well for domination-type problems.

There is one crucial difference to classical kernelization algorithms, though: while in kernelization, it is highly exceptional when certain rules are applied that increase the size of the instance (see the discussion in [18]), this could well be the case at the end of a cascade of rule applications designed for approximation purposes. In our case study, the size of the graph even doubled. More generally speaking, the decrease of size of the instance is usually less important here compared to the goal of getting a better structure.

Some more examples Minimum Dominating Set is as bad to approximate as Minimum Total Dominating Set. What about maximizing the complement of a dominating set instead, also known as an *enclaveless set* [19]? This maximization problem is known as the Maximum Star Forest problem, see [20–22]. By way of contrast, the parameterized complexity community addresses the corresponding decision problem as Nonblocker [23]. It was shown in [24] that the approach of Nguyen *et al.* from [22] fits into the framework of using reduction rules for approximation, although other approaches [20, 21] provide better approximation factors in this case. Another example is provided by Maximum Differential, which was proven to correspond to the maximization version of Minimum Roman Domination in [25]. Based on but also improving on earlier works, using [26, 27], a factor-$3.\overline{3}$ approximation was shown in [4].

4. Self-monitoring approximation algorithms for Minimum Vertex Cover

Minimum Vertex Cover is probably one of the best studied graph problems both in terms of parameterized complexity and in terms of approximation. Despite all efforts, the best known constant factor approximation algorithm for general graphs is still (basically) a factor-2 approximation. More specifically, G. Karakostas [28] derived an approximation factor of $2-\Theta(\frac{1}{\sqrt{\log n}})$. I. Dinur and S. Safra [29] showed that there is no polynomial-time approximation algorithm for MINVC achieving an approximation ratio better than $10\sqrt{5}-21 \approx 1.36067$, unless $\mathsf{P} = \mathsf{NP}$. Moreover, assuming that the Unique Games Conjecture is true, no approximation factor of the form $2-\varepsilon$ is possible for any $\varepsilon > 0$, as shown by S. Khot and O. Regev [30]. There are better results for some specific classes of graphs, most notably graphs with bounded degree ($2 - \frac{2}{\Delta}$ [31] and $2 - \frac{5}{\Delta+3} + \varepsilon$ [32]), and 4-colorable graphs (1.5) [31]. A good (although a bit out-dated) survey is contained in the third chapter of D. Hochbaum's book [33]. The above-mentioned result for degree-bounded graphs was improved by E. Halperin [34] towards a performance factor of $(2-(1-o(1)))\cdot\frac{2\ln\ln\Delta}{\ln\Delta}$. Even for graphs with maximum degree 3, no approximation factor arbitrarily close to one (*i.e.*, no PTAS) can be achieved in polynomial time, as shown by P. Alimonti and V. Kann [35]. Another interesting result is due to T. Imamura and K. Iwama [36]: they showed that graphs $G = (V, E)$ of maximum degree

$\Delta \in \Omega\left(|V| \cdot \frac{\log\log(|V|)}{\log(|V|)}\right)$ and average degree d_{avg} can be approximated with a factor of $\frac{2}{1+\frac{d_{avg}}{2\Delta}}$, provided that $|E| \leq \Delta \cdot (|V| - \Delta)$. This yields factor-$\frac{4}{3}$ approximations for dense regular graphs (*i.e.*, $d_{avg} = \Delta$) and an approximation factor of $\frac{3}{2}$, provided that $d_{avg} \geq \frac{2}{3}\Delta$ for dense graphs.

4.1. *A bird's eye view on local ratio techniques*

The local ratio technique has been developed by R. Bar-Yehuda and his colleagues, we refer to [37–40] as the most relevant papers for our exposition. One of the easiest approaches to approximating MINIMUM VERTEX COVER is, starting with a graph G and an empty set of vertices C, to repeatedly remove isolates or grab an edge and place both endpoints into C (the cover to be constructed) until the graph is empty. Alternatively, also fixing notations, we could formulate this procedure recursively. Let $G_0 = G$ be the input graph. We initialize $C_0 = \emptyset$. Assume that in step i of our algorithm, we are left with the instance $G_i = (V_i, E_i)$. If $E_i = \emptyset$, we stop and return $C = C_i$ as the (approximative) solution. If $E_i \neq \emptyset$, then pick some edge $e_i = u_i v_i$ and define $C_{i+1} = C_i \cup \{u_i, v_i\}$, $G_{i+1} = G_i - \{u_i, v_i\}$ and recurse. By the halting condition, C will be a vertex cover of G. Moreover, assuming that $C = C_k$, *i.e.*, the algorithm is executed k recursive steps, then for each $0 \leq i < k$ and any vertex cover C' of G, $u_i \in C'$ or $v_i \in C'$. Hence, in particular for a minimum vertex cover C^*, $|C^*| \geq k$. Conversely, $|C| = 2k$, which explains that the algorithm computes a factor-2 approximation. Notice that this algorithm is usually considered in the setting of vertex-weighted graphs when analyzing it with the local ratio technique. Let us stick with the unweighted setting here, though. Also, the weighted setting suggests a kind of reasoning (based on the modification of weight functions) that we like to replace by more combinatorial arguments in what follows.

From the perspective of FPT, it is interesting to rephrase the algorithm in terms of reduction rules. Let us call this the *edge rule* in the following.

> Select an edge $e = uv$ in G, put both endpoints u, v into C and remove them from G.

Now, the algorithm simply consists in, starting with input G and $C = \emptyset$, and then executing this reduction rule as long as possible. As argued above, this is a factor-2 approximation algorithm. The main reason for this fact is that whenever u, v are put into C, $u \in C^*$ or $v \in C^*$ holds true. This property can also be viewed as a local factor-2 approximation property of this reduction rule. Using the terminology we introduced above, the rule is a 2-preserving reduction with factor proof $(a, b) = (1, 2)$. In particular, our argument shows that an optimum vertex cover of instance G is at least $a = 1$ bigger than an optimum solution to instance $G - \{u, v\}$. Moreover, we will add $b = 2$ vertices to the approximate solution recursively obtained for $G - \{u, v\}$ to produce our approximate solution for G.

Are there any better rules? Notice that this question is different from and does not necessarily contradict the previous lower-bound discussion for the approximability of MINIMUM VERTEX COVER. One well-known rule will be called the *triangle rule* in the following.

> Select a triangle $t = uvw$ in G, put all its vertices u, v, w into C and remove them from G.

Observe that out of these three vertices, at least two must be in any vertex cover, so in particular in any vertex cover of minimum size. This shows that, at least intuitively speaking, the triangle rule looks better than the edge rule, because it is 1.5-preserving (with factor proof $(2, 3)$), much better than the factor 2 implied by the edge rule. Of course, we can mirror this intuition by executing the triangle rule as long as the graph is not triangle-free. We expect this to help getting a better approximation when actually running the algorithm, but we cannot guarantee that this will happen. However, we can easily compute a better approximation guarantee *a posteriori* by the following observation, which can be deduced from Corollary 2.8.

Lemma 4.1: *If the reduction-based algorithm terminated after $k_{3/2}$ executions of the triangle rule and k_2 executions of the edge rule, then $|C| = 3k_{3/2} + 2k_2$ holds for the output cover C, while for the size of an optimum cover, we know that $|C^*| \geq 2k_{3/2} + k_2$. Hence, we know that the a posteriori approximation factor of C can be upper-bounded by $\frac{3k_{3/2}+2k_2}{2k_{3/2}+k_2}$.*

Hence, if the graph contains many triangles, which is in particular the case if the graph is dense, then the approximation factor will be rather close to 1.5. But in reality, we often expect rather sparse instances (for instance, if our graphs arise from social network applications). Can we say anything in this case, as well? This will be treated in the next subsection.

Remark 4.2: It is tempting to conjecture that for the two rules presented so far and the corresponding rule-based approximation algorithm yielding a solution C, it is even true that there exists a minimum vertex cover $C^* \subseteq C$.

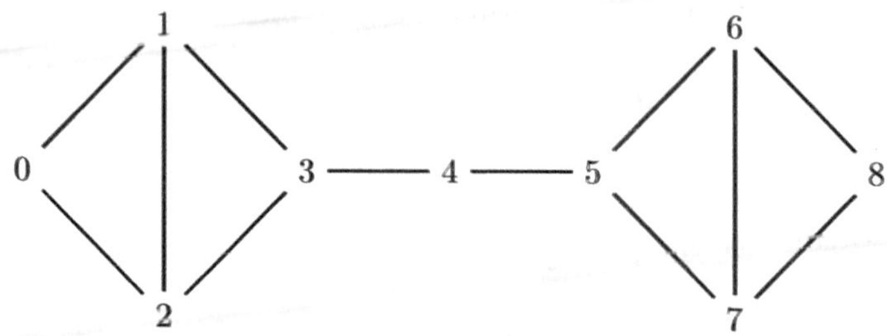

Fig. 4. Our rules might produce a solution not containing an optimum one.

However, this intuition is wrong, as can be seen by looking at the example graph in Fig. 4. Here, the triangle rule might produce $C = \{1, 2, 3, 5, 6, 7\}$. However, the (unique) minimum solution is $C^* = \{1, 2, 4, 6, 7\}$. Yet, by observing our algorithm and noticing that we only applied rules that guarantee a factor of 1.5, this would not be clear.

There is another case we want to discuss first, namely that of two well-known exact reduction rules (*i.e.*, rules that by themselves give an approximation factor of one). We start with the *isolate rule*:

> If G contains a vertex x of degree zero, then remove x from G; do not change the set C.

Actually, we have integrated this rule implicitly in our first version of our algorithm. As it does not make sense to integrate isolated vertices in any cover, this rule is optimal. In our terminology, the isolate rule is α-preserving for any α, as it has the factor proof $(0, 0)$.

Let us now consider vertices of degree one with the *leaf rule*.

> If G contains a vertex x of degree one, *i.e.*, $N(x) = \{y\}$, then remove x, y from G and put y into the set C.

This is an exact rule in the sense of being 1-preserving with a factor proof $(1, 1)$. Algorithmically, this is very desirable, so a rule-based approximation algorithm should prefer executing leaf and isolate rules over the triangle rule, which should again take priority over the edge rule, which seems to be the worst choice. If we had executed in particular the leaf rule k_1 times, then we could re-write the formula for the *a posteriori* approximation factor of the obtained solution as

$$\frac{3k_{3/2} + 2k_2 + k_1}{2k_{3/2} + k_2 + k_1}. \tag{4.1}$$

The meaning of the indices of the counters in Formula (4.1) should have become clear by now: it is reflecting the local approximation factor of the rule. Hence, if we add more rules in the following, we need not introduce any more variables, unless we encounter different local approximation factors. We could even keep track of these variables by simple increments upon executing the corresponding rules.

Let us move back to our little example from Fig. 4. First, no isolate and no leaf rules trigger, so we could still start by selecting the triangle $\{1, 2, 3\}$. Removing these three vertices results in a new situation, though. First, the isolated vertex 0 is removed by the isolate rule, and then we find that 4 has degree one, which puts 5 into the cover. We are left with the triangle $\{6, 7, 8\}$, so that altogether seven vertices are included into the solution, which is worse than before. So even additional exact rules do not necessarily lead to an improved outcome of the approximation algorithm. Yet, something has improved, as the *a posteriori* analysis provides a ratio of $\frac{6+1}{4+1} = 1.4$, which does improve on the previous *a posteriori* analysis yielding a factor of 1.5. How did we arrive at this conclusion, although the output of this run of our new algorithm is in fact worse than the previously studied run? Actually, we now have a proof (by running the algorithm) that a minimum solution has a size of at least five: This is exactly what the denominator of Formula (4.1) means. As we have seen above, there is in fact a minimum solution of that size for our small instance, but of course we cannot tell this in the general situation. Yet, there is a further lesson we can take away with our small example: by running different versions of our algorithm, or also by (sometimes) randomly choosing the next rule to be executed, we could improve both lower and upper bounds of the solutions, and we could even use the sharpest lower bound in any of the runs and combine it with the sharpest upper bound to improve the *a posteriori* analysis. In our case, it is now clear that the solution of size six obtained by the first run of the approximation algorithm is in fact a factor-1.2 approximation (or even better), as we know by the second run of the algorithm that any solution has to have size at least five.

We will discuss more reduction rules later in this chapter but switch over to special graph classes for now.

4.2. *Looking at bounded average-degree graphs*

One of the easiest ways to model sparse graphs is via bounding the degree. As explained in the introduction to this section, better approximation ratios for MINIMUM VERTEX COVER can be obtained in this scenario. However, having in mind again graphs that occur in practice, like social networks, we cannot expect these graphs to have a low maximum degree. Rather, we have relatively few vertices of high degree that are kind of counter-balanced by a large number of vertices of very low degree. Therefore, it is far more realistic to consider graphs of bounded average degree. Interestingly, while the case of bounded maximum degree has been thoroughly studied for nearly every graph parameter, much less has been done with respect to bounded average degree. So, as a general research agenda, we propose to study this class of graphs, as they might be easier to handle compared to other notions of sparse graphs. Yet, they have their particularities, as we will explain next with the current approach in mind.

One of the most obvious observations is the fact that some of the rules we considered so far for approximating this problem could in fact increase the average degree that we are aiming at. To make this thought more concrete, let us consider (as in [7]) the case of graphs of average degree bounded by four. For a graph with n vertices and m edges, this means that $m \leq 2n$. Now, consider the border case of a graph $G = (V, E)$, where the average degree satisfies $\frac{2|E|}{|V|} = 4$. Now, if $v \in V$ is an isolate, then our reduction rule would produce the graph $G - v$ with average degree $\frac{2|E|}{|V|-1} > 4$. Similarly, the leaf rule would lead to an average degree of $\frac{2(|E|-1)}{|V|-1} > 4$. In order to maintain an average degree of four, we have to delete (at least) two edges along with one vertex. Therefore, the algorithm presented in [7] is tailored towards working with graphs of minimum degree (at least) two, so that the mentioned problems can be avoided. This also means we have to be careful when incidentally creating low-degree vertices (and hence triggering, for instance, the leaf rule).

Let us first consider the triangle rule with respect to the average degree. If each of the vertices had a degree of at least three, then one would, by deleting these three vertices x, y, z, also remove at least six edges from the graph. As can be easily verified, this is exactly a scenario that preserves an average degree of at most four, *i.e.*, if G had average degree at most four, then also $G - \{x, y, z\}$ has average degree at most four. If at least one of the vertices in the triangle formed by x, y, z has degree two, say, $N(x) = \{y, z\}$, then notice that the *special triangle rule* that puts (only)

y, z into the cover (but not x) is a 1-preserving reduction with factor proof $(2, 2)$. Now, consider the following idea for the case when y and z both are of degree three (which would mean that the graph $G - \{x, y, z\}$ could have average degree above four), an idea that was inspired in [7] by the previously presented work on a parameterized approximation algorithm for MINIMUM VERTEX COVER: Select an edge such that the sum of the degrees of the incident vertices is maximum in G. As removing x, y, z was critical in the sense that the average degree of the new graph is above four, there must be an edge uv, such that $|N(u)| + |N(v)| \geq 9$. Hence, by removing x, y, z, u, v from G (and putting y, z, u, v in the approximate cover), we will remove at least three edges for the triangle plus eight edges (by deleting u, v), so that $G - \{x, y, z, u, v\}$ has average degree at most four if G does. In [7], the algorithmic step that deleted an edge uv with high-degree vertices u, v was called a *worsening step*, as it actually worsens the quality of the approximation. In our concrete example, we know that out of the vertices $\{y, z, u, v\}$ put into the cover, three should be also in some optimum cover, which means that this combined rule is in fact 4/3-preserving, so even better than our aim of 3/2. However, the worsening step serves the purpose of maintaining the average degree bound.

Let us mention one subtlety that could happen in particular when removing triangles xyz: if at least two of the three vertices are adjacent to the same vertex w, then it could be that the degree of w drops to 1 or even to 0, again triggering some low-degree rule. This should be avoided or at least taken into account. In particular, in combination with a worsening step, this is not that bad, because we gained enough edges to compensate for one more application of the leaf rule. Otherwise, consider removing the triangle xyw instead of xyz.

A combinatorial simple lemma that is often handy for this type of analysis is the following one.

Lemma 4.3: [7] *In a graph G with average degree d_{avg}, if one deletes x vertices and at least $\frac{d_{avg}}{2}x$ edges, the remaining graph G' has average degree $d'_{avg} \leq d_{avg}$.*

Another well-known reduction rule (developed in Parameterized Algorithms) is the *folding rule*. It applies to vertices x of degree two, where the two neighbors y, z of x do not form a triangle with x. If $z_1, \ldots, z_r$ are all neighbors of z but x, then this rule would delete x, z but add edges between y and all vertices $z_1, \ldots, z_r$ to create graph G' from G. Altogether, this rule deletes two vertices and two edges. This is again a problem for the bounded average degree of four, but incorporating a worsening step could boost this to deleting four vertices altogether and ten edges. The idea of correctness is to determine the solution only after returning from the recursion on G'. Now, if y had been put into a solution, z will be put into the solution, as well, but when y is not put into a solution, z will not be put there either, but rather x is put into the solution then. In any case, one additional vertex is put into the solution, and this is done both for the approximate solution (that our algorithm actually computes) and for an optimum solution (that we try to estimate). Hence, the folding rule can be seen as a 1-preserving rule.

There are only two more ingredients needed to understand the basics of the factor-3/2 algorithm for MINIMUM VERTEX COVER on graphs of bounded average degree (at most four) proposed in [7]:

- We can process degree-3 vertices by looking for one such vertex, x, that has a neighbor u that is part of an edge uv, $v \neq x$, such that the sum of the degrees of u, v is maximum. Now first remove u, v (plus incident edges) and then execute the folding rule (or the special triangle rule). Together, this idea of having a worsening step precede a 1-preserving rule gives a 3/2-preserving rule that also maintains an average degree of at most four.
- Four-regular graphs will be encountered only once in one recursion and could be hence processed, say, by again doing a worsening step first.

It is shown in [7] that an algorithm that processes first all degree-2 vertices, then all degree-3 vertices and finally (possibly) a degree-4 vertex would always resolve a given graph of average degree bounded by four. This leads to the following result:

Theorem 4.4: [7] *For any graph G with average degree $d_{avg} \leq 4$ and minimum degree $\delta \geq 2$, there exists a polynomial-time factor-*1.5 *approximation algorithm for* MINIMUM VERTEX COVER.

We can optimize the algorithm with the following idea in mind: Why should we (immediately) execute worsening steps to balance the average degree condition? This can be easily deferred to the time when there are no small-degree vertices available to execute our reduction rules. Now, it becomes important that all our reduction rules come with a local approximation factor, so that we can monitor the approximation behavior of our overall algorithm run, so that we know when we can still allow ourselves to finally execute worsening steps. Let us clarify this idea by working through

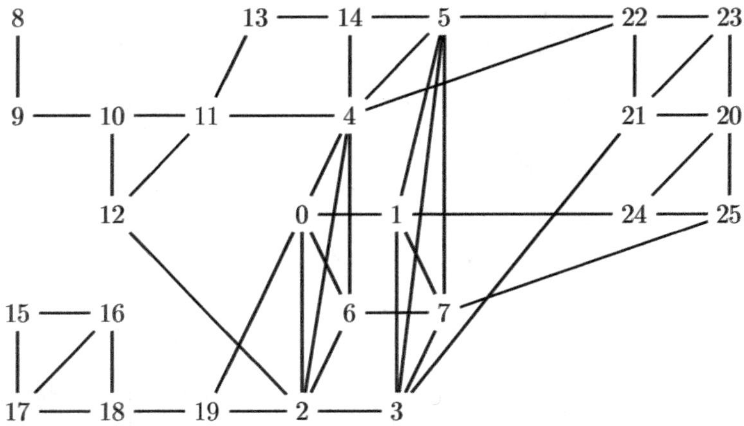

Fig. 5. An example graph for which we compute a small vertex cover.

an example step-by-step; along the way, we also explain further reduction rules. Our example graph is shown in Fig. 5. The reader might wish to look for a good vertex cover solution, without reading further first.

The graph is not fulfilling the minimum-degree requirement of Theorem 4.4, but it is clearly not forbidden to start with our reduction rules, working from low degrees up to higher degrees. As shown on the left side of Fig. 6, we can find one vertex of degree one (colored red) that puts one vertex (colored blue) into the vertex cover C that we are constructing. At the same time, our previous considerations show that the blue vertex should indeed be put in (some) minimum vertex cover C^* that we are studying for comparison. In the following, we will grow both C and C^*, also focussing on their ratio. At present, we see $|C| = 1$, $|C^*| \geq 1$. After removing vertices $8, 9$ from the graph (actually, on the right-hand side of Fig. 6, these vertices are not deleted, but shown in light orange color), no more vertices of degree one are around, but three vertices of degree two. Now, it depends on the implementation which of the vertices are chosen for continuation. We follow the track what occurs when selecting vertex 13. This folding rule also involves the neighboring vertices $11, 14$, so that we highlight all this gadgetry by coloring. Color blue is again chosen to indicate that one of the vertices will also be put into an optimum cover. Observing the sets C and C^*, we see that $|C| = 2$, $|C^*| \geq 2$, because we (again) executed a 1-preserving rule. The reader is encouraged to look into the course of the algorithm if, say, the degree-2 vertex 10 had been chosen instead of 13 at this stage.

In Fig. 7, on the left-hand side, one first notices the result of applying the

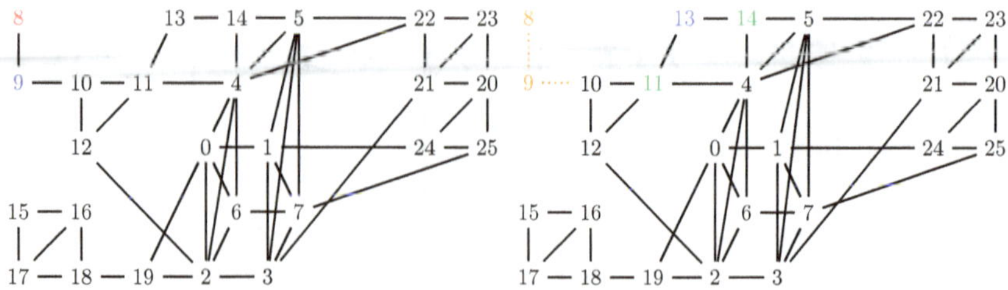

Fig. 6. Triggering first leaf, then folding rules.

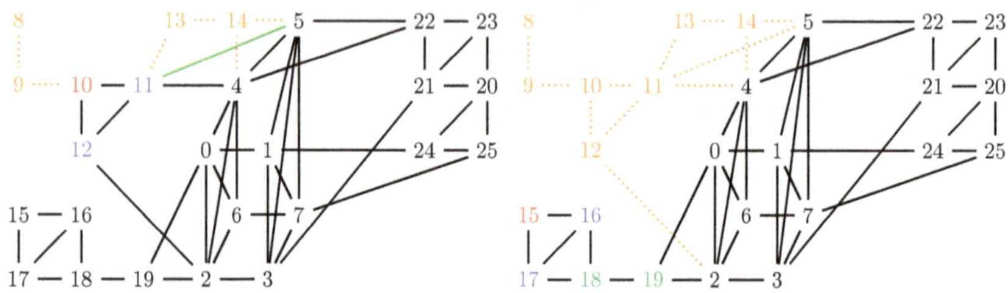

Fig. 7. Executing the folding rule and special triangle rules.

folding rule, marking the seemingly new edge between 5 and 11 in green. We now pick the triangle with degree-2 vertex 10. This is again a 1-preserving rule, so after its execution, we have collected four vertices into C, but also $|C^*| \geq 4$ is clear. On the right-hand side, the finding of the other triangle with the degree-2 vertex 15 is shown. Notice that after executing the special triangle rule, the degree of vertex 18 drops to one, which triggers a leaf rule. Altogether, after executing all these exact rules, we see that $|C| = 7$ and $|C^*| \geq 7$.

Now, there are no more vertices of degree smaller than three in the graph. This seems to be an end to using 1-preserving rules, but we have one more rule upon our sleeves that we did not yet disclose. It is generalizing the special triangle rule, but we will call it *special edge rule*.

> If there is an edge uv, such that $N(v) \subseteq N[u]$, then delete u and its incident edges and put u into the cover.

We claim that this rule is 1-preserving. Namely, consider any solution that does not contain u. As the solution is a vertex cover, all neighbors of u must be in the cover, including v. But as $N(v) \subseteq N[u]$, we could produce another solution of the same size by replacing v by u in the solution. For instance, v could be a vertex of degree two in a triangle and u any other

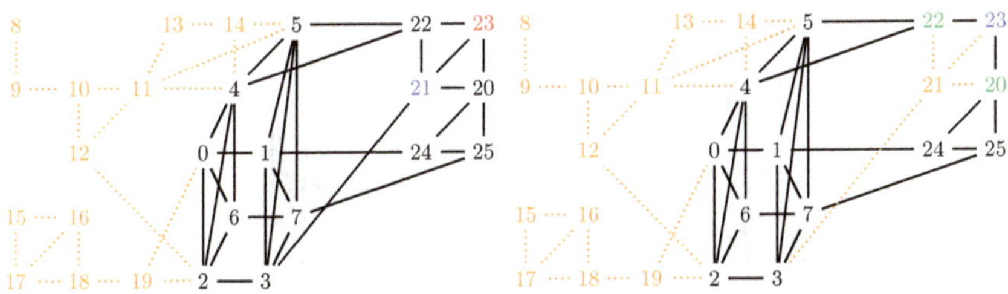

Fig. 8. Executing the special edge and then the folding rules.

vertex in the triangle. Then, the special edge rule would put u in the cover, but after removing u from the graph, v would become a leaf, putting the third triangle vertex in the cover, as well.

In our example, we find $v = 23$ and $u = 21$ in the context of the special edge rule, so that u is put into the cover, and we arrive at $|C| = 8$, $|C^*| \geq 8$. Afterwards, observe that a vertex of degree two was created, so that a folding rule applies. Now, $|C| = 9$, $|C^*| \geq 9$. These rules are highlighted in Fig. 8.

Now, we finally arrived at a situation where none of the 1-preserving reduction rules applies. Hence, we have to look for possibilities to apply rules that are not exact. Clearly, the graph is full of triangles, so that the triangle rule is applicable. But is there any heuristic telling us where to apply this rule? One idea would be to apply it to such a triangle that guarantees the creation of many vertices of degree at most two after removing the selected triangle. We hence choose the triangle $1, 5, 7$. In the new situation (see Fig. 9 on the left-hand side), we know that $|C| = 12$, $|C^*| \geq 11$. The right-hand side of Fig. 9 shows a possibility to apply a leaf rule, yielding $|C| = 13$, $|C^*| \geq 12$. Now, two special triangle rule apply, so that we get away with a vertex cover C with $|C| = 17$ for the original graph, knowing that an optimum solution can possess no less than 16 vertices. This proves (*a posteriori*) that the solution we found has an approximation ratio no worse than $\frac{17}{16} = 1.0625$. We believe that in many practical situations, a guarantee saying that the obtained solution is about 5% off from the optimum is good enough. By observing that $0, 2, 4, 6$ and $1, 3, 5, 7$ both induce complete graphs K_4, one can see: the remaining graph shown in Fig. 9 needs 8 vertices to be covered, which is exactly what the approximation produces.

Let us now reconstruct a concrete instantiation of our solution; this is necessary, because we had several nondeterministic choices and even worse, the recursively determined solutions will resolve some situations in the

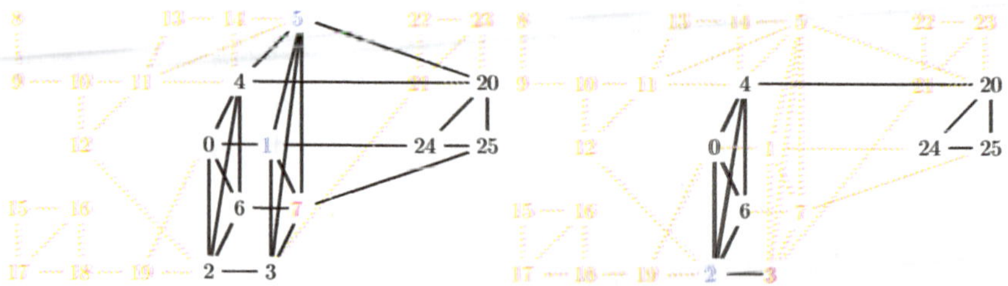

Fig. 9. Applying a triangle rule to create a vertex of degree two.

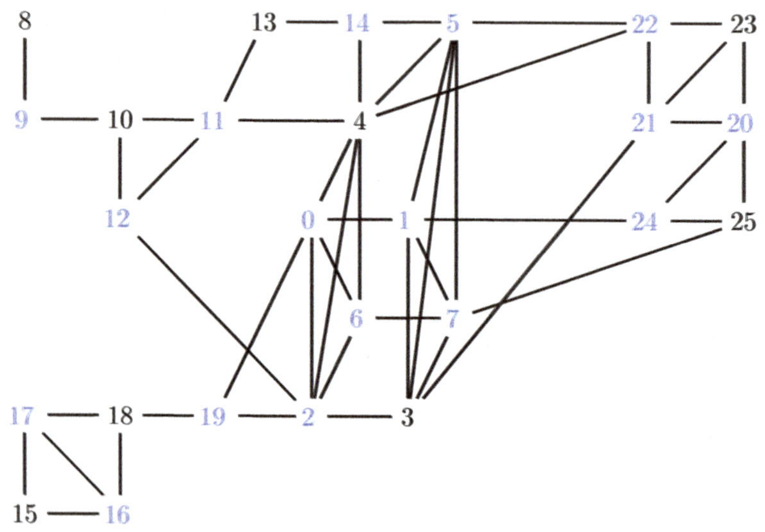

Fig. 10. An example graph for which we have computed a small vertex cover (in blue).

folding rule. Assume we chose in the very last step (when resolving the last triangle) to put vertices 0, 6 into the cover, and for the penultimate triangle, we picked 20, 24 for the solution. Working backwards, we put 2 into the cover. Then, we selected 1, 5, 7. Because we have put 20 into the cover, we should also put 22 in the cover by the folding rule. We already decided to put 21 in the cover. Working further backwards, we see that 19, 16, 17 and then 11, 12 were deterministic decisions. Since 11 went into the cover, 14 will go there, as well (folding rule). Then, 9 went into the cover (actually, this was done as the very first step). Our solution is displayed in Fig. 10.

Experimental insights A student of ours, F. Kruschewski-Kursawe, has implemented and tested the algorithm from [7] for solving MINIMUM VERTEX COVER on graphs of average degree at most four. Let us summarize the experimental insights from thousands of random graphs; also confer [41]:

- Graphs of average degree two are always optimally solved.
  This is no surprise, but a good test for the implementation.
- Graphs of average degree 2.5 are always optimally solved.
  This is more of a surprise, as counterexamples can be constructed.
- Graphs of average degree 3 are mostly optimally solved.
  All factors that the algorithm could guarantee were below 1.01.
- Graphs of average degree 3.5 are never optimally solved (according to the computed approximation guarantee).
  Nearly all factors that the algorithm could guarantee were below 1.05.
- Graphs of average degree 4 give solutions with a guaranteed factor below 1.1.
  Recall that we could only warrant an approximation factor of 1.5 in this case.
- Some reduction rules never (or hardly ever) apply.

Let us mention two concrete example cases, the so-called 60-cell and 120-cell graphs, which are well-known 4-regular graphs with 300 vertices and 600 vertices, respectively. They are known to pose hard instances for VERTEX COVER. Optimum solutions are known to contain 190 and 380 vertices, respectively. We refer to the discussions in [42–44]. Variations of our algorithm (described in more detail in the next subsection) computed the results listed in Fig. 11 within less than one millisecond on standard hardware (with a program written in C++). We list the sizes of the solutions we found, as well as the computed lower bound estimate on the size of an optimum solution. One can see that this estimate is actually quite pessimistic, as we know the optimum size of a vertex cover solution. Therefore, the best (correct) approximation factors that we get with our methods are $197/190 \approx 1.03684$ for the 60-cell and $398/380 \approx 1.04737$ for the 120-cell. It is possible to combine the results of the three versions, taking the minimum of the solution sizes and the maximum of the lower bounds to improve, in particular, on the approximation guarantee that our approach can offer. Although this value is still somewhat pessimistic, it gives a far better guarantee compared to the standard estimate, which would only give a promise of a factor of 1.5. It is also worth mentioning that the solution and approximation guarantees that our algorithm found have been obtained within fractions of milliseconds with everyday hardware, while the

Graph	Type	Solution size	Lower bound	Factor
60-cell	Version 1	202	161	1.25466
	Version 2	**197**	157	1.25478
	Version 3	202	**165**	1.22424
	Combination	197	165	1.19394
120-cell	Version 1	**398**	314	1.26752
	Version 2	402	323	1.24458
	Version 3	412	**333**	1.23724
	Combination	398	333	1.19520

Fig. 11. Results of running our Vertex Cover approximation algorithm on the 60-cell and 120-cell graphs.

exact solution (for the 60-cell graph) has been computed on one thousand cores (each with four hardware threads) within three quarters of an hour (see [43]). The proof showing optimality of the 380-vertex solution for the 120-cell graph required quite some combinatorial arguments in [44].

4.3. *Oblivious application of our algorithm*

So far, we proved that our algorithm always works for graphs satisfying some minimum degree condition and some upper-bound on the average degree. What would happen if we feed a graph into our algorithm that does not satisfy these conditions? As our algorithm does not make any explicit use of these graph properties, this is possible. In other words, our algorithm is oblivious of these conditions; they are required only for the worst-case analysis regarding our algorithm (both regarding running time and approximation ratio).

We tested this idea and collected some more experimental insights, again from thousands of random graphs. First, we stopped running the algorithm once we can no longer guarantee a factor of 1.5.

- The algorithm was also run on graphs of higher average degree.
- Our theorem does not guarantee that such graphs can be solved.
- However, in practice, graphs up to average degree of eight are completely solved by the algorithm.
- The only observable disadvantage is that the approximation factor guaranteed by the algorithm gets closer to our cut-off value of 1.5.
- One could alternatively see this approach as a still probably quite nice preprocessing for parameterized approximation algorithms for all realistic graphs, as they are often of bounded average degree.

Algorithm Type	Criterion	Aver. degree 6 Factor	Time	Aver. degree 8 Factor	Time	Aver. degree 10 Factor	Time
Version 1	Maximum	1.240	2355	1.343	1352	1.419	888
	Average	1.233	2203	1.337	1255	1.413	827
	Minimum	1.228	2107	1.331	1164	1.407	766
Version 2	Maximum	1.238	89	1.342	52	1.417	157
	Average	1.232	55	1.335	45	1.412	144
	Minimum	1.225	48	1.327	42	1.406	132
Version 3	Maximum	1.236	84	1.337	78	1.824	15
	Average:	1.229	63	1.331	68	1.760	10
	Minimum:	1.228	60	1.331	68	1.407	7

Fig. 12. Reporting on approximation factors and running times on random graphs of average degree six, eight and ten; times are in milliseconds.

So, the algorithm can be run on graphs that violate the average degree condition formulated in Theorem 4.4, but the *a priori* analysis leading to factor of 1.5 is no longer valid. Yet, the *a posteriori* analysis often proves that on a concrete instance, the guaranteed approximation factor is better than the theoretical *a priori* bound, which is then just two. But also observe that the worsening step (edge rule) serves as a fall-back, so that no worse scenario than factor two can occur.

We are now reporting on experiments with three times 250 random graphs, each with 16384 vertices and with 49152 edges (in case of average degree six). With these experiments, we also continued to run the algorithm if (theoretically) it would no longer achieve a factor of 1.5. The average degrees violate the condition we have set in the correctness proofs (regarding the approximation factor). Yet, version 1 is the algorithm described above. In particular, it will process all vertices of degree one before moving on to process vertices of degree two and so forth. It also implemented a certain order between rules applying to vertices of the same degree. Simply by dropping this strict condition and by moving on to the next finding of a lowest degree vertex, we arrive at version 2, which offers a considerable speed-up, as proven by Fig. 12. Rather unexpectedly, also the quality ensured by the *a posteriori* analysis improved a bit. The third variant plays with one technicality of the triangle rule: it might be that its application creates vertices of very low degree that trigger rules associated to vertices of very low degrees, which in turn necessitates calling worsening steps.

Ignoring this (because we are violating the degree conditions anyway) means to further relax the original processing. Yet, on larger average degrees, this seems to lead to worse factor estimates, traded off for faster running times.

5. Conclusions

In this chapter, we have presented the idea to use reduction rules as a main tool to come up with approximation algorithms that allow both for an *a priori* and an *a posteriori* analysis; the latter idea is also dubbed self-monitoring. It is based on the observation that the classical worst-case *a priori* analysis typically provides far too pessimistic approximation factor guarantees. This was already observed by Graham in his first-ever approximation paper (as discussed in the introduction). Let us quote from the last page of this paper [1]:

> *. . . one might ask how "typical" the examples are for which ω'/ω_0 [the approximation factor] is close to the upper bound $2 - (1/n)$. While very little work has been done in this respect, empirical results . . . indicate that examples in which $\omega'/\omega_0 \geq 1.1$ are quite common.*

Some comments may be in order here:

- Presumably, the author meant $\omega'/\omega_0 \leq 1.1$ or $\omega'/\omega_0 \approx 1.1$ instead of $\omega'/\omega_0 \geq 1.1$, as such a lower bound would not give rise to any surprise.
- It is not said how these empirical results were obtained (a technical report of Manacher is cited), as now (with hindsight) this seems to be a major task: because the underlying decision problem is NP-hard, a fact Graham could not know at that time, it is quite hard to come up with exact solutions to compare the obtained approximate solution with.

The latter item is still true today, although computer technology has advanced considerably during the last 50 years. Therefore, the *a posteriori* analysis that we are proposing is not trying to really find the best solution, but only some good locally justified bound. Still, the *a posteriori* factors tend to be much better than the *a priori* ones, which is no surprise, as they can really look at the instance and not only at some abstract worst-case scenario. Hence, we believe that this approach can help boost the application of approximation algorithms with guarantees (as opposed to mere heuristics), because knowing that a certain solution is only five percent off the optimum should often suffice.

Acknowledgments

The content of this chapter is based on a series of lectures given at IMS of National University of Singapore in August 2017. The author thanks IMS for providing such a nice opportunity.

This work was partially supported by MEXT Grant-in-Aid for Scientific Research 13224040. Also, some of the work of Florian Kruschewski-Kursawe was undertaken after finishing his bachelor thesis within a student project supported by the DFG project FE 560 / 6-1.

References

1. R. L. Graham, Bounds for certain multiprocessing anomalies, *The Bell Systems Technical Journal.* **XLV** (1966), 1563–1581.
2. R. G. Downey and M. R. Fellows, *Parameterized Complexity.* Springer, 1999.
3. F. V. Fomin, D. Lokshtanov, S. Saurabh, and M. Zehavi, *Kernelization: Theory of Parameterized Preprocessing.* Cambridge University Press, 2019. doi: 10.1017/9781107415157.
4. F. N. Abu-Khzam, C. Bazgan, M. Chopin, and H. Fernau, Data reductions and combinatorial bounds for improved approximation algorithms, *Journal of Computer and System Sciences.* **82** (2016), 503–520.
5. G. Ausiello, P. Creczenzi, G. Gambosi, V. Kann, A. Marchetti-Spaccamela, and M. Protasi, *Complexity and Approximation; Combinatorial Optimization Problems and Their Approximability Properties.* Springer, 1999.
6. H. Shachnai and M. Zehavi, A multivariate framework for weighted FPT algorithms, *Journal of Computer and System Sciences.* **89** (2017), 157–189.
7. L. Brankovic and H. Fernau, A novel parameterised approximation algorithm for MINIMUM VERTEX COVER, *Theoretical Computer Science.* **511** (2013), 85–108.
8. M. Hennings and A. Yeo, *Total Domination in Graphs.* Springer, 2013.
9. G. Ausiello, C. Bazgan, M. Demange, and V. T. Paschos, Completeness in differential approximation classes, *International Journal of Foundations of Computer Science.* **16** (2005), 1267–1295.
10. C. Bazgan, J. Monnot, V. Th. Paschos, and F. Serriére, On the differential approximation of MIN SET COVER, *Theoretical Computer Science.* **332** (2005), 497–513.
11. M. Demange, P. Grisoni, and V. Th. Paschos, Differential approximation algorithms for some combinatorial optimization problems, *Theoretical Computer Science.* **209** (1998), 107–122.
12. C. Bazgan and M. Chopin. The robust set problem: Parameterized complexity and approximation. In eds. B. Rovan, V. Sassone, and P. Widmayer, *Mathematical Foundations of Computer Science 2012 – 37th International Symposium, MFCS*, vol. 7464, *LNCS*, pp. 136–147, Springer, 2012.
13. E. J. Cockayne, R. M. Dawes, and S. T. Hedetniemi, Total domination in graphs, *Networks.* **10** (1980), 211–219.

14. L. Sun, An upper bound for total domination number, *J. Beijing Inst. Technol., Engl. Ed.* **4** (1995), 111–114.
15. D. Archdeacon, J. A. Ellis-Monaghan, D. Fisher, D. Froncek, P. C. B. Lam, S. M. Seager, B. Wei, and R. Yuster, Some remarks on domination, *Journal of Graph Theory.* **46** (2004), 207–210.
16. C. Bazgan and M. Chopin, The complexity of finding harmless individuals in social networks, *Discrete Optimization.* **14** (2014), 170–182.
17. P. C. B. Lam and B. Wei, On the total domination number of graphs, *Utilitas Mathematica.* **72** (2007), 223–240.
18. H. Fernau, T. Fluschnik, D. Hermelin, A. Krebs, H. Molter, and R. Niedermeier, Diminishable parameterized problems and strict polynomial kernelization. In eds. F. Manea, R. G. Miller, and D. Nowotka, *Sailing Routes in the World of Computation, 14th Conference on Computability in Europe, CiE*, vol. 10936, *LNCS*, pp. 161–171, Springer, 2018.
19. P. J. Slater, Enclaveless sets and MK-systems, *Journal of Research of the National Bureau of Standards.* **82** (1977), 197–202.
20. S. Athanassopoulos, I. Caragiannis, C. Kaklamanis, and M. Kyropoulou, An improved approximation bound for spanning star forest and color saving. In eds. R. Královic and D. Niwinski, *Mathematical Foundations of Computer Science 2009, 34th International Symposium, MFCS*, vol. 5734, *LNCS*, pp. 90–101, Springer, 2009.
21. N. Chen, R. Engelberg, C. T. Nguyen, P. Raghavendra, A. Rudra, and G. Singh, Improved Approximation Algorithms for the Spanning Star Forest Problem, *Algorithmica.* **65** (2013), 498–516.
22. C. T. Nguyen, J. Shen, M. Hou, L. Sheng, W. Miller, and L. Zhang, Approximating the spanning star forest problem and its application to genomic sequence alignment, *SIAM Journal on Computing.* **38** (2008), 946–962.
23. F. Dehne, M. Fellows, H. Fernau, E. Prieto, and F. Rosamond, NONBLOCKER: parameterized algorithmics for MINIMUM DOMINATING SET. In eds. J. Štuller, J. Wiedermann, G. Tel, J. Pokorný, and M. Bielikova, *Software Seminar SOFSEM*, vol. 3831, *LNCS*, pp. 237–245, Springer, 2006.
24. C. Bazgan, L. Brankovic, K. Casel, and H. Fernau, On the complexity landscape of the domination chain. In eds. S. Govindarajan and A. Maheshwari, *Algorithms and Discrete Applied Mathematics - CALDAM*, vol. 9602, *LNCS*, pp. 61–72, Springer, 2016.
25. S. Bermudo, H. Fernau, and J. M. Sigarreta, The differential and the Roman domination number of a graph, *Applicable Analysis and Discrete Mathematics.* **8** (2014), 155–171.
26. S. Bermudo and H. Fernau, Computing the differential of a graph: hardness, approximability and exact algorithms, *Discrete Applied Mathematics.* **165** (2014), 69–82.
27. S. Bermudo and H. Fernau, Combinatorics for smaller kernels: The differential of a graph, *Theoretical Computer Science.* **562** (2015), 330–345.
28. G. Karakostas, A better approximation ratio for the vertex cover problem, *ACM Trans. Algorithms.* **5** (2009), 41:1–41:8.
29. I. Dinur and S. Safra, On the hardness of approximating minimum vertex cover, *Annals of Mathematics.* **162** (2005), 439–485.

30. S. Khot and O. Regev, Vertex cover might be hard to approximate to within $2 - \varepsilon$, *Journal of Computer and System Sciences.* **74** (2008), 335–349.
31. D. S. Hochbaum, Efficient bounds for the stable set, vertex cover and set packing problems, *Discrete Applied Mathematics.* **6** (1983), 243–254.
32. P. Berman and T. Fujito, On the approximation properties of independent set problem in degree 3 graphs. In eds. S. G. Akl, F. K. H. A. Dehne, J.-R. Sack, and N. Santoro, *Algorithms and Data Structures, 4th International Workshop, WADS*, vol. 955, *LNCS*, pp. 449–460, Springer, 1995.
33. D. S. Hochbaum, ed., *Approximation Algorithms for NP-Hard Problems.* PWS Publishing Company, Boston, MA, 1996.
34. E. Halperin, Improved approximation algorithms for the vertex cover problem in graphs and hypergraphs, *SIAM Journal on Computing.* **31** (2002), 1608 1623.
35. P. Alimonti and V. Kann, Some APX-completeness results for cubic graphs, *Theoretical Computer Science.* **237** (2000), 123–134.
36. T. Imamura and K. Iwama, Approximating vertex cover on dense graphs. In *Proceedings of the Sixteenth Annual ACM-SIAM Symposium on Discrete Algorithms, SODA*, pp. 582–589, SIAM, 2005.
37. R. Bar-Yehuda and S. Even, A local-ratio theorem for approximating the weighted vertex cover problem, *Annals of Discrete Mathematics.* **25** (1985), 27–45.
38. R. Bar-Yehuda, One for the price of two: a unified approach for approximating covering problems, *Algorithmica.* **27** (2000), 131–144.
39. R. Bar-Yehuda, K. Bendel, A. Freund, and D. Rawitz, Local ratio: a unified framework for approximation algorithms, *ACM Surveys.* **36** (2004), 422–463.
40. R. Bar-Yehuda and D. Rawitz, On the equivalence between the primal-dual schema and the local-ratio technique, *SIAM Journal of Discrete Mathematics.* **19** (2005), 762–797.
41. F. Kruschewski-Kursawe, Datenreduktionsbasierte Approximationsalgorithmen für das Knotenüberdeckungsproblem: Implementierung und Analyse. Bachelorarbeit, Abteilung Informatikwissenschaften, Fachbereich IV, Universität. Trier, Germany, 2016.
42. F. N. Abu-Khzam, M. A. Langston, A. E. Mouawad, and C. P. Nolan, A hybrid graph representation for recursive backtracking algorithms. In eds. D.-T. Lee, D. Z. Chen, and S. Ying, *Frontiers in Algorithmics, 4th International Workshop, FAW*, vol. 6213, *LNCS*, pp. 136–147, Springer, 2010.
43. F. N. Abu-Khzam, K. Daudjee, A. E. Mouawad, and N. Nishimura, On scalable parallel recursive backtracking, *J. Parallel Distrib. Comput.* **84** (2015), 65–75.
44. S. Debroni, E. Delisle, W. Myrvold, A. Sethi, J. Whitney, J. Woodcock, P. Fowler, B. de La Vaissière, and M. Deza, Maximum independent sets of the 120-cell and other regular polytopes, *ARS MATHEMATICA CONTEMPORANEA.* **6** (2013), 197–210.

https://doi.org/10.1142/9789811278631_0010

Disjoint NP-Pairs and Propositional Proof Systems

Christian Glaßer
University at Würzburg
glasser@informatik.uni-wuerzburg.de

Andrew Hughes
University at Buffalo, New York
ahughes6@buffalo.edu

Alan L. Selman
University at Buffalo, New York
selman@buffalo.edu

Nils Wisiol
Technische Universität Berlin
nils.wisiol@tu-berlin.de

This chapter surveys results on disjoint NP-pairs, propositional proof systems, function classes, and promise classes—including results that demonstrate close connections that bind these topics together. We illustrate important links between the questions of whether these classes have complete objects and whether optimal proof systems may exist.

1. Introduction

Let Σ be the alphabet $\{0, 1\}$. In the following, all sets are subsets of Σ^*. Depending on the context, these might be sets of natural numbers or strings that represent efficient codings of data structures. Given two nonempty, disjoint sets A and B, a *separator* of the pair (A, B) is a set S, such that $A \subseteq S$ and $B \subseteq \overline{S}$. We say that S *separates* A from B. In general, we are concerned with the computational difficulty of separators. The study of separability (rather, inseparability) arose over the years in different contexts.

Separability in Descriptive Set Theory and Computability Theory. The notion of separability first appeared in descriptive set theory, where Luzin [Luz30] proved the separation theorem for analytic sets. It says that any two disjoint analytic sets are separable by a Borel set. In 1950, researchers began to study separability in the context of computability theory. Here a disjoint pair (A, B) is *computably separable* if there is a decidable separator of (A, B). Otherwise, the pair is *computably inseparable*. Kleene [Kle50] showed the existence of a disjoint pair (A, B) of computably enumerable (c.e.) sets, such that (A, B) is computably inseparable. So with respect to separability, we have different situations in descriptive set theory and computability theory.

Trakhtenbrot [Tra53] proved that in first-order logic, the set of finitely satisfiable formulas, the set of unsatisfiable formulas, and the set of only infinitely satisfiable formulas are pairwise computably inseparable. Shoenfield [Sho58] showed that every undecidable, computably enumerable degree contains disjoint sets A and B, such that (A, B) is computably inseparable.

Smullyan [Smu58] showed for Peano arithmetic (PA), the standard theory of arithmetic, that the set of provable formulas and the set of refutable formulas are computably inseparable.

Promise Problems and Disjoint NP-Pairs. Even, Selman, and Yacobi [ESY84] defined a promise problem to be a partial decision problem consisting of two predicates (unary relations) (Q, R), so that for an input word x, if the *promise* predicate Q holds, then we want to know whether the property $R(x)$ holds. Formally, a *promise problem* is a pair of predicates (Q, R). A deterministic Turing machine M *solves* (Q, R) if

$$\forall x \in \Sigma^*[Q(x) \Rightarrow [M(x) \text{ stops } \wedge [M(x) = \text{“}yes\text{”} \Leftrightarrow R(x)]]].$$

Of course, a decision problem is a promise problem with the condition that $Q(x)$ holds for all inputs x.

Promise problems are completely characterized by disjoint pairs [GS88]: Let A be the set of “yes” instances, i.e., instances x so that $Q(x) \wedge R(x)$ and let B be the set of instances x so that $Q(x) \wedge \neg R(x)$. Now a deterministic Turing machine to solve the promise problem must accept all strings in A, must reject all strings in B, and may behave arbitrarily for all other strings in Σ^*. This is the formulation most used in current studies [Gol06].

Both Even, Selman, and Yacobi [ESY84] and Grollmann and Selman [GS88] were interested in analyzing issues in public-key cryptography, and technical considerations led them to be especially interested in disjoint pairs (A, B), where A and B are in the class NP. These are called disjoint NP-pairs, and we let DisjNP denote the collection of all disjoint NP-pairs.

Separability in Computational Complexity. The interesting question now is whether there is a separator of (A, B) that belongs to P. If there is such a separator, then (A, B) is P-*separable.* Otherwise, the pair is P-*inseparable.*

We know that such pairs do exist if P $\neq$ UP or P $\neq$ NP $\cap$ coNP (see Proposition 2.1). Further evidence for the existence of P-inseparable disjoint NP-pairs comes from cryptography. Even, Selman, and Yacobi [ESY84] formulated the problem of cracking a public-key cryptosystem as a promise problem, and Grollmann and Selman [GS88] then observed that secure public-key cryptosystems imply the existence of P-inseparable disjoint NP-pairs. Also, Even, Selman, and Yacobi [ESY84] formulated a conjecture that would imply that no public-key cryptosystem is NP-hard to crack. The conjecture is that every disjoint NP-pair has some separator that is not NP-hard.

Observe the analogies with the concepts from computability theory, where we replace c.e. by NP and decidable by membership in P. It is a theorem that every disjoint pair of c.e. sets has some separator, whose Turing degree is strictly less than the degree $0'$ of the c.e.-complete sets [Sho60]. Similarly, we have already noted that computably-inseparable disjoint pairs of c.e. sets do exist.

Disjoint NP-pairs also relate naturally to the theory of proof systems for propositional calculus [Pud03,Raz94], and that is the connection we will primarily explore here. Before we do so, here is an interesting example of a disjoint NP-pair known as the Clique-Coloring pair.

Example 1.1: Consider the following sets.

$$CC_0 = \{(G, k) \mid G \text{ is a graph that has a clique of size } k\}.$$
$$CC_1 = \{(G, k) \mid G \text{ is a graph that can be colored with } k - 1 \text{ colors}\}.$$

The sets are disjoint, since a clique of size k cannot be colored with $k-1$ colors. Moreover, CC_0 and CC_1 are NP-complete, and hence (CC_0, CC_1) is a disjoint NP-pair. Surprisingly, this pair is P-separable [Pud03], a result that is based on deep combinatorial ideas by Lovász [Lov79] and Tardos [Tar88].

2. Separability and Reducibility

The following proposition gives evidence for the existence of P-inseparable disjoint NP-pairs.

Proposition 2.1:

(1) If $\mathrm{P} \neq \mathrm{NP} \cap \mathrm{coNP}$*, then* DisjNP *contains* P*-inseparable pairs.*
(2) If $\mathrm{P} \neq \mathrm{UP}$*, then* DisjNP *contains* P*-inseparable pairs* [GS88].

Proof: If $A \in (\mathrm{NP} \cap \mathrm{coNP}) - \mathrm{P}$, then $(A, \overline{A})$ is a disjoint NP-pair. The only separator is $A \notin \mathrm{P}$.

Now assume $\mathrm{P} \neq \mathrm{UP}$. Let $L \in \mathrm{UP} - \mathrm{P}$ and let N be a nondeterministic, polynomial-time Turing machine that accepts L, that runs in (polynomial) time p, and that has at most one accepting path. Hence the following sets are disjoint sets that belong to NP.

$$A = \{(x, i) \mid \text{there exists an accepting path } w \text{ of } N \text{ on } x, \text{ such that the } i\text{-th bit of } w \text{ is } 0\}.$$

$$B = \{(x, i) \mid \text{there exists an accepting path } w \text{ of } N \text{ on } x, \text{ such that the } i\text{-th bit of } w \text{ is } 1\}.$$

Assume that (A, B) is separable by a separator $S \in \mathrm{P}$. Consider the algorithm that on input x computes in polynomial time the string $w = \chi_S(x,1)\,\chi_S(x,2) \cdots \chi_S(x, p(|x|))$ and that accepts if and only if w is an accepting path of N on x.

If $x \in L$, then w is the unique accepting path of N on x, and hence the algorithm accepts. If $x \notin L$, then there is no accepting path of N on x, and hence the algorithm rejects. So the algorithm decides L in polynomial time, which contradicts our assumption. □

Reducibilities between disjoint NP-pairs are as important to our study as are reductions between sets in order to properly formulate notions such as NP-completeness and NP-hardness. Also, the reduction notions we define here are identical to those of promise problems in general [ESY84, GS88]. We define polynomial-time many-one and Turing reductions.

Definition 2.2: Let (A, B) and (C, D) be disjoint pairs.

(1) (A, B) is *many-one reducible in polynomial-time* to (C, D), $(A, B) \leq_m^{pp} (C, D)$, if for every separator T of (C, D), there exists a separator S of (A, B), such that $S \leq_m^p T$.
(2) (A, B) is *Turing reducible in polynomial-time* to (C, D), $(A, B) \leq_T^{pp} (C, D)$, if for every separator T of (C, D), there exists a separator S of (A, B), such that $S \leq_T^p T$.

The definitions tell us that for every separator of (C, D), there is a separator of (A, B) that is no more complex. If (C, D) is P-separable, then so is (A, B). However, these definitions are nonuniform. Consider $(A, B) \leq_m^{pp} (C, D)$. If S_1 is a separator of (C, D), then for an appropriate polynomial-time computable function f_1, we have $f_1(S_1)$ is a separator of (A, B). However, for a different separator S_2 of (C, D), there might be a different function f_2, so that $f_2(S_2)$ is a separator of (A, B). This nonuniformity make these definitions extremely difficult to work with. Fortunately, we have the following equivalent uniform definitions [GS88].

Definition 2.3: Let (A, B) and (C, D) be disjoint pairs.

(1) (A, B) is *uniformly many-one reducible in polynomial time* to (C, D), $(A, B) \leq_{um}^{pp} (C, D)$, if there exists a polynomial-time computable function f, such that for every separator $T \in \mathrm{Sep}(C, D)$, there exists a separator $S \in \mathrm{Sep}(A, B)$, such that $S \leq_m^p T$ via f.
(2) (A, B) is *uniformly Turing reducible in polynomial time* to (C, D), $(A, B) \leq_{uT}^{pp} (C, D)$, if there exists a polynomial-time oracle Turing machine M, such that for every separator $T \in \mathrm{Sep}(C, D)$, there exists a separator $S \in \mathrm{Sep}(A, B)$, such that $S \leq_T^p T$ via M.

Razborov [Raz94] provided yet another definition of many-one reductions.

Definition 2.4: Let (A, B) and (C, D) be disjoint pairs. (A, B) is Razborov reducible[a] to (C, D), if there exists a polynomial-time computable function λ, such that $\lambda(A) \subseteq C$ and $\lambda(B) \subseteq D$.

The following result is fortuitous. A comprehensive proof is in the paper of Glaßer *et al.* [GSSZ04, Theorems 2.8, 2.10, 2.14].

Lemma 2.5: *Let (A, B) and (C, D) be disjoint pairs. Then the following assertions are equivalent.*

(1) $(A, B) \leq_m^{pp} (C, D)$.
(2) $(A, B) \leq_{um}^{pp} (C, D)$.
(3) There exists a polynomial-time computable function λ, such that $\lambda(A) \subseteq C$ and $\lambda(B) \subseteq D$.

[a]Razborov reducible is not a term commonly used in the literature. We will use it only until we prove its equivalence to many-one reducibility.

Similarly, the following result shows that uniform and nonuniform Turing reductions are equivalent. This result was first proven by Grollmann and Selman [GS88].

Lemma 2.6: *Let (A, B) and (C, D) be disjoint pairs. Then the following two assertions are equivalent:*

(1) $(A, B) \leq_T^{pp} (C, D)$.
(2) $(A, B) \leq_{uT}^{pp} (C, D)$.

Therefore, we will only use the notation for many-one and Turing reducibility, but always intend the uniform version. If $(A, B) \leq_m^{pp} (C, D)$ and $(C, D) \leq_m^{pp} (A, B)$, then we write $(A, B) \equiv_m^{pp} (C, D)$. Similarly we define the relation $\equiv_T^{pp}$. Obviously, $\equiv_m^{pp}$ and $\equiv_T^{pp}$ are equivalence relations. The *degree* of a pair $(A, B) \in \text{DisjNP}$ is defined as $\mathbf{d}(A, B) = \{(C, D) \in \text{DisjNP} \mid (A, B) \equiv_T^{pp} (C, D)\}$, i.e., the class of NP-pairs that are Turing equivalent to (A, B). In a canonical way, Turing reductions extend from pairs to Turing-degrees: $\mathbf{d}(A, B) \leq_T^{pp} \mathbf{d}(C, D)$ if $(A, B) \leq_T^{pp} (C, D)$. The *degree structure* of disjoint NP-pairs is the structure of the partial ordering $(\{\mathbf{d}(A, B) \mid (A, B) \in \text{DisjNP}\}, \leq_T^{pp})$.

Razborov raised the question of whether DisjNP contains complete disjoint NP-pairs. His interest was primarily in the question of whether many-one complete pairs exist, but because of the conjecture of Even, Selman, and Yacobi mentioned in the introduction, the question is also interesting for Turing-completeness. We will discuss this in another section. Here we state the formal definitions.

Definition 2.7: Let (A, B) be a disjoint NP-pair.

(1) (A, B) is $\leq_m^{pp}$-hard (resp., $\leq_T^{pp}$-hard) for NP if every separator of (A, B) is $\leq_m^p$-hard (resp., $\leq_T^p$-hard) for NP.
(2) (A, B) is $\leq_m^{pp}$-complete (resp., $\leq_T^{pp}$-complete) if for every $(C, D) \in \text{DisjNP}$ it holds that $(C, D) \leq_m^{pp} (A, B)$ (resp., $(C, D) \leq_T^{pp} (A, B)$).

We conclude this section with a short description of two complexity classes of partial functions. We will mention these briefly in coming sections.

A nondeterministic transducer T computes a value y on an input x if there is an accepting computation of T on x, for which y is the final contents of the output tape of T. Such transducers compute multivalued functions.

Definition 2.8: We consider the following function classes (cf. [BLS84, Sel94]).

(1) NPMV is the set of all partial, multivalued functions computed by nondeterministic polynomial-time-bounded transducers.
(2) NPSV is the set of all $f \in$ NPMV that are single-valued.
(3) $\text{NPSV}_{\{0,1\}}$ is the class of all 0,1-valued functions in NPSV [KM00] (functions that only take on the value 0 or 1).

Given partial multivalued functions f and g, define g to be a *refinement* of f if domain(g) = domain(f) and for all $x \in$ domain(g) and all y, if $g(x) \mapsto y$, then $f(x) \mapsto y$. Let $\mathcal{F}$ and $\mathcal{G}$ be classes of partial multivalued functions. If f is a partial multivalued function, we define $f \in_c \mathcal{G}$ if $\mathcal{G}$ contains a refinement g of f, and we define $\mathcal{F} \subseteq_c \mathcal{G}$ if for every $f \in \mathcal{F}$, $f \in_c \mathcal{G}$.

Let *sat* be the multivalued function defined by $sat(x) \mapsto y$ if and only if x encodes a propositional formula and y encodes a satisfying assignment of x.

3. Propositional Proof Systems

The resolution principle [Rob65] provides a way to prove the unsatisfiability of Boolean formulas. From the correctness and completeness of the resolution principle it follows that

$$\overline{\text{SAT}} = \{\varphi \mid \text{there exists a resolution proof for } \varphi\}, \text{ and} \tag{3.1}$$

$$\text{TAUT} = \{\varphi \mid \text{there exists a resolution proof for } \neg\varphi\}. \tag{3.2}$$

Given a formula φ and a resolution proof w, we can easily check whether w is a valid resolution proof for $\neg\varphi$. So the following function is polynomial-time computable.

$$f(\langle \varphi, w \rangle) = \begin{cases} \varphi, & \text{if } w \text{ is a valid resolution proof for } \neg\varphi \\ \text{true}, & \text{otherwise.} \end{cases}$$

By (3.2), f maps from Σ^* onto TAUT. Hence we can think of the resolution proof system as a polynomial-time computable function that maps onto TAUT. Cook and Reckhow [CR79] observed that this yields an elegant definition for general propositional proof systems.

Definition 3.1: A *propositional proof system* (*proof system* for short) is a partial, polynomial-time-computable function f from Σ^* onto TAUT.

For any w, we say w is an *f-proof for φ* if $f(w) = \varphi$. If there is a polynomial p, such that for all φ and all f-proofs w of φ we have $|w| \leq p(|\varphi|)$, then f is called *polynomially bounded*.

Cook and Reckhow started a line of research [CR79] that studies the length of the shortest proof of a propositional tautology relative to the length of the tautology. To this end they introduced the notion of simulation.

Definition 3.2: Let f and g be proof systems. We say f *simulates* g if there is a function h and a polynomial p, such that for all w it holds that $f(h(w)) = g(w)$ and $|h(w)| < p(|w|)$. We call h a *translation function*. If h is polynomial-time computable, we say f *p-simulates* g. A proof system that simulates (resp., p-simulates) every other proof system is called *optimal* (resp., *p-optimal*).

So f simulates g if for every g-proof for some φ there exists an f-proof for φ that is at most polynomially longer. An optimal proof system then has, up to a polynomial factor, the shortest proofs for tautologies.

Theorem 3.3: [CR79] *There exist polynomially-bounded proof systems if and only if* NP = coNP.

Proof: Assume NP = coNP, and let N be an NP-machine accepting TAUT $\in$ coNP. Let

$$f(\langle\varphi, w\rangle) = \begin{cases} \varphi, & \text{if } w \text{ is an accepting path of } N \text{ on input } \varphi \\ \text{true}, & \text{otherwise.} \end{cases}$$

This function is polynomial-time computable and it outputs only tautologies. Moreover, for every tautology φ there exists an accepting path w, such that N on φ accepts along w. So $f(\langle\varphi, w\rangle) = \varphi$, which shows that f is onto TAUT. Hence f is a proof system, which is even polynomially-bounded, since $|w|$ is polynomially-bounded in $|\varphi|$.

To prove the converse, suppose there is a polynomially-bounded proof system f. Since TAUT is coNP-complete, it suffices to show TAUT $\in$ NP. Let N be a nondeterministic Turing machine that on input φ first guesses a polynomial-length f-proof w and then accepts if and only if $f(w) = \varphi$. This is an NP-machine that accepts TAUT. □

Corollary 3.4: *If* NP = coNP, *then optimal proof systems exist.*

Proof: Assume NP = coNP. By Theorem 3.3, there exists a polynomially-bounded proof system f. We show that f simulates every proof system g. Let

$$h(w) = \text{the lexicographically smallest } v\text{, such that } f(v) = g(w).$$

Hence $f(h(w)) = g(w)$. Since f is a polynomially-bounded proof system and g is polynomial-time computable, there are polynomials p, q, such that $|h(w)| \leq q(|g(w)|) \leq p(|w|)$. So f simulates g. □

Köbler, Meßner, and Torán [KMT03] proved sufficient conditions for the existence of optimal proof systems under the weaker (but still unlikely) hypothesis that NEE $\cap$ TALLY $\subseteq$ coNEE. As they stated in their paper, "the sufficient conditions show however that it would be very hard to prove that optimal proof systems do not exist, since this would imply a separation of complexity classes."

4. Disjoint NP-Pairs that Are Complete or NP-Hard

One of the most interesting open questions about disjoint NP-pairs is whether there exist complete pairs [Raz94].[b] A proof of the nonexistence implies NP $\neq$ coNP, and we will relate complete pairs to propositional proof systems [Raz94]. Also, it is an open question as to whether disjoint NP-pairs can be NP-hard [ESY84].[c] Even, Selman, and Yacobi [ESY84] conjectured the following:

- ESY-T: No disjoint NP-pair is $\leq_T^{pp}$-hard for NP.

This conjecture would imply that no public-key cryptosystem has an NP-hard cracking problem. Moreover, the conjecture has fascinating consequences for computational complexity: ESY-T implies that NP $\neq$ coNP, NP $\neq$ UP, and satisfying assignments of Boolean functions cannot be computed by single-valued NP-machines (i.e., NPMV $\not\subseteq_c$ NPSV) [ESY84,GS88, Sel94].

Recently, Hughes *et al.* [HPRS12] have studied variants of this conjecture, of which we mention here the following:

- ESY-*tt*: No disjoint NP-pair is $\leq_{tt}^{pp}$-hard for NP.
- ESY-*m*: No disjoint NP-pair is $\leq_m^{pp}$-hard for NP.

For a reduction r, the ESY-r conjecture says that for every disjoint NP-pair there exists a separator that is not r-hard for NP. Note that ESY-T

[b]For c.e. sets, this question has an affirmative answer: There exist many-one complete pairs of disjoint c.e. sets [Rog67, Theorem XII(c)].

[c]For c.e. sets, this question has a negative answer: Every pair of disjoint c.e. sets has a separator, whose Turing-degree is strictly less than the degree $0'$ of the c.e.-complete sets [Sho60].

implies ESY-tt, and ESY-tt implies ESY-m. Moreover, the ESY-conjectures immediately relate to the existence of complete pairs.

Theorem 4.1: *If ESY-r does not hold, then there exists an r-complete disjoint* NP *pair.*

Proof: Assume that ESY-r does not hold, then there is a pair (A, B) of disjoint sets in NP, such that every separator is r-hard for NP. We claim (A, B) is r-complete for DisjNP. Let $(C, D) \in$ DisjNP, and let S be any separator for (A, B). Then since $C \in$ NP and S is r-hard for NP, we have $C \leq_r S$. This proves C, which is a separator of (C, D), reduces to any separator of (A, B). By Definition 2.2, we have $(C, D) \leq_r^{pp} (A, B)$ and thus (A, B) is r-complete for DisjNP. □

For the ESY-tt conjecture we know the same consequences for computational complexity as for ESY-T.

Theorem 4.2: [HPRS12] *If ESY-tt holds, then* NP $\neq$ coNP, NP $\neq$ UP, *and* NPMV $\not\subseteq_c$ NPSV.

For this reason, ESY-tt is probably as difficult to settle as is ESY-T. We should note in this regard that NPMV $\subseteq_c$ NPSV holds only if the polynomial hierarchy collapses to $\mathrm{ZPP}^{\mathrm{NP}}$ [HNOS96] (and indeed even to $S_2^{\mathrm{NP}\cap\mathrm{coNP}}$ [CCHO05]). However, the following result settles ESY-m as well as can be expected.

Theorem 4.3: [GSSZ04] *ESY-m holds if and only if* NP $\neq$ coNP.

Proof: Assume ESY-m does not hold. Let $(A, B) \in$ DisjNP, such that all separators are many-one-hard for NP. The set $\overline{B}$ is a separator of (A, B), and therefore SAT $\leq_m^p \overline{B}$. Thus $\overline{\mathrm{SAT}} \leq_m^p B$, which means that $\overline{\mathrm{SAT}} \in$ NP. It follows by the completeness of SAT that NP = coNP.

Now assume NP = coNP. The pair $(\mathrm{SAT}, \overline{\mathrm{SAT}})$ has only separators that are many-one-hard for NP. □

5. Optimal Propositional Proof Systems

In this section, we present a proof of a result by Köbler, Meßner, and Torán [KMT03]. Here we show that the existence of an optimal proof system implies the existence of a complete set for NP $\cap$ SPARSE, the sparse sets that belong to NP. We tend to believe that there are no complete sets for

NP $\cap$ SPARSE. Therefore, this important result by Köbler, Meßner, and Torán provides evidence that there are no optimal proof systems.

Theorem 5.1: [KMT03] *If optimal proof systems exist, then* NP$\cap$SPARSE *has many-one complete sets.*

Proof: We define the set SP, containing descriptions of nondeterministic Turing machines that do not accept too many different strings if we restrict the runtime and the length of the input.

$$\begin{aligned} SP = \{\langle N, 0^l, 0^n\rangle \mid\ & (1)\ N \text{ is a nondeterministic Turing machine} \\ & (2) \text{ there are at most } l \text{ pairs } (x_i, y_i), \text{ such that} \\ & \quad \text{(a) all } x_i \text{ are distinct} \\ & \quad \text{(b) } |x_i| \le n, |y_i| \le l \\ & \quad \text{(c) } N \text{ accepts } x_i \text{ on path } y_i\}. \end{aligned}$$

A triple $\langle N, 0^l, 0^n\rangle$ does not belong to SP if and only if N is not the encoding of a nondeterministic Turing machine or there exist $l+1$ pairs (x_i, y_i), such that (a)–(c) hold. This shows $SP \in$ coNP.

Now we define polynomial-time computable subsets of SP. Assume N is a nondeterministic Turing machine with polynomial runtime q, such that for every n, N accepts at most $q(n)$ strings of length at most n. Hence $\langle N, 0^{q(n)}, 0^n\rangle \in SP$ for all $n \ge 1$. So the set $SP_N = \{\langle N, 0^{q(n)}, 0^n\rangle \mid n \ge 1\}$ is a subset of SP. Moreover, SP_N is polynomial-time decidable: On input $\langle M, 0^l, 0^n\rangle$, the program checks whether $M = N$ and $l = q(n)$, where N and q are coded into the program.

We now define a set $S \in$ NP $\cap$ SPARSE and use the sets SP_N to prove the completeness of S. Let h be an optimal proof system and let γ be a many-one reduction from SP to TAUT. Define

$$\begin{aligned} S = \{\langle 0^N, 0^j, 0^l, x\rangle \mid\ & \text{(I) } N \text{ is a nondeterministic Turing machine} \\ & \text{(II) there exists a string } w \text{ of length at most } j, \text{ such that} \\ & \quad \text{(a) } h(w) = \gamma(\langle N, 0^l, 0^{|x|}\rangle) \\ & \quad \text{(b) } N \text{ accepts } x \text{ in at most } l \text{ steps}\}. \end{aligned}$$

S belongs to NP. To see that S is sparse, we fix N, j, l, and $m = |x|$. If $\langle 0^N, 0^j, 0^l, x\rangle \in S$ where $|x| = m$, then by (II)(b), N accepts x in at most l steps. By (II)(a), $\gamma(\langle N, 0^l, 0^{|x|}\rangle) \in$ TAUT and hence $\langle N, 0^l, 0^{|x|}\rangle \in SP$. So there are at most l inputs of length at most $|x| = m$ that are accepted by N in at most l steps. Thus for fixed N, j, l, and m, there are at most l triples

$\langle 0^N, 0^j, 0^l, x\rangle \in S$, such that $|x| = m$. For triples of length n, there is only a polynomial number of possibilities for the values N, j, l, and $m = |x|$, because of the tally encoding of N, j, and l. Hence S is sparse.

We argue that every $S' \in \mathrm{NP} \cap \mathrm{SPARSE}$ many-one reduces to S. Let q be a polynomial, such that S' contains at most $q(n)$ strings of length at most n, and S' is accepted by N in time q. Since we know that SP_N is polynomial-time decidable, we can define a polynomial-time-computable function

$$g_N(w) = \begin{cases} \gamma(z) & \text{if } w = 1z \text{ and } z \in SP_N, \\ h(z) & \text{if } w = 0z, \\ h(\varepsilon) & \text{otherwise,} \end{cases}$$

with range TAUT. So g_N is a proof system and thus is simulated by the optimal proof system h. Hence, there exists a translation function λ and a polynomial r, such that for all v, we have $h(\lambda(v)) = g_N(v)$ and $|\lambda(v)| \leq r(|v|)$. Let $r'(n) = r(|1\langle N, 0^{q(n)}, 0^n\rangle|)$. We show that S' reduces to S via the polynomial-time-computable function $x \mapsto \langle 0^N, 0^{r'(|x|)}, 0^{q(|x|)}, x\rangle$.

Assume $x \in S'$. Let $z = \langle N, 0^{q(|x|)}, 0^{|x|}\rangle$. By definition, $z \in SP_N$. Thus, $g_N(1z) = \gamma(z)$, and therefore $h(\lambda(1z)) = \gamma(z)$. Define $w = \lambda(1z)$, $j = r'(|x|)$, and $l = q(|x|)$. So $|w| = |\lambda(1z)| \leq r(|1z|) = r'(|x|) = j$ and $h(w) = \gamma(z)$, which shows that w satisfies condition (II)(a). Moreover, N accepts x in at most l steps, since $x \in S'$. This shows $\langle 0^N, 0^{r'(|x|)}, 0^{q(|x|)}, x\rangle \in S$.

For $x \notin S'$ condition (II)(b) is not satisfied and hence $\langle 0^N,\ 0^{r'(|x|)},\ 0^{q(|x|)},\ x\rangle \notin S$. □

Köbler, Meßner, and Torán [KMT03] show that this proof can be generalized to other promise classes, such as UP, Few, FewP, NP ∩ SPARSE, and NP ∩ coNP.

Theorem 5.2: [KMT03]

(1) If p-optimal proof systems exist, then UP *has many-one complete sets.*
(2) If optimal proof systems exist, then UP *contains sets that are complete under nonuniform many-one reducibility.*

Another remarkable consequence of the existence of optimal proof systems is the existence of complete NP-pairs, first proven by Razborov [Raz94]. Our proof requires preparation, which we provide in the next section.

6. Canonical Disjoint NP-Pairs for Proof Systems

In the introduction we mentioned the inseparability of Peano Arithmetic. The analogous question for propositional calculus is extremely difficult, since the set of provable formulas SAT and the set of refutable formulas $\overline{\text{SAT}}$ are P-separable if and only if P = NP. Razborov [Raz94] studied the variant of this question that asks for *short* proofs that refute a formula. For a propositional proof system f, he defined the *canonical pair* $(\text{SAT}^*, \text{REF}_f)$, where

$$\text{SAT}^* = \{(\varphi, 1^m) \mid \varphi \in \text{SAT and } m \geq 0\} \text{ and}$$
$$\text{REF}_f = \{(\varphi, 1^m) \mid \neg\varphi \in \text{TAUT and } \exists y, |y| \leq m, \text{ such that } f(y) = \neg\varphi\}.$$

These sets are disjoint and they belong to NP. We know that there exists a propositional proof system f, such that SAT^* and REF_f are P-inseparable if and only if NP contains a pair of disjoint, P-inseparable sets [GSZ07]. Moreover, the notion of canonical pairs is closely related to the notion of simulation of proof systems.

Proposition 6.1: [Raz94] *Let f and g be propositional proof systems. If f simulates g, then* $(\text{SAT}^*, \text{REF}_g) \leq_m^{pp} (\text{SAT}^*, \text{REF}_f)$.

Proof: Since f simulates g, there is a function h and a polynomial p, such that $g(w) = f(h(w))$ and $|h(w)| \leq p(|w|)$. Define $\lambda(\varphi, 0^n) = (\varphi, 0^{p(n)})$, we show $\lambda(\text{SAT}^*) \subseteq \text{SAT}^*$ and $\lambda(\text{REF}_g) \subseteq \lambda(\text{REF}_f)$.

If $(\varphi, 0^n) \in \text{SAT}^*$, then $(\varphi, 0^{p(n)}) \in \text{SAT}^*$. If $(\varphi, 0^n) \in \text{REF}_g$, then $\neg\varphi \in \overline{\text{TAUT}}$ and there is some y, such that $|y| \leq n$ and $g(y) = \neg w$. So $\neg w = g(y) = f(h(y))$ and $|h(y)| \leq p(|y|)$. Hence $(w, 0^{p(n)}) \in \text{REF}_f$. □

The next theorem shows the close relation between disjoint NP-pairs and canonical pairs.

Theorem 6.2: [GSZ07] *For every* $(A, B) \in \text{DisjNP}$, *there exists a proof system f, such that* $(A, B) \equiv_m^{pp} (\text{SAT}^*, \text{REF}_f)$.

Proof: Let $\langle \cdot, \cdot \rangle$ be a polynomial-time computable, polynomial-time invertible pairing function, such that $|\langle v, w\rangle| = 2|vw|$. Choose a function g that is polynomial-time computable and polynomial-time invertible, such that $A \leq_m^p \text{SAT}$ via g (such a g exists, since SAT is a paddable NP-complete set). Let M be an NP-machine that accepts B in time p. Define the following function f:

$$f(z) \stackrel{def}{=} \begin{cases} \neg g(x) & \text{if } z = \langle x, w \rangle, |w| = p(|x|), M(x) \text{ accepts along path } w \\ x & \text{if } z = \langle x, w \rangle, |w| \neq p(|x|), |z| \geq 2^{|x|}, x \in \text{TAUT} \\ \text{true} & \text{otherwise.} \end{cases}$$

The function f is polynomial-time computable, since in the second case $|z|$ is large enough so that $x \in$ TAUT can be decided by the brute-force algorithm in deterministic time $O(|z|^2)$. (Note that in the second case, the condition $|z| \geq 2^{|x|}$ is equivalent to the condition $\log |z| \geq |x|$.) In the first case of f's definition, $x \in B$ and so $g(x) \notin$ SAT. It follows that $f : \Sigma \to$ TAUT. The mapping is onto, since for every tautology y,

$$f(\langle y, 0^{2^{|y|}} \rangle) = y.$$

Therefore, f is a propositional proof system.

Claim 1: $(\text{SAT}^*, \text{REF}_f) \leq_m^{pp} (A, B)$.

Choose elements $a \in A$ and $b \in B$. Define the reduction function h as follows.

```
1: Input (y, 0^n)
2: if n ≥ 2^|y| then
3:     if y ∈ SAT then output a else output b
4: endif
5: if g^{-1}(y) exists then output g^{-1}(y)
6: output a.
```

The condition in line 2 is equivalent to $\log n \geq |y|$. The exhaustive search in line 3 is possible in quadratic time in n. So h is computable in polynomial time.

Assume $(y, 0^n) \in \text{SAT}^*$. If we reach line 3, then we output $a \in A$. Otherwise we reach line 5. If $g^{-1}(y)$ exists, then it belongs to A. Therefore, in either case (output in line 5 or in line 6) we output an element from A.

Assume $(y, 0^n) \in \text{REF}_f$ (in particular $\neg y \in$ TAUT). So there exists z, such that $|z| \leq n$ and $f(z) = \neg y$. If we reach line 3, then we output b. Otherwise we reach line 5 and so it holds that $|z| \leq n < 2^{|y|}$ and $\neg y$ syntactically differs from the expression true. Therefore, $f(z) = \neg y$ must be due to line 1 in the definition of f. It follows that $g^{-1}(y)$ exists and belongs to B (again by line 1 of f's definition). This shows Claim 1.

Claim 2: $(A, B) \leq_m^{pp} (\text{SAT}^*, \text{REF}_f)$.

The reduction function is $h'(x) \stackrel{def}{=} (g(x), 0^{2(|x|+p(|x|))})$. If $x \in A$, then $g(x) \in$ SAT and therefore, $h'(x) \in \text{SAT}^*$. Otherwise, let $x \in B$. Let w be an accepting path of $M(x)$ and define $z \stackrel{def}{=} \langle x, w \rangle$. So $|w| = p(|x|)$ and $|z| = 2(|x| + p(|x|))$. By line 1 in f's definition, $f(z) = \neg g(x)$. Therefore, $h'(x) \in \text{REF}_f$. This proves Claim 2 and finishes the proof of Theorem 6.2. □

Corollary 6.3: *Disjoint* NP*-pairs and canonical pairs for proof systems have identical degree structure.*

Regarding this degree structure, Glaßer, Selman, and Zhang [GSZ07] proved that between any two comparable and inequivalent disjoint NP-pairs (A, B) and (C, D), there exist P-inseparable, incomparable NP-pairs (E, F) and (G, H), whose degrees lie strictly between (A, B) and (C, D). This is an analogue of Ladner's result for NP [Lad75]. The proof is based on previous work by Schöning [Sch82] and Regan [Reg83, Reg88]. Thus, assuming that P-inseparable disjoint NP-pairs exist, the class DisjNP has a rich, dense, degree structure and each of these degrees contains a canonical pair.

From Proposition 6.1 and Theorem 6.2, we obtain an important connection between optimal proof systems and complete NP-pairs.

Corollary 6.4: *If f is an optimal proof system, then* $(\text{SAT}^*, \text{REF}_f)$ *is complete for* DisjNP.

Proof: For any $(A, B) \in$ DisjNP, there is a proof system g, such that $(A, B) \equiv_m^{pp} (\text{SAT}^*, \text{REF}_g)$. Since f simulates g, we have $(A, B) \leq_m^{pp} (\text{SAT}^*, \text{REF}_g) \leq_m^{pp} (\text{SAT}^*, \text{REF}_f)$. □

It is an open question whether the converse of Corollary 6.4 holds: Does the existence of many-one complete disjoint NP-pairs imply the existence of optimal proof systems? Section 8 suggests that it is difficult to answer this question, since there exist oracles demonstrating that both answers are possible.

7. Uniform Enumerations

The notion of uniform enumeration—and most particularly linking the existence of complete sets to the existence of enumerations of machines echoing the flavor of the class—was first explored in the 1980s in a flurry of papers on the (possible non)existence of complete sets [Sip82, Kow84, HI85, HH88]. In this section, we discuss such a link for the classes DisjNP and NPSV.

Definition 7.1: Suppose $\{N_i\}_{i\geq 0}$ is an effective enumeration of nondeterministic polynomial-time machines. DisjNP is *uniformly enumerable* if there is a total computable function $f : \Sigma^* \to \Sigma^* \times \Sigma^*$, such that

(1) $\forall(i,j) \in \text{range}(f)[(L(N_i), L(N_j)) \in \text{DisjNP}]$, and
(2) $\forall(C,D) \in \text{DisjNP}\exists(i,j)[(i,j) \in \text{range}(f) \wedge C = L(N_i) \wedge D = L(N_j)]$.

Definition 7.2: Suppose $\{N_i\}_{i\geq 0}$ is an effective enumeration of nondeterministic polynomial-time transducers and h_i denotes the partial, multivalued function computed by N_i. NPSV is *uniformly enumerable* if there is a total computable function $f : \Sigma^* \to \Sigma^*$, such that

(1) $\forall i \in \text{range}(f)[h_i \in \text{NPSV}]$, and
(2) $\forall g \in \text{NPSV}\ \exists i\ [i \in \text{range}(f) \wedge h_i = g]$.

Definition 7.3: Let $f, g \in \text{NPSV}$. g is *polynomial-time many-one reducible* to f ($g \leq_m^p f$) if there is a deterministic polynomial-time function h, such that $f(h(x)) = g(x)$.

Definition 7.4: Let $f \in \text{NPSV}$. f is $\leq_m^p$-complete for NPSV if for all $g \in \text{NPSV}$, $g \leq_m^p f$.

The following theorem is extracted from Glaßer *et al.* [GSS05], where seven distinct open questions are proved to be logically equivalent.

Theorem 7.5: [GSS05] *The following are equivalent.*

(1) There is a $\leq_m^{pp}$-complete disjoint NP*-pair.*
(2) DisjNP *is uniformly enumerable.*
(3) NPSV *has a $\leq_m^p$-complete function.*
(4) NPSV *is uniformly enumerable.*

Proof: Glaßer *et al.* [GSS05] have shown the equivalence of statements (1)–(3). We will give the proof for the equivalence of statements (3) and (4).

Let $\{M_i\}_{i\geq 0}$ (resp., $\{N_i\}_{i\geq 0}$) be a standard enumeration of deterministic (resp., nondeterministic) polynomial-time transducers. In the proof below, we identify a transducer with the function that is computed by this transducer.

To show that statement (3) implies statement (4), assume $f \in \text{NPSV}$ is a $\leq_m^p$-complete function. Then define the enumerating function $h : \Sigma^* \to \Sigma^*$ as $h(i) = \langle f \circ M_i \rangle$. Suppose $g \in \text{NPSV}$. Then $g \leq_m^p f$ via some M_i. So $g(x) = f(M_i(x))$ for all x and $h(i) = \langle f \circ M_i \rangle$. For all i, $h(i)$ outputs $f \circ M_i$

and $f \in$ NPSV, so $h(i) \in$ NPSV. Therefore h is a uniform enumeration for NPSV.

To show that statement (4) implies statement (3), assume that f is a uniform enumeration for NPSV. Let M compute f. Then we will define the following function h,

$$h(z) = N_i(x) \qquad \text{if } z = 0^n 10^t 1x \text{ and } M(n) = i \text{ within } t \text{ steps.}$$

Given $z = 0^n 10^t 1x$, $h(z)$ uses a polynomial amount of time to compute $M(n)$ for t steps and to simulate $N_i(x)$. Since N_i computes a function in NPSV, this gives that h is also single-valued. So $h \in$ NPSV.

Suppose $g \in$ NPSV. Then there exists a nondeterministic transducer N_ℓ such that $g = N_\ell$. So there exists an n and t, such that $M(n) = \ell$ in t steps. Then the function $\lambda(x) = 0^n 10^t 1x$ is a polynomial-time reduction from g to h. This gives that h is $\leq_m^p$-complete for NPSV. □

It seems improbable that items (2) and (4) hold. Therefore, it seems that DisjNP does not have $\leq_m^{pp}$-complete pairs. Hence, by Corollary 6.4 it seems that optimal proof systems do not exist.

8. Relativization

Several questions regarding disjoint NP-pairs and propositional proof systems remain open. For some of them, we know that an answer to the question solves prominent open problems in computational complexity. For example, proving or disproving the ESY-m conjecture immediately solves the NP versus coNP problem (see Theorem 4.3). This is a satisfactory explanation for the difficulty of the question. For some other open questions, we do not have such a connection to prominent open problems. In this situation, the construction of oracles relative to which the question is answered in each direction can give some evidence for the difficulty of the question.

8.1. *Existence of Optimal Proof Systems*

Krajíček and Pudlák [KP89] constructed an oracle relative to which optimal proof systems do not exist. Buhrman *et al.* [BFFvM00] constructed an oracle relative to which NP ∩ SPARSE has no many-one complete sets. Köbler, Meßner, and Torán [KMT03] show with a relativizing proof that this implies that optimal proof systems do not exist (see Theorem 5.1).

For an oracle relative to which optimal proof systems do exist, one can use the simple oracle by Baker, Gill, and Solovay [BGS75], relative to which P = NP. By Corollary 3.4, it follows that optimal proof systems exist relative to this oracle as well.

8.2. *Do Complete Pairs Imply Optimal Proof Systems?*

By Corollary 6.4, we know the following implication: If optimal proof systems exist, then DisjNP has many-one complete pairs. It is open whether the converse of this implication holds. Here the oracles O_1 and O_2 by Glaßer *et al.* [GSSZ04] give evidence for the difficulty of the question. Relative to both oracles, complete NP-pairs exist, but relative to O_1 (resp., O_2), optimal proof systems exists (resp., do not exist).

8.3. *P-Inseparable Pairs versus Optimal Proof Systems*

The following is known regarding the relationship between the existence of optimal proof systems and the existence of P-inseparable NP-pairs. Rackoff [Rac82] constructs an oracle relative to which P $\neq$ NP = coNP. By Proposition 2.1 and Corollary 3.4, relative to this oracle there exist optimal proof systems and P-inseparable NP-pairs. Baker, Gill, and Solovay [BGS75] construct an oracle relative to which P = NP. By Corollary 3.4, relative to this oracle there exist optimal proof systems but no P-inseparable NP-pairs. Glaßer *et al.* [GSSZ04] construct an oracle relative to which optimal proof systems do not exist, but P-inseparable NP-pairs exist. We do not know of an oracle for the remaining possibility that optimal proof systems and P-inseparable NP-pairs do not exist. Such an oracle would tell us that the implication "if all NP-pairs are P-separable, then optimal proof systems exist" is difficult.

Another open question is inspired by Shoenfield's result [Sho58] that every undecidable, computably enumerable degree contains disjoint sets A and B, such that (A, B) is computably inseparable. Is there a complexity-theoretic analog? We know that if P-inseparable NP-pairs exist, then the degree of NP-complete sets contains disjoint sets A and B that are P-inseparable. Do we find such sets in all degrees in NP – P?

8.4. *Separation of ESY-m, ESY-tt, and ESY-T*

Clearly, ESY-T implies ESY-tt, and ESY-tt implies ESY-m. What about the converse implication? Glaßer and Wechsung [GW03] constructed an oracle relative to which UP = NP and NP $\neq$ coNP. It follows from Sec. 4 that relative to this oracle, ESY-T and ESY-tt do not hold, but ESY-m holds. This gives evidence for the difficulty of proving (ESY-m $\Longrightarrow$

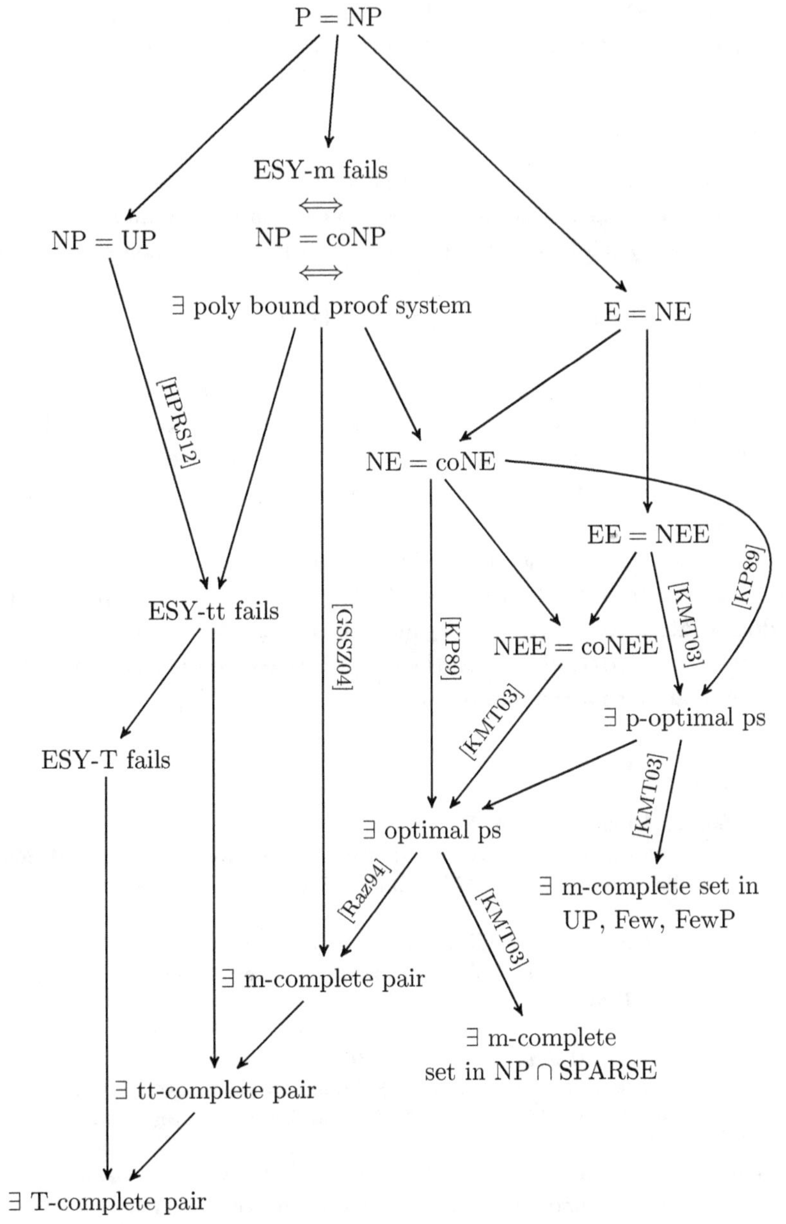

Fig. 1. Known Implications for proof systems, disjoint pairs and the ESY conjecture.

ESY-tt) or (ESY-$m \implies$ ESY-T). It is an open question whether there exists an oracle relative to which ESY-tt holds, but ESY-T does not.

9. Conclusion and Open Questions

For each assertion shown in Fig. 1, it is not known if the assertion is true or false. For each of the implications shown in the chart, it would be interesting to find oracles, such that the converse of the implication holds or does not hold.

For the previously mentioned oracle D by Glaßer and Wechsung [GW03], it is an interesting question to see if there exist many-one complete pairs relative to D, since oracle D implies that NP $\neq$ coNP.

The most significant converse to study is the question of whether the existence of many-one complete disjoint NP-pairs implies the existence of optimal proof systems. As stated in Sec. 8, this question cannot be answered with a relativizable proof.

Our final question concerns random oracles. It is known that NP $\neq$ coNP holds relative to a random oracle [BG81]. So ESY-m holds relative to a random oracle. Also, NP $\neq$ UP and NPMV $\not\subseteq_c$ NPSV hold relative to random oracles [NRRS98]. Nevertheless, it is open whether ESY-T or ESY-tt holds relative to a random oracle.

References

BFFvM00. H. Buhrman, S. A. Fenner, L. Fortnow, and D. van Melkebeek. Optimal proof systems and sparse sets. In Horst Reichel and Sophie Tison, editors, *STACS*, volume 1770 of *Lecture Notes in Computer Science*, pages 407–418. Springer, 2000.

BG81. C. Bennett and J. Gill. Relative to a random oracle A, $P^A \neq NP^A \neq$ co-NP^A with probability 1. *SIAM Journal on Computing*, 10(1):96–113, 1981.

BGS75. T. Baker, J. Gill, and R. Solovay. Relativizations of the P=NP problem. *SIAM Journal on Computing*, 4:431–442, 1975.

BLS84. R. Book, T. Long, and A. Selman. Quantitative relativizations of complexity classes. *SIAM Journal on Computing*, 13(3):461–487, 1984.

CCHO05. Jin-yi Cai, Venkatesan T. Chakaravarthy, Lane A. Hemaspaandra, and Mitsunori Ogihara. Competing provers yield improved Karp-Lipton collapse results. *Inf. Comput.*, 198(1):1–23, 2005.

CR79. S. A. Cook and R. A. Reckhow. The relative efficiency of propositional proof systems. *J. Symb. Log.*, 44(1):36–50, 1979.

ESY84. S. Even, A. L. Selman, and Y. Yacobi. The complexity of promise problems with applications to public-key cryptography. *Information and Control*, 61(2):159–173, May 1984.

Gol06. O. Goldreich. On promise problems: A survey. In O. Goldreich, A. L. Rosenberg, and A. L. Selman, editors, *Essays in Memory of Shimon Even*, volume 3895 of *Lecture Notes in Computer Science*, pages 254–290. Springer, 2006.

GS88. J. Grollmann and A. L. Selman. Complexity measures for public-key cryptosystems. *SIAM J. Comput.*, 17(2):309–335, 1988.

GSS05. C. Glaßer, A. L. Selman, and S. Sengupta. Reductions between disjoint NP-pairs. *Inf. Comput.*, 200(2):247–267, 2005.

GSSZ04. C. Glaßer, A. L. Selman, S. Sengupta, and L. Zhang. Disjoint NP-pairs. *SIAM Journal on Computing*, 33(6):1369–1416, 2004.

GSZ07. C. Glaßer, A. L. Selman, and L. Zhang. Canonical disjoint NP-pairs of propositional proof systems. *Theor. Comput. Sci.*, 370(1-3):60–73, 2007.

GW03. C. Glaßer and G. Wechsung. Relativizing function classes. *J. UCS*, 9(1):34–50, 2003.

HH88. J. Hartmanis and L. Hemachandra. Complexity classes without machines: On complete languages for UP. *Theoretical Computer Science*, 58(1–3):129–142, 1988.

HI85. J. Hartmanis and N. Immerman. On complete problems for NP$\bigcap coNP$. In *Proceedings of the 12th International Colloquium on Automata, Languages, and Programming*, pages 250–259. Springer-Verlag *Lecture Notes in Computer Science #194*, July 1985.

HNOS96. L. A. Hemaspaandra, A. V. Naik, M. Ogihara, and A. L. Selman. Computing solutions uniquely collapses the polynomial hierarchy. *SIAM Journal on Computing*, 25:697–708, 1996.

HPRS12. A. Hughes, A. Pavan, N. Russell, and A. L. Selman. A thirty year old conjecture about promise problems. In Artur Czumaj, Kurt Mehlhorn, Andrew M. Pitts, and Roger Wattenhofer, editors, *International Colloquium on Automata, Languages, and Programming*, volume 7391 of *Lecture Notes in Computer Science*, pages 473–484. Springer, 2012.

Kle50. S. C. Kleene. A symmetric form of Gödel's theorem. In *Proceedings of the section of sciences*, volume 12 of *Koninklijke Nederlandse Akademie van Wetenschappen*, pages 800–802. Springer Verlag, 1950.

KM00. J. Köbler and J. Meßner. Is the standard proof system for SAT P-optimal? In Sanjiv Kapoor and Sanjiva Prasad, editors, *Foundations of Software Technology and Theoretical Computer Science (FSTTCS)*, volume 1974 of *Lecture Notes in Computer Science*, pages 361–372. Springer, 2000.

KMT03. J. Köbler, J. Meßner, and J. Torán. Optimal proof systems imply complete sets for promise classes. *Inf. Comput.*, 184(1):71–92, 2003.

Kow84. W. Kowalczyk. Some connections between representability of complexity classes and the power of formal reasoning systems. In *Proceed-*

ings of the 11th Symposium on Mathematical Foundations of Computer Science, pages 364–369. Springer-Verlag *Lecture Notes in Computer Science #176*, 1984.

KP89. J. Krajíček and P. Pudlák. Propositional proof systems, the consistency of first order theories and the complexity of computations. *J. Symb. Log.*, 54(3):1063–1079, 1989.

Lad75. R. Ladner. On the structure of polynomial-time reducibility. *Journal of the ACM*, 22:155–171, 1975.

Lov79. L. Lovász. On the Shannon capacity of a graph. *IEEE Transactions on Information Theory*, 25(1):1–7, 1979.

Luz30. N. N. Luzīn. *Leçons sur les ensembles analytiques et leurs applications.* Gauthier-Villars, 1930.

NRRS98. A. V. Naik, J. D. Rogers, J. S. Royer, and A. L. Selman. A hierarchy based on output multiplicity. *Theor. Comput. Sci.*, 207(1):131–157, 1998.

Pud03. P. Pudlák. On reducibility and symmetry of disjoint NP-pairs. *Theoretical Computer Science*, 295:323–339, 2003.

Rac82. C. Rackoff. Relativized questions involving probabilistic algorithms. *J. ACM*, 29(1):261–268, 1982.

Raz94. A. A. Razborov. On provably disjoint NP-pairs. *Electronic Colloquium on Computational Complexity (ECCC)*, 1(6), 1994.

Reg83. K. Regan. On diagonalization methods and the structure of language classes. In *Proceedings Foundations of Computation Theory*, volume 158 of *Lecture Notes in Computer Science*, pages 368–380. Springer Verlag, 1983.

Reg88. K. Regan. The topology of provability in complexity theory. *Journal of Computer and System Sciences*, 36:384–432, 1988.

Rob65. J. A. Robinson. A machine-oriented logic based on the resolution principle. *J. ACM*, 12(1):23–41, 1965.

Rog67. H. Rogers Jr. *Theory of Recursive Functions and Effective Computability.* McGraw-Hill, New York, 1967.

Sch82. U. Schöning. A uniform approach to obtain diagonal sets in complexity classes. *Theoretical Computer Science*, 18:95–103, 1982.

Sel94. A. L. Selman. A taxonomy of complexity classes of functions. *Journal of Computer and System Sciences*, 48(2):357–381, 1994.

Sho58. J. R. Shoenfield. Degrees of formal systems. *Journal of Symbolic Logic*, 23:389–392, 1958.

Sho60. J. R. Shoenfield. Degrees of models. *Journal of Symbolic Logic*, 25(3):233–237, 1960.

Sip82. M. Sipser. On relativization and the existence of complete sets. In *Proceedings of the 9th International Colloquium on Automata, Languages, and Programming*, pages 523–531. Springer-Verlag *Lecture Notes in Computer Science #140*, July 1982.

Smu58. R. M. Smullyan. Undecidability and recursive inseparability. *Zeitschr. f. math. Logik und Grundlagen d. Math.*, 4:143–147, 1958.

Tar88. E. Tardos. The gap between monotone and non-monotone circuit complexity is exponential. *Combinatorica*, 8(1):141–142, 1988.

Tra53. B. Trakhtenbrot. On recursive separability. *Dokl. Akad. Nauk SSSR*, 88:953–955, 1953.

https://doi.org/10.1142/9789811278631_0011

How to Verify Computations with a Rational Network

Sanjay Jain[1,†], Prateek Saxena[1,‡], Frank Stephan[2] and Jason Teutsch[3]
[1] *School of Computing, National University of Singapore*
[†] *sanjay@comp.nus.edu.sg*, [‡] *prateeks@comp.nus.edu.sg*
[2] *Department of Mathematics, National University of Singapore*
fstephan@comp.nus.edu.sg
[3] *TrueBit Establishment, jt@truebit.io*

This chapter introduces a practical protocol for provably secure, outsourced computation. Our protocol minimizes overhead for verification by requiring solutions to withstand an interactive game between a prover and challenger. For optimization problems, the best or nearly best of all submitted solutions is expected to be accepted by this approach. Financial incentives and deposits are used in order to overcome the problem of fake participants.[a]

Contents

[a]Author order is alphabetical. P. Saxena and J. Teutsch research has been supported by Singapore Ministry of Education Grant No. R-252-000-560-112. S. Jain and F. Stephan have been supported in part by NUS grant No. R252-000-534-112 and R146-000-181-112. S. Jain has also been supported by NUS grant No. C252-000-087-001. The authors thank Christian Reitwießner for discussions.

1. Introduction

In the classical Byzantine General's Problem [9], three parties try to agree on whether to "attack" or "retreat" by passing messages back and forth. Here neither "attack" nor "retreat" represents a semantic truth. The generals simply want to agree on one plan or the other, and neither plan is *de jure* more correct than the other. In other consensus situations, such as those involving mathematics, a true answer may in fact exist. A simple way a group may achieve agreement about a known computational problem is for each party to compute the solution locally. With local computation, the communication cost between parties is zero and therefore beats any Byzantine consensus protocol in terms of the number of messages sent. In practice, however, this local computational approach may not be feasible due to its heavy computational burden.

A machine with limited resources may wish to outsource a computational problem to an external system. Today cloud services exist which provide outsourcing service to businesses and individuals. These services generally require the individual to trust the cloud, which may or may not behave correctly for any reason, including hardware failures or server cheating, in order to save computational resources.

One promising approach to ensuring a correct answer from an outsourced computation involves forcing the cloud to provide a short proof which witnesses the correctness of its computation. Researchers have achieved much progress on this method in recent years [18], however the cryptographic setup costs and computational overhead for state-of-the-art systems make these methods impractical for many applications. We take a different approach. In Sec. 5.1, we contrast our approach with this probabilistically checkable proof method.

Let us assume that the individual, who wants to outsource his computation, offers to pay for a correct answer. Suppose that a peer-to-peer distributed system of *rational* servers exists, who want to solve computational problems in exchange for monetary rewards. In this chapter, we show

how to harness the power of economics and consensus in order to incentivize correct computation from such a network with minimal overhead costs.

1.1. *The Verifier's Dilemma*

The Bitcoin network provides evidence that systems can economically incentivize correct computation. Bitcoin miners correctly solve cryptographic hashing problems in exchange for monetary rewards in the form of electronic currency. The scope of problems correctly computed on Bitcoin does not, however, extend beyond "inverting" SHA2 hashes. The new cryptocurrency Ethereum [3], which runs essentially the same consensus protocol as Bitcoin, permits one to achieve an outsourced computation protocol through a series of transactions [12]. Anyone can publicly offer a reward for a puzzle, and the first person to announce a correct solution gets the reward.

Under the assumption that Ethereum miners are rational, the Ethereum network may not correctly distribute rewards for solutions whose verification requires more than minimal computational work [12]. As in the protocol for Bitcoin, the Ethereum protocol dictates that every miner should check every transaction that appears on the cryptocurrency's *blockchain*, or public ledger. While Ethereum miners receive rewards for being the first to solve mining puzzles, they do not receive rewards for checking transactions. Thus miners may skip *heavy* transactions, which take a long time to verify, in order to get ahead in the mining race. But whether a rational miner, who is trying to maximize his mining rewards, benefits from verification depends on whether other miners are honestly verifying, building on top of the current blockchain without verifying, or backtracking along the blockchain and extending with a block that doesn't include the heavy transaction. When a heavy transaction appears on the blockchain, a rational miner doesn't know whether to verify it or not. This conundrum is known as the *Verifier's Dilemma* [12], and Ethereum's `gas limit` system does not entirely mitigate this problem.

As a result of the Verifier's Dilemma, the Ethereum blockchain may contain unverified transactions. An example of a heavy transaction would be a solution to an outsourced computational puzzle which takes a long time to verify. Thus, even though the Ethereum protocol achieves consensus on the blockchain, it may not achieve consensus on the *correct* blockchain. In particular, assuming rational miners, Ethereum's *consensus computer*, or consensus-based outsourced computation system, yields incorrect results.

1.2. *Contribution of this chapter*

We describe a modification to the Ethereum consensus protocol, which incentivizes against the Verifier's Dilemma described above. Under the assumption that computational entities on the network are both:

- *non-lazy* in the sense that one can convince them to perform computations for the right price and
- *rational* in the sense that they will try to maximize their individual payoffs from participation,

then our modified Ethereum system yields correct results for any feasible function.

In the following discussion, a prover who provides a solution to a puzzle must withstand challenges from other parties on the network. These challenges result in an authenticated, public transcript of interaction between the prover and challenger, which may be communicated on the Ethereum blockchain or through some other means. We assume throughout this chapter that the Ethereum smart contract mechanism can effectively enforce the intended penalty and reward consequences of such transcripts. The network identifies each party in the protocol by its Ethereum wallet address.

We remark that the Verifier's Dilemma may extend to other cryptocurrencies besides Ethereum, such as Bitcoin, when the number of transactions included becomes large enough that verifying all of them requires significant computational effort. One can also use the verification game presented in this chapter to counteract the Verifier's Dilemma in Bitcoin by appropriate modification to the Bitcoin protocol. Assuming a modification to the Bitcoin in which miners can challenge blocks with wrong transactions, either by posting "challenge" transactions or through some other means and with appropriate rewards for finding errors and penalties for false alarms, one can achieve not only consensus on the Bitcoin blockchain but, in fact, *correct* consensus.

1.3. *Incentives from cryptocurrencies*

Luu, Teutsch, Kulkarni and Saxena [12] proposed to use cryptocurrencies, in particular Ethereum, for outsourcing computations. The network verifies outsourced computations, either deterministically or in a statistical way. This chapter improves on this by using a stricter verification protocol by using a verification game between a prover and a challenger in order to check a proposed solution to a computation task. The role of the network then becomes to verify that the players (prover and challenger) stick to the rules of the verification game. Provers and challengers have to pay deposits in order to participate. These deposits are refunded if the players are honest; however, the deposits of dishonest provers and verifiers are distributed in order to compensate the honest participants of a verification game for their effort and the task giver for the additional waiting time.

Under the assumption that at least one participant of the pool of provers is honest and takes the effort to push his solution through, the proposed framework leads to the acceptance of a correct solution at the end. In the case that no one solves the problem, but at least one challenger is honest and invests the time to challenge the wrong solutions, no solution is accepted by the framework and the task giver receives back the offered prize money. This model resembles the situation of grand problems in mathematics, such as the seven millennium problems [7], where one million dollar prize money has been offered for seven important mathematical problems to be solved and the main effort of checking the solutions was shifted to the mathematical community.

Consider the case that only a few places in the world own supercomputers — such as Quantum computers — and someone without direct access to these computers wants to solve a difficult problem, say factorizing a large integer. He could then outsource the computation and the verification to such a network. This example also shows why one should permit challengers who need not contribute a solution: some owners of traditional supercomputers might not be able to factorize, but are still able to run the algorithm of Agrawal, Kayal and Saxena [1] to check whether all factors in a proposed solutions are prime.

1.4. *Systems that outsource computation*

Humans today process more information than they ever have in the past. One typically associates scientific data processing with "supercomputers," but even your average consumer invokes computationally-intensive processes every time they do a search query on Google, receive news feeds on Facebook, or ask Mathematica to compute Gröbner bases. The utility of mobile devices derives not only from their connectivity to the Internet but also from their connection to more powerful machines.

Cloud computing offers a simple way to outsource computation. A user who wants to compute something submits a query to the cloud in the form of a program and then waits for the cloud's answer. In this model, the user must trust that the machine hardware, software, and administrator all function the way she expects them to. In some situations, one or more of these may not be reasonable assumptions. Hardware failures occur, software bugs abound, and what's to stop a dishonest cloud that wants to save CPU cycles from spewing a random answer when he knows that the user can't tell the difference from a real one?

Due to the possibility of such errors, one might ask the cloud, in addition to providing the desired computation, to also provide a proof of its correctness. Several recent papers, including [17], follow this approach. Since reading a complete proof of correctness might take as much time as simply doing the calculation oneself, the authors employ the machinery of interactive proofs [11] so that users need not read the entire proof but only sample a few bits to confirm its correctness. Although constructions have drastically improved efficiency in recent years [18], general-purpose methods at the time of this writing still carry impractically high computational overheads for the cloud server.

1.5. *Our approach*

The approach to outsourced computation described herein achieves provably correct computation through financial incentives rather than proof systems. We describe a distributed system in which anyone connected to the network can offer a reward for a puzzle, and anyone connected to the network may obtain that reward in exchange for a correct solution to the puzzle. In order to maximize speed and minimize redundant work, we consider two types of protocols. In computations where it is easy to randomly guess and check a solution, for example finding a satisfying assignment to a SAT instance, provers can immediately submit their solutions to a puzzle. This approach gives good speed, and if all the provers guess solutions randomly, the amount of duplicated work is minimal. In computations which require more verification work, for example multiplying extremely large numbers, the puzzle giver may decide to hold a preliminary bidding round to select a single prover. This bidding reduces the price the puzzle pay for a correct solution because each prover expects to get paid a fixed amount per unit work, and the bidding process reduces the total work done among all provers.

This chapter introduces a practical, new consensus protocol especially suited for outsourced computation. The consensus protocol offers a mechanism by which the network can check the correctness of a prover's solution with minimal overhead. If the given solution has an error, anyone may challenge it (for a reward), then the prover may respond to the challenge, then the challenger may challenge back again, etc. in such a way that verifying that the prover and challenger follow the predefined rules of their interaction game requires only a small amount of computational work. In case the finite interaction game reaches its last possible round, and up to that point no rules have been broken, then the trivial final check is passed to an authority, for example the set of Ethereum miners [3]. At this point, the authority either accepts the last challenge, in which case the original solution was wrong, or rejects, in which case it was right.

Our consensus protocol compromises between minimal communication complexity and minimal redundant work. On one extreme, you have the Bitcoin network in which a single announcement of a new block results in immediate consensus among all parties as to which new transactions are acceptable [16]. Consensus in Bitcoin, called *Nakamoto consensus* [13], achieves low communication complexity in exchange for an extremely high amount of local work from miners. In the other extreme, Byzantine consensus [4] avoids local work, but requires a considerable amount of message passing among parties. Our verification game below requires nontrivial interaction and some local work, but not nearly as much of either as Byzantine or Nakamoto consensus, respectively. Perhaps consequently, it permits more robust verification than the purely Nakamoto-based outsourced computation system proposed in [12].

2. The Verification Game

The main idea of this chapter — compared to prior work on consensus computing — is that there is an explicit game for verifying the solution between a prover and a challenger. The basic idea is similar to interactive proof systems, except that both prover and challenger have limited resources.

Nodes on the network have four types of computational Tasks:

(1) solving puzzles,
(2) challenging puzzle solutions,
(3) checking that provers and challengers play the verification game according to the established rules, and
(4) doing proof-of-work as required for participation in the blockchain consensus.

The presence of Task (4) implicitly assumes that we embed our verification game system inside of a cryptocurrency, such as Ethereum. One need not have a cryptocurrency system in order to make use of the verification game protocol below, however, employing an existing cryptocurrency construction simplifies the implementation. Task (4) encompasses Task (3) in the following sense. Nakamoto consensus dictates that miners always mine on the longest blockchain in which all transactions are correct. Thus miners can vote for or against the validity of a set of transactions by choosing which fork of the blockchain to extend.

Without loss of generality, we may assume that network nodes engage either in some combination of Task (1) and Task (2) or in Task (4), but not a combination of these two sets. Moreover, we identify Task (3) and Task (4) into a single task. This disjointness of computational resources may materialize in practice since miners use specialized hardware to achieve Task (3) whereas Task (1) and Task (2) require general-purpose computers. Therefore one cannot use machines for one of the sets of Tasks on the other. We summarize our assumed network properties as follows.

Assumption 1: In order to guarantee correctness of the protocol, we assume the following:

(i) for each submitted puzzle solution, either:

  a. at least one node on the network is able and sufficiently non-lazy to both properly check the full solution and raise alerts for mistakes as needed, or
  b. a fixed fraction of the nodes of the network are able and sufficiently non-lazy to check a share of the solution and raise alerts for mistakes as needed, and

(ii) more than half of the nodes who verify the rules of the verification game are ϵ-*rational* [12], in the sense that they will do the verification correctly without receiving a reward, so long as their computational burden in doing so does not exceed some small threshold ϵ.

Whether we choose Assumption (i)a or Assumption (i)b above depends on the verification game contracts. We discuss two possible protocols in the next section, called consensus-competition or consensus-contract

We shall also assume that, in any verification game, the computations of both the prover and the challenger run in time polynomial in the size n of the input. The puzzle giver himself, who wants the problem to be solved, is sufficiently wealthy to pay for the computation, but not in possession of the necessary computational resources. The network (miners) consists of many participants who verify that both sides in the verification game stick to the rules and that the winner is determined correctly; however, each single step which they verify should be of small computational complexity.

Definition 2: In general, a (f, g, h)-*verification game* consists of

(i) $f(n)$ rounds on the block chain where
(ii) the players can write up to $O[g(n)]$ bits of information in every round, and
(iii) a miner in the network spends up to poly$[h(n)]$ time for verifying each step.

Here the time is measured for random access computation, and thus a miner need not read the full information provided in a round but only the relevant parts needed for verification.

Normally, one would expect $f(n)$ to be small, say constant or $\log n$. The function $h(n)$ is expected to be around $\log n$ or poly $\log n$, as this is the time that the miners use for verification. The function $g(n)$ should ideally be around n (say $n \log n$); however, in some cases, such as in Proposition 9, $g(n)$ may be a polynomial in n. Note that the space $g(n)$ used need not always occupy a blockchain. One may put these $g(n)$ bits on a publicly readable board, where they cannot be altered.

As the above verification game has costs, every participant in the consensus computation has to place a deposit. This deposit of a participant is split among the hurt parties, in the case that the verification game shows that the participant has been cheating, for example, by submitting a false solution or challenging a correct solution.

Example 3: Matrix multiplication of $n \times n$ matrices over a fixed finite field $\mathbb{F}$ has a $(2, n, \log n)$-verification game.

The prover has solution C for the product of A and B.

In the first round, the challenger outputs coordinates i, j, such that

$$c_{i,j} \neq \sum_k a_{i,k} \cdot b_{k,j}$$

with corresponding evidence consisting of a sequence $d_0, d_1, \ldots, d_n$ of elements in $\mathbb{F}$ with

$$d_m = \sum_{k=1}^{m} a_{i,k} \cdot b_{k,j}.$$

The network verifies that i, j are coordinates and that $d_n \neq c_{i,j}$ and that $d_0 = 0$. The challenger loses if this verification does not go through.

In the second round, the prover tries to defend himself by providing a k, such that $d_k \neq d_{k-1} + a_{i,k} \cdot b_{k,j}$ in $\mathbb{F}$. If the network can verify this claim then the prover wins, otherwise the challenger wins.

Remark 4: One can combine this idea together with the idea to verify remainders in order to make a $(O(1), n \log n, \log n)$-verification game, which permits to check whether the product of two $n \times n$ matrices of n-digit numbers gives a claimed $n \times n$ matrix of $2n$-digit numbers.

In the example above, as in the subsequent examples below, we do not specify the means by which the challenger discovers the witness for his challenge. The challenger might be one of the other provers who has lower priority for receiving the reward. In this case, he has already computed the product himself and can therefore check the correctness of the entries in C quickly. The puzzle giver might also try to check the prover's answer himself by computing a random handful of entries from C. If the puzzle giver finds an error, he can proceed with a challenge. Alternatively, one could modify the verification game for matrix multiplication, so as to have the challenges proceed according to Freivalds' probabilistic $O(n^2)$-time algorithm for matrix multiplication verification [5, 15].

Example 5: Consider the problem of finding intersection of two sets A and B, which are a subset of some universal set U. This problem has a $(2, O(\log n), O(\log n))$-verification game.

Assume that sets A, B, and the solution C are represented in array form. Furthermore, it is assumed that equality of two elements from U can be checked fast enough by the network. The proof can be generalized to various other representations. Suppose A and B are the input sets. The prover gives a solution C. Let n_a, n_b, n_c, n_u be the size of the arrays for A, B, C, and U, respectively.

Round 1. The challenger challenging the solution provides a counter-example by giving either

(i) positive indices $i_a \leq n_a$ and $i_b \leq n_b$ (denoting that $A(i_a) = B(i_b)$ but $A(i_a)$ not in C), or

(ii) a positive index $j_c \leq n_c$ (denoting that $C(j_c) \notin A \cap B$).

The miners verify that indeed $i_a \leq n_a, i_b \leq n_b$ and $A(i_a) = B(i_b)$ in case (i) and $j_c \leq n_c$ in case (ii). If the above test fails, then challenger loses the game.

Round 2. To defend himself, the prover gives in case (i) above a positive index $i_c \leq n_c$ intended to show that $A(i_a) = C(i_c)$ and in case (ii) two positive indices $j_a \leq n_a$ and $j_b \leq n_b$ intended to witness that $C(j_c) = A(j_a) = B(j_b)$.

The miners verify in case (i) that

- $i_c \leq n_c$ and
- $A(i_a) = C(i_c)$

or in case (ii) that

- $j_a \leq n_a$,
- $j_b \leq n_b$ and
- $C(j_c) = A(j_a) = B(j_b)$.

If the above test fails then the challenger wins the game else the prover wins the game.

Note the assumption that the comparison of the set elements (such as $A(i_a) = B(i_b)$ etc. in the protocol above) can be done in "small time" by the network. In case the representation of the elements is large (say taking r bits), then we can introduce another round of challenge/verification by prover or challenger to reduce the time needed by the miners.

In case the prover wants to object that $A(i_a) = B(i_b)$, as provided by challenger in Round 1, he can provide a position where $A(i_a)$ and $B(i_b)$ differ. The miners then just verify whether the bits of $A(i_a)$ and $B(i_b)$ at the corresponding positions are equal or not. Similar method can be used in case challenger wants to object that $A(j_a) = C(i_c)$ or $B(j_b) = C(i_c)$. Note that this would take additional $O(\log r)$ bits of space. Similar comparison method can also be used for checking the following inequalities:

$$
\begin{array}{ll}
i_a \leq n_a, & j_a \leq n_a, \\
i_b \leq n_b, & j_b \leq n_b, \\
i_c \leq n_c, & j_c \leq n_c.
\end{array}
$$

Example 6: Consider the problem of sorting n numbers $A(1), A(2), \ldots, A(n)$ each of at most r bits in non-decreasing order. This has a $(2, \log(n + r), \log n)$-verification game.

The prover provides a solution $f(1), \ldots, f(n)$, each being a number of $\log(n)$ bits, where this sequence is supposed to be a sorted permutation of the given items $1, \ldots, n$.

Round 1. The challenger provides one of the following challenges:

(i) a positive number $j \leq n$, such that
- $f(j) > n$ or
- $f(j) < 1$;

(ii) two positive numbers $i, j \leq n$, such that
- $i \neq j$ and
- $f(i) = f(j)$;

(iii) positive numbers $j \leq n$ and $b \leq r$, where the intention is that

$$A[f(j)] > A[f(j+1)]$$

and the least bit where they differ is b-th bit.

In case (i) or case (ii), the miners just check whether the claim holds. If so, then challenger wins the game. In case (iii), the miners check whether the b-th bit in $A[f(j)]$ is 1 and b-th bit in $A[f(j+1)]$ is 0. If not, then prover wins the game. Otherwise, the game proceeds to Round 2.

Round 2. The prover provides a bit b', with $1 \leq b' < b$. The intention is that $b' < b$ is the bit-position, where $A[f(j)]$ and $A[f(j+1)]$ differ.

The miners then just check whether the above claims hold, namely that the b'-th bits of $A[f(j)]$ and $A[f(j+1)]$ are different. If so, then prover wins the game. Otherwise, challenger wins the game.

Example 7: Computing the greatest common divisor of two n-digit numbers has an $(3, n \log n, \log n)$-verification game.

For the solution, the prover provides the greatest common divisor c of input numbers a, b plus signed n-digit integers d, d', e, e', such that the following equations hold:

$$c = d \cdot a + e \cdot b,$$
$$a = d' \cdot c,$$
$$b = e' \cdot c.$$

Alternatively, one could view this also as a $(4, n \log n, \log n)$-verification game, where the prover provides d, d', e, e' in the initial round of verification.

Let $a\%p$ denote the remainder of a by p, which is a number in $\{0, 1, \ldots, p-1\}$, even for negative numbers.

Round 1. The challenger claims that the solution is wrong modulo a $2\log n$ digit number p; if the solution is indeed false, such a number p exists by the Chinese Remainder Theorem and the fact that almost all n-digit numbers are smaller than the product of primes smaller than n^2 (which can be represented by $2\log n$ digits). The challenger provides p, $a\%p$, $b\%p$, $c\%p$, $d\%p$, $d'\%p$, $e\%p$, $e'\%p$, and the network checks that

- these numbers all have at most $2\log n$ digits (plus a suitable constant to cover the finitely many exceptions for small values of n), and
- the equations above hold for these values modulo p.

Round 2. The prover, to defend the solution, will select a $k \in \{a, b, c, d, e, d', e'\}$ and claim that $k\%p$ does not have the value claimed by the challenger. To substantiate the claim, the prover provides two lists of numbers:

- an array of n bits $k_0, \ldots, k_n$, which supposedly is the binary representation of k,
- an array of numbers $\tilde{k}_0, \ldots, \tilde{k}_n$, each at most $2\log n$ bits long, which allegedly gives the binary representation of $k_0 \ldots k_m \mod p$.

The network will verify that $\tilde{k}_n$ differs from what the challenger had claimed to be $k\%p$, otherwise the prover loses the game.

Round 3. The challenger selects a number m, such that either

(i) $m = 0$ and $\tilde{k}_0 \neq k_0$, or
(ii) $m > 0$ and $\tilde{k}_m \neq (2\tilde{k}_{m-1} + k_m)\%p$.

This can be verified by each miner involved in time poly $\log n$. If the complaint is justified, then the challenger wins the game else the prover wins the game.

Remark 8: One could replace the checking of the remainder computation — which involves the usage of $O(n \log n)$ space of the block chain — by an iterated game between the two players, which compare their computations with some type of interval search for a mistake in the computation of the remainder. This has $2\log n + O(1)$ rounds and would result in a $(2\log n + O(1), \log n, \log n)$-verification game for remainder. In this scenario, it is necessary that the solution c is accompanied by the numbers d, d', e, e'.

Proposition 9: *Let ε be any positive rational number, and let k and h be positive constants. If a computational puzzle can be solved in time n^k and space n^h, then it has a $(O(1), n^{h+\varepsilon}, \log n)$-verification game.*

Proof: Without loss of generality, the model of computation is a Turing machine. The basic idea is to check the Turing machine configurations (instantaneous descriptions). Either the prover or challenger's computation sequence must be wrong. Therefore, for finding the cheater, it is enough to find a t, such that the configuration at time t is same for both but the configurations are different at time $t+1$, and then check which transition is valid based on the Turing machine.

When challenging the result of the prover, the challenger wants to convince the network that the Turing machine computation is different than the result. So he publishes i (with $i = n^\varepsilon$) configurations spaced out equally over the computation time n^k and named $y_{0,1}, \ldots, y_{0,i}$. The prover can now challenge either that

(i) the input is wrongly encoded in $y_{0,1}$ (with a pointer to the wrongly coded symbol), or
(ii) the output encoded in $y_{0,i}$ is the same as the prover provided, or
(iii) there is a j_0, such that y_{0,j_0+1} is not obtained from y_{0,j_0} by running the machine $n^{h-\varepsilon}$ steps.

For the first complaint, the network can compare the two symbols at the two positions and decide whether the complaint on y_0 is justified. For a complaint of the third type, the challenger can give a pointer to the position of difference and the network can decide who is right. For a complaint of y_{j_0} not leading to y_{j_0+1}, the process can now be iterated with the challenger providing a new set $y_{1,1}, \ldots, y_{1,i}$ of configurations spaced out by $n^{k-2\varepsilon}$ computation steps and again the prover can either challenge the copying of y_{0,j_0} into $y_{1,1}$ by providing a pointer to the point of difference or challenge the copying of y_{0,j_0+1} into $y_{1,i}$ by providing a pointer or challenging the computation from y_{1,j_1} to y_{1,j_1+1} for some j_1. In general, for each $u < k/\varepsilon$, provided that the copying is correct, the game goes from analysing the configurations $y_{u,1}, \ldots, y_{u,i}$ to $y_{u+1,1}, \ldots, y_{u+1,i}$, where $y_{u+1,1} = y_{u,j_u}$ and $y_{u+1,i} = y_{u,j_u+1}$ for some j_u selected by the prover.

In the last round, the computations are each progressing by at most one time step and the prover has to pinpoint the positions at which the two subsequent configurations did not conform with the rules of updating the Turing machine configurations in one step. The difference between the two configurations only is the binary coding of the head positions, the state and the content of the cell at the old head position; the network can check whether, at the position pointed by prover, the computation conforms with the update rules of the Turing machine. If so, the challenger wins the game; if not, the prover wins the game by successfully refuting the alternative computation of the challenger. Thus the overall number of rounds is approximately k/ε, which is a constant, as h, k, ε are constants. □

Example 10: While finding a prime factorization in polynomial time needs a quantum computer (based on any algorithm known up to now), prime factorization itself has a $(O(1), n^{4+\varepsilon}, \log n)$-verification game, which can be played by polynomial time players on both sides of the game. So let $b_1 \cdots b_m$ be the proposed prime factorization of a given n-digit number a.

Recall that by the algorithm of Agrawal, Kayal and Saxena [1], a number b_k, having up to n bits, is a prime iff there are no numbers r, c, d satisfying one of the following conditions:

(i) $c^d = b_k$ and $1 < d \leq 2n$;
(ii) $b_k = c \cdot d$ and $1 < c < n^4$ and $d \neq 1$;
(iii) r is the least number with
- $b_k, b_k^2, \ldots, b_k^{n^2} \not\equiv 1 \mod r$,
- $c < \sqrt{r} \cdot n$, and
- $(x+c)^{b_k} \not\equiv x^{b_k} + c \mod b_k, x^r - 1$,

where the second modular equivalence is over a polynomial in the formal variable x.

In the game the challenger would either claim that $b_1 \cdots b_m$ is not equal to a or that some b_k is not a prime. There is a bound of $O(n^3)$ on the size of r [10] and thus the space used by the algorithm is $O(n^4)$, which is the space needed to represent the polynomials $(x+c)^u$ modulo $b_k, x^r - 1$, when computing the power $(x+c)^{b_k}$ modulo $b_k, x^r - 1$ fast. The existence of the verification game now follows from Proposition 9.

Definition 11: In an optimization problem where one is trying to find a solution maximized on some measure, the *quality* of a solution refers to the magnitude of that measurement. Thus a high quality solution is closer to optimal than a low quality one.

Example 12: In machine learning, given a sample $(x_1, c_1), \ldots, (x_n, c_n)$ of data items x_k to be classified as c_k by some classifier f (like a perceptron or decision tree), whose parameters are to be learned by a learning algorithm, one can phrase it as an optimization task. Here each solution is given of the form $(\sigma, k_0, \ldots, k_n)$, where σ is the set of parameters of f and k_m says how many of the first m solutions are classified correctly; the value k_n is the quality of the solution. A solution is valid iff

(i) $0 = k_0 \leq \cdots \leq k_n$, and
(ii) $k_{m+1} = k_m + 1$ in the case that the classifier given by σ satisfies $f(x_m) = c_m$ and $k_{m+1} = k_m$ if f satisfies $f(x_m) \neq c_m$.

The network first verifies that $k_0 = 0$ and the remaining verification game runs in constantly many rounds. The first round is a challenge of the form $m \in \{0, \ldots, n-1\}$ indicating that k_{m+1} is wrongly updated from k_m. Subsequent rounds then investigate whether the underlying computation of $f(x_m)$ gives c_m, so that one can check whether the challenge to the update of k_m is justified.

For example, in the case of separation by a hyperplane, σ will be the coefficients of the hyperplane and $f(x_m)$ gives the side (0 or 1) of the hyperplane in which x_m lies in. In case of classification by decision trees, σ is a particular decision tree. Then, $f(x_m)$ gives the value of classification as given by the decision tree when the input is x_m.

Remark 13: Similarly, one can also formulate the factorization game as an optimization task. In this case, any solution of an n-bit number a to be factorized consists of a list $(m, b_1, \ldots, b_k)$, such that

- $m = k$,
- $2 \leq b_1 \leq \cdots \leq b_k$, and
- $a = b_1 \cdots b_k$.

Any solution not obeying these rules is invalid and the larger the m the better the solution. Note that the number m can be written down using $\log(n)$ bits (as $m \leq n$) and one can verify solutions in LOGSPACE. Thus the game has, for every $\varepsilon > 0$, a $(O(1), n^{\varepsilon}, \log n)$-verification game.

3. Frameworks of Consensus-Competition and Consensus-Contract

These verification games can be used to put together a framework to carry out consensus computations, where the bulk of the work is done by the task giver who wants a task to be solved, the provers who hand in possible solutions to be checked and the challengers who are incentivized to check the handed-in solutions in order to eliminate faulty ones.

We present below two verification game variants, which we call consensus-competition (Protocol 14) and consensus-contract (Protocol 15). While the consensus-competition protocol has the advantage that one can verify some puzzles more quickly and that the work of the challenger may be reduced, however we will focus primarily on the consensus-contract protocol for its ease of implementation. In a consensus-competition, challengers race to check a prover's solution using probabilistic or any ad-hoc methods, and the first one to find an error wins a prize. In a consensus-contract, the challengers collectively decide ahead of time, by bidding or otherwise, who should check the given solution first. When puzzles do not lend themselves to guess-and-check error finding, consensus-contracts have the advantage that they reduce the amount of redundant work done by the network.

Protocol 14: The framework of a *consensus-competition* is the following. For covering the network costs, it is assumed that all participants (task-givers, provers and challengers) provide some fees to the network for the costs incurred by the network. These costs are constants and are left out from the following description. The deposits mentioned below do not include these fees.

(a) The task giver T specifies a task to be solved, an algorithm for the verification process and the time permitted for phases (b) and (d) listed below (where the time of (d) might be a formula depending on the number m of solutions), and the run time permitted for each step in the verification game. The task giver also specifies the amount of the prize to be awarded for the solution and also the minimum deposit to be paid by each prover and challenger. Additionally, for optimization problems, the task giver may provide a minimal quality of the solution needed to enter the competition.

(b) The provers provide encrypted pairs (S, q, d), where S is the solution and d the amount of deposit which they offer, and in case of optimization problems, q is the quality of the solution. Note that d and q must be above the minimum as specified in (a). Higher amount of deposit is used to get higher priority as mentioned below. Both numbers d, q are assumed to have only small number of digits so that the network working bound of the protocol is met.

(c) Each prover decrypts his solution and deposit in a verifiable way and the network sorts the pairs (S, q, d) — solution, quality (of optimization problems) and deposit — in an order such that higher quality solutions are before lower quality solutions and, for equal quality solutions, higher deposits go before lower deposits. The network furthermore eliminates solutions not meeting the minimum deposit or quality criterion. Let $S_1, S_2, \ldots, S_m$ be the solutions in the above order.

(d) In the next round, each prover and each member of the network can challenge solutions. The priority among the challenges are again given by encrypted deposits, where one deposit works for all challenges by the same user. A prover is allowed to top up its deposit on the solution in order to get a higher priority for challenges. The network orders the challenges according to the amount of the deposit given by them.

(e) The network maintains list of pairs of solutions and challenges in lexicographic order (first all pairs involving S_1, then those involving S_2 and so on). In each round, one processes the least pair (S_i, C_j) in the above order, where neither S_i nor C_j have been disqualified so far, until the list is exhausted. For each such pair (S_i, C_j), the players S_i and C_j play the verification game and the network decides who wins and who loses the game. The loser loses his deposit, which is split into half/half between the winner and T; furthermore, the loser will now be disqualified from further games. This process terminates when for all pairs (S_i, C_j) in the list, either the game has been decided or one of the parties has been disqualified.

(f) If no solution survives, then T gets his prize refunded. If one or several solutions survive, then the least i such that S_i survives receives the prize. Furthermore, the deposits of all surviving provers and challengers get refunded to the party who paid the deposit.

Protocol 15: A *consensus-contract* is a contract between a task-giver T and a client S where T gets the solution of S checked in a way similar to a consensus-competition. S has to pay a deposit for his solution to be checked. If within the agreed time-frame no challengers successfully challenge the solution, then the deposit is refunded to S together with the prize for solving the problem; the deposits of the challengers are split between task-giver and prover in order to compensate them for the delays and work caused by the checking. If a challenger successfully challenges the solution S, then the deposit of S is split between task-giver and challenger and the prize is recycled for the next contractor willing to take up the task. Here each challenge is handled in the same way as before by a verification game and challengers are put in an order based on the deposits that they provide.

It can be easily verified that among the solutions, the prover S_i providing a correct solution wins (with i being minimal, in case of several correct solutions), provided that T gives enough time for provers and challengers to compute the winning strategies for the corresponding verification games: Each S_j with $j < i$ will be defeated by a sufficiently well-equipped challenger and S_i will defend its solutions successfully to all challenges. So the task should be solved successfully provided that T gives enough time to compute solutions, make the moves in the verification games and provides enough incentive for an honest and sufficiently capable player to participate in the competition for the solution. This is essentially the content of the following theorem. A similar result can also be shown for optimization tasks.

Theorem 16: *Assume that the following conditions hold:*

- *A is a task which has exactly one output for every input;*
- *The task giver provides a prize money which is at least two times the cost of computing the solution and the network fees paid by the provers, for every sufficiently powerful prover of the network;*
- *The deposit required is at least four times the maximum of following costs:*
  - *the network fees to be downpaid when submitting a solution or challenge, and*
  - *the expected local computation costs for a sufficiently powerful prover or challenger to do the computations to carry out all rounds of the verification game;*
- *The task giver provides enough time for phase (b) so that a member of the network with sufficient computation power can solve the problem within the specified time;*
- *The task giver provides enough time for phase (d) so that a member of the network with sufficient computation power can compare each solution with its own solution and, in the case that it differs, submit a challenge;*
- *The task giver provides enough time for each step in the verification game so that the provers and challengers with sufficient computation power can run their winning strategy for the case that their solution or challenge, respectively, is correct.*

Then, in the case that one prover with sufficient computation power participates in the game, this prover is incentivized to submit a correct solution and to win the prize money and one such prover will eventually win the prize money. Furthermore, the deposits motivate challengers to detect false solutions and provers to defend correct solutions and thus deter provers and challengers from putting up false solutions or challenges.

Proof: Note that if the provers and challengers have enough time for the verification game, then they can challenge any wrong solution. The costs involved for them is the network cost paid and the cost of computation. If they win the challenge, then they get twice the above costs as payback (as the remaining deposit of the loser goes to the task giver). Thus, there is enough incentive for the challengers to challenge wrong solutions.

If the prover S has enough time to compute a solution, then its costs are just the network fees and the cost of computation. If the prize money is double the above costs, then there is enough incentive for S to enter the contest. As mentioned above, S also has enough time/incentive to challenge a higher priority wrong solution. Thus, either S wins, or only correct solutions with priority higher than S survive. In either case, a correct prover wins the prize money. □

Note that if the provers/challengers solve the problem not just based on reward/deposit gained but also based on expected chance that they win the reward in case of presence of large number of provers, then the above incentive structure need to be appropriately updated. For example, if the number of expected correct solutions posted is 5, then for a prover expected return is 1/5-th of the reward. Thus, the reward needs to be at least 5 times the costs. However, as it is not possible to estimate the number of provers who will attempt the problem, we have used twice the costs as enough incentive.

Remark 17: A consensus-contract has the advantage that contract-taker knows that the prize will be awarded to him, provided that a correct solution is given (and thus the verification goes through). The task-giver has the advantage that the prize money does not need to be too high: the provers will only try to solve if their expected award is more than the work done — for example in the consensus competition case, if the expected number of correct submitted solutions is three, then each prover knows that he gets the prize only with probability 1/3 and therefore he would only work if either the prize money is at least three times the expected costs or if he has the means to bypass the other correct solutions through a higher deposit. For tasks where many users spend a small amount of effort to solve the problem and where solving the problem is a game of chance — like guessing next week's lottery numbers or, in a computational setting, finding the solutions of some satisfiability problem — the consensus-competition might be the more adequate way to go. On the other hand, where the task is deterministic and effort needed is well-defined, just large, consensus-contract is the more adequate way to go.

Remark 18: For the choice of the parameters in a consensus-competition, the following ideas might be helpful as a guideline.

- The encrypted phase (b) is there to avoid the case that some users read the solutions of the other users and cooperate with miners in order to get their solution into the blockchain with priority (timestamp) ahead of the others. Furthermore, paying a deposit is like having some confidence that the solution will survive a verification game; hence solutions with higher deposits are given priority over solutions with lower deposit.
- Deposits might not deincentivise malicious players on the network if they have enough money. They might, however, compensate the persons involved for the work they did (such as provers playing a verification game to justify the solution, challengers playing a game to eliminate fraudulent provers and the task-giver suffering a delay from ongoing fraudulent behavior of provers and challengers).
- The task giver has to select the amount of the prize and the minimum deposits for provers and challengers and the time given for the competition carefully. Too low minimum deposits attract cheaters who try to get the prize unjustifiably; too high minimum deposits might prevent honest provers or challengers from participating due to the lack of funds. Too high prize money or too long solving time might attract too many solutions, which then makes the whole process lengthy and results in many verification games to be played. Too short solving time might prevent any honest prover from being able to submit a solution.
- Similarly the task giver has to decide about the time-frame in which players have to move in the verification game or other parts of the process. If players stop to move, they must get disqualified; however, in order to avoid honest players getting disqualified due to the time-complexity of computing the right move, the time-out must be sufficient generous. On the other hand, a too high time-out might slow down the whole computation and decision process.
- Malicious players might, under various names, put up several solutions. A player providing a correct solution is allowed to use his

single deposit to challenge all other solutions in order to prevent that he goes bankrupt when challenging a multitude of malicious incorrect solutions of higher priority; similarly a challenger who does not provide a solution does also pay only one deposit, though he might challenge all solutions.

- In order to avoid that the system is abused by a player making multiple false challenges, all challenges and the solution of a player are thrown out when he loses a verification game — honest players are supposed to win them under all circumstances. Thus each run of the verification game is paid by the deposit of the loser (a prover or a challenger).
- Players might circumvent the rule that false solutions or challenges throw them out of the competition by using different identities; however, then they have to also pay multiple deposits, and thus this does not pay off financially.
- The verification game will make the best solution win phase (e) of the protocol for the consensus computation provided that there is a good solution and that its prover challenges all false solutions which pretend to be better.
- Phase (d) allows additional challengers to enter the game. For example, a large factorization problem might only be solvable using quantum computers, while primality testing algorithms [1] permit sufficiently powerful traditional supercomputers to check for the primality of the factors of these large numbers; hence they are allowed to enter the game. As indicated, the verification games then use only $O(n^{4+\varepsilon})$ space on the block chain entries and this makes it possible for T to get the factorization task verified through the network rather than by using its own limited resources. Furthermore, phase (d) also allows T to challenge incorrect solutions and avoid the situation where it is forced to award a prize unjustifiably.

4. A verification protocol with incentives

We assume that individuals on the network are rational, in the sense of Nash equilibrium, and wish to exchange computational resources for financial rewards. We want to incentivize correct computations. At a high level, our protocol is as follows.

Protocol 19: A solution to a puzzle is presented, followed by challenges.

Step 1. A puzzle giver G presents a puzzle.
Step 2. A prover P proposes a solution S to G's puzzle.
Step 3. A small set of randomly selected challengers $\mathsf{C}_1, \mathsf{C}_2, \ldots$ checks whether S is correct.

(a) If there are no objections from the challengers, then
- the network *accepts* S,
- P pays G for his solution, and
- each of the challengers receives a nominal reward for participating.

(b) If some C_i reports a mistake in S, then:
- C_i checks the solution S by playing the verification game with P, and
- the loser of this verification game pays a large penalty to the winner.

(c) Next:
- In case this game continues until the last possible round, the community V determines the winner by verifying the correctness of the final response in the game.
- In case V determines that C_i made a false alarm, after C_i pays a penalty we repeat Step 3.
- Otherwise, C_i receives her reward for correctly raising an alarm, and the network *rejects* S.

We now give more details.

4.1. *Step 1: Presenting the puzzle*

The puzzle giver posts his puzzle in a public place. The puzzle may be a known problem, such as checking whether all the transactions in a Bitcoin block are valid, or a puzzle that an individual independently wants to solve. External to the system we describe, G must commit both his puzzle and its prize to this system so that in the end the system can either reward or penalize the prover for his solution. For instance, the problem and its reward could be committed to some public blockchain. G's puzzle is converted into a verification game either manually or via Proposition 9.

4.2. *Step 2: Committing a solution*

In this step, the network selects a candidate solution S for the puzzle. The prover P commits a significant amount of capital to an escrow which, according to some contract, he gets back later iff his solution S turns out to be correct. The prover wants to be sure that no one else can steal his answer and claim a reward for his solution, and the community wants to be sure that the prover can't change his solution once it has been committed.

The paper [12] describes a commitment scheme in which a prover's solutions might be securely broadcast and selected using the Ethereum blockchain. First, everyone races to solve the puzzle, and then an Ethereum blockchain lottery elects a prover from among the first solutions to appear. The designated prover then hashes his solution onto the blockchain, thereby committing his answer, and once the blockchain decides that the prover P has been elected, he then commits his solution in plaintext to the blockchain along with a smart contract carrying his deposit.

4.3. *Selecting a subcommittee*

Next we select a random subcommittee of challengers to check P's solution S. The number of challengers selected should be large enough that at least one among them is honest enough to check the solution S in exchange for the potentially large reward for finding a mistake, and the subcommittee size is not too large so as to be prohibitively expensive. The expense here derives from the fact that each member of the subcommittee must be compensated financially for his computational effort in checking, so larger subcommittees are more expensive. The rewards for challengers should be at least enough to compensate for checking effort in case the solution turns out to be correct, and *significantly* more in case the challenger finds an error. The "significantly" part is needed in order to incentivize the challenger to actually do the checking, particularly in the case where the challenger believes that his chances of finding an error are small.

We select the subcommittee of challengers via a lottery. When the solution is announced, everyone who is interested in participating in the subcommittee does a proof-of-work. Ideally the proof-of-work puzzle should be *sequential* rather than *parallelizable*, as defined in [8], so that each party, regardless of computational power, takes the same amount of time to complete the proof-of-work task. Then everyone with a minimum amount of computational power has a chance to win the race and therefore gets a lottery ticket. Sequential proof-of-work differs from the parallelizable work used in Bitcoin and other cryptocurrencies. An ideal candidate for sequential proof-of-work has yet to appear [8], nevertheless, but we need only imagine that everyone who can demonstrate a minimum amount of computational power should have a chance to be selected. Finally, we randomly sample a subset of those who successfully completed the proof-of-work task. If the proof-of-work task had several solutions, we could, for example, elect those members to subcommittee whose hash of their solution ends in some prescribed sequence of digits.

In addition to submitting a proof-of-work, each candidate for the subcommittee must deposit some money into an escrow account. This money will be used to impose a penalty in case the candidate is selected for the subcommittee and subsequently raises a false alarm (or an incorrect check). We omit precise details of the subcommittee construction as we anticipate embedding this game inside of another forthcoming system, which can securely appoint the subcommittee [2, 6].

4.4. *Playing the verification game*

Each challenger privately checks the solution S. The checking round ends as soon as one of the following things happens:

- some challenger C_i announces a mistake in S, or
- the predetermined time limit for challenging has been reached.

Due to the non-laziness assumption, at least one challenger will respond within the given time limit. In the latter case, each C_j gets paid for verification in proportion to the *gas limit*, or predetermined CPU cycles, for the problem. If, on the other hand, some challenger C_i announces a mistake in C_i, then C_i and P play a verification game. In this case, C_i will either receive a substantial reward for detecting a mistake or face a substantial penalty for raising a false alarm.

The community vote V only gets invoked if the final round of the verification game occurs. Hence, if the challengers agree that P's solution is correct or if C_i and P can settle their verification game amicably, then the community need not do any work. In this way, we minimize the computational burden on the community which, as in Bitcoin, we expect does a small amount of verification work for free.

4.5. *Correctness and efficiency*

We assume that challengers in the network are rational, and that there are sufficiently many of them that are non-lazy so that at least one non-lazy challenger belongs to each randomly selected subcommittee, thereby satisfying Assumption 1(i)a and Assumption 1(ii). In Sec. 3, we discussed two flavors of verification games, based on either consensus-competition or consensus-contract protocols. Consensus-competitions cost less because the challengers only sample the output and do not necessarily check the whole thing. Consensus-contracts, on the other hand, cost more because challengers are expected to replicate the prover's entire work. For simplicity, we will discuss only consensus-contracts here. In general, challengers have incentive to be both non-lazy and verify honestly in this protocol because they gain large prizes for discovering errors, and they are discouraged from reporting errors falsely by monetary penalties.

For example, suppose that the challenger is supposed to check whether a matrix multiplication is correct, and suppose that a consensus-competition compensates challengers for the expected work required to find an error assuming 10% of the entries are wrong, and that the prize for finding an error during this check is 50 times more than the compensation for checking. If the challenger believes that the actual chances of finding a mistake in the prover's solution are only 1/5000, then the rational challenger views this 50× prize as inadequate and therefore lacks incentive to do the check. This discrepancy between the challenger's belief regarding the probability of finding an error and his actual probability of finding an error exists whether we employ a consensus-contract or a consensus-competition.

The difference between these methods is that in a consensus-competition, the challenger might honestly do his verification, find no mistakes in his samples, and yet not receive a reward because the overall submitted solution is actually incorrect. This outcome unfairly penalizes the challenger, and so a rational challenger might not accept such a contract in the first place. Whether or not a challenger will accept such a contract for matrix multiplication depends on what the challenger believes is the true probability of finding an error. Since the challenger does not *a priori* know how likely an anonymous prover is to make an error, no rational argument can say whether or not the challenger should agree to the consensus-competition. In contrast, a rational challenger should accept an adequately compensated consensus-contract, and if the reward for finding a bug suffices to convince the challenger to be non-lazy, then the challenger will do the intended work and give the correct answer. We incentivize all of the challengers on the subcommittee to respond by only paying compensation once sufficiently many challengers have responded.

It is not unreasonable to expect that the community will do a negligible amount of work for free. We see this commonly today in various distributed systems. The challengers on the network may function also as verifiers.

5. Practical considerations

5.1. *Outsourced computation*

Mobile, scientific, and other various big data applications often require local machines to outsource computations to an external cloud computer. How can one be sure that the computation done by a cloud is correct given only limited local resources? Computations can contain errors due to hardware failures, software errors, or even laziness to expend resources on the part of the cloud. The past five years have witnessed a rising interest in secure, outsourced computation protocols, in which the cloud service provides, in addition to the computation itself, some certificate of the computation's correctness, which is easy for the client to verify [18]. Ideally, time required to verify the computation should be small compared to the time required to simply execute the computation directly, while the cloud's extra overhead in producing the certificate should be minimal. The techniques applied thus far involve PCP-type encodings and SNARK-like cryptographic protocols.

One can use the protocol in Sec. 4 to outsource computation. The puzzle giver G puts forth the problem and then some prover P puts forth a solution. The economics of the verification system ensure that G receives a correct answer. In Walfish and Blumberg's survey article "Verifying Computations without Reexecuting Them" [18], the authors argue against replicating computations as a means to achieving secure computation. Potential solutions to this problem, they say, suffer from either systemic errors which appear in repetitions across cloud platforms, wrongly assume that auditors can detect rare errors, or rely on assuming that a particular piece of hardware will execute correctly. Since anyone can participate in the verification network, we make no assumptions about how a particular cloud sets up its hardware, and we employ powerful economic forces to not only detect, but reduce, errors.

Our present system offers some advantages over the PCP/SNARK outsourced verification systems to date. Firstly, it doesn't use PCPs or SNARKS. Our protocol is simple in the sense that we do not require any complicated encoding schemes or cryptographic protocols. Secondly, our protocol can be used to correctly compute any polynomial-time functions, whereas the PCP-style systems proposed thus far generally operate effectively on smaller classes of functions [18]. Finally, as we argue below, our system has low computational overheads to the point of being practical.

Our system has negligible setup cost. In contrast, many of the proof-based systems to-date, including Zaatar, Pinocchio, Ginger, and TinyRAM, require one to run thousands of instances of a single function before breaking even on the overall verification time on a 128×128 matrix multiplication puzzle [18]. Some systems, only run on restricted circuits or try to amortize setup costs. CMT, Allspice, and Thaler effectively slash verification time with negligible overhead costs. The prover overhead for Thaler in particular, under matrix multiplication, has the same order of magnitude worker overhead as the protocol we have presented here.

In many practical cases, a user might wish to reuse input, rather than reuse a function, in order to execute several analyses on the same data. In such cases, proof-based systems with overhead generally become impractical on the verification side. Our incentive-based system, however handles data-analysis cases well as there is no overhead in switching between functions.

In general, the prover side of these proof-based systems is the bottleneck which prevents them from being practical. With the exception of Thaler, all of the systems benchmarked in [18] require a factor of 10^5 worker overhead for the prover when performing a 128×128 matrix multiplication. Thus the cloud processing power being used should be significantly more than $100,000$ times the verifier could do on his own hardware in order for outsourcing to make any sense. Matrix multiplication, in our protocol, can be done with a 3-round verification game plus the verification time from the community (see Example 3). So if we assume that the communication time for posting the problem and electing a prover to give a solution takes no more time than solving the problem itself, we can conclude that the prover overhead for matrix multiplication, in terms of both computational time and effort, is generously at most a factor of 5. This is much cheaper than the above systems in terms of computation and hence in terms of dollars as well. In general, an arbitrary polynomial-time function can be done with $\log n$ overhead (Proposition 9).

5.2. *Outsourced verification*

Cryptocurrencies, like Bitcoin, maintain a public ledger, called a blockchain, which includes all transactions that take place over the network. Miners on the network are responsible for maintaining the integrity of these transactions through two steps:

(1) they select which transactions to include in the blockchain, and
(2) they verify that blockchain transactions are valid.

As the Verifier's Dilemma in [12] illustrates, miners do not always verify transactions correctly. We can use our protocol in Sec. 4 as a means to allow miners to offload their verification burden, thereby relieving them of the second task. Here, each miner plays the role of the Puzzle Giver in Protocol 19, who provides a block of transactions to be verified, along with a financial contract which guarantees funding for the network to perform the verification. The source of these funds may come either from the miners themselves, or in the case of resource-heavy transactions, from the party who posts the transactions. The system now places a $\surd$ or a $\times$ next to each transaction (or block of transactions), depending on whether it is correct or incorrect.

Since miners can rely on the $\surd$ and $\times$ marks for correctness, they do not need to verify these transactions themselves. Therefore this verification protocol reduces the amount of redundant work done on the cryptocurrency. Only a few challengers in Protocol 19 check each block, hence the total amount of work spent in verification across both the cryptocurrency and verification network is less than what would be spent if each miner on the cryptocurrency verified each transaction for himself. In this way, a verification network can offer positive ecological impact.

6. Conclusion

By restricting Ethereum to transactions which require little verification time, we achieve a consensus computer, which achieves correct outsourced computation results for feasible functions, when the network's computational entities are rational. The verification game described herein permits us to verify, through financial incentives, interaction, and Nakamoto consensus, any polynomial-time puzzle with only poly log effort. We show how to embed this game into real systems and argue that the computational overhead of the implementation is less than the overhead for state-of-the-art for probabilistically-checkable-proof systems.

Christian Reitwießner independently developed a similar concept as the authors of this chapter. Vitalik Buterin brought Teutsch and Reitwießner together and they worked out the Truebit whitepaper [14]. The Truebit protocol adds an incentive layer, which replaces the assumption of an honest verifier with a rational one. Teutsch went on to found an organization [19] which built an implementation [20] based on this later work. This system now allows to outsource computation, including a game-based verification of the results found as described above. The Truebit system can, currently, support tasks written in the languages used by the system WebAssembly which are C, C++, and Rust. In summary, the ideas outlined in this chapter are one of the very few theoretical ideas which found an application in some real start-up.

References

1. Manindra Agrawal, Neeraj Kayal and Nitin Saxena. PRIMES is in P. *Annals of Mathematics*, 160(2):781–793, 2004.
2. Vitalik Buterin, Jeff Coleman and Matthew Wampler-Doty. Notes on scalable blockchain protocols. (Version 0.3).
3. Ethereum Foundation, *Ethereum's White Paper*. `https://ethereum.org/en/whitepaper/`, 2014.
4. Michael J. Fischer. The consensus problem in unreliable distributed systems (a brief survey). *Foundations of Computation Theory*, ser. Lecture Notes in Computer Science, M. Karpinski, Ed. Springer Berlin Heidelberg, 1983, vol. 158, pp. 127–140.
5. Rūsiņš Freivalds. Fast probabilistic algorithms. *Mathematical Foundations of Computer Science*, MFCS 1979. *Springer LNCS* 74:57–69, 1979.
6. Rachid Guerraoui, Floriang Huc Anne-Marie Kermarrec. Highly dynamic distributed computing with Byzantine Failures. *Proceedings of the 2013 ACM Symposium on Principles of Distributed Computing*, PODC 2013, ACM, pages 176–183, 2013.
7. Arthur M. Jaffe. The Millenium Grand Challenge in mathematics. *Notices of the American Mathematical Society*, 53(6):652–660, 2006.
8. Jonathan Katz, Andrew Miller and Elaine Shi. *Pseudonymous broadmast and secure computation from cryptographic puzzles.* Technical Report on `http://eprint.iacr.org/2014/857.pdf`, 2014.
9. Leslie Lampert, Robert Shostak and Pease Marshal. The Byzantine Generals Problem. *ACM Transactions on Programming Languages and Systems*, 4(3):382–401, 1982.
10. Hendrik W. Lenstra and Carl Pomerance. Primality testing with Gaussian periods. *Journal of the European Mathematical Society*, Available Online at DOI: 10.4171/JEMS/861, 2019.
11. Carsten Lund, Lance Fortnow, Howard Karloff and Noam Nisan. Algebraic methods for interactive proof systems. *Journal of the ACM*, vol. 39, no. 4, pp. 859–868, Oct. 1992.
12. Loi Luu, Jason Teutsch, Raghav Kulkarni and Prateek Saxena. Demystifying incentives in the consensus computer: How to repurpose cryptocurrencies for outsourced computation. *ACM Conference on Computer and Communications Security*, CCS 2015, pp. 206–710, 2015.
13. Satoshi Nakamoto. *Bitcoin: A Peer-to-Peer Electronic Cash System.* `http://bitcoin.org/bitcoin.pdf`, 2009.
14. Jason Teutsch and Christian Reitwießner. *A scalable verification solution for blockchains.* Truebit White Paper, 50 pages, http://www.truebit.io, 2017.
15. Justin Thaler. Time-optimal interactive proofs for circuit evaluation. *Advances in Cryptology*, CRYPTO 2013. *Springer LNCS* 8043:71–89, 2013.

16. Florian Tschorsch and Björn Scheuermann. *Bitcoin and beyond: A technical survey on decentralized digital currencies.* http://eprint.iacr.org/2015/464.pdf.
17. Riad S. Wahby, Srinath Setty, Zuocheng Ren, Andrew J. Blumberg and Michael Walfish. Efficient RAM and control flow in verifiable outsourced computation. *Twentysecond Annual Network and Distributed System Security Symposium*, NDSS 2015, San Diego, California, USA, 8–11 February 2015. The Internet Society, 2015.
18. Michael Walfish and Andrew J. Blumberg. Verifying computations without reexecuting them. *Communications of the ACM*, 58(2):74–84, 2015.
19. Truebit: The road to Veritas. https://medium.com/truebit/the-road-to-veritas-3399bc72a8c8, Medium, 2018.
20. Jason Teutsch and Christian Reitwießner. A Scalable Verification Solution for Blockchains. Institute for Mathematical Sciences, National University of Singapore, Lecture Notes Series, Vol. 42, Ch. 15, N. Greenberg, S. Jain, K. M. Ng, S. Schewe, F. Stephan, G. Wu and Y. Yang (eds.), https://doi.org/10.1142/9789811278631_0015.

https://doi.org/10.1142/9789811278631_0012

Punctual Degrees and Lattice Embeddings

Iskander Kalimullin

Kazan Federal University, ikalimul@gmail.com

Alexander Melnikov

Victoria University of Wellington, Wellington, New Zealand
sasha.melnikov@vuw.ac.nz

Maxim Zubkov

Kazan Federal University, maxim.zubkov@kpfu.ru

We systematically investigate into the online content of finitely generated (f.g.) structures. We show that a countable lattice L is embeddable into the punctual degrees of a rigid f.g. structure if, and only if, L is distributive. The right-to-left implication is achieved by embedding the atomless Boolean algebra below any non-zero punctual degree of a given f.g. rigid structure. In contrast, we show that a lower cone is never a Boolean algebra, for any (not necessarily rigid) f.g. structure. We also prove that the punctual degrees of the standard arithmetic are neither a lower semi-lattice nor an upper semi-lattice.

Keywords: finitely generated structure; isomorphism; primitive recursive function.

Contents

1. Introduction

It is well-known that decision procedures in finitely generated algebraic structures are intrinsic, in the following sense. For example, if G is a finitely presented group and it has an algorithmically decidable word or conjugacy problem, then every finite presentation $H \cong G$ will also have a decidable word problem. Indeed, fix some generators $\bar{g}$ of G and their isomorphic counterparts $\bar{h}$ in H. Then every element of G is a word in the alphabet of $\bar{g}$, and it can be naturally mapped to the respective word in the alphabet of $\bar{h}$. This observation is heavily exploited in combinatorial group theory [12, 19], often without explicit reference. In the terminology of computable structure theory [1, 8], finitely generated structures are *relatively computably categorical.* Notice, however, that the observation above uses a rather general notion of a Turing computable process. Imagine that you are given an algorithmic description $C = \langle c \rangle$ of the infinite cyclic group. Your task is to associate every $x \in C$ with an integer in the standard representation of $(\mathbb{Z}, +)$. The naive algorithm waits for a late enough stage at which $mc = x$. However, how long will you have to wait? It is easy to diagonalise against all polynomial time, all exponential, or even all hyper-exponential (etc.) algorithms. In other words, this procedure uses an instance of a *truly unbounded search*; this is the same as to say that the algorithm is not *primitive recursive* [25].

It is natural to ask what happens if we forbid unbounded search. Kalimullin, Melnikov, and Ng [16] proposed that an "online" algebraic structure must minimally satisfy the following general definition.

Definition 1.1: A countable structure is *punctually computable* or simply *punctual* if its domain is $\mathbb{N}$ and the operations and predicates of the structure are (uniformly) primitive recursive.[a]

Of course, one naturally seeks to understand the truly efficient algorithms, in the spirit of, e.g., [5,7,17]. Kalimullin, Melnikov and Ng [16] observed that producing a punctual presentation is often sufficient to obtain a polynomial-time one; e.g., [11]. Therefore it makes sense to develop a general theory of punctual algebraic and combinatorial structures; see survey [4] for the foundations of this new emerging theory. See also Alaev [2,3] for an alternative approach.

Recall that the inverse of a primitive recursive function does not have to be primitive recursive. Fix a punctual structure A. The collection of all punctual presentations of A carries a natural *reduction*, as defined below.

Definition 1.2: Let A be a punctual structure. Then, for punctual C, B isomorphic to A,

$C \leq_{pr} B$ if there exists a surjective primitive isomorphism $f : C \rightarrow_{onto} B$.

This leads to an equivalence relation $\cong_{pr}$ and *the punctual degrees* on the equivalence classes, which will be denoted $PR(A)$. What does $PR(A)$ reflect? If $C \leq_{pr} B$ then, in a way, B has more online content than C does. For example, the standard copy of $(\mathbb{Q}, <)$ punctually embeds any other punctual copy of the rationals. The standard copy has a primitive recursive Skolem function, while a "non-standard" punctual copy of $(\mathbb{Q}, <)$ will have intervals in which points appear arbitrarily late. The punctual degrees serves as a punctual invariant of a structure. A reader familiar with the terminology of computable structures should compare punctual degrees with degree spectra and categoricity spectra of structures. Note that non-trivial degree or categoricity spectra are typically realised by unnatural structures, which have to be specifically constructed; see, e.g., [9, 10, 15, 21]. In contrast, the punctual degrees of natural and common algebraic structures, such as the dense linear order or the random graph seem to possess remarkably non-trivial properties, which are yet to be understood. Therefore, it makes sense to study the punctual degrees of specific and natural algebraic structures and compare them with the punctual degrees

[a]To include finite structures, we also allow initial segments of $\mathbb{N}$ to serve as domains. We will never consider infinite languages in this chapter. Also, Kalimullin, Melnikov and Ng called punctually computable structures "fully primitive recursive". This term is a bit too technical and does not reflect the intuition behind the notion.

of other structures. For example, the punctual degrees of the dense linear order, the random graph, and the universal countable abelian p-group are pairwise non-isomorphic; see [22]. It is not known whether the punctual degrees of $(Q, <)$ and the atomless Boolean algebra are isomorphic [4].

We go back to the finitely generated case. Suppose A is a punctual f.g. structure. What can be said about $PR(A)$? Obviously, $PR(A)$ will have the least element $\mathbf{0}$, which is the "natural" term-representation of A. It is known that the punctual degrees of a f.g. structure can be a singleton [16]. Bazhenov, Kalimullin, Melnikov and Ng [6] have showed that the punctual degrees of any f.g. structure are dense. They also showed that there exists a f.g. rigid A, such that $PR(A)$ is infinite, yet $PR(A)$ has a greatest element. However, apart from the above-mentioned density, no result in [6] gives any information about the algebraic structure of the punctual degrees of common f.g. structures. For instance, how do the PR-degrees of the arithmetic $(\omega, +, \times, <, 0, 1)$ look like?

Our initial goal was to attack the following problem:

Describe the punctual degrees of $\mathbb{N} = (\omega, +, \times, <, 0, 1)$.

Our first result below seems to suggest that the punctual degrees of $\mathbb{N}$ are too ill-behaved to be of any interest.

Theorem 1.3: *Let* $\mathbb{N} = (\omega, +, \times, <, 0, 1)$ *be the ordered semiring of the standard arithmetic. Then* $PR(\mathbb{N})$ *is neither an upper semi-lattice nor a lower semi-lattice.*

In fact, the theorem above follows from the stronger Theorem 3.1 saying that there is a pair of punctual copies of $\mathbb{N}$ having no infimum and no supremum.

Nonetheless, we decided to proceed and look at the punctual degrees of $\mathbb{N}$ from the perspective of lattice embeddings. More specifically, we asked for an algebraic description of finite lattices, which are embeddable into $PR(\mathbb{N})$ preserving sup and inf. This led to the unexpected main result of this chapter.

Theorem 1.4: *Let L be a countable lattice and A a finitely generated rigid punctual structure. Then the following are equivalent:*

(1) L is embeddable into $PR(A)$;
(2) L is distributive.

We note that A could be *any* f.g. rigid structure; this clearly covers the case when A is $\mathbb{N}$. In fact, we could weaken the assumption on rigidity by, say, allowing all automorphisms in any copy to be primitive recursive; see also Remark 4.1 for a further discussion. For example, finitely generated *complete groups* have this property. (Recall that a group is complete if each automorphism of the group is inner, i.e., is a conjugation by some fixed element of the group. See [23] for many examples of infinite f.g. complete groups.[b]) We do not know if the theorem above holds without any assumption on the automorphism group.

The theorem is proved in two steps. The first, and the simpler of the two steps, is to show that each lower cone $\breve{A} = \{X : X \leq_{pr} A\}$ has to be a distributive lattice if the structure is rigid and f.g.; this is Theorem 2.1. The second step requires more care. More specifically, in Theorem 4.1 we will use a subtle Lemma 4.2 to embed the atomless Boolean algebra $\beta = Int(\eta)$ below any degree $A >_{pr} \mathbf{0}$. These two results combined together clearly imply Theorem 1.4.

We know that each $\breve{\mathbf{a}} = \{x : x \leq_{pr} a\}$ is distributive and embeds the atomless Boolean algebra β. It is natural to ask whether $\breve{\mathbf{a}} = \{x : x \leq_{pr} a\}$ can be *isomorphic* to β or at least to *some* Boolean algebra. To finish off this chapter, we give a definite negative answer to this question.

Theorem 1.5: *Suppose G is finitely generated, and assume $A \in PR(G)$. Then $\breve{A} = \{X : X \leq_{pr} A\}/\cong_{pr}$ is not a Boolean algebra.*

Note that there is no rigidity assumption on G in the theorem above. The proof is contained in the last section of this chapter. It combines the above-mentioned Lemma 4.2 with a diagonalisation attempt.

We briefly discuss the techniques of our proofs. Although the injury is only finite, various effects specific to primitive recursive analysis make some of our arguments quite subtle. For example, we invite the reader to try to eliminate Lemma 4.2 from the proof of the main result Theorem 1.4.

Although we have learnt much about the punctual degrees of f.g. structures, there are many questions that remain open. For example, the only case when we have a full understanding of the punctual degrees of a f.g. structure is when it has exactly one degree. We do not know if the rigidity assumption in Theorem 1.4 can be dropped, but we suspect that there exists a f.g. structure, whose punctual degrees have a non-distributive lower cone. Also, we still do not know enough about punctual degrees of

[b]Many thanks to Prof. Marston Conder for this reference.

common algebraic structures. For example, are the punctual degrees of all f.g. infinite abelian groups isomorphic? What about the n-generated free groups? It seems that new techniques and insights will be required to answer questions of this sort.

2. Distributivity of lower cones

Theorem 2.1: *For every $A \in PR(G)$ for a f.g. rigid G, the lower cone below A is a distributive lattice under $\leq_{pr}$.*

Proof: Then proof will be split into several lemmas.

Lemma 2.2: *For every $X, Y \leq_{pr} A$, $\sup(X, Y)$ exists.*

Proof: Consider the union Z of the images of X and Y within A. Note that Z itself does not have to be punctual, in the sense that a natural number can be kept outside of the domain of Z for quite long. However, Z contains a natural image of the least copy $\mathbf{0}$ of the structure; this natural image has a rapid enumeration. Using $\mathbf{0}$, it is easy to pass to a punctual copy of Z. This copy has a 1-1 punctual correspondence with Z, and it has domain ω. We identify Z with this copy. It remains to observe that, since G is rigid, any upper bound of X and Y punctually embeds Z as well. □

Lemma 2.3: *For every $X, Y \leq_{pr} A$, $\inf(X, Y)$ exists.*

Proof: Consider the intersection of the images of X and Y within A. Pass to a punctual copy of it (using a copy of $\mathbf{0}$). Using that G is rigid, observe that every lower bound of X and Y punctually embeds the natural punctual copy of the intersection. □

It remains to check distributivity.

Lemma 2.4: *Suppose $Z \leq \sup(X, Y)$. Then there exist $X_1 \leq_{pr} X$ and $Y_1 \leq Y$, such that $Z = \sup(X_1, Y_1)$.*

Proof: Consider $X_1 = im(Z) \cap im(X)$ and $Y_1 = im(Z) \cap im(Y)$, where $im(X)$ stands for the image of X in the supremum of Y and X under some fixed punctual reduction. (As before, use $\mathbf{0}$ to produce copies upon the domain of ω if necessary.) Clearly, $X_1 \leq_{pr} X$ and $Y_1 \leq_{pr} Y$. Also, by the proof of the lemma for sup, the definition of Z coincides with the definition of the supremum of X_1 and Y_1. □

Then lemma above excludes both M_3 and N_5, and therefore implies the theorem.

3. No infimum

Theorem 3.1: *There is a pair of punctual presentations of* $\mathbb{N} = (\omega, +, \times, <, 0, 1)$*, such that neither* $\sup(A_1, A_2)$ *nor* $\inf(A_1, A_2)$ *exists.*

Proof: We will construct a pair of punctual copies A_1 and A_2 of $\mathbb{N}$ without infimum. It follows from Theorem 2.1 that $\sup(A_1, A_2)$ does not exist either.

We are building A_1 and A_2 of $\mathbb{N}$ and an infinite sequence of $B_i, i = 1, 2, \ldots$, and satisfy the requirements:

U_i: $B_i \leq_{pr} A_1$ and $B_i \leq_{pr} A_2$,
R_i: $B_i <_{pr} B_{i+1}$,
S_e: if P_e is below both A_1 and A_2, then P_e is reducible to B_i, for some i.

(These will also imply that A_1 and A_2 are incomparable.)

3.1. *Intuition*

Let $\mathbf{0}$ denote the least punctual presentation of $\mathbb{N}$ under $\leq_{pr}$. It is easy to describe what $\mathbf{0}$ looks like, up to $=_{pr}$. More specifically, at stage s we have $\mathbf{0}[s] = \{\underline{k} : k \leq s\}$, where

$$\underline{k} = 1 + 1 + \ldots (k \text{ times}) \ldots + 1.$$

What does an $A > \mathbf{0}$ look like? In A, there will be infinitely many stages at which some elements x will be "far-far away" from the origin 0 for a long time. Such elements will look "non-standard" in the sense that $x \neq \underline{k}$ for any $\underline{k}$ generated so far, and they will stay "non-standard" for a long time. More formally, if there was a uniform bound on the stages at which every element is expressed as a term in 0 and 1, then we would have $A =_{pr} \mathbf{0}$. The collection of those elements, which already received expressions in terms of 0 and 1, will be called *the standard part of* A. If we construct A, then we can keep x disconnected from the standard part for as long as we like, and in this case, we say that we *keep* x *non-standard.* We will also use variations of this informal terminology throughout the proof; we believe that in each specific instance the meaning of these terms will be self-explanatory.

Now, when we have gained a bit of intuition on temporarily non-standard elements, we are ready to explain the basic diagonalisation strategy. Then we will finally explain the proof idea which, among other things, will be using the strategy below.

3.1.1. *The elementary diagonalisation strategy*

Suppose we are given a punctual $Y \cong \mathbb{N}$ and we would like to build an $X >_{pr} Y$. To diagonalise against $p_e : X \to Y$, do the following:

(1) Copy Y into X.
(2) In X, create a new x and keep it non-standard.
(3) Calculate $p_e(x) \in Y$.
(4) Wait for a stage and a k, such that $p_e(x) = \underline{k}$ in Y.
(5) In X, define $x = \underline{m}$, where m is larger than k. (This stage may require a bounded delay on the enumeration of X. The delay might be necessary to compute $\underline{m}$ for all z currently present in X, and thus in Y.)

If Y is the least copy $\mathbf{0}$ of $\mathbb{N}$, then the delay in (5) will not be necessary: just pick m very large. However, in the general case, Y will contain some other non-standard elements, and since $X[s]$ essentially is $Y[s]$ together $\langle x\rangle[s]$ we must be a bit careful in how we incorporate x into the standard part of X. For that, evaluate $+$ and $\times$ in Y for a sufficient number of steps and find a freshly generated substructure Ξ, which could accommodate $\langle x\rangle[s]$ without any overlap with $Y[s]$; there is clearly a primitive recursive bound on the number of steps necessary to find such a substructure. Only after such an interval is found, we can resume the enumeration of X and set the image of Ξ equal to $\langle x\rangle[s]$ under the natural embedding of Y into X.

3.1.2. *An informal description of the construction*

Recall that we are building two copies, A_1 and A_2, of $\mathbb{N}$. At every stage, both A_1 and A_2 will always have *at most one* currently non-standard generator present in them. For example, A_1 will consist of a long enough initial segment of the natural numbers and also a generator x, such that every element of A_1 at that stage will be seen as a term in 1 and x. These generators in A_1 and A_2 will be thought as equal, unless prescribed otherwise by some requirement.

We will have infinitely many kinds of such non-standard generators which will appear at some stages in both A_1 and A_2. When the time comes, we will once in a while create fresh non-standard-looking generators in both A_1 and A_2 (one in each) and call them blue. There will be infinitely many stages of the construction at which we create blue generators in both A_1 and A_2. They will be promised to be identical as numbers of the form $\underline{k}$, i.e., they will be matched under the unique isomorphism between the structures. They will stay detached from the standard part for an equal number of steps (long enough), and at a late enough stage they will be incorporated identically into the standard part of both A_1 and A_2. Only after the blue generators are declared standard, we will be allowed to put a fresh non-standard generator into A_1 or A_2 (or both), if we need to. (The colour of this fresh generator will not have to be blue.) Then, at some much later stage, we will again create blue generators, and then later again more blue generators, etc. At intermediate stages, we will also create generators of colours that are not necessarily blue, as will be explained below.

The structure B_1 will be consisting of the standard part and the blue generators (only). This property will make B_1 below both A_1 and A_2. The speed with which we connect the blue generators is dictated by the requirements making B_1 not reducible to the standard "bottom" copy $\mathbf{0}$ of $\mathbb{N}$, according to the simple strategy described at the subsection above. These will be the only strategies working with the blue generators.

Later, when we are ready, we will once in a while introduce red generators. No blue generator can be red. They will be handled similarly to how blue generators were dealt with, i.e., they will appear in the structures simultaneously. We also monitor P_0, which is the first punctual structure in the total listing $(P_i)_{i\in\omega}$ of all punctual structures. More specifically, we actually look at the triple (P_0, p_0, q_0), where p_0 and q_0 are potential isomorphisms from P_0 onto A_1 and A_2, respectively.

If no action needs to be performed (to be explained), the red generators will be similar to the blue ones in the way they will be handled. They will be created and then, when necessary, connected to the standard part. The structure B_2 will incorporate the standard initial segment, the red and the blue generators, and *nothing else*. We will use the red generators to make $B_2 > B_1$, using the elementary strategy from the previous subsection.

If either p_0 or q_0 is not an isomorphism then it will be eventually discovered. If no point of P_0 is ever mapped into a red (in the general construction, non-blue) generator via q_0 and p_0, then P_0 must be reducible to B_1. Since B_2 is above B_1, and both are below A_1 and A_2, P_0 is not the infimum of A_1 and A_2.

Thus, we assume p_0 does eventually map some $x \in P_0$ into a red (in general, non-blue) generator of A_1. But if it does send x into a red generator of A_1, we can make $p(x)$ not equal to $q(x)$ when we finally decide on the terms for the red generators.

Note that if both p_0 and q_0 map x into red generators of A_1 and A_2, respectively, then we will have to connect the red generators differently in these two structures. Thus, we will potentially upset the definition of B_2

which assumes that the red generators behave identically in both A_1 and A_2. If this ever happens, we will have to restart B_2. We make a new copy of B_2, and proceed to the next requirements. Use the speed with which the red generators are connected to make B_1 strictly below B_2, and we will never use the red generators again for any other purpose.

The case of all B_i is handled using simple priority; each colour can be upset by only finitely many P_e, thus making the definition of the respective B_j eventually correct.

3.2. *Formal proof*

We describe the basic strategies, and then we describe the construction. In the construction, we will be introducing generators into A_1 and A_2 and assign colours to the generators. The generators will be kept non-standard and then connected to the standard part, according to the strategies below. When a generator becomes standard, it no longer has a colour.

3.2.1. *The strategy for U_i*

The strategy is responsible for building B_i a pair of isomorphisms $v_i : B_i \to A_1$ and $u_i : B_i \to A_2$. At every stage, we identify all generators of the same color in both structures. More formally, the isomorphisms will be defined as identical pullback-copying of A_1 (resp., A_2) into B_i on all currently standard elements (which will be the same in A_1 and A_2) and all generators of *colours* j, $j \leq i$. Unless injured by higher priority requirements (to be clarified), the automorphism types of the j-coloured generators will be set equal in both A_1 and A_2. Therefore, assuming the strategy is never initialised, the definition of B_1 and the primitive recursive isomorphic reductions u_i and v_i will be consistent and correct.

3.2.2. *The strategy for R_i*

The strategy is responsible for illustrating $B_i <_{pr} B_{i+1}$. The basic strategies for P_i and P_{i+1} will guarantee $B_i \leq_{pr} B_{i+1}$. To ensure $B_{i+1} \not\leq_{pr} B_i$, implement the basic diagonalisation strategy (§3.1.1) on the generators having colour $i+1$ to defeat every potential isomorphism $p_e : B_{i+1} \to B_i$ throughout the construction. (We note that the coloured generators are formally introduced in A_1 and A_2, but they are naturally copied into B_i and B_{i+1}, and thus the respective generators can be identified.) In particular, the strategy will define the stages at which the generators having colour $j+1$ will be assigned with a term, i.e., will become "standard".

3.2.3. *The strategy for S_e*

The strategy monitors P_e. It will be split into infinitely many substrategies $S_{e,i,j}$, each working with the respective triple (P_e, p_i, q_j). Here p_i and q_j are potential isomorphisms onto A_1 and A_2, respectively. Recall that initially we intend to make A_1 and A_2 naturally isomorphic, with generators naturally matched with their natural images of the same colour. The instructions for the substrategy $S_{e,i,j}$ are as follows:

(1) Wait for $q_j(x)$ and $p_i(x)$ to converge. If $q_j(x)$ is unequal to $p_i(x)$ as a term of the respective naturally isomorphic generators, then the strategy for R_c will guarantee that $q_j(x)$ is not equal to $p_i(x)$ (as terms of $\underline{1}$) in the respective copies of the arithmetic. In this case, do nothing.

(2) Wait for an $x \in P_e$, such that both $p_i(x)$ and $q_j(x)$ are equal to a term which non-trivially involves a generator of colour $c > \langle e, i, j \rangle$. If such an x is never found, then do nothing. In this case, P_e is below $B_{\langle e,i,j \rangle}$.

(3) Suppose $p_i(x) = \underline{m}_0 + \underline{m}_1 z + \ldots + \underline{m}_s z^s$ and $q_j(x) = \underline{m}_0 + \underline{m}_1 z' + \ldots + \underline{m}_s z'^s$, where z and z' are generators which are currently kept non-standard in A_1 and A_2, respectively, have colour $c > \langle e, i, j \rangle$, and such that we initially intended to declare z and z' equal as terms of $\underline{1}$ at some later stage. We must ensure these linear combinations will become unequal (as terms of $\underline{1}$ in the respective copies of the arithmetic) when we are ready to agree on the values for z and z'. Note that, initially, the strategy for U_c intended to assign identical terms to both z and z', and therefore U_c will be *injured*.

3.2.4. *Construction*

In the construction, we build A_1 and A_2 punctually, so that for every colour k there are infinitely many stages at which a new currently non-standard generator of color k is introduced to both A_1 and A_2.

We then let the U_i-strategies build the respective B_i, and we let the other strategies act according to their instructions. At stage s, we monitor P_i with $i < s$, and we let $S_{e,i,j}$ with $\langle e, i, j \rangle < s$ to act if necessary, with the priority given to the strategy having the smallest index. If U_i is *injured* due to an action of some $S_{e,i,j}$-strategy (and there can be only finitely many such), then it restarts its definition of embeddings u_e and v_e onto A_1 and A_2, respectively, correcting the asymmetry in A_1 and A_2 introduced by the $S_{e,i,j}$-strategy.

3.2.5. *Verification*

It is clear that the structures $A_1, A_2, B_i, i = 0, 1, \dots$ are all punctual, and each of them is isomorphic to $\mathbb{N}$.

Note that there are at most finitely many S-substrategies, which can possibly upset the definition of the embeddings of B_i onto A_1 and A_2. Furthermore, every such substrategy acts at most once. Thus, for every i, U_i is eventually met. It follows at once from the description of the basic diagonalisation strategy (§3.1.1) that U_i is met for every i.

Also, the sub-strategy $S_{e,i,j}$ will either ensure that either p_i and p_j are not isomorphisms form P_e onto A_1 and A_2, respectively, or P_e is reducible to $B_{\langle e,i,j\rangle}$. We claim that S_e is met for every e. Indeed, suppose $P_e \leq_{pr} A_1$ and $P_e \leq_{pr} A_2$, as witnessed by, say, p_i and p_j. Then P_e must be pr-below B_c. However, $B_{c+1} >_{pr} B_c$, and therefore $B_{c+1} >_{pr} B_c \geq_{pr} P_e$. Since B_{c+1} is below both A_1 and A_2, P_e cannot be their infimum. □

4. Embedding $Int(\eta)$

Theorem 4.1: *Suppose X is a finitely generated rigid punctual structure, and suppose $A >_{pr} \mathbf{0}$ is any punctual presentation of X. Then the lower cone $\check{A} = \{Z : Z \leq_{pr} A\}$ embeds the atomless Boolean algebra $Int(\eta)$ preserving the supremums and the infimums.*

Proof: The lemma below seems to be crucial for the proof.

Lemma 4.2: *Suppose $A > \mathbf{0}$ is finitely generated (there is no assumption on rigidity). There is a $A^* > \mathbf{0}$ below A with the property: At every stage s we can pick at most one currently non-standard generator in $A^*[s]$ which, together with the currently standard part of $A^*[s]$, s-generates $A^*[s]$.*

Proof: Using a reduction from $\mathbf{0}$ onto A, fix a collection of generators $\bar{d}$ in A, which is the isomorphic image of some fixed tuple $\bar{c}$ of generators in $\mathbf{0}$, under the fixed reduction.

We will copy the standard part $\langle \bar{d} \rangle [s]$ into A^*, as well as some other elements (to be explained). Let $\bar{d}'$ be the natural image of $\bar{d}$ in A^*. We will view each p_e as a map from A^* to $\mathbf{0}$. For every e, we will wait for $p_e(\bar{d}')$ to generate $\bar{c}$. This non-primitive recursive waiting will be performed at the background and will not be a part of the strategy below. If we ever see that the elements $p_e(\bar{c})$ generate $\bar{d}$, we declare p_e *ready*. At every stage there will be some finite — perhaps, empty — collection of such p_e, which have already been declared ready.

We have not yet defined A^*; the construction of A^* will also be performed based on the strategy below. As we have mentioned above, A^* will always be copying the standard part from A. The further elements of A^* are picked as follows.

(1) Wait for at least one p_e to be declared ready, and pick e least such.
(2) Choose x with the smallest index outside the standard part $\langle \bar{d} \rangle[s]$ of A, and copy x into A^*.
(3) Keep copying only elements from $\langle \bar{d} \cup x \rangle$ into A^* and monitor $p_e(x)$.
(4) If p_e is discovered to be not an isomorphism, then wait for x to be incorporated into $\langle \bar{d}' \rangle$, and then go to (1) and wait for the next primitive recursive map to be declared ready.
(5) Otherwise, when $p_e(x) \downarrow$, we will see $p_e(x) \in \langle \bar{c} \rangle$ in $\mathbf{0}$. Since p_e had been declared ready, we will also see $p_e(x) \in \langle p_e(\bar{d}') \rangle$ on that stage. Since p_e insists on being an isomorphism, we will also see $x \in \langle \bar{d}' \rangle$ in A^*.
(6) Define $q : x \to \mathbf{0}$ on x as follows. Map $\bar{d}$ to the natural pre-image $\bar{u}$ of $\bar{d}'$ in $\mathbf{0}$, map x to the $\bar{u}$-term obtained from the $\bar{d}'$-term witnessing $x \in \langle \bar{d}' \rangle$ using the straightforward replacement. Since each element $\langle \bar{d}' \cup x \rangle$ is also a term in $\bar{d}'$ and x, extend this map naturally to every such element seen so far. Go to (2).

Note that at every stage s, A^* contains an element x which, together with the currently standard part of A^*, generates the whole of $A^*[s]$. Also, suppose $A^* \leq \mathbf{0}$ as witnessed by p_e, and e is the smallest such. Then p_e will eventually be declared ready, and each p_j, $j < e$, will either show that they are not isomorphisms in finite time or will never be declared ready. (We note here that the outcome when p_j is not onto is simply impossible. If p_j has been declared ready then it has no choice but to copy the structure along the generators or show a local disagreement.)

At some stage, p_e will have the smallest index among the maps which are ready, and thus it will be used by the strategy in the definition of x. If p_e never shows a local disagreement, then this means that every x that we pick will be incorporated into the standard part of A^*, and this leads to a primitive recursive isomorphism $q : A \to \mathbf{0}$, whose convergence speed can be calculated using p_e. This contradicts the choice of $A >_{pr} \mathbf{0}$. □

Lemma 4.3: *Let $A^* > \mathbf{0}$ satisfy the property from Lemma 4.2, and additionally assume that the finitely generated structure under consideration is rigid. Then there exist incomparable P_0, P_1, such that* $\sup(P_0, P_1) = A^*$ *and* $\inf(P_0, P_1) = \mathbf{0}$.

Proof: The proof is non-technical, thus there is no point in giving a formal stage-by-stage construction. The difficulties which occur in the proof are related purely to the timing at which the (elementary) actions below will be performed. We give a detailed explanation of what needs to be done, and why the assumptions on A^* make the proof work.

To build P_0, switch back and forth between copying $\mathbf{0}$ and copying A^*. The elementary formal details of the switching procedure can be found in [6]; we will also give more details below. The other structure P_1 will also be switching, but in the opposite way. That is, P_1 will be copying $\mathbf{0}$ when P_0 copies A^*, and vice versa. When P_0 copies A^* and P_1 copies $\mathbf{0}$, we diagonalise against $p_e : P_0 \to P_1$, which will be achieved in finite time. Indeed, if p_e was an isomorphism, we would be able to define a primitive recursive isomorphism of A^* onto $\mathbf{0}$, contradicting the choice of A^*. As in the proof of Lemma 4.2, we do not even attend p_e unless it has mapped the generators to the generators (recall the structure is rigid). Since the generators will be mapped to their natural images, p_e will have no choice but to either copy the structure locally or to show a local disagreement, which will happen in finite time. When we see p_e die, we declare that we are ready to *switch sides.*

When we switch sides, we wait for (the natural image of) $\mathbf{0}$ (in A^*) to catch up on the unique non-standard generator, which may currently be in the structure. Once we see this happen, we declare that P_0 now copies $\mathbf{0}$, and we also declare that P_1 copies A^*. At that stage, we instantly switch sides and allow P_1 to copy A^*. *It is crucial that as soon as the least non-standard generator is incorporated into the standard part, everything in P_0 immediately becomes standard.* While we wait for the switch to be performed, there will be no further non-standard elements in A^*, which would not be generated by a specifically chosen special non-standard generator. Thus, while we are waiting, P_0 is still identical to A^*, and P_1 can be kept identical to $\mathbf{0}$. Then we meet another diagonalisation requirement and then switch back similarly, etc.

We claim that $\sup(P_0, P_1) = A^*$. If $H \geq P_0, P_1$, then we define $h : A^* \to H$ as follows. At stage s, see which of the two structures copies A^*. Assume this is P_0. To define $h(x)$, see where x is mapped into P_0, and use the reduction from P_0 to H to match x with the respective image in H. Since the structure is rigid, the composition of the maps coming through the P_0-side will agree with the composition of the maps coming from the P_1-side, thus there will be no problem in extending the isomorphism even after we switch sides.

Remark 4.1: We (strongly) conjecture that rigidity can be replaced by some more relaxed assumption, such as: Every automorphism in a punctual copy of the structure is primitive recursive. We do not know if the lemma holds without any assumptions on the automorphisms of the structure, and we suspect that, in fact, it does not.

We claim that $\inf(P_0, P_1) = \mathbf{0}$. Again, if there is some $B \leq_{pr} P_0, P_1$, then we use either the P_0-side or the P_1-side to match every element of B with the respective element in $\mathbf{0}$.

Remark 4.2: It is crucial that, when we are ready to switch sides, there is at most one non-standard generator in T. Otherwise we could run into the scenario when we have connected some non-standard parts in P_0, but some other non-standard parts stay disconnected from the generators for much longer. Then there is a possibility of constructing another infimum (or supremum), which is tighter than the one claimed in the lemma. □

Note that both P_0 and P_1 satisfy the property of Lemma 4.2. Iterate the process to embed $Int(\eta)$, as follows. Split P_0 and P_1 over $\mathbf{0}$, and then split each of the two, etc. The process can, in fact, be made simultaneous and effective, but this is not important for establishing the theorem.

5. Not a Boolean algebra. Proof of Theorem 1.5

Fix $A >_{pr} \mathbf{0}$ of a finitely generated structure. We must show that the lower cone below A is not a Boolean algebra. (If $A =_{pr} \mathbf{0}$ then there is nothing to prove, because every Boolean algebra contains at least two elements.) The plan is to build a copy $A^* \leq_{pr} A$ and construct a $C <_{pr} A^*$, such that C does not have the relative complement below A^*. Use Lemma 4.2 to produce an $A^* > \mathbf{0}$ below A, which has at most one special non-standard generator at every stage. (Recall that Lemma 4.2 did not need any assumption on $Aut(A)$.)

We will build a C, such that $\mathbf{0} <_{pr} C <_{pr} A^*$ and satisfy the requirements:

$$R_e : P_e \text{ is not the complement of } C \text{ relative to } A^*,$$

where $(P_e)_{e\in\omega}$ is the total enumeration of all punctual structures.

The strategy for making C strictly in-between $\mathbf{0}$ and A^* is exactly the same as in the previous theorem. That is, we will attempt to switch back and forth between copying $\mathbf{0}$ and copying A^*. When C copies $\mathbf{0}$, we diagonalise against $p_e : A^* \to C$, where p_e has already been declared ready (i.e., the generators and their images are bi-spanned), and e is the least such. When C copies A^*, we diagonalise against isomorphisms from C to $\mathbf{0}$ similarly. In any of these cases, the diagonalisation process, once initiated, will finish in finite time. Note that the nice property of A^* allows to switch without any extra delay; see Remark 4.2.

The strategy for R_e is less straightforward. We split it into substrategies, each monitoring a triple (P_e, q_e, l_e), where $q_e : P_e \to A^*$ and $l_e : \mathbf{0} \to P_e$ are potential isomorphisms. The sub-strategy computes $q_e(l_e(\bar{c}))$, where $\bar{c}$ is some fixed tuple of generators in $\mathbf{0}$, and waits for the image of $\bar{c}$ in A^* and $q_e(l_e(\bar{c}))$ to generate each other. If this ever happens, then the substrategy will be declared ready.

5.1. *The definition of $N(C, P_e)$*

Let $p : C \to A^*$ be the primitive recursive isomorphism, which will be constructed by stages. At every stage, its value will be well-defined. Let $N(C, P_e)$ be the punctual re-enumeration of $\langle p(C) \cup q_e(P_e)\rangle \subseteq A^*$. The re-enumeration uses that the fixed tuple of generators of A^* will be contained in $p(C) \cup q_e(P_e)$, and therefore we can punctually list the "standard" substructure of A^*, which will also be contained in $\langle p(C) \cup q_e(P_e)\rangle$; the other elements can be incorporated with a bounded delay. Although $N(C, P_e)$ is not really a subset of A^*, but rather its retract via a punctual bijection, we abuse our notation and at every stage identify $N(C, P_e)$ with a subset of A^*. Note that, since q_e was declared ready, the standard part of $q_e(P_e)$ can be essentially identified with the standard part of A^* and of $p(C)$.

If q_e and l_e are indeed isomorphisms, we will have that $N(C, P_e)$ is a punctual copy of A^* with the property $N(C, P_e) \leq_{pr} A^*$; the reduction is essentially the identity. (Strictly speaking, the reduction will be the bijection defined for the sake of making the domain of $N(C, P_e)$ equal to ω). Also, under the same assumptions, we will have $C \leq_{pr} N(C, P_e)$ and $P_e \leq N(C, P_e)$.

In other words, we have constructed *some* upper bound $N(C, P_e)$ of P_e and C, but only under the assumption that all the maps under consideration are indeed isomorphisms.

Remark 5.1: Since we have assumed that the generators have been seen to span each other, we do not have to worry about the unpleasant case of a proper isomorphic embedding. In other words, if something goes wrong with the potential isomorphisms, we will see a confirmation at a finite stage.

5.2. *Pressing P_e*

If A^* is indeed the supremum of C and P_e, then so is $N(C, P_e) \leq_{pr} A^*$. In particular, there should exist some isomorphism $v_e : A^* \to N(C, P_e)$. For every e, we shall test whether $v_e : A^* \to N(C, P_e)$ is an isomorphism. We further split the sub requirements, each monitoring its own primitive recursive v_e. As usual, we wait for $v_e(\bar{c})$ and $\bar{c}$ to generate each other. Again, if this never happens then we never attend the sub-requirement monitoring v_e.

The value of v_e is that it will allow us to press P_e to show us non-standard elements when they appear in A^*, by intentionally keeping them out of C.

5.3. *The definition of $L(C, P_e)$*

The subrequirement working with (P_e, q_e, l_e, v_e) will attempt to define some lower bound $L(C, P_e)$ below C and P_e, such that $L(C, P_e) >_{pr} \mathbf{0}$. The definition of $L(C, P_e)$ will be dynamic, but it will be similar to $N(C, P_e)$, in the sense that it will be "living" inside A^*, but up to a punctual bijection whose only role is to make the domain equal to ω.

We split the subrequirement even further, for the last time. Each individual one will be trying to diagonalise against $u_e : L(C, P_e) \to \mathbf{0}$, and again will act only if it is declared ready.

When the subrequirement working with $(P_e, q_e, l_e, v_e, u_e)$ is ready to act, it goes over the following stages. At every stage it waits for a disagreement showing that one of the maps is not an isomorphism. If it discovers such a disagreement, it instantly stops. Recall that at every stage A^* has at most one special non-standard generator.

(1) Let C copy $\mathbf{0}$.
(2) Use $v_e : A^* \to N(C, P_e)$ to compute the image of a special generator in P_e, when it is introduced in A^*.
(3) Copy the special generator into C and define a (perhaps, not optimal) lower bound $L(C, P_e)$, which contains the standard part as well as the special generator, and matches these with their natural images in P_e and C in accordance with p, q_e, v_e. More specifically, we copy A^* into $L(C, P_e)$ naturally, using p. We define the image of the special generator (thus, of the rest) in $N(C, P_e)$ using v_e, and then we use the definition of $N(C, P_e)$ to punctually find the respective pre-images under q_e. (Later in the construction, while no sub strategy of the strategy

for $(P_e, q_e, l_e, v_e,)$ is active, we will copy C into $L(C, P_e)$ and will not wait for any diagonalisation to occur.)

(4) Wait for either $u_e : L(C, P_e) \to \mathbf{0}$ to die or $v_e : A^* \to N(C, P_e)$ to die. Repeat 1–4 until either we diagonalise u_e or one of the monitored maps proves to be not an isomorphism.

The strategy is finitary in the following sense. Either at least one of the maps observed by the substrategy must prove itself wrong, or we will eventually diagonalise against u_e. It is crucial that the substrategy acts only when all the maps observed by it have proved themselves ready (in the sense of bi-spanning of the generators and their images under the map). If all the maps except for u_e are never wrong, then we will eventually diagonalise against u_e, for otherwise we would be able to use u_e to build a reduction from A^* onto $\mathbf{0}$.

5.4. *Construction*

We list all subrequirements $(P_e, q_e, l_e, v_e, u_e)$ in some priority order and wait for them to be declared ready. Meanwhile, we construct C switching back and forth between copying $\mathbf{0}$ and A^*, according to its basic strategy. As soon as a fresh diagonalisation subrequirement is declared ready, we pick the highest priority such. We then allow C to switch to copying $\mathbf{0}$ and act according to the instructions of the substrategy. After the action of the substrategy is finished, we allow C, to switch at least twice back and forth between $\mathbf{0}$ and A^*. Then repeat.

5.5. *Verification*

A large portion of the verification was incorporated into the description of the basic strategies. We put everything together.

The construction is finitary, and there is no explicit injury or initialisation involved. It is crucial that, as soon as a map is declared ready, we no longer have to worry about the Π^0_2-outcome in which the map is a proper isomorphic embedding. In the construction, we simply never get to work with any map, which is never declared ready.

Once a map is declared ready, all the diagrams will be maintained commutatively until we see some disagreement. Either a disagreement must occur, or we succeed in diagonalising against u_e. For either A^* is not the real supremum, and then q_e or v_e must prove to be wrong, or we will punctually copy A^* into the lower bound $L(C, P_e)$ and rely on the fact that

$A^* > \mathbf{0}$. Indeed, if we keep successfully copying A^* into $L(C, P_e)$, then $u_e : L(C, P_e) \to \mathbf{0}$ cannot be an isomorphism.

Assume q_e, l_e, v_e are all actual isomorphisms and A^* is a supremum. Then each u_e will either never be declared ready, or will eventually become ready. In the former case, it cannot be an onto isomorphism $u_e : L(C, P_e) \to \mathbf{0}$, and in the latter case, we will eventually diagonalise against it, and furthermore will see a finitary disagreement witnessing that $u_e : L(C, P_e) \to \mathbf{0}$ is not an isomorphism. Either way, we will succeed in constructing $L(C, P_e)$ strictly above $\mathbf{0}$.

Acknowledgments

The work of the first and the third author is supported by RSF grant no. 18-11-00028 and performed under the development program of Volga Region Mathematical Center (agreement no. 075-02-2023-944). The second author is partially supported by Marsden Fund of New Zealand and Rutherford Discovery Fellowship (Wellington) RDF-MAU1905, Royal Society Te Aparangi.

References

1. C. Ash and J. Knight. *Computable structures and the hyperarithmetical hierarchy*, volume 144 of *Studies in Logic and the Foundations of Mathematics.* North-Holland Publishing Co., Amsterdam, 2000.
2. P. E. Alaev. Structures computable in polynomial time. I. *Algebra Logic*, 55(6):421–435, 2017.
3. P. E. Alaev. Structures computable in polynomial time. II. *Algebra Logic*, 56(6):429–442, 2018.
4. N. Bazhenov, R. Downey, I. Kalimullin, and A. Melnikov. Foundations of online structure theory. Submitted.
5. Achim Blumensath and Erich Grädel. Automatic structures. In *15th Annual IEEE Symposium on Logic in Computer Science (Santa Barbara, CA, 2000)*, pages 51–62. IEEE Comput. Soc. Press, Los Alamitos, CA, 2000.
6. N. Bazhenov, I. Kalimullin, A. Melnikov, and K. M. Ng. Punctual presentations of finitely generated structures. Submitted.
7. D. Cenzer and J. B. Remmel. Complexity theoretic model theory and algebra. In Yu. L. Ershov, S. S. Goncharov, A. Nerode, and J. B. Remmel, editors, *Handbook of recursive mathematics, Vol. 1*, volume 138 of *Stud. Logic Found. Math.*, pages 381–513. North-Holland, Amsterdam, 1998.
8. Y. Ershov and S. Goncharov. *Constructive models.* Siberian School of Algebra and Logic. Consultants Bureau, New York, 2000.
9. A. Frolov, I. Kalimullin, V. Harizanov, O. Kudinov, and R. Miller. Spectra of high$_n$ and non-low$_n$ degrees. *J. Logic Comput.*, 22(4):755–777, 2012.

10. Ekaterina B. Fokina, Iskander Sh. Kalimullin, and Russell Miller. Degrees of categoricity of computable structures. *Arch. Math. Log.*, 49(1):51–67, 2010.
11. Serge Grigorieff. Every recursive linear ordering has a copy in $DTIME$-$SPACE(n, log(n))$. *J. Symb. Log.*, 55(1):260–276, 1990.
12. G. Higman. Subgroups of finitely presented groups. *Proc. Roy. Soc. Ser. A*, 262:455–475, 1961.
13. H. A. Kierstead. An effective version of Dilworth's theorem. *Trans. Am. Math. Soc.*, 268:63–77, 1981.
14. H. A. Kierstead. On line coloring k-colorable graphs. *Israel J. Math.*, 105(1):93–104, 1998.
15. I. Kalimullin, B. Khoussainov, and A. Melnikov. Limitwise monotonic sequences and degree spectra of structures. *Proc. Amer. Math. Soc.*, 141(9):3275–3289, 2013.
16. Iskander Kalimullin, Alexander Melnikov, and Keng Meng Ng. Algebraic structures computable without delay. *Theoret. Comput. Sci.*, 674:73–98, 2017.
17. Bakhadyr Khoussainov and Anil Nerode. Automatic presentations of structures. In *Logic and Computational Complexity (Indianapolis, IN, 1994)*, volume 960 of *Lecture Notes in Comput. Sci.*, pages 367–392. Springer, Berlin, 1995.
18. H. A. Kierstead, S. G. Penrice, and W. T. Trotter Jr. On-line coloring and recursive graph theory. *SIAM J. Discrete Math.*, 7:72–89, 1994.
19. Roger C. Lyndon and Paul E. Schupp. *Combinatorial group theory*. Classics in Mathematics. Springer-Verlag, Berlin, 2001. Reprint of the 1977 edition.
20. L. Lovász, M. Saks, and W. T. Trotter Jr. An on-line graph coloring algorithm with sublinear performance ratio. *Discrete Math.*, 75:319–325, 1989.
21. R. Miller. The Δ_2^0-spectrum of a linear order. *J. Symbolic Logic*, 66(2):470–486, 2001.
22. A. G. Melnikov and K. M. Ng. The back-and-forth method and computability without delay. Preprint.
23. Viatcheslav N. Obraztsov. On infinite complete groups. *Comm. Algebra*, 22(14):5875–5887, 1994.
24. J. B. Remmel. Graph colorings and recursively bounded Π_1^0-classes. *Ann. Pure Appl. Logic*, 32:185–194, 1986.
25. H. Rogers. *Theory of recursive functions and effective computability*. MIT Press, Cambridge, MA, second edition, 1987.

https://doi.org/10.1142/9789811278631_0013

Maximal Automatic Complexity and Context-Free Languages

Bjørn Kjos-Hanssen

Department of Mathematics
University of Hawai'i at Mānoa
2565 McCarthy Mall, Honolulu, HI 96822, U.S.A.
bjoernkh@hawaii.edu

Let A_N denote nondeterministic automatic complexity and

$$L_{k,c} = \{x \in [k]^* : A_N(x) > |x|/c\}.$$

In particular, $L_{k,2}$ is the language of all k-ary words for which A_N is maximal, while $L_{k,3}$ gives a rough dividing line between complex and simple. Let **CFL** denote the complexity class consisting of all context-free languages. While it is not known that $L_{2,2}$ is infinite, Kjos-Hanssen (2017) showed that $L_{3,2}$ is **CFL**-immune but not **coCFL**-immune. We complete the picture by showing that $L_{3,2} \notin$ **coCFL**.

Turning to Boolean circuit complexity, we show that $L_{2,3}$ is $\mathbf{SAC}^0$-immune and $\mathbf{SAC}^0$-coimmune. Here $\mathbf{SAC}^0$ denotes the complexity class consisting of all languages computed by (non-uniform) constant-depth circuits with semi-unbounded fanin.

As for arithmetic circuits, we show that $\{x : A_N(x) > 1\} \notin \oplus\mathbf{SAC}^0$. In particular, $\mathbf{SAC}^0 \not\subseteq \oplus\mathbf{SAC}^0$, which resolves an open implication from the Complexity Zoo.

Contents

1. Introduction

Automatic complexity is a computable form of Kolmogorov complexity that was introduced by Shallit and Wang in 2001 [17]. It was studied by Jordon and Moser in 2021 [10], and in a series of papers by the author and his coauthors. Roughly speaking, the automatic complexity $A(x)$ of a string is the minimal number of states of an automaton accepting only x among its equal-length peers. The nondeterministic version A_N was introduced by Hyde [9]:

Definition 1: Let Σ be finite a set called the *alphabet* and let Q be a finite set whose elements are called *states*. A *nondeterministic finite automaton* (NFA) is a 5-tuple $M = (Q, \Sigma, \delta, q_0, F)$. The *transition function* $\delta : Q\times\Sigma \to \mathcal{P}(Q)$ maps each $(q, b) \in Q\times\Sigma$ to a subset of Q. Within Q we find the *initial state* $q_0 \in Q$ and the set of *final states* $F \subseteq Q$. The function δ is extended to a function $\delta^* : Q \times \Sigma^* \to \mathcal{P}(Q)$ by[a]

$$\delta^*(q, \sigma i) = \bigcup_{s\in\delta^*(q,\sigma)} \delta(s, i).$$

Overloading notation we also write $\delta = \delta^*$. The *language accepted by M* is

$$L(M) = \{x \in \Sigma^* : \delta(q, x) \cap F \neq \emptyset\}.$$

A *deterministic finite automaton* (DFA) is also a 5-tuple $M = (Q, \Sigma, \delta, q_0, F)$. In this case, $\delta : Q \times \Sigma \to Q$ is a total function and is extended to δ^* by $\delta^*(q, \sigma i) = \delta(\delta^*(q, \sigma), i)$. If the domain of δ is a subset of $Q \times \Sigma$, M is a *partial DFA*. Finally, the set of words accepted by M is

$$L(M) = \{x \in \Sigma^* : \delta(q, x) \in F\}.$$

Definition 2: ([9, 17]) Let $L(M)$ be the language recognized by the automaton M. The nondeterministic automatic complexity $A_N(w)$ of a word w is the minimum number of states of an NFA M, such that M accepts w and the number of paths along which M accepts words of length $|w|$ is 1.

The fundamental upper bound on A_N was discovered by Kayleigh Hyde in 2013:

Theorem 3: (Hyde [9]) *Let x be a word of length $n \in \mathbb{N}$. Then*

$$A_N(x) \leq \lfloor n/2 \rfloor + 1.$$

[a]We denote the concatenation of two words σ and τ by juxtaposition, $\sigma\tau$.

The sharpness of Theorem 3 over a ternary alphabet is also known [9]. Therefore, if $A_N(x) < \lfloor n/2 \rfloor + 1$, we say that x is A_N-*simple* (see Theorem 8).

The question whether the computation of automatic complexity is itself computationally tractable was raised by Shallit and Wang and also by Allender [2]. Here we show that a natural decision problem associated with automatic complexity does not belong to the complexity class $\mathbf{SAC}^0$ consisting of languages accepted by constant-depth Boolean circuits with semi-unbounded fan-in. We also confirm [11, Conjecture 11] by showing that the set of maximally complex words over a ternary alphabet is not co-context free.

2. Context-free languages

A word is called *squarefree* if it cannot be written as $xyyz$ where y is nonempty. In ruling out context-freeness of the set of words of nonmaximal nondeterministic automatic complexity, our method of proof goes back to the 1980s. It consists of analyzing the proof in [14] in the form given in Shallit 2008 [16], to conclude that it pertains to all languages L with $\mathsf{SQ}_3 \subseteq L \subseteq \mathsf{REP}_3$. Here SQ_k is the set of square words over the alphabet $[k]$ and REP_k is the set of words over $[k]$ that are repetitive, i.e., not squarefree.

Our key tool will be the Interchange Theorem 4, originally due to Ehrenfeucht and Rozenberg [6]; see also Berstel and Boasson [3]. We denote the length of a word x by $|x|$, and let $[k] = \{0, 1, \ldots, k-1\}$ for a nonnegative integer k.

Lemma 4: (Interchange Lemma [14]) *For each $L \in \mathbf{CFL}$ there is an integer $c > 0$, such that for all integers $n \geq 2$, all subsets $R \subseteq L \cap \Sigma^n$, and all integers with $2 \leq m \leq n$, there exists a subset $Z \subseteq R$, $Z = \{z_1, z_2, \ldots, z_k\}$, such that $k \geq \frac{|R|}{c(n+1)^2}$ and such that there exist decompositions $z_i = w_i x_i y_i$, $1 \leq i \leq k$, such that for all $1 \leq i, j \leq k$,*

(a) $|w_i| = |w_j|$,
(b) $|y_i| = |y_j|$,
(c) $\frac{m}{2} < |x_i| = |x_j| \leq m$; *and*
(d) $w_i x_j y_i \in L$.

Remark 5: When establishing Theorem 6 and Theorem 8 we needed a word with a certain property. We first found an example over a 6-letter alphabet, $(\mathtt{123})^2\mathtt{0}(\mathtt{12345})^2$. We were able to reduce the alphabet size from 6 to 5 by using the slightly modified word $(\mathtt{123})^2\mathtt{0}(\mathtt{12341})^2$. We expect that the alphabet size, and word length, can be reduced further, but we do not pursue it here.

Lemma 6: *Let* $w = (\texttt{123})^2\texttt{0}(\texttt{12341})^2 \in \{\texttt{0},\texttt{1},\texttt{2},\texttt{3},\texttt{4}\}^{17}$. *Then* $A_N(w) = 8$.

Proof: Let M be the following NFA.

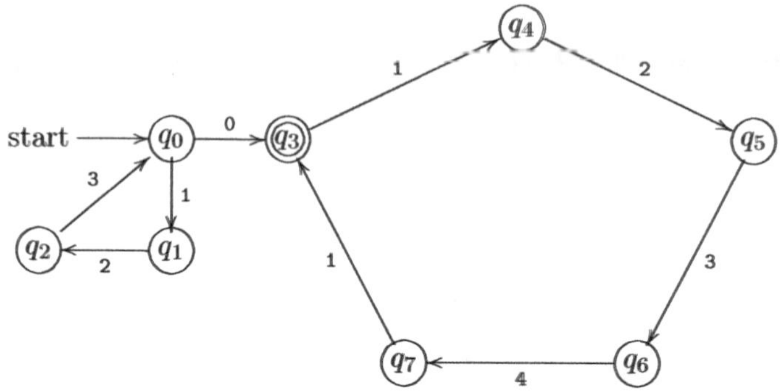

Since the equation $3x + 1 + 5y = 17$ has the unique solution $(x, y) = (2, 2)$ over $\mathbb{N}$, we see that M accepts w uniquely via the sequence of states

$$(q_0, q_1, q_2, q_0, q_1, q_2, q_0, q_3, q_4, q_5, q_6, q_7, q_3, q_4, q_5, q_6, q_7, q_3).$$

Since M has only 8 states, it follows that $A_N(w) \leq 8$. Conversely, the inequality $A_N(w) \geq 8$ was verified with a *brute force* computerized search in Python 3.7 on July 7, 2021. □

Definition 7: For a word x and positive integers p, q, we define the power x^α, $\alpha = p/q$, to be the prefix of length n of $x^\omega = xxx\cdots$, where $n = p|x|/q$ is an integer, if it exists. The word x^α is called an α-power. If a word w has a (consecutive) subword of the form x^α then x is said to occur with exponent α in w. The word $\mathbf{w}$ is α-power-free if it contains no subwords which are α-powers. If $\alpha = 2$, this is read as squarefree, if $\alpha = 3$, cubefree. Finally, w is *overlap-free* if it is α-power-free for all $\alpha > 2$.

As an example of Theorem 7, we have $(\texttt{0110})^{3/2} = \texttt{011001}$.

Lemma 8: *There exists an A_N-simple overlap-free word of odd length.*

Proof: Consider the word w of Theorem 6, of length 17. □

Lemma 9: *Let Σ be a finite alphabet. Let $\alpha \in \mathbb{Q}$, $\alpha \geq 1$ and let $x \in \Sigma^*$ such that x is an α-power. Then $A_N(x) \leq |x|/\alpha$.*

Proof: Let $x = (x_1 \cdots x_v)^\alpha$, where $m, v \in \mathbb{N}$ are such that the fractional part of α is m/v and $x_1, \ldots, x_v \in \Sigma$. Let M be the following NFA, whose digraph is a cycle:

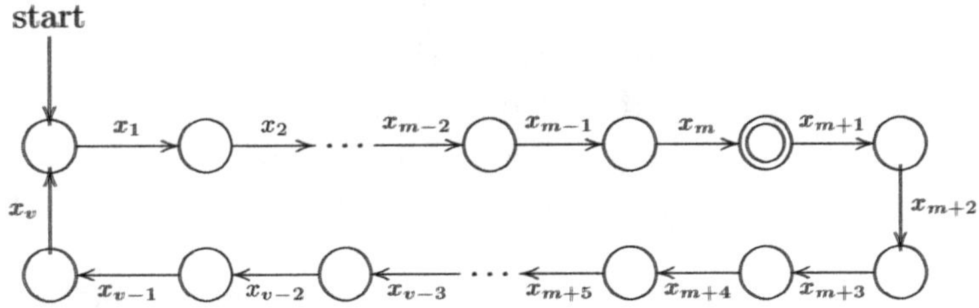

The number of states is v, so $A_N(x) \leq v = |x|/\alpha$. □

Lemma 10: *Let Σ be a finite alphabet. Consider the following implication for a word $x \in \Sigma^*$.*

$$A_N(x) \leq |x|/\alpha \implies x \text{ contains an } \alpha\text{-power}. \tag{2.1}$$

(i) Let $\alpha \geq 1$ be an integer. Then (2.1) *holds.*
(ii) (2.1) *fails for $\alpha = 2 + \frac{1}{8}$.*

Proof:
Proof of (i): Suppose that $A_N(x) \leq |x|/k$ as witnessed by an NFA M. Then the $|x| + 1$ visiting times of M during its computation on input x are distributed among the at most $|x|/k$ states. Consequently, some state is visited at least $\frac{|x|+1}{|x|/k} = k(1 + \frac{1}{|x|}) > k$, and hence at least $k + 1$, times. However, [9, Theorem 16] says: If an NFA M uniquely accepts x of length n, and visits a state p at least $k + 1$ times, where $k \geq 2$, then x contains a kth power.
Proof of (ii): consider the word x of Theorem 6 of length 17 and let $\alpha = 2+\frac{1}{8}$. Note that $|x|/\alpha = 17/(2 + \frac{1}{8}) = 8$. □

Let Σ, Δ be finite alphabets. A *morphism* $\varphi : \Sigma^* \to \Delta^*$ is a function satisfying $\varphi(xy) = \varphi(x)\varphi(y)$ for all $x, y \in \Sigma^*$. Note that, in order to define a particular morphism φ, it suffices to define $\varphi(a)$ for each $a \in \Sigma$.

A function f is called squarefree-preserving if for each squarefree x, the word $f(x)$ is squarefree.

Theorem 11: (Brandenburg [5]) *The following morphism h is squarefree-preserving:*

$$
\begin{aligned}
0 &\to 010201202101210201\text{0}212\\
1 &\to 010201202120121012\text{0}212\\
2 &\to 010201210120210121\text{0}212\\
3 &\to 010201210120212012\text{1}012\\
4 &\to 010201210201021012\text{0}212\\
5 &\to 010201210212021012\text{0}212.
\end{aligned}
$$

Definition 12: The perfect shuffle of two words x, y of the same length a is $Ш(x, y) = x_1y_1 \cdots x_ay_a$.

We follow Shallit's presentation [16, Theorem 4.5.4] in Theorem 13, which is a slight strengthening thereof.

Theorem 13: *For each language L,*

$$\mathsf{SQ}_6 \subseteq L \subseteq \mathsf{REP}_6 \implies L \notin \mathbf{CFL}.$$

Proof: Assume that $L \in \mathbf{CFL}$. Let c be the constant in the Interchange Theorem 4 and choose n divisible by 8 and sufficiently large so that

$$\frac{2^{n/4}}{c(n+1)^2} > 2^{n/8}.$$

Let r' be a squarefree word over the alphabet $\{0, 1, 2\}$ of length $\frac{n}{4} - 1$ and let $r = 3\,r'$, so that $|r| = \frac{n}{4}$. Let

$$A_n = \{Ш(rr, s) : s \in \{4, 5\}^{n/2}\} \subseteq \{0, \dots, 5\}^n.$$

As shown in [16],

(1) if $z_i = w_ix_iy_i \in A_n$ for $i \in \{1, 2\}$ with $|w_1| = |w_2|$, $|x_1| = |x_2|$, and $|y_1| = |y_2|$, then $w_1x_2y_1 \in A_n$ and $w_2x_1y_2 \in A_n$, too; and
(2) if $z \in A_n$, then z contains a square if and only if z is a square.

Now let $B_n = A_n \cap \mathsf{SQ}_6 = A_n \cap \mathsf{REP}_6 = \{Ш(rr, ss) : s \in \{4, 5\}^{n/4}\}$ and note that $|B_n| = 2^{n/4}$. By the Interchange Theorem 4 with $m = n/2$ and $R = B_n$, there is a subset $Z \subseteq B_n$, $Z = \{z_1, \dots, z_k\}$ with $z_i = w_ix_iy_i$ satisfying the conclusions of that lemma. In particular,

$$k = |Z| \geq \frac{|R|}{c(n+1)^2} = \frac{2^{n/4}}{c(n+1)^2} > 2^{n/8}.$$

Case 1: There exist indices g, h, such that $x_g \neq x_h$. By the Interchange Theorem 4, $w_g x_h y_g \in L$. Moreover, one of the 4 or 5's was changed going from x_g to x_h. Since $|x| \leq m = n/2$, the corresponding 4 or 5 in the other half was not changed. So $w_g x_h y_g$ is not a square and hence does not contain a square. So $w_g x_h y_g \notin L$, contradiction.

Case 2: Case 1 fails. Then all the x_i are the same and have length at least $m/2 = n/4$. Therefore there are at least $n/4$ positions in which all the z_i are the same. Because of the shuffle and since $n/4$ is even, at least half of these, i.e., at least $n/8$ positions, contain 4's and 5's, which means that these $n/8$ positions have constant values within Z. This leaves at most $n/8$ positions, where both choices $4, 5$ are available within Z. Thus $|Z| \leq 2^{n/8}$, which is a contradiction. □

Theorem 14: ([16, Theorem 4.1.4]) *Let h be a morphism and $L \in$ **CFL**. Then $h^{-1}(L) = \{x \mid h(x) \in L\} \in$ **CFL**.*

Theorem 15: *For each language L,*

$$\mathsf{SQ}_3 \subseteq L \subseteq \mathsf{REP}_3 \implies L \notin \mathbf{CFL}.$$

Proof: Assume that $\mathsf{SQ}_3 \subseteq L \subseteq \mathsf{REP}_3$. Let h be the squarefree-preserving morphism of Theorem 11. Since h is a morphism, for all x, w we have

$$w = xx \implies h(w) = h(x)h(x) \in \mathsf{SQ}_3.$$

Thus $\mathsf{SQ}_6 \subseteq h^{-1}(\mathsf{SQ}_3)$. Also, the statement that h is squarefree-preserving is equivalent to: $h^{-1}(\mathsf{REP}_3) \subseteq \mathsf{REP}_6$. Thus

$$\mathsf{SQ}_6 \subseteq h^{-1}(\mathsf{SQ}_3) \subseteq h^{-1}(L) \subseteq h^{-1}(\mathsf{REP}_3) \subseteq \mathsf{REP}_6.$$

By Theorem 13, $h^{-1}(L) \notin$ **CFL**. Hence by Theorem 14, $L \notin$ **CFL**. □

Corollary 16: $\{x \in \{0, 1, 2\}^* : A_N(x) \leq |x|/2\} \notin$ **CFL**.

Proof: Let $L = \{x \in \{0, 1, 2\}^* : A_N(x) \leq |x|/2\}$. By Theorem 9 and Theorem 10 with $\alpha = 2$, $\mathsf{SQ}_3 \subseteq L \subseteq \mathsf{REP}_3$. By Theorem 15, $L \notin$ **CFL**. □

3. Sensitivity of automatic complexity

A natural decision problem associated with automatic complexity can be defined as follows.

Definition 17: Let $k, c \geq 1$ be integers, and let A_N denote nondeterministic automatic complexity. We define the language $L_{k,c}$ by

$$L_{k,c} = \{x \in [k]^* : A_N(x) > |x|/c\}.$$

The intent is that the decision problem $L_{2,3}$ captures the "random or not" characteristic of a word x. Shallit and Wang [17] asked whether automatic complexity can be computed in polynomial time. This remains open, in particular so does the question whether $L_{2,3} \in \mathbf{P}$. It is easy to see that $L_{2,3} \in \mathbf{coNP}$ [9]. Our state of knowledge is shown in Fig. 1.

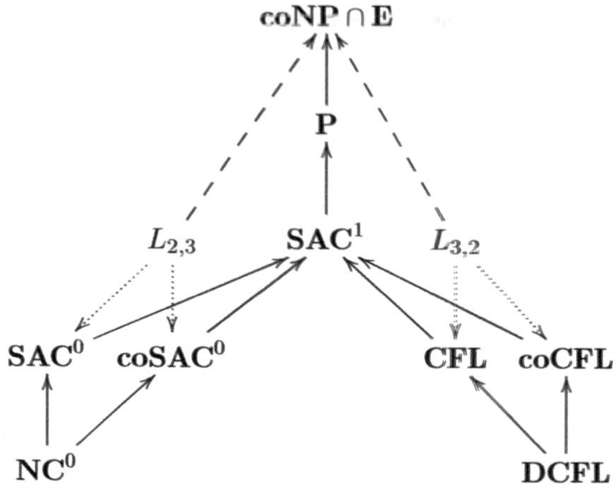

Fig. 1. Known implications ($\subseteq$) denoted by solid arrow, containments ($\in$) by dashed arrow, non-containments ($\notin$) by dotted arrow. The non-containments are demonstrated in this chapter for the first time except for the one from $L_{3,2}$ to **CFL**.

In order to prove the remaining main result of this chapter, Theorem 33, we shall use Theorem 27 on the abundance of high-complexity words, and a new result Theorem 25. The latter states that for all $y \in \{0,1\}^n$, and all c-tuples $a_1 < a_2 < \cdots < a_c < n$, there exists $x \in \{0,1\}^n$, such that $x(a_i) = y(a_i)$ for each $1 \le i \le c$, and $A_N(x) = O(c^2 \log n)$. Here, the constant O is independent of c and n. Before starting the proof, the reader is invited to consider Table 1.

We now prove some lemmas that will culminate in Theorem 25. Theorem 18 provides a small but valuable refinement of what we would otherwise get more crudely in Theorem 20.

Lemma 18: (Max Alekseyev [8]) *Let $c \ge 2$ be an integer and let $a_1 < a_2 < \cdots < a_c$ be real numbers. The average of the values $a_j - a_i$, $i < j$, is at most $\frac{c}{2(c-1)}\ell$, where $\ell = a_c - a_1$.*

Table 1. The best bound on $A_N(x)$ as a function of c and n for some small values of c and n.

$c \setminus n$	0	1	2	3	4	5	6
0	1	1	1	1	1	1	1
1	-	1	1	1	1	1	1
2	-	-	2	2	2	3	3
3	-	-	-	2	3	3	4
4	-	-	-	-	3	3	4
5	-	-	-	-	-	3	4
6	-	-	-	-	-	-	4

Proof: The average of the values $a_j - a_i$, $i < j$ is

$$\binom{c}{2}^{-1} \sum_{i<j} |a_j - a_i| = \frac{2}{c^2 - c} \sum_{i<j} a_j - a_i$$

$$= \frac{2}{c(c-1)} \sum_{i=0}^{(c-1)/2} (c - 1 - 2i)(a_{c-i} - a_{i+1}) \quad (3.1)$$

$$\leq \frac{2\ell}{c(c-1)} \sum_{i=0}^{(c-1)/2} (c - 1 - 2i) \quad (3.2)$$

$$\leq \frac{c}{2(c-1)} \ell. \quad (3.3)$$

For (3.2), the "large" values $a_c, a_{c-1}, \ldots, a_{c-i}, \ldots, a_{c-(c-1)/2}$ are the larger of the two numbers in a difference $c - 1 - i$ times and the smaller of the two i times. The "small" values a_{i+1} are the smaller $c - 1 - i$ times and the larger i times.

For (3.3), if $c \in \{2d, 2d+1\}$, the sum is

$$\sum_{i=0}^{d-1} (c - (2i+1)) = cd - d^2 = \begin{cases} d^2, & c = 2d \\ d^2 + d, & c = 2d + 1 \end{cases}$$

$$\leq (c/2)^2 = \begin{cases} d^2, & c = 2d \\ (d + \frac{1}{2})^2, & c = 2d + 1. \end{cases} \qquad \square$$

Definition 19: Let p_n denote the nth prime number: $p_1 = 2, p_2 = 3$, etc. The *primorial function* $x\#$ is defined by $x\# = \prod_{i=1}^{q} p_q$, where q is maximal such that $p_q \leq x$.

Lemma 20: *Let* $0 \leq a_1 < \cdots < a_c < n$, *all integers. Let* $q \in \mathbb{N}$ *be such that* $n \leq \frac{2(c-1)}{c} (p_q\#)^{1/\binom{c}{2}}$. *There exists a modulus* m *and an integer* $j \leq q$ *such that* $m = p_j \leq p_q$ *and all the* a_i *are distinct mod* m.

Proof: Suppose that for each $1 \le m = p_j \le p_q$, there exist $i \ne j$ with $a_i \equiv a_j \pmod{m}$. Then $\prod_{i=1}^q p_j$ divides $\prod_{i<j}(a_j - a_i)$. Since $i \ne j \implies a_i \ne a_j$ this gives, using Theorem 18 and the AM-GM inequality,

$$\begin{aligned} p_q\# = \prod_{i=1}^q p_j &\le \prod_{1\le i<j\le c} a_j - a_i \\ &\le (\text{average}(a_j - a_i))^{\binom{c}{2}} \\ &\le \left(\frac{c}{2(c-1)}\ell\right)^{\binom{c}{2}}, \end{aligned}$$

where $\ell = a_c - a_1 < n$. So

$$\frac{2(c-1)}{c}(p_q\#)^{1/\binom{c}{2}} < n.$$

Contrapositively, if all the $a_i < n \le \frac{2(c-1)}{c}(p_q\#)^{1/\binom{c}{2}}$ then there exists a prime $p_j \le p_q$, such that all the a_i are non-congruent mod p_j. □

Let log with no subscript denote the natural logarithm.

Definition 21: The first Chebyshev function $\vartheta : \mathbb{N} \to \mathbb{R}$ is defined by $\vartheta(x) = \sum_{i=1}^q \log p_i = \log(x\#)$ for $p_q \le x < p_{q+1}$.

Theorem 22: ([15, Equation (3.16)])

$$x(1 - 1/\log x) < \vartheta(x), \quad \textit{for all } x \ge 41.$$

Corollary 23: *For each $\varepsilon > 0$, and all sufficiently large x,*

$$(1-\varepsilon)\exp(x) \le x\#.$$

Proof: Theorem 22 is equivalent to

$$\frac{\exp(x)}{\exp(x/\log x)} < x\#, \quad \text{for } 41 \le x.$$ □

Corollary 24: *Let $\varepsilon > 0$ and $c \in \mathbb{N}$. Let $0 \le a_1 < \cdots < a_c < n$, all integers, with n sufficiently large. Let $q \in \mathbb{N}$ be such that $n \le \frac{2(c-1)}{c}((1-\varepsilon)\exp(p_q))^{1/\binom{c}{2}}$. There exists a modulus m and an integer $j \le q$, such that $m = p_j \le p_q$ and all the a_i are distinct mod m.*

Proof: From Theorem 23 and Theorem 20. □

Theorem 25: *Let $c \in \mathbb{N}$ and let n be sufficiently large. For all $a_1 < \cdots < a_c < n$ and each $y \in \{0,1\}^n$, there is an $x \in \{0,1\}^n$, such that $x(a_i) = y(a_i)$ for each $1 \le i \le c$, and $A_N(x) \le \binom{c}{2}\log n < n/3$.*

Proof: Notice that if we take $n = \frac{2(c-1)}{c}((1-\varepsilon)\exp(p_q))^{1/\binom{c}{2}}$ (within ± 1) in Theorem 24 then solving for p_q, if $c \geq 2$ we have

$$\begin{aligned} p_q &= \log\left\{\left(\frac{c}{2(c-1)}n\right)^{\binom{c}{2}} \cdot \frac{1}{1-\varepsilon}\right\} \\ &= \binom{c}{2}\left(\log\frac{n}{2} + \log\frac{c}{(c-1)}\right) - \log(1-\varepsilon) \\ &\leq \binom{c}{2}\log n, \end{aligned}$$

and so

$$m \leq p_q = \binom{c}{2}\log n. \tag{3.4}$$

Take x to be a (non-integral, in general) power of z, $x = z^{n/m}$, $|z| = m$ as in (3.4), where $z(a_i \mod m) = y(a_i)$ for each i. In other words, given $y \in \{0,1\}^n$, consider y on $\{a_1, \ldots, a_c\}$, copy those values $(y(a_1), \ldots, y(a_c))$ to a $z \in \{0,1\}^m$, then extend that z to $x \in \{0,1\}^n$ by repetition. $\square$

Remark 26: Let us consider an example of the construction in Theorem 25. Let $c = 12$, $m = 14$ (admittedly not a prime, but optimal in this example), and let a_i and $b_i = a_i \mod m$ for each i be as follows.

$$\begin{aligned} (a_1, \ldots, a_{12}) &= (3,4,5,7,8,11,20,23,24,26,27,28) \\ (b_1, \ldots, b_{12}) &= (3,4,5,7,8,11,6,9,10,12,13,0). \end{aligned}$$

We let y be an interesting word from [12, Theorem 20] and let z be obtained as in Theorem 25. We underline the positions a_i in x and y.

$$\begin{aligned} y &= \texttt{000}\underline{\texttt{101}}\texttt{0}\underline{\texttt{11}}\texttt{01}\underline{\texttt{0}}\texttt{00011001}\underline{\texttt{0}}\texttt{01}\underline{\texttt{11}}\texttt{1}\underline{\texttt{101}}\texttt{11} \\ x &= \texttt{122}\underline{\texttt{101}}\texttt{0}\underline{\texttt{11}}\texttt{11}\underline{\texttt{0}}\texttt{10122101}\underline{\texttt{0}}\texttt{11}\underline{\texttt{11}}\texttt{0}\underline{\texttt{101}}\texttt{22} = z^{31/14} \\ z &= \texttt{12210101111010}. \end{aligned}$$

The occurrences of 2 in z can be replaced by either 0 or 1 as they are unconstrained. The NFA showing that $A_N(x) \leq m$ is as follows

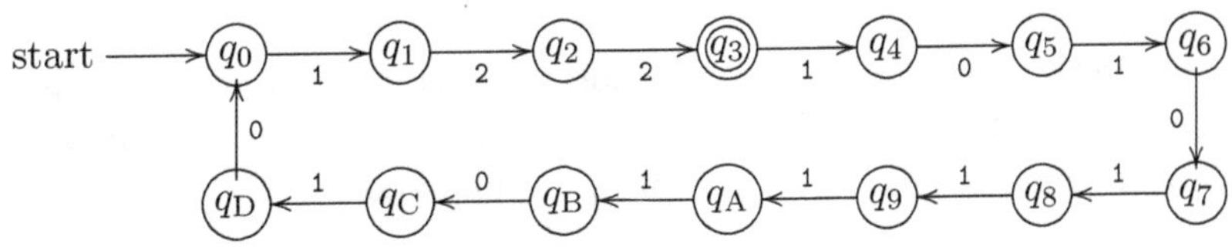

where the final state, q_3, is indicated by a double circle.

Theorem 25 showed that it is possible for a low-complexity word to agree with a given high-complexity word in given positions. Theorem 27 will be used for a converse problem: showing that a high-complexity word can agree with a given low-complexity word in given positions.

Theorem 27: ([13]) *Let $\mathbb{P}_n$ denote the uniform probability measure on words $x \in \Gamma^n$, where Γ is a finite alphabet of cardinality at least 2. For all $\epsilon > 0$,*

$$\lim_{n\to\infty} \mathbb{P}_n\left(\left|\frac{A_N(x)}{n/2} - 1\right| < \epsilon\right) = 1.$$

3.1. *Defining the class* SAC^0

The main fact about **SAC**^{0} that we shall use is Theorem 28.

Theorem 28: *If $A \subseteq \Sigma^*$ is in* **SAC**^{0} *then there is a constant c, such that for each n, there is a formula $\psi = \bigvee_{i=1}^k \varphi_i$, such that each φ_i mentions at most c variables, and for $x = (x_1, \ldots, x_n) \in \Sigma^n$, $x \in A$ iff $\psi(x)$ holds.*

Remark 29: The literature is not consistent on whether **SAC**^{0} is by definition uniform, i.e., uniformly efficiently computable. Aaronson [1] states that "a uniformity condition may also be imposed". Our result does not require the circuit families to be uniform.

Remark 30: The requirement of polynomial size circuits, while presumably important for **SAC**^{k} with $k \geq 1$, is redundant for **SAC**^{0}. When the depth is bounded by c, the number of variables in each φ_i is bounded by another constant $d = d(c)$, and there are only $e = e(d)$ formulas in d variables, so that the φ_i are chosen from a set of size at most $\binom{n}{d}e(d)$, which is a polynomial in n of degree d.

Theorem 28 can be considered already well-known. Borodin *et al.* [4, page 560], referring back to Venkateswaran [18, page 383] describe the semi-unbounded fan-in circuit model as follows.

> [...] in this model, we allow OR gates with arbitrary fan-in, whereas all AND gates have bounded fan-in. Input variables and their negations are supplied, but negations are prohibited elsewhere.

An example of a circuit computing the function $f(x_1, x_2) = x_2$, with some redundant gates:

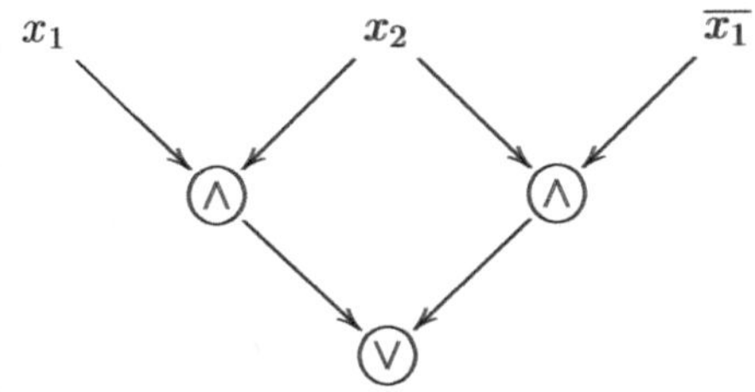

It is important for us to note that for the described model, all unbounded fan-in OR gates may be assumed to be at the maximum depth, i.e., the OR gates is the output gate and is the furthest removed from the input variables and their negations. This is shown by repeated use of the distributive law for $\wedge$ and $\vee$.

3.2. *Immunity to* $\mathbf{SAC^0}$

Definition 31: Let $L \subseteq \Sigma^*$ and $\mathsf{C} \subseteq \mathcal{P}(\Sigma^*)$, where $\mathcal{P}$ denotes power set. Let $\mathsf{Inf} = \{L : L \text{ is an infinite set}\}$.

- The set L is C-immune if for all $C \in \mathsf{Inf}$, $C \in \mathsf{C} \implies C \not\subseteq L$.
- The set L is C-coimmune if $\Sigma^* \setminus L$ is C-immune.

Writing $\mathsf{coC} = \{L \subseteq \Sigma^* : L \notin \mathsf{C}\}$, we then also have:

- The set L is coC-immune if for all $C \in \mathsf{Inf}$, $C \notin \mathsf{C} \implies C \not\subseteq L$.
- The set L is coC-coimmune if for all $C \in \mathsf{Inf}$, $C \notin \mathsf{C} \implies C \not\subseteq \Sigma^* \setminus L$.

Remark 32: We can think of immune sets as "small" and coimmune sets as "big" in some sense. Conversely, if a set L is both C-immune and C-coimmune then it points to an ability to avoid having subsets in C, i.e., sets in C may tend to be "big" in some sense.

From this point of view, Theorem 33 says that $L_{2,3}$ is neither big nor small (or if the reader prefers, both big and small), and many of the sets in the class $\mathbf{SAC^0}$ are rather large.

Theorem 33: *$L_{2,3}$ is $\mathbf{SAC^0}$-immune and $\mathbf{SAC^0}$-coimmune.*

Proof: Let φ_n be a $\mathbf{SAC^0}$ formula, representing a circuit family. Then $\varphi_n = \bigvee_i \varphi_{i,n}$, where $\varphi_{i,n}$ depends only on c bits of the input, where c does not depend on n but the choice of c bits can depend on both n and i.

Suppose some $\varphi_{i,n}$ determines membership in $L_{2,3}$ based on the bits $p_1 < \cdots < p_c$ of the input.

To show $L_{2,3}$ is $\mathbf{SAC}^0$-coimmune, we show that the bits p_i cannot guarantee low complexity, i.e., $x \notin L_{2,3}$. Let n be sufficiently large. Fix values $x(p_i)$ and select the values $x(k)$, $k \notin \{p_1, \dots, p_c\}$, $1 \le k \le n$, randomly (uniformly and independently). The probability of the values $x(p_i)$ is 2^{-c}. Let $\epsilon = 1/3$. Then

$$\left|\frac{A_N(x)}{n/2} - 1\right| < \epsilon \implies A_N(x) \ge \frac{\lfloor n/2 + 1 \rfloor}{2} > n/3 \implies x \in L_{2,3}.$$

By Theorem 27, for large enough n,

$$\mathbb{P}_n\left(\left|\frac{A_N(x)}{n/2} - 1\right| < \epsilon\right) \ge 1 - 2^{-(c+1)}.$$

Therefore, by the union bound, the probability of either disagreeing with some $x(p_i)$, or having low complexity, is at most

$$(1 - 2^{-c}) + 2^{-(c+1)} < 1,$$

so the probability of agreeing with the $x(p_i)$ and also having high complexity is positive.

Since this occurs with positive probability, in particular it occurs for at least one x. For that x, $x \in L_{2,3}$, as desired. This shows that no single $\varphi_{i_0,n}$ can imply that $x \notin L_{2,3}$. Since $\varphi_{i_0,n}$ implies $\bigvee_i \varphi_{i,n}$, neither can $\bigvee_i \varphi_{i,n}$ imply that $x \notin L_{2,3}$. So $L_{2,3}$ is $\mathbf{coSAC}_0$-immune.

To show $L_{2,3}$ is $\mathbf{SAC}^0$-immune, we show that the bits p_i cannot guarantee high complexity, i.e., $x \in L_{2,3}$.

This follows from Theorem 25, which shows how in the limit we can force $x \notin L_{2,3}$. □

4. The class $\oplus\mathbf{SAC}^0$

Gál and Wigderson [7] considered arithmetic circuits with gates from the basis $\{+, -, \times\}$ over fields such as $\mathbb{Z}/2\mathbb{Z}$, i.e., $GF(2)$. All constants of the field may be used. Boolean circuits have the standard Boolean basis $\{\wedge, \vee, \neg\}$. *Semi-unbounded fan-in* circuits have constant fan-in $\times$ (resp. $\wedge$) gates and unbounded fan-in $+$ (resp. $\vee$) gates. Semi-unbounded fan-in Boolean circuits may have negations only at the input level. $\mathbf{SAC}^k$ denotes the class of languages accepted by polynomial size, depth $O((\log n)^k)$ semi-unbounded fan-in Boolean circuits. $\oplus\mathbf{SAC}^k$ denotes the class of languages accepted by polynomial size, depth $O((\log n)^k)$ semi-unbounded fan-in arithmetic circuits over $GF(2)$.

Gál and Wigderson showed that $\mathbf{SAC}^1 \subseteq \oplus\mathbf{SAC}^1$. We note that $\mathbf{SAC}^0 \neq \mathbf{coSAC}^0$ was shown in [4]. However, $\mathrm{co}\oplus\mathbf{SAC}^0 = \oplus\mathbf{SAC}^0$ since if φ is an multilinear polynomial representing L then $\varphi + 1$ represents $\Sigma^* \setminus L$. Analogously to Theorem 28 for $\mathbf{SAC}^0$, the main fact about $\oplus\mathbf{SAC}^0$ that we shall use is Theorem 34.

Theorem 34: *If $A \subseteq \Sigma^*$ is in $\oplus\mathbf{SAC}^0$ then there is a constant c, such that for each n, there is a formula $\psi = \bigoplus_{i=1}^k \varphi_i$, such that each φ mentions at most c variables, and for $x = (x_1, \ldots, x_n) \in \Sigma^n$, $x \in A$ iff $\psi(x)$ holds.*

Proof: A formula of bounded depth with bounded fan-in can only mention a bounded number of variables. All the unbounded fan-in occurrences of $\bigoplus$ can be folded into one using the distributive law of $\cdot$ and $+$. $\square$

Remark 35: In Theorem 34 it follows that the formulas ψ have size polynomial in n, much as in Theorem 30.

We can show that $\{x : A_N(x) > 1\} \in \mathbf{SAC}^0 \setminus \oplus\mathbf{SAC}^0$ and hence $\mathbf{SAC}^0 \not\subseteq \oplus\mathbf{SAC}^0$. Theorem 39 resolves an open implication from the Complexity Zoo [1].

Definition 36: We define the *degree* of a multilinear polynomial by

$$\deg\left(\sum_{F\in\mathcal{F}}\prod_{i\in F} x_i\right) = \max\{|F| : F \in \mathcal{F}\}.$$

Lemma 37: *A multilinear polynomial over the ring $(\mathbb{Z}/2\mathbb{Z}, +\cdot)$ is identically 0 as a function over $\mathbb{Z}/2\mathbb{Z}$ only if all the coefficients are 0.*

Proof: Let P be the set of (formal) multilinear polynomials in n variables with coefficients in $\{0, 1\}$. Let F_P be the set of Boolean functions computed by elements of F. It suffices to show that two multilinear polynomials are equal as functions only if they are equal as (formal) polynomials, i.e., the map sending a formal polynomial to its function is one-to-one. Thus, it suffices to show that $|F_P| \leq |P|$.

Let B be the set of all Boolean functions in n variables. Each Boolean function may be expressed as a multilinear monomial over $\mathbb{Z}/2\mathbb{Z}$. To wit,

$$\bot = 0, \quad \top = 1, \quad \neg a = a + 1, \quad \text{and} \quad a \wedge b = a \cdot b.$$

Thus $B = F_P$.

For each set $S \subseteq \{1, \ldots, n\}$ there is a multilinear monomial $\prod_{i \in S} x_i$. Thus there are 2^n multilinear monomials in n variables. Each subset $\mathcal{S}$ of these may be included in a multilinear polynomial

$$\sum_{S \in \mathcal{S}} \prod_{i \in S} x_i.$$

Thus $|P| = 2^{2^n}$. As it is well-known that $|B| = 2^{2^n}$, we conclude

$$|F_P| = |B| = 2^{2^n} = |P|. \qquad \square$$

Remark 38: Theorem 37 is a known result, but it should not be confused with the similar representation of Boolean functions as multilinear polynomials over $\mathbb{R}$. For instance, XOR is $x + y - xy$ over $\mathbb{R}$, but $x + y$ over $GF(2)$. Also, Theorem 37 is somewhat sharp in that if we go beyond linear polynomials we quickly get a counterexample: $x^2 + x$ is identically 0 as a function.

Theorem 39: $\mathbf{SAC^0} \not\subseteq \oplus\mathbf{SAC^0}$.

Proof: We will show that the family of disjunction functions

$$\bigvee_n (x_1, \ldots, x_n) = x_1 \vee \cdots \vee x_n$$

is in $\mathbf{SAC^0} \setminus \oplus\mathbf{SAC^0}$. To see that $\bigvee_n$ is in $\mathbf{SAC^0}$ is trivial: it is computed by the following constant-depth circuit.

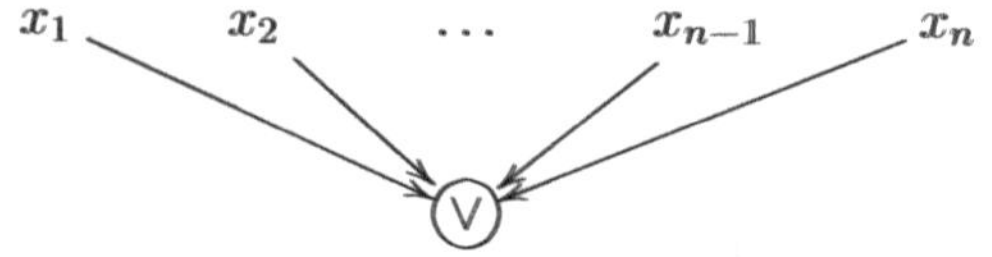

It remains to show that the function $\bigvee_n$ is not in $\oplus\mathbf{SAC^0}$. By de Morgan's law,

$$\bigvee_n (x_1, \ldots, x_n) = 1 + \prod_{i=1}^{n} (1 + x_i) = 1 + \sum_{F \subseteq [n]} x_F = \sum_{\emptyset \neq F \subseteq [n]} x_F \qquad (4.1)$$

where $x_F = \prod_{i \in F} x_i$, $x_\emptyset = 1$. For instance, $\bigvee_2 (x, y) = 1 + (xy + x + y + 1) = xy + x + y$.

By Theorem 37, if $\bigvee_n$ is equal to the function expressed by a $\oplus\mathbf{SAC^0}$ formula (circuit) then that formula expands to the polynomial in (4.1). By Theorem 34, an $\oplus\mathbf{SAC^0}$ formula is a sum of terms $\sum_i \varphi_i \pmod 2$, where each φ_i depends on a bounded number of variables. So when we express φ_i as a multilinear polynomial, it will have bounded degree since

$$\deg\left(\sum_i \varphi_i\right) \leq \max_i \deg(\varphi_i).$$

Since $\bigvee_n$ has degree n by (4.1), the result follows. □

Corollary 40: $\{x \in \{0,1\}^* : A_N(x) = 1\} \notin \oplus\mathbf{SAC}^0$.

Proof: We have $A_N(x) = 1$ iff $x \in \{0^n, 1^n\}$ (where $n = |x|$), iff

$$\prod_{i\in[n]} x_i + \prod_{i\in[n]} (x_i + 1) = \sum_{\emptyset\neq F\subsetneq[n]} \prod_{i\in F} x_i = 1.$$

Since $\sum_{\emptyset\neq F\subsetneq[n]} \prod_{i\in F} x_i$ has degree $n-1$, which is unbounded as $n \to \infty$, the result follows as in Theorem 39. □

While we have not settled whether $L_{2,3} \in \oplus\mathbf{SAC}^0$, Theorem 40 does show that $\oplus\mathbf{SAC}^0$ is too limited to capture the computational complexity of A_N.

Acknowledgments

This work was partially supported by a grant from the Simons Foundation (#704836 to Bjørn Kjos-Hanssen).

References

1. Scott Aaronson. The complexity zoo. `https://complexityzoo.net/Complexity_Zoo`, 2022.
2. Eric Allender. The complexity of complexity. In *Computability and Complexity*, volume 10010 of *Lecture Notes in Comput. Sci.*, pages 79–94. Springer, Cham, 2017.
3. Jean Berstel and Luc Boasson. Context-free languages. In *Handbook of Theoretical Computer Science, Vol. B*, pages 59–102. Elsevier, Amsterdam, 1990.
4. Allan Borodin, Stephen A. Cook, Patrick W. Dymond, Walter L. Ruzzo, and Martin Tompa. Two applications of inductive counting for complementation problems. *SIAM J. Comput.*, 18(3):559–578, 1989.
5. Franz-Josef Brandenburg. Uniformly growing kth power-free homomorphisms. *Theoret. Comput. Sci.*, 23(1):69–82, 1983.
6. Andrzej Ehrenfeucht and Grzegorz Rozenberg. On the separating power of EOL systems. *RAIRO Inform. Théor.*, 17(1):13–22, 1983.
7. Anna Gál and Avi Wigderson. Boolean complexity classes vs. their arithmetic analogs. In *Proceedings of the Seventh International Conference on Random Structures and Algorithms (Atlanta, GA, 1995)*, volume 9, pages 99–111, 1996.

8. Max Alekseyev (https://mathoverflow.net/users/7076/max-alekseyev). Least modulus distinguishing some integers. MathOverflow. URL:https://mathoverflow.net/q/416457 (version: 2022-02-18).
9. Kayleigh K. Hyde and Bjørn Kjos-Hanssen. Nondeterministic automatic complexity of overlap-free and almost square-free words. *Electron. J. Combin.*, 22(3), 2015. Paper 3.22, 18.
10. Liam Jordon and Philippe Moser. Normal sequences with non maximal automatic complexity. In *Symposium on Fundamentals of Computation Theory*, FCT '21, 2021.
11. Bjørn Kjos-Hanssen. On the complexity of automatic complexity. *Theory Comput. Syst.*, 61(4):1427–1439, 2017.
12. Bjørn Kjos-Hanssen. Automatic complexity of shift register sequences. *Discrete Math.*, 341(9):2409–2417, 2018.
13. Bjørn Kjos-Hanssen. An incompressibility theorem for automatic complexity. *Forum of Mathematics, Sigma*, 9:e62, 2021.
14. William Ogden, Rockford J. Ross, and Karl Winklmann. An "interchange lemma" for context-free languages. *SIAM J. Comput.*, 14(2):410–415, 1985.
15. J. Barkley Rosser and Lowell Schoenfeld. Approximate formulas for some functions of prime numbers. *Illinois J. Math.*, 6:64–94, 1962.
16. Jeffrey Shallit. *A Second Course in Formal Languages and Automata Theory*. Cambridge University Press, New York, NY, USA, 1 edition, 2008.
17. Jeffrey Shallit and Ming-Wei Wang. Automatic complexity of strings. *J. Autom. Lang. Comb.*, 6(4):537–554, 2001. 2nd Workshop on Descriptional Complexity of Automata, Grammars and Related Structures (London, ON, 2000).
18. Hari Venkateswaran. Properties that characterize LOGCFL. *J. Comput. System Sci.*, 43(2):380–404, 1991.

https://doi.org/10.1142/9789811278631_0014

Roots of Polynomials in Fields of Generalized Power Series

Julia Knight
Department of Mathematics, University of Notre Dame
j1knight@nd.edu

Karen Lange
Department of Mathematics, Wellesley College
klange2@wellesley.edu

Reed Solomon
Department of Mathematics, University of Connecticut
david.solomon@uconn.edu

We consider roots of polynomials in fields of generalized power series. Newton [7] and Puiseux [8,9] showed that if K is an algebraically closed field of characteristic 0, then the field $K\{\{t\}\}$ of *Puiseux series* over K is algebraically closed. Mac Lane [5] showed that if K is an algebraically closed field of characteristic 0 and G is a divisible ordered Abelian group, then the *Hahn field* $K((G))$ is algebraically closed. Our goal is to measure the complexity of the roots and the root-taking process in these fields. For a polynomial over $K\{\{t\}\}$, the roots are computable in K and the coefficients. In fact, knowing that a polynomial is non-constant, and given K and the coefficients, we can apply a uniform effective procedure to find a root. Puiseux series have length at most ω. Hahn series may be longer, and the complexity of the roots goes up with the length. In [3] and [4], there are results bounding the lengths of roots of a polynomial over $K((G))$ in terms of the lengths of the coefficients. Using these results, we show that the generalized Newton-Puiseux Theorem holds in any admissible set. We set bounds on the complexity of initial segments of a root. The bounds are sharp for initial segments of length less than $\omega + \omega$.

1. Introduction

An *integer part* of a real closed field R is a discrete ordered subring Z, such that for every $r \in R$, there is a unique $z \in Z$, for which $z \leq r < z + 1$. Mourgues and Ressayre [6] proved that every real closed field R has an integer part. In their proof, they constructed a particular embedding of R into a Hahn field $K((G))$, a field of series of ordinal length, where K is the residue field and G is the value group of R. (We define Hahn fields below.) Constructing this embedding requires extracting appropriate roots in $K((G))$. In [3] and [4], Knight and Lange studied the complexity of these roots in terms of their ordinal lengths. In this chapter, we consider the complexity of finding roots in Hahn fields and in the simpler setting of Puiseux series from the perspective of computability theory. Rather than working with real closed fields, we focus on the case when the field K is algebraically closed of characteristic 0.

The *Puiseux series* over a field K are formal power series of the form $\sum_{i \in \omega} a_i t^{q_i}$, where $a_i \in K$ and $(q_i)_{i \in \omega}$ is an increasing sequence of rationals with an upper bound on the denominators. Newton [7], in 1676, and Puiseux [8, 9], in 1850–1851, showed that if K is an algebraically closed field of characteristic 0, then the set $K\{\{t\}\}$ of Puiseux series over K is an algebraically closed field. Our goal is to understand the computability theoretic complexity of the roots of a polynomial with coefficients in $K\{\{t\}\}$ in terms of the complexity of the coefficient series.

For a divisible ordered Abelian group G and a field K, the *Hahn series* have the form $\sum_{i<\alpha} a_i t^{g_i}$, where $a_i \in K$ and $(g_i)_{i \in \alpha}$ is an increasing sequence of elements of G indexed by an ordinal α. If G has cardinality κ, the length of the sequence may be any ordinal $\alpha < \kappa^+$. Mac Lane proved the analogue of the Newton-Puiseux Theorem for Hahn fields, showing that if K is algebraically closed field of characteristic 0, and G is a divisible ordered Abelian group, then the set $K((G))$ of Hahn series is a field that is again algebraically closed.

We are interested in complexity in these fields. In Sec. 2, we give some background on fields of Puiseux series. In Sec. 3, we consider complexity in Puiseux series. We show that if K is algebraically closed of characteristic 0, with universe a subset of ω, and $p(x) = A_0 + A_1 x + \ldots + A_n x^n$, where $A_i \in K\{\{t\}\}$, then the roots of $p(x)$ are all computable in K and the A_i's. Moreover, we have a uniform effective procedure that, given K with universe a subset of ω, and the coefficients of a non-constant polynomial $p(x)$, produces a root. In Sec. 4, we give some background on Hahn fields. In

Sec. 5, we consider complexity. We show that Mac Lane's Theorem holds in any admissible set M. That is, if G and K are in M, then for any polynomial over $K((G))$ with coefficients in M, the roots are all in M. For this, we use results from [3,4] bounding the length of roots of polynomials over $K((G))$ in terms of the lengths of the coefficients. In the case where G and K have universe a subset of ω, we consider the complexity of initial segments of the roots of polynomials.

2. Puiseux series

Definition 2.1: The *Puiseux series* over a field K are the formal sums of the form $s = \sum_{i \geq k} a_i t^{\frac{i}{n}}$, where n is a positive integer, k is an integer, and for each integer $i \geq k$, $a_i \in K$. The *support* of s, denoted by $Supp(s)$, is the set of rationals $\frac{i}{n}$, such that $a_i \neq 0$.

It is helpful to think of t as infinitesimal. Thus, t^q is infinite if $q < 0$, and t^q is infinitesimal if $q > 0$. We write $K\{\{t\}\}$ for the set of Puiseux series with coefficients in K. We define addition and multiplication on $K\{\{t\}\}$ in a natural way.

- In the sum $s + s'$, the coefficient of t^q is the sum of the coefficients of t^q in s and in s' (if there is no term corresponding to q, we take the coefficient to be 0).
- In the product $s \cdot s'$, the coefficient of t^q is the sum of the products $b \cdot b'$, such that for some $r, r' \in \mathbb{Q}$ with $r + r' = q$, b is the coefficient of t^r in s and b' is the coefficient of $t^{r'}$ in s'.

There is a natural valuation w on $K\{\{t\}\}$ defined as follows.

$$w(s) = \begin{cases} \min(Supp(s)) & \text{if } s \neq 0 \\ \infty & \text{if } s = 0. \end{cases}$$

Note that if $s \neq 0$, then $Supp(s) \neq \emptyset$ and so has a least element. We sometimes refer to $w(s)$ as the weight of s. If K is ordered, there is also a natural order on $K\{\{t\}\}$ defined by s is positive if $s \neq 0$ and the coefficient of $t^{w(s)}$ is positive.

Newton [7] and Puiseux [8,9] showed the following.

Theorem 2.2: *(Newton-Puiseux) If K is algebraically closed of characteristic 0, then $K\{\{t\}\}$ is also algebraically closed.*

For a proof, see Basu, Pollack, and Roy [2], or Walker [11]. Below, we say a little about the root-taking process.

Lemma 2.3: *Let $p(x)$ be a non-constant polynomial over $K\{\{t\}\}$. For suitably chosen g, the coefficients of $t^g p(x)$ all have non-negative valuation, and the least valuation is 0. The roots are the same as for $p(x)$.*

Proof: For $p(x) = A_0 + A_1x + \ldots + A_nx^n$, take the least $w(A_i)$, and let $g = -w(A_i)$. □

Lemma 2.4: *Let $p(x)$ be a non-constant polynomial over $K\{\{t\}\}$. Assuming that $p(x)$ has a non-zero root, then for suitably chosen g, $p(t^{-g}x)$ has a root with positive valuation.*

Proof: Let $p(x) = A_0 + A_1x + \ldots + A_nx^n$, where $w(A_i) \geq 0$. If r is a non-zero root of $p(x)$ and $-g < w(r)$, then $t^g r$ is a root of $p(t^{-g}x)$ with positive valuation. We will see soon how to choose g just by looking at the coefficients A_i. □

We can now sketch the proof of the Newton-Puiseux Theorem.

Proof: Consider a polynomial $p(x) = A_0 + A_1x + \ldots + A_nx^n$, where $w(A_i) \geq 0$ for all i. If $A_0 = 0$, then 0 is a root of $p(x)$. We suppose that $A_0 \neq 0$. We draw the *Newton polygon*. We start by graphing the points $(i, w(A_i))$, for $0 \leq i \leq n$, omitting any points such that $A_i = 0$. The polygon is convex, with certain of the points $(i, w(A_i))$ lying on the edges, and other points lying above. Since $A_0 \neq 0$, the first vertex is $(0, w(A_0))$. We rotate a ray through the vertex $(0, w(A_0))$ counter-clockwise from a starting position pointing downwards until it touches another point $(j, w(A_j))$. It may hit more than one point at the same time, in which case, we let $(j, w(A_j))$ be the point with the largest index that it hits. The segment from $(0, w(A_0))$ to $(j, w(A_j))$ is the first side of the Newton polygon. We repeat this process with A_j and continue until every point is either on or above a side of the Newton polygon.

To find the valuation of a root r of $p(x)$, we consider a side of the Newton polygon. For a side with first point $(i, w(A_i))$ and last point $(j, w(A_j))$, $\nu = \frac{w(A_i) - w(A_j)}{j - i}$ is the valuation of at least one root. (We will not prove this fact, but see Sec. 2.6 of [2] or Sec. IV.3 of [11].) The negatives of the slopes of the sides of the polygon are exactly the valuations of the roots. Convexity means that the slopes are increasing. Thus, the greatest valuation of a root is the one we get from the first side, the one with least slope.

Fix a side L of the Newton polygon, and let ν be the negative of the slope. The *carrier* Δ_ν is the set of points $(k, w(A_k))$ that lie on the side L.

The *ν-principal part* is the polynomial $\sum_{k\in\Delta_\nu} c_k z^{k-i}$, where c_k is the first non-zero coefficient in A_k. This is a polynomial over K. If b is a root of the ν-principal part, then b is the coefficient of t^ν in a root of $p(x)$. Any root of the ν-principal part serves. Then bt^ν is the first term of a root. Note that if the constant term c_i in the ν-principal part is non-zero, then the root b is also not zero. Considering the ν's corresponding to all sides of the Newton polygon, and all roots b of the ν-principal parts counted with multiplicity, we have found the first non-zero terms of all non-zero roots of $p(x)$.

Let $r_1 = b_0 t^{\nu_0}$ be a possible first term of a root, where $\nu_0 = \nu$ and $b_0 = b$ are determined above. To find the second term of a root with r_1 as the first term, we consider the new polynomial $q(x) = p(r_1 + x) = B_0 + B_1 x + \ldots + B_n x^n$. If $B_0 = 0$, then r_1 is a root of $p(x)$ because $q(0) = 0$. Otherwise, we find the first term of a root of $q(x)$, using the method above. Say this is $b_1^{\nu_1}$, and let $r_2 = b_0 t^{\nu_0} + b_1 t^{\nu_1}$. Given r_n, the sum of the first n terms of a root r, we consider $p_n(x) = p(r_n + x)$. If the constant term is 0, then r_n is a root of $p(x)$. Otherwise, we find the first term of a root $b_{n+1} t^{\nu_{n+1}}$ of $p_n(x)$. Adding this to r_n, we have r_{n+1}. Continuing, we get a root of form r_n or r_ω. □

3. Complexity in fields of Puiseux series

In this section, our goal is to measure complexity in a field of Puiseux series $K\{\{t\}\}$. We must say exactly how we are representing Puiseux series.

3.1. *Representing Puiseux series*

We suppose that the field K has universe ω, and we fix a computable copy of $\mathbb{Q}$ with universe ω. We represent elements of $K\{\{t\}\}$ by functions $f : \omega \to K \times \mathbb{Q}$, such that if $f(n) = (a_n, q_n)$, then

- q_n increases with n, and
- there is a uniform bound on the denominators of the q_n terms.

To be explicit, the function $f : \omega \to K \times \mathbb{Q}$, such that $f(n) = (a_n, q_n)$ corresponds to the series $\sum_{n\in\omega} a_n t^{q_n}$. Throughout this section, we assume Puiseux series are representing this way. Note that for $f(n) = (a_n, q_n)$ representing a Puiseux series, q_n is defined for all n. This fact, together with the fact that there is a bound on the denominators of the q_n, implies that $\lim_{n\to\infty} q_n = \infty$.

3.2. *Complexity of basic operations*

Lemma 3.1: *Given K and $s, s' \in K\{\{t\}\}$, we can effectively compute $s+s'$ and $s \cdot s'$.*

Proof: Say that $s(n) = (a_n, q_n)$ and $s'(n) = (a'_n, q'_n)$. We compute the value $(s+s')(n) = (b_n, r_n)$ as follows. Let r_0 be the smaller of q_0, q'_0. If $q_0 < q'_0$, then $b_0 = a_0$, if $q'_0 < q_0$, then $b_0 = a'_0$, and if $q_0 = q'_0$, then $b_0 = a_0 + a'_0$. Given (b_n, r_n), we find the least m, such that $r_n < q_m$ and the least m', such that $r_n < q'_{m'}$. Then r_{n+1} is the smaller of $q_m, q'_{m'}$. If $q_m < q'_{m'}$, then $b_{n+1} = a_m$, if $q'_{m'} < q_m$, then $b_{n+1} = a'_{m'}$, and if $q_m = q'_{m'}$, then $b_{n+1} = a_m + a'_{m'}$.

For simplicity, we suppose that $q_0, q'_0 \geq 0$ and we compute the value $(s \cdot s')(n) = (b_n, r_n)$ and the finite set $S_n = \{(m, m') : q_m + q'_{m'} = r_n\}$ as follows. Let $r_0 = q_0 + q'_0$, and let $b_0 = a_0 a'_0$. We have $S_0 = \{(0,0)\}$. Given (b_n, r_n) and S_n, we compute (b_{n+1}, r_{n+1}) and S_{n+1} as follows. Let M be least integer greater than all m for $(m, m') \in S_n$, and let M' be the least integer greater than all m' for $(m, m') \in S_n$. Let r_{n+1} be the least rational greater than r_n, such that for some $m \leq M$ and $m' \leq M'$, $r_{n+1} = q_m + q'_{m'}$. Let S_{n+1} be the set of pairs (m, m'), such that $q_m + q'_{m'} = r_{n+1}$. (For all of these pairs, $m \leq M$ and $m' \leq M'$.) The coefficient b_{n+1} is equal to $\sum_{(m,m') \in S_{n+1}} a_m a'_{m'}$. □

Using similar arguments, one can show the following lemma.

Lemma 3.2: *Given K and s, it is Π^0_1, but not computable, to say that $s = 0$. Given that $s \neq 0$, we can effectively find $w(s)$. Moreover, regardless of whether s is non-zero, we can effectively determine the ordering between $w(s)$ and q for any $q \in \mathbb{Q}$.*

Here is an effective version of Lemma 2.3.

Lemma 3.3: *Given a non-constant polynomial $p(x)$ over $K\{\{t\}\}$, we can effectively find a polynomial $q(x) = t^g p(x)$ with the same roots, such that the coefficients of $q(x)$ all have non-negative valuation and the sides of the Newton polynomials for $p(x)$ and $q(x)$ have the same slope.*

Proof: Let $p(x) = A_0 + A_1 x + \ldots + A_n x^n$. Since $p(x)$ is not constant, there is some $i > 0$, such that $A_i \neq 0$. We can find $w(A_i)$ for some such i. We can effectively tell whether $w(A_j) < w(A_i)$ for each $j \neq i$ by Lemma 3.2. Thus, we can find the least value of $w(A_i)$, say $-g$, and we can find the least k, such that $w(A_k) = -g$. Then $q(x) = t^g p(x)$ has the same roots as $p(x)$, and if $q(x) = B_0 + B_1 x + \ldots + B_n x^n$, then $w(B_i) = w(A_i) + g$. Therefore, its coefficients have non-negative valuations and the Newton polygon for $q(x)$ is the result of shifting the one for $p(x)$ vertically by g. It follows that the slopes of the sides of the polygons for $p(x)$ and $q(x)$ are the same. □

Lemma 3.4: *Given a non-constant polynomial $p(x)$, such that 0 is not a root and all coefficients have non-negative valuation, we can effectively find the first side of the Newton polygon, including identifying all of the points $(i, w(A_i))$ on the first side of the polygon.*

Proof: Let $p(x) = A_0 + A_1 x + \ldots + A_n x^n$. Since 0 is not a root, we can find $w(A_0)$. Since $p(x)$ is non-constant, there exists $i > 0$, such that $A_i \neq 0$. Therefore, we can find $w(A_i)$ and the equation of the line through $(0, w(A_0))$ and $(i, w(A_i))$. For each j, we can determine whether $(j, w(A_j))$ lies strictly below or on this line by checking whether $w(A_j) < q$ or $w(A_j) = q$ for the point (j, q) on the line. If we find $(j, w(A_j))$ lies strictly below this line, we try the line through $(0, w(A_0))$ and $(j, w(A_j))$. Again, we look for a point below this line. Continuing in this way, we eventually find the first side of the Newton polygon and all of the points on this side. □

The next lemma is an effective version of Lemma 2.4

Lemma 3.5: *Given a non-constant polynomial $p(x)$, such that 0 is not a root and all coefficients have non-negative valuation, we can effectively find g and h, such that the first side of the Newton polygon for $t^h p(t^{-g} x)$ has negative slope, and all coefficients have non-negative valuation.*

Proof: Suppose the first side of the Newton polynomial for $p(x)$ joins $(0, w(A_0))$ to $(i, w(A_i))$. If the slope is positive, we consider

$$q(x) = p(t^{-g} x) = B_0 + B_1 x + \ldots + B_n x^n,$$

where $B_i = t^{-gi} A_i$. The Newton polygon for $q(x)$ will also join $(0, w(B_0))$ to $(i, w(B_i))$. The slope of the first side is at most $\frac{w(B_i) - w(B_0)}{i} = \frac{w(A_i) - ig - w(A_0)}{i} \leq \frac{w(A_i)}{i} - g$. If $gi > w(A_i)$, then this is negative. The first side may have slope that is even more negative.

This transformation may yield coefficients B_i, some of which have negative valuation. We can make another transformation as in Lemma 3.3 to determine h, so that the coefficients of the polynomial have non-negative valuation, and the first side of the Newton polynomial has negative slope. □

Note that if the first side of the Newton polygon has a negative slope, then the term in the root constructed from this side (as described above in the sketch of the proof of the Newton-Puiseux theorem) will have positive weight. Furthermore, the Newton polygon is convex and therefore the term corresponding to the first side has larger weight than the terms determined by any other side. In particular, if we start with a polynomial with only non-zero roots and we use the first side of the Newton polygon to compute the first term of a root, then this root will have the largest valuation of all the roots. This fact provides the key step in the following lemma.

Lemma 3.6: *If $p(x)$ is a non-constant polynomial and $n \in \omega$, then we can effectively tell whether there is a root r with $w(r) > n$. If there is no such root, then we know that $A_0 \neq 0$.*

Proof: Let $p(x) = A_0 + A_1 x + \ldots + A_n x^n$. We may suppose that $w(A_i) \geq 0$ for all i. Not knowing whether $A_0 = 0$, we do not know whether 0 is a root. Since $p(x)$ is non-constant, there is some $i > 0$, such that $A_i \neq 0$. We can find such an i and then effectively check whether $w(A_0) > ni + w(A_i)$.

If $w(A_0) \leq ni + w(A_i)$, then we know $A_0 \neq 0$. By Lemma 3.4, we can find the first side of the Newton polygon. Using this side, we can calculate the greatest valuation of a root and hence determine whether there is a root r with $w(r) > n$.

Otherwise, suppose $w(A_0) > ni + w(A_i)$. If $A_0 \neq 0$, then the first side of the Newton polygon has slope at most $\frac{w(A_i) - w(A_0)}{i} \leq -\frac{ni}{i} = -n$. For any root r corresponding to this side, $w(r) > n$. Of course, if $A_0 = 0$, then 0 is a root with valuation greater than n. □

Lemma 3.7: *Let $p(x) = B_0 + B_1 x + \cdots + B_n x^n$ and let $0 < r < n$ be such that $w(B_i) > 0$ for $i < r$, $w(B_r) = 0$, and $w(B_i) \geq 0$ for $i > r$. If $w(B_0) = q$ and the slope of the first side of the Newton polygon is δ, then $\delta \leq -q/r$.*

Proof: Assume $w(B_0) = q$. By the assumptions on the weights of the coefficients, the first side of the Newton polygon connects $(0, B_0)$ to (i, B_i) for some $0 < i \leq r$. If it ends at (r, B_r), then the slope is $-q/r$. If it connects to a point (i, B_i) for $0 < i < r$, then the slope will be more steeply negative. □

3.3. *Complexity of the root-taking process*

Let K be an algebraically closed field of characteristic 0 with universe a subset of ω and let $p(x) = A_0 + A_1x + \ldots + A_nx^n$ be a non-constant polynomial with coefficients in $K\{\{t\}\}$. Following the sketched proof of the Newton-Puiseux Theorem, we want to uniformly compute a root r of $p(x)$ of the form

$$r = b_0t^{\nu_0} + b_1t^{\nu_1} + b_2t^{\nu_2} + \cdots$$

by inductively computing b_i and ν_i with $b_i \in K$ and $\nu_i \in \mathbb{Q}$, such that $\nu_0 < \nu_1 < \cdots$ and the denominators of the ν_i bounded.

There are two main problems to doing this computation uniformly in the field K and the coefficients A_i. The first problem is that we cannot effectively tell whether a coefficient A_i is 0. In particular, this means we cannot determine whether 0 is a root of a given polynomial by uniformly checking whether the constant term is 0. Also, we cannot uniformly determine the valuations $w(A_i)$ of the coefficients and hence cannot uniformly compute the Newton polygon.

The second problem in giving a uniform version of the standard proof is that the root r may be given by a finite summation. This problem is related to the previous one in the sense that the sum will be finite when 0 is the root of a certain polynomial in the inductive construction. In the classical setting, there is no problem with viewing a finite sum as an element of $K\{\{t\}\}$ because we can think about appending additional terms with coefficient 0. However, we have to represent our root as a total function and hence we need to explicitly append these terms with coefficient 0. The danger is that we must append these terms while checking whether the constant term of a given polynomial is 0. If we discover that the given polynomial does not have 0 as a root, we have to construct the next term of r in a way that is consistent with the terms we appended during our search process.

In the proof of Theorem 3.8 below, we will follow the outline and the notation of the proof of the Newton-Puiseux Theorem in Walker [11]. Our focus will be on overcoming the two problems noted above. We will note various algebraic properties of the defined parameters, but we leave it to the reader to consult [11] for the proofs of these properties.

Theorem 3.8: *There is a uniform effective procedure that, given K and the sequence of coefficients for a non-constant polynomial over $K\{\{t\}\}$, yields a root.*

Proof: Let $p(x) = A_0 + A_1x + \ldots + A_nx^n$ be a non-constant polynomial over $K\{\{t\}\}$. By Lemma 3.3, we may suppose that for all i, $w(A_i) \geq 0$. Our initial plan is to start approximating a root of $p(x)$ that looks like 0 until (if ever) we see evidence that $A_0 \neq 0$. At that point, we switch to the general case of producing a root of the form

$$c_1t^{\gamma_1} + c_2t^{\gamma_1+\gamma_2} + c_3t^{\gamma_1+\gamma_2+\gamma_3} + \ldots$$

where each $c_i \in K$ and each $\gamma_i \in \mathbb{Q}$, with $\gamma_i > 0$, whenever $i > 1$.

Not knowing whether $A_0 = 0$, we use Lemma 3.6 to test whether there is a root r with $w(r) > 0$. If so, we give our root r the first term $0t^0$. We test whether there is a root r with $w(r) > 1$, and, if so, we add the term $0t^1$. We continue, testing whether there is a root r with $w(r) > n$, and, if so, adding the term $0t^n$. If the answer is always positive, then $A_0 = 0$ and we have produced the root $0 = 0t^0 + 0t^1 + \cdots$ of $p(x)$. If for some positive n, we find that there is no root r with $w(r) > n$, then we know that $A_0 \neq 0$. In this case, we switch to trying to find a root of the form above.

Suppose $p(x)$ has no root r with $w(r) > n$ and hence $A_0 \neq 0$. By Lemma 3.5, we can effectively find g and h, such that the first side of the Newton polygon for $p_0(x) = t^hp(t^{-g}x)$ has negative slope and all the coefficients of $p_0(x)$ have non-negative valuations. We switch to finding a root y of $p_0(x)$ and then translating y back to a root r of $p(x)$. We need to check that this process is consistent with having already determined that $0t^0 + \cdots + 0t^{n-1}$ is an initial segment of our root r for $p(x)$. (We can assume $n > 0$ since if $n = 0$ then we have not specified any initial terms of our root r for $p(x)$.)

Note that y is a root of $p_0(x)$ if and only if $r = t^{-g}y$ is a root of $p(x)$. Using the first side of the Newton polygon for $p_0(x)$ to find the first term in a root y, we will eventually obtain a root y of $p_0(x)$ with maximum weight. Therefore, $t^{-g}y$ will be a root of $p(x)$ with maximum weight, and hence $w(t^{-g}y) > n - 1$ because $p(x)$ has a root with weight $> n - 1$. It follows that we can put the terms in the root $t^{-g}y$ after our declared initial segment $0t^0 + \cdots + 0t^{n-1}$ of the root of $p(x)$ without conflict. (More formally, each time we add a term ct^{γ} to y below, we add the corresponding term $ct^{\gamma-g}$ to r. To simplify the notation, we focus only on the root y of $p_0(x)$.)

We have now reduced our algorithm to determining the parameters c_i and γ_i in a root y of $p_0(x)$ of the form

$$y = c_1t^{\gamma_1} + c_2t^{\gamma_1+\gamma_2} + c_3t^{\gamma_1+\gamma_2+\gamma_3} + \ldots$$

where each $c_i \in K$ and each $\gamma_i \in \mathbb{Q}$, with $\gamma_i > 0$ whenever $i > 1$.

Furthermore, we need to use the first side of the Newton polygon to determine γ_1, as noted in the previous paragraph. We will determine γ_i and c_i inductively, with c_i and γ_i (as well as the additional parameters L_i, β_i and r_i) determined in the ith round of computation.

Having performed this transformation from $p(x)$ to $p_0(x)$ using Lemma 3.5, we have fixed a rational g, such that $r = t^{-g}y$ is our desired root of $p(x)$. It is important that we do not need to perform similar transformations in later rounds of computation in this algorithm. After describing the first round of computation, we list properties that hold inductively at the end of each round of computation and which ensure that we do not need to apply Lemma 3.5 again in this process. In describing the computation, we ignore this transformation from $p(x)$ to $p_0(x)$ and we abuse notation by letting $p_0(x) = A_0 + A_1x + \ldots + A_nx^n$.

1st round of computation

(1) Find γ_1 and the parameters L_1 and β_1:

- By Lemma 3.4, we can find the negative slope of the first side of the Newton polygon of $p_0(x)$ and the points $P_i = (i, w(A_i))$ that lie on this side. Since $A_0 \neq 0$, we know $(0, w(A_0))$ is one endpoint of this line segment.
- Let L be the line forming the first side of the Newton polygon. Viewing the polygon as lying in the u-v plane, the equation for L has the form $v + \gamma u = \beta$ with $\gamma, \beta \in \mathbb{Q}$. Note that $\beta = w(A_0)$ and that $\gamma > 0$ because it is the negative of the slope of L.
- Set $L_1 = L$, $\gamma_1 = \gamma$ and $\beta_1 = \beta$.

(2) Find c_1 and the parameter r_1.

- Let I be the set of indices i, such that P_i is on L_1 and let $k > 0$ be the greatest index in I. For each $i \in I$, let $a_i \in K$ be the coefficient of the leading term in A_i. Since we know the finite value of $w(A_0)$, we can assume $a_0 \neq 0$ by ignoring any initial terms in A_0 with 0 coefficients.
- Let $\psi(z) \in K[x]$ be the polynomial $\psi(z) = \sum_{i\in I} a_iz^i$ of degree k. Since $a_0 \neq 0$, $\psi(z)$ has a non-zero constant term, so $\psi(0) \neq 0$. Because K is algebraically closed, $\psi(z)$ has a non-zero root in K.
- Let c_1 be a root of $\psi(z)$ of minimal multiplicity and let r_1 be the multiplicity of the root c_1.

This completes the first round of computation to determine c_1 and γ_1.

Recall that our goal is to compute a root of the form

$$y = c_1 t^{\gamma_1} + c_2 t^{\gamma_1+\gamma_2} + c_3 t^{\gamma_1+\gamma_2+\gamma_3} + \ldots = t^{\gamma_1}(c_1 + c_2 t^{\gamma_2} + c_3 t^{\gamma_2+\gamma_3} + \ldots).$$

If we think of y_1 as the undetermined part $y_1 = c_2 t^{\gamma_2} + c_3 t^{\gamma_2+\gamma_3} + \ldots$, then y_1 is a root of the polynomial $p_1(x) = t^{-\beta_1} p_0(t^{\gamma_1}(c_1 + x))$. If $p_1(0) = 0$, then $c_1 t^{\gamma_1}$ is already a root of $p_0(x)$. Otherwise, we need to determine c_2 and γ_2. The key fact about $p_1(x)$ that will allow us to continue uniformly is that when we write $p_1(x)$ as $p_1(x) = B_0 + B_1 x + \ldots + B_n x^n$, we have the following three properties (see [11]).

(P1) $w(B_i) \geq 0$ for all $i \leq n$,
(P2) $w(B_i) > 0$ for all $i \leq r_1 - 1$, and
(P3) $w(B_{r_1}) = 0$.

By these properties and Lemma 3.7, if $B_0 \neq 0$, then the slope of the first side of the Newton polygon for $p_1(x)$ is at most $-w(B_0)/r_1$, and in particular, is negative.

The second round of computation forms the basis for the continuing induction. We do not know which of the new coefficients $B_0, \ldots, B_n$ are non-zero, so we can only determine if their valuations are finite in a c.e. manner. In particular, at the start of this round of computation, we need to expand our root by adding zero terms $c_1 t^{\gamma_1} + 0t^{2\gamma_1} + 0t^{3\gamma_1} + \cdots$ until we see evidence that $B_0 \neq 0$. If $B_0 = 0$, then we obtain a representation of the root $c_1 t^{\gamma_1}$ of $p_0(x)$. If $B_0 \neq 0$, then once we see this fact, we need to compute the next parameters γ_2 and c_2. However, we need to add these zero terms in a careful manner so that we do not add a term $0t^{k\gamma_1}$, such that our eventual value γ_2 satisfies $\gamma_1 + \gamma_2 \leq k\gamma_1$. Under such circumstances, we cannot include a term of the form $c_2 t^{\gamma_1+\gamma_2}$ in a computable fashion after appending $0t^{k\gamma_1}$. To handle this difficulty, we use Lemma 3.7 and the fact that if $w(B_0)$ is finite, the γ_2 will be the absolute value of the (negative) slope of the first side of the Newton polygon for $p_1(x)$.

2nd round of computation

(1) Expand y with terms of the form $0t^{k\gamma_1}$ as we check whether $B_0 \neq 0$.

- Set $k = 2$ and start the following loop.
- Determine $q \in \mathbb{Q}$, such that $q/r_1 \geq k\gamma_1$ and check whether $w(B_0) > q$.
- If $w(B_0) > q$, then by Lemma 3.7, we know that if $w(B_0)$ is finite, then the absolute value of the slope of the first side of the Newton

polygon is greater than $k\gamma_1$. We add the term $0t^{k\gamma_1}$ to y, increment k and repeat the loop.

- If $w(B_0) \leq q$, then we know the exact value of $w(B_0)$. We move on to Step 2 to start our calculation of the next term of y.

(2) Find γ_2 and the parameters L_2 and β_2:

- By Lemma 3.4, we can find the line L forming the first side of the Newton polygon for $p_1(x)$ and the points P_i that lie on this side. The equation for L is $v + \gamma u = \beta$ with $\beta = w(A_0)$ and $\gamma > 0$.
- Set $L_2 = L$, $\gamma_2 = \gamma$ and $\beta_2 = \beta$.

(3) Find c_2 and the parameter r_2.

- Let I be the set of indices i, such that P_i is on L_2, let k be the greatest index in I and for each $i \in I$, let $a_i \in K$ be the coefficient of the leading term in B_i. Note that $0 < k \leq r_1$ by (P1)–(P3). Because $w(B_0)$ is finite, we can assume $a_0 \neq 0$.
- Let $\psi(z) \in K[x]$ be the polynomial $\psi(z) = \sum_{i \in I} a_i z^i$ of degree k. Since $a_0 \neq 0$ and K is algebraically closed, $\psi(z)$ has a non-zero root in K.
- Let c_2 be a root of $\psi(z)$ of minimal multiplicity and let r_2 be the multiplicity of the root c_2. Since $\psi(z)$ has degree $k \leq r_1$, we have $0 < r_2 \leq r_1$.

This completes the second round of the computation. We are now in a position to iterate the construction by repeating the second round of computation (and adjusting the indices accordingly) with the polynomial

$$p_2(x) = t^{-\beta_2} p_1(t^{\gamma_2}(c_2 + x)) = t^{-\beta_2 - \beta_1} p_0(c_1 t^{\gamma_1} + c_2 t^{\gamma_1 + \gamma_2} + t^{\gamma_1 + \gamma_2} x).$$

Writing $p_2(x)$ in the form $p_2(x) = B_0 + B_1 x + \cdots + B_n x^n$ (with new coefficients that more formally ought to be labeled $B_{i,2}$), the properties (P1)–(P3) hold.

In general, to determine the value of c_{n+1} and γ_{n+1}, we start by adding terms of the form $0t^{\gamma_1+\gamma_2+\cdots+\gamma_{n-1}+k\gamma_n}$, while we run a search procedure to see if the constant term of $p_{n+1}(x)$ satisfies $B_0 \neq 0$. We use Lemma 3.7 to make sure that we do not get into a situation in which we have added such a term for which we might have $\gamma_{n+1} \leq k\gamma_n$.

If this search does not halt, then $B_0 = 0$ and hence $p_{n+1}(0) = 0$. In this case, we have computed a representation of the root $c_1 t^{\gamma_1} + \ldots + c_n t^{\gamma_1 + \cdots + \gamma_n}$ of $p_0(x)$. Otherwise, we eventually determine that $B_0 \neq 0$ and we find the finite value of $w(B_0)$. We use Lemma 3.4 to determine the first side of the

Newton polygon, and let γ_{n+1} be the absolute value of the slope of this line and c_{n+1} be the root of the appropriate polynomial in $K[x]$.

This completes the construction of the root y of $p_0(x)$ and hence the root of $p(x)$. The proof that y is an element of $K\{\{t\}\}$ (i.e. there is a bound on the denominators of the γ_i terms) and that y is a root of $p_0(x)$ is exactly as in [11] because we have produced the same root. □

Corollary 3.9: *For a polynomial $p(x) = A_0 + A_1x + \ldots + A_nx^n$ over $K\{\{t\}\}$, all roots are computable in K and the coefficients A_i.*

Proof: The polynomial $p(x)$ has at most n distinct roots. Some of these may be finite, so they are computable. For an infinite root r, fix a finite initial segment $r_k = b_0t^{\nu_0} + \cdots b_kt^{\nu_k}$ of r that does not extend to any other root of $p(x)$. Let $r = r_k + y$ with $w(y) > \nu_k$. Consider the polynomial $q(x) = p(r_k + x)$ and note that y is a root of $q(x)$. We claim that y is the root of $q(x)$ with greatest valuation. Suppose $s \neq y$ were a root of $q(x)$ with $w(s) > w(y)$. Then $w(s) > \nu_k$, so $r_k + s$ is a root of $p(x)$ that extends r_k, contradicting the fact that r is the unique root of $p(x)$ extending r_k.

Applying the procedure in Theorem 3.8 to $q(x)$ gives us the root of $q(x)$ with maximal weight, and so gives us the root y, such that $r = r_k+y$. Therefore, y is computable from K and the coefficients of $q(x)$, or equivalently, it is computable from K, the coefficients of $p(x)$ and the finite (non-uniform) sum r_k. It follows that r is (non-uniformly) computable from K and the coefficients of $p(x)$. □

Theorem 3.8 yields the following.

Theorem 3.10: *Let I be a family of subsets of ω, such that if $X_1, \ldots, X_k \in I$ and Y is computable from $X_1, \ldots, X_n$, then $Y \in I$. Then the Newton-Puiseux Theorem holds in I.*

Proof: Suppose K is an algebraically closed field of characteristic 0 computable from some $X \in I$, and let R be the set of Puiseux series over K represented in I. If $p(x) = A_0 + \ldots + A_nx^n$ is a non-constant polynomial for which all $A_i \in R$, then $p(x)$ has a root in R. □

4. Hahn series

Let K be a field, and let G be a divisible ordered Abelian group.

Definition 4.1: The *Hahn series* (obtained from K and G) are formal power series of the form $s = \sum_{g \in S} a_g t^g$, where S is a well-ordered subset of G and $a_g \in K$. The *support* of s is $Supp(s) = \{g \in S : a_g \neq 0\}$ and the *length* of s is the order type of $Supp(s)$.

We write $K((G))$ for the set of Hahn series with coefficients in K and terms corresponding to elements of G. The operations and the valuation on $K((G))$ are defined in the natural way.

Definition 4.2: Let $s = \sum_{g \in S} a_g t^g$, $s' = \sum_{g \in S'} a'_g t^g$, where S, S' are well-ordered.

- $s + s' = \sum_{g \in S \cup S'} (a_g + a'_g) t^g$, where $a_g = 0$ if $g \notin S$, and $a'_g = 0$ if $g \notin S'$.
- $s \cdot s' = \sum_{g \in T} b_g t^g$, where $T = \{g_1 + g_2 : g_1 \in S \ \&\ g_2 \in S'\}$, and for each $g \in T$, we set $B_g = \{(g_1, g_2) \in S \times S' : g_1 + g_2 = g\}$ and $b_g = \sum_{(g_1, g_2) \in B_g} b_{g_1} \cdot b_{g_2}$.

Definition 4.3: For $s \in K((G))$, the valuation (or weight) $w(s)$ is the least $g \in Supp(s)$, if $s \neq 0$, and $w(0) = \infty$.

Here is the generalized Newton-Puiseux Theorem of Mac Lane [5].

Theorem 4.4: *Let G be a divisible ordered Abelian group, and let K be a field that is algebraically closed of characteristic* 0*. Then $K((G))$ is also algebraically closed.*

We sketch the root taking process, which is essentially a transfinite version of the Newton-Puiseux process.

Proof: Let $p(x) = A_0 + A_1 x + \ldots + A_n x^n$ be a polynomial over $K((G))$, where $p(x)$ is non-constant. We may suppose that for all i, $w(A_i) \geq 0$—if this is not so initially, we replace $p(x)$ by $t^g p(x)$ for suitable positive g. We may also suppose that there is an infinitesimal root—if this is not so initially, we replace $p(x)$ by $p(t^{-g} x)$ for suitable positive g. We define a sequence of initial segments of a root. Let $r_0 = 0$ and let $p_0(x) = p(r_0 + x)$. If the constant term is 0, then r_0 is a root. Otherwise, we consider the Newton polygon, and find the first term of a root as in the case of Puiseux series. Let r_1 consist of just this first term and form $p_1(x) = p(r_1 + x)$. Given r_α an initial segment of a root having length α, we form the polynomial $p_\alpha(x)$. If the constant term is 0, then 0 is a root of $p_\alpha(x)$, which means that r_α is a root of $p(x)$. Otherwise, we consider the Newton polygon for $p_\alpha(x)$

and we find the first term of a root. We let $r_{\alpha+1}$ be the result of adding this to r_α. For limit α, having determined r_β for $\beta < \alpha$, we let r_α consist of all of the terms of all r_β. Eventually, the process stops. The length must be less than κ^+, where κ is the cardinality of G. □

We can bound the lengths of roots of a polynomial in terms of the lengths of the coefficients. The result below is proved in [3, 4].

Theorem 4.5: *Let K be an algebraically closed field of characteristic 0, and let G be a divisible ordered Abelian group. Let $p(x) = A_0 + \ldots + A_n x^n$ be a polynomial over $K((G))$. We suppose that the coefficients all have countable length. If γ is a countable limit ordinal greater than the lengths of all A_i, then the roots of $p(x)$ all have length less than ω^{ω^γ}.*

We can extend Theorem 4.5 to the case where γ is uncountable.

Corollary 4.6: *Let K be an algebraically closed field of characteristic 0, and let G be a divisible ordered Abelian group. Let $p(x) = A_0 + \ldots + A_n x^n$ be a polynomial over $K((G))$. If γ is a limit ordinal (possibly uncountable) greater than the lengths of all A_i, then the roots of $p(x)$ all have length less than ω^{ω^γ}.*

Proof: Take a transitive set M satisfying some set theory (enough to prove Mac Lane's Theorem) and containing the field K, the group G, and the coefficients A_i. Say that α_i is the length of A_i, these may be uncountable ordinals. Let γ be the first limit ordinal greater than all α_i. Let $r_1, \ldots, r_n$ be the roots of $p(x)$ in M (listed with multiplicity). Say that β_i is the length of r_i. The coefficients A_i and the roots r_i are each identified with a function defined on all of G, giving the coefficients in K. For each coefficient A_i and each root r_i, there is an isomorphism between the support (with the ordering from G) and the ordinal length.

Let M' be a countable elementary substructure of M that contains the following elements: K, G, the coefficients A_i, the roots r_i, the ordinals α_i, γ, and β_i, and the functions mapping the supports isomorphically onto the lengths. Note that since M' is countable, the elements of G may not all be present. Similarly, the ordinals α_i, γ, and β_i may not have all of their predecessors in M'. Applying a Mostowski collapse f, we get M'', such that $M' \cong_f M''$. M'' is again a transitive set, and the ordinals are actual ordinals. If G and γ were uncountable in M, then $f(G)$ and $f(\gamma)$ are countable (viewed in the real world), even though they are uncountable in M''.

Being a root, and having a specific ordinal length are absolute between M'' and the real world. In M'', and in the real world, $f(r_i)$ is a root of $f(p)(x)$. The coefficients $f(A_i)$ have length $f(\alpha_i) < f(\gamma)$, where $f(\gamma)$ is a limit ordinal. The roots $f(r_i)$ have length $f(\beta_i)$. Since the ordinals are countable in the real world, we can see in the real world that the roots $f(r_i)$ of $f(p)$ have length $f(\beta_i)$, which must be less than $\omega^{\omega^{f(\gamma)}}$. By absoluteness, in M'', $f(r_i)$ has length $f(\beta_i)$, which is less than $\omega^{\omega^{f(\gamma)}}$. Then in M' and M, r_i has length β_i, which is less than $\omega^{\omega^{\gamma}}$. □

5. Complexity in Hahn fields

We want to measure complexity of basic operations and root-taking in a Hahn field $K((G))$. We first say how we are representing Hahn series. To represent $s \in K((G))$, we could use a function f from an ordinal α to $K \times G$, such that the second component of $f(\beta)$ increases with $\beta < \alpha$. Alternatively, we could use a function σ from G to K, such that $\{g \in G : \sigma(g) \neq 0\}$ is well-ordered. We use the first approach to get a result on Mac Lane's Theorem in admissible sets. We then use the second approach to say more precisely how complicated it is to compute the basic operations on Hahn series, and to set bounds on the complexity of the root-taking process.

5.1. *Mac Lane's Theorem in admissible sets*

An *admissible* set A is a transitive set that satisfies the axioms of Kripke-Platek set theory. We consider admissible sets that contain ω. What is important for us is that we can define functions F by induction on the ordinals, provided that we have a Σ_1 formula describing the way we pass from $F|\alpha$ to $F(\alpha)$. We will show that the generalized Newton-Puiseux Theorem holds in admissible sets. We consider an admissible set A that contains the algebraically closed field K and the divisible ordered Abelian group G. We consider those elements of $K((G))$ that are represented by a function s in A from an ordinal α to $K \times G$, such that if $s(\beta) = (b_\beta, g_\beta)$, then $\beta < \gamma < \alpha$ implies $g_\beta < g_\gamma$.

Lemma 5.1: *Let A be an admissible set containing the algebraically closed field K and the divisible ordered Abelian group G. If s, s' are elements of $K((G))$ in A, then $s + s'$, $s \cdot s'$, Supp(s) and the length of s are all in A.*

Lemma 5.2: *Let A be an admissible set. The function $\alpha \to \omega^\alpha$ is Σ_1-definable on A.*

Theorem 5.3: *Let A be an admissible set. Then the generalized Newton-Puiseux Theorem holds in A.*

Proof: We sketch the proof which follows the standard root-taking process, indicating where the bounds in Corollary 4.6 are used. Let K be an algebraically closed field of characteristic 0, and let G be a divisible ordered Abelian group, both in A. Let $p(x)$ be a polynomial over $K((G))$, with coefficients A_i in A. Let α_i be the length of A_i. We must show that $p(x)$ has a root r in A. We define, by Σ_1-induction on ordinals α, a sequence of initial segments r_α of a root r, where r_α has length α until/unless there is some $\beta < \alpha$, such that r_β is a root. We let $r_0 = 0$. Given r_α, we form $p_\alpha(x) = p(r_\alpha + x) = B_0 + B_1 + \ldots + B_n x^n$. By Taylor's Theorem, $B_i = \frac{p^{(i)}(r_\alpha)}{i!}$. These coefficients are in A. If $B_0 = 0$, then r_α is a root of $p(x)$. Otherwise, we consider the Newton polygon for $p_\alpha(x)$. This polygon will have at least one side, such that if ν is the negative of the slope of this side, then ν is greater than the elements in Supp$(p_\alpha(x))$. Fix such a value ν and let b be a root of the ν-principal part. Then $r_{\alpha+1} = r_\alpha + bt^\nu$. For limit α, we let r_α be the element of $K((G))$, whose initial segments are the r_β for $\beta < \alpha$.

We have Σ_1 formulas saying how $r_\alpha + 1$ is obtained from r_α and, for limit α, how r_α is obtained from the r_β's for $\beta < \alpha$. Thus, we have a Σ_1 formula defining the function F on ordinals α in A, such that $F(\alpha) = r_\alpha$ if no proper truncation of r_α is a root, and $F(\alpha) = r_\beta$ if $\beta < \alpha$ and r_β is a root. Using our bounds on lengths of roots, we can show that some r_β is a root. Let γ be the maximum of $\alpha_i + \omega$. Then γ is a limit ordinal in A, and the ordinal ω^{ω^γ} is also in A. By the results above, the roots of $p(x)$ all have length less than ω^{ω^γ}. So, for some $\beta \in A$, r_β is a root of $p(x)$. □

5.2. *More precise results*

Let K be an algebraically closed field and G be a divisible ordered Abelian group, both with universe a subset of ω. To represent an element of $K((G))$, we take a function s from G to K, such that $Supp(s) = \{g \in G : s(g) \neq 0\}$ is well-ordered. We want to know how many jumps it takes to compute the basic operations on $K((G))$, and to find initial segments of a root of a polynomial over $K((G))$.

5.3. *Complexity of basic operations*

We begin with the complexity of the basic operations.

Lemma 5.4:

(1) *There is a uniform effective procedure that, given K, G, and elements s, s' of $K((G))$, yields $s + s'$.*
(2) *There is a uniform effective procedure that, given n, K, G, and elements $s_1, \ldots, s_n$ of $K((G))$, yields $s_1 + \ldots + s_n$.*

Proof: Both parts follow since $(s_1 + \ldots + s_n)(g) = s_1(g) + \ldots + s_n(g)$. □

Lemma 5.5: *We have a Δ_2^0 procedure that, given K, G, and $s \in K((G))$ determines whether $s = 0$. (In fact, the set of representations of 0 is effective Π_1^0.)*

Proof: We suppose that G has universe ω. Given s, we ask whether there is some g, such that $s(g) \neq 0$. □

Lemma 5.6: *We have a Δ_2^0 procedure that, given K, G, and $s \in K((G))$, determines $w(s)$.*

Proof: If X computes K, G, and s, then using X', we find $w(s)$ as follows. We first ask whether $Supp(s) = \emptyset$. If so, then $w(s) = \infty$. If not, we can effectively find some $g_1 \in Supp(s)$. We then ask whether $Supp(s) \cap pred(g_1) = \emptyset$. If so, then $w(s) = g_1$. If not, we find $g_2 \in Supp(s)$ with $g_2 < g_1$. We continue until we find $g_k \in Supp(s)$, such that $Supp(s) \cap pred(g_k) = \emptyset$. Then $w(s) = g_k$. □

Lemma 5.7:

(1) *There is a uniform Δ_2^0 procedure that, given K, G, and elements s_1, s_2 of $K((G))$, yields $s_1 \cdot s_2$.*
(2) *There is a uniform Δ_2^0 procedure that, given n, K, G, and elements $s_1, \ldots, s_n$ of $K((G))$, yields $s_1 \cdot \cdots \cdot s_n$.*

Proof: For a given $g \in G$, we ask whether there exist g_1 and g_2, such that $g_1 + g_2 = g$ and $s_i(g_i) \neq 0$. If we have one pair, we ask whether there is another. Because $Supp(s_i)$ are well-ordered, there are only finitely many such pairs. Therefore, after finitely many Δ_2^0 questions, we have the full set S of pairs (g_1, g_2), such that $g_1 + g_2 = g$ and $s_i(g_i) \neq 0$. The coefficient of g is the sum over the pairs $(g_1, g_2) \in S$ of the products $s_1(g_1)s_2(g_2)$. The uniform process for multiplying n terms is similar. □

Lemma 5.8: *There is a Δ_2^0 procedure that, given K, G, a polynomial $p(x_1, \ldots, x_n) \in K[x_1, \ldots, x_n]$, and elements $s_1, \ldots, s_n \in K((G))$, yields $p(s_1, \ldots, s_n)$.*

Proof: The polynomial p is a sum of monomials $ax_1^{i_1} \cdot \ldots \cdot x_n^{i_n}$. Given $s_1, \ldots, s_n \in K((G))$, we can find the value of any monomial $as_1^{i_1} \cdot \ldots \cdot s_n^{i_n}$, for $a \in K$, in the same way that we found the product of two elements s, s'. We can then take the sum effectively. □

Taking products becomes simpler when one of the sums is finite. This observation will help lower the complexity of the initial segments in the root taking process later.

Lemma 5.9: *Let $s \in K((G))$ and let $r = b_0 t^{\nu_0} + \cdots b_k t^{\nu_k}$ with $b_i \in K$, $\nu_i \in G$ and $\nu_0 < \nu_1 < \cdots < \nu_k$. There is a uniform procedure that, given s and r, yields $s \cdot r$.*

Proof: First, consider the case of $s \cdot b_0 t^{\nu_0}$, where $s = \sum_{g \in S} a_g t^g$. We have

$$\left(\sum_{g \in S} a_g t^g\right) \cdot b_0 t^{\nu_0} = \sum_{g \in S} a_g b_0 t^{g+\nu_0} = \sum_{g \in \widehat{S}} \widehat{a}_g t^g$$

where $g \in \widehat{S}$ if and only if $g - \nu_0 \in S$ and $\widehat{a}_g = a_{g-\nu_0} b_0$. For the general case, we take the sum of the uniformly computed terms $s \cdot b_i t^{\nu_i}$. □

Lemma 5.10: *Let $p(x) = A_0 + \cdots + A_n x^n$ be a polynomial over $K((G))$ and let $r = b_0 t^{\nu_0} + \cdots + b_k t^{\nu_k}$ be a finite sum as above. There is a uniform procedure that, given $p(x)$ and r, yields $p(r)$.*

Proof: Expanding r^i, we can uniformly calculate $A_i r^i$ by Lemma 5.9 and then take a sum to obtain $p(r)$. □

Lemma 5.11: *Given K, G, the coefficients of a polynomial $p(x)$ over $K((G))$, and a finite sum $r = b_0 t^{\nu_0} + \cdots + b_k t^{\nu_k}$ as above, we can effectively compute $p(r + x)$.*

Proof: To find the coefficients of $p(r+x)$, it suffices, by Taylor's Theorem, to be able to compute $p^{(i)}(r)$ for $i \leq n$. We can do these computations uniformly by Lemma 5.10. □

5.4. *Initial segments of roots*

We consider the complexity of a procedure that, given K, G, and a polynomial $p(x) = A_0 + A_1x + \ldots + A_nx^n$ over $K((G))$, at step α determines an initial segment r_α of a root of $p(x)$, where

(1) $r_0 = 0$
(2) for $\alpha > 0$, either r_α has length α and extends r_β for all $\beta < \alpha$ or else there is some $\beta < \alpha$, such that r_β is already root and $r_\alpha = r_\beta$.

Note that r_α is an element of $K((G))$ and hence is a function $r_\alpha : G \to K$ with well-ordered support. In the case when none of the r_β for $\beta < \alpha$ are a root, $Supp(r_\alpha)$ has order type α. The totality of r_α will be important in determining the complexity at limit levels because, to present r_α as a total function that has each r_β as an initial segment, we need to be able to determine the sequence r_β uniformly and we need to be able to determine for any element $g \in G$ whether there is a $\beta < \alpha$, such that $g \in Supp(r_\beta)$.

We describe a function f, such that for computable ordinals α, it is $\Delta^0_{f(\alpha)}$ in K, G, and p to carry out step α of the procedure. Here are the first few values of f.

(1) At Step 0, we write 0. This is computable, so $f(0) = 1$.
(2) At Step 1, we must decide whether $r_0 = 0$ is a root, and if not, we must give r_1 of length 1. To determine whether 0 is a root of $p(x) = A_0 + \ldots + A_nx^n$, it is enough to check whether $A_0 = 0$. This is Δ^0_2. If $A_0 \neq 0$, then we find $w(A_0)$. This is also Δ^0_2. In fact, we have a uniform procedure Δ^0_2 in K, G, and $p(x)$, for determining which A_i are non-zero, and finding $w(A_i)$ for each such i. Then, proceeding as Newton did, we can effectively compute (b_0, ν_0), such that $r_1 = b_0t^{\nu_0}$. So, $f(1) = 2$.
(3) At Step $n+1$, having found r_n of length n, we must decide whether r_n is a root, and if not, we must extend to r_{n+1}. By Lemma 5.11, we can compute effectively the new polynomial $p_n(x) = p(r_n + x)$. As in Step 1, deciding whether r_n is a root of $p(x)$, or finding r_{n+1} is Δ^0_2 in K, G, and p. So, $f(n) = 2$ for all finite $n \geq 1$.
(4) At Step ω, we must decide whether r_n is a root for some finite n, and if not, we must give r_ω as a function. To determine whether there is some finite n, such that r_n is a root of $p(x)$ is Δ^0_3. Assuming that this is not the case, the sequence $(r_n)_{n\in\omega}$ is Δ^0_2 in K, G, and $p(x)$, but giving r_ω as a total function is Δ^0_3. So, $f(\omega) = 3$.
(5) At Step $\omega + 1$, we must determine whether r_ω is a root, and if not, we

must give $r_{\omega+1}$. To determine whether r_ω is a root of $p(x)$, we could either determine directly whether $p(r_\omega) = 0$, or find $p_\omega(x) = p(r_\omega + x)$ and then check whether the constant term is 0. If we find $p_\omega(x)$, and the constant term is not 0, then we could continue to find $r_{\omega+1}$. To find p_ω, we use the sequence version of r_ω, which is Δ^0_2. The new coefficients have the form $B_i = \frac{p^{(i)}(r_\omega)}{i!}$. Each B_i is a finite sum of products of a coefficient A_j and a power of r_ω. For $g \in G$, to find the coefficient of t^g in B_i, we ask whether there exist $h_0 \in Supp(A_j)$ and some n, with a tuple of h's in $Supp(r_n)$ for which the sum is g. If we find one such tuple, we ask whether there is another. We continue until we have the full finite set S of tuples. For each tuple in S, we take the product of the coefficients. The sum of these products is the coefficient of t^g in B_i. Thus, finding $p_\omega(x)$ is Δ^0_3 in K, G, and p. Then it is Δ^0_4 to say whether the constant term is 0, and if not, to find the first term of a root. So, $f(\omega + 1) = 4$.

(6) At Step $\omega + 2$, we must determine if $r_{\omega+1}$ is a root, and if not, we must find the next term to give $r_{\omega+2}$. Writing $r_{\omega+1} = r_\omega + b_\omega t^{\nu_\omega}$, we have $p_{\omega+1}(x) = p(r_{\omega+1} + x) = p_\omega(b_\omega t^{\nu_\omega} + x)$. Since $p_\omega(x)$ is Δ^0_3, it follows from Lemma 5.11 that $p_{\omega+1}(x)$ is also Δ^0_3. Therefore, it is Δ^0_4 to determine whether the constant term is 0, and if not, to find the next term. So, $f(\omega+2) = 4$. Following the same pattern, $f(\omega+n) = 4$ for all $n \geq 1$.

These results are sharp.

Proposition 5.12: *For finite $n \geq 1$, Step n is Δ^0_2-hard, Step ω is Δ^0_3-hard, and for finite $n \geq 1$, Step $\omega + n$ is Δ^0_4-hard.*

Proof: At Step 1, we must decide whether r_0 is a root, and if not, we must give r_1. We focus on the first part of Step 1, deciding whether 0 is a root. Now, 0 is a root iff the constant term is 0, and this is Π^0_1. Let K be a computable copy of the field of algebraic numbers, and let G be a computable copy of the additive group of rationals. For Step 1, we show that for an arbitrary c.e. set S, there is a uniformly computable sequence of polynomials $p_e(x) = B_e - x$, such that $e \in S$ iff 0 is not a root of $p_e(x)$. We let $Supp(B_e) = \emptyset$ if $e \notin S$, and if for some first s, $e \in S_s$, then we put the pair $\langle e, s\rangle$ into $Supp(B_e)$. We have $e \in S$ iff 0 is not a root of $p_e(x)$.

Similarly, for any finite $n \geq 1$, we show that for an arbitrary c.e. set S, there is a uniformly computable sequence of polynomials $p_e(x) = B_e - x$, such that $e \in S$ iff $Supp(B_e)$ has more than n elements. We start by including $0, 1, \ldots, n-1$ in $Supp(B_e)$. At step s, we define more of χ_{B_e}, keeping elements out of B_e, until/unless e appears in S. If this happens, then we add one more element into $Supp(B_e)$. We have $e \in S$ iff r_n is not a root of $p_e(x)$ because $|Supp(r_n)| = n$.

At Step ω, we must decide whether r_n is a root for some finite n, and if not, then we must combine the r_n's to form r_ω. Consider the m-complete Σ^0_2 set $Fin = \{e : W_e \text{ is finite}\}$. We produce a uniformly computable sequence of polynomials $p_e(x) = B_e - x$, such that $e \in Fin$ iff $Supp(B_e)$ is finite. At step s, we keep the next element g of G (in the $\leq_\omega$ order) out of $Supp(B_e)$ unless s is first, such that some $k \in W_{e,s}$, and in this case, we put g into $Supp(B_e)$. Then $e \in Fin$ iff some r_n is a root.

At Step $\omega + 1$, we must decide whether r_ω is a root, and if not, then we must find $r_{\omega+1}$. Consider an arbitrary Π^0_3 set S. We define a uniformly computable sequence of polynomials $p_e(x) = B_e - x$, such that if $e \in S$, then $Supp(B_e)$ has order type ω, and otherwise, $Supp(B_e)$ has order type at least $\omega + \omega$. We have uniformly c.e. sets $W_{f(e,k)}$, such that $e \in S$ iff for all k, $W_{f(e,k)}$ is finite. We have a computable sequence of rationals $s_0 < s_1 < \ldots s_n < \ldots$. We start with the plan to have $Supp(B_e)$ consist of these s_i. At step s, the characteristic function of $Supp(B_e)$ is defined on all s_i plus the first s elements of G (in the $<_\omega$-ordering). Whenever a new element enters $W_{f(e,n)}$, we choose a new q in the interval (r_n, r_{n+1}) to the right of any we have chosen earlier, and we add it to the support. Consider the polynomial $B_e - x$, with the unique root $r = B_e$. If $e \in S$, then $r = r_\omega$. If $e \notin S$, then r has length at least $\omega + \omega$, so r_ω is not a root.

Similarly, we can show that for finite $n > 1$, for any Π^0_3 set S, there is a uniformly computable sequence of polynomials $p_e(x) = B_e - x$, such that $e \in S$ iff $Supp(B_e)$ has order type $\omega + n - 1$. We let B_e be as above, except that we add to the support of B_e an extra $(n-1)$-tuple, greater than all s_i. If $e \in S$, then B_e has length $\omega + n - 1$, and if $e \notin S$, then B_e has length at least $\omega + \omega$. As above, we take polynomials $B_e - x$. If $e \in S$, then $r_{\omega+n-1}$ is a root. If $e \notin S$, then $r_{\omega+n-1}$ is not a root. □

Continuing the complexity pattern, we have the following general result about the values of f.

Proposition 5.13: *Let α be a computable limit ordinal.*

(1) $f(\alpha) = sup_{\beta<\alpha} f(\beta) + 1$.
(2) for $n \geq 1$, $f(\alpha + n) = f(\alpha) + 1$.

Determining $r_{\omega+\omega}$ as a function is Δ^0_5, but we do not know whether this upper bound is sharp. However, there is a weak hardness result showing that the complexity continues to go up with length.

Proposition 5.14: *For each computable ordinal α, Step ω^α is $\Pi^0_{2\alpha}$-hard.*

Proof: Let S be a $\Pi^0_{2\alpha}$ set. We can build a uniformly computable sequence of orderings $\mathcal{A}_n$, such that $\mathcal{A}_n$ has order type ω^α if $n \in S$ and some $\gamma < \omega^\alpha$ otherwise. We get a uniformly computable sequence C_n of subsets of $\mathbb{Q}$, all contained in the interval $(0,1)$, such that $C_n \cong \mathcal{A}_n$. Let B_n be the sum of t^q for $q \in C_n$, and consider the polynomial $B_n - x$, with the unique root $r = B_n$. For these polynomials, if $n \in S$, then $r = r_{\omega^\alpha}$, and if $n \notin S$, then $r = r_\gamma$ for some $\gamma < \omega^\alpha$. Thus, S is reducible to Step ω^α, applied to these polynomials. □

References

1. J. Barwise, *Admissible Sets and Structures: An Approach to Definability*, Perspectives in Math. Logic, vol. 7, Springer-Verlag, 1975.
2. S. Basu, R. Pollack, and M.-F. Roy, *Algorithms in Real Algebraic Geometry*, Springer, 2011.
3. J. F. Knight and K. Lange, "Lengths of developments in $K((G))$", *Selecta Mathematica*, vol 25(2019).
4. J. F. Knight and K. Lange, "Lengths of roots of polynomials in a Hahn field," *Algebra and Logic*, vol. 60(2021), pp. 95–107.
5. S. Mac Lane, "The universality of formal power series fields", *Bull. Amer. Math. Soc.*, vol. 45(1939), pp. 888–890.
6. M.H. Mourgues and J.-P. Ressayre, "Every real closed field has an integer part," *J. Symb. Logic*, vol. 58 (1993), pp. 641–647.
7. I. Newton, "Letter to Oldenburg dated 1676 Oct 24", *The Correspondence of Isaac Newton II*, 1960, Cambridge University Press, pp. 126–127.
8. V. A. Puiseux, "Recherches sur les fonctions algébriques", *J. Math. Pures Appl.*, vol. 15(1850), pp. 365–480.
9. V. A. Puiseux, "Nouvelles recherches sur les fonctions algébriques", *J. Math. Pures Appl.*, vol. 16(1851), pp. 228–240.
10. H. Rogers, *Theory of Recursive Functions and Effective Computability*, McGraw-Hill, 1967.
11. Robert J. Walker, *Algebraic Curves*, 1950, Dover.

https://doi.org/10.1142/9789811278631_0015

A Scalable Verification Solution for Blockchains

Jason Teutsch[1] and Christian Reitwießner[2]

[1] *TrueBit Establishment, jt@truebit.io*

[2] *Ethereum Foundation, chris@ethereum.org*

Bitcoin and Ethereum, whose miners arguably collectively comprise the most powerful computational resource in the history of mankind, offer no more power for processing and verifying transactions than a typical smart phone. The system described herein bypasses this bottleneck and brings scalable computation to Ethereum. Our new system consists of a financial incentive layer atop a dispute resolution layer, where the latter takes the form of a versatile "verification game." In addition to secure outsourced computation, immediate applications include decentralized mining pools, whose operator is an Ethereum smart contract, a cryptocurrency with scalable transaction throughput, and a trustless means for transferring currency between disjoint cryptocurrency systems.

Contents

1. Securing computations with economics

Every Nakamoto consensus-based cryptocurrency (e.g. Bitcoin or Ethereum) offers something that everyone in the world agrees on: a definitive, public ledger of financial transactions, or *blockchain*. This consensus technology enables basic Bitcoin transactions, which transfer currency from one party to another, while Ethereum transactions perform financial and database operations contingent on the evaluation of more complex computational scripts. Anonymous miners, who freely join and leave cryptocurrency networks, determine the validity of these transactions and thereby establish who owns which coins. Remarkably, this verification scheme requires no central authority to operate securely.

In practice, miners can successfully maintain blockchain integrity so long as the computational burden of verification remains minimal. Indeed, Bitcoin and Ethereum, whose miners arguably collectively comprise the most powerful computational resource in the history of mankind, offer no more power for processing and verifying transactions than a typical smart phone. One cannot simply increase the volume or complexity of transactions flowing into the blockchain without risking inclusion of invalid transactions due to the so-called *Verifier's Dilemma* [1] (see Sec. 3.1). In 2015, the

Verifier's Dilemma manifested itself in the form of the July 4 Bitcoin fork [2], which temporarily derailed the entire network. The Ethereum community also witnessed a related exploit in its 2016 denial-of-service attacks [3].

A *consensus computer* [1] permits users to outsource computations to the Ethereum network and receive *correct* answers in exchange for payment, so long as the effort required to verify solutions does not exceed the threshold induced by the Verifier's Dilemma. Thus, the trustless consensus computer offers a small but reliable kernel of semantic truth.

Our contribution. We present here a system, called *TrueBit*, which amplifies the consensus computer's capabilities. TrueBit enables trustless smart contracts, in theory, to securely perform any computation task. Moreover, TrueBit vastly reduces the number of redundant network node computations used in traditional Ethereum smart contracts. Presently, every Ethereum miner has to independently replicate each smart contract action in its entirety, whereas TrueBit outsources most computation work to a handful of entities. In this way, TrueBit makes secure computing affordable.

1.1. *Outsourced computation*

Let us ponder for a moment why cryptocurrencies might offer an especially convenient framework for secure outsourced computation, the core application of TrueBit. In the traditional cloud models, users must trust that the machine hardware, software, and cloud administrator all perform as expected. A wide range of things can go wrong, particularly when one wishes to tie the results of such computations to monetized entities, such as smart contracts. Proper economic incentives, the cornerstone of any cryptocurrency, can deter many types of errors from occurring in ways that simple task repetition cannot. Furthermore, in contrast to a cloud whose configuration details may not be visible to users, any systemic network errors that might occur on a blockchain-based system like TrueBit would appear in plain sight to the entire community. Cryptocurrencies, indeed, provide an excellent starting point as they already achieve several desirable properties.

(1) As witnessed by the Ethereum consensus computer, Nakamoto consensus grants networks the ability to trustlessly perform small computations correctly.
(2) Blockchain public ledgers provide perfect transparency and immutability. Ethereum smart contracts inherit these characteristics.
(3) Currency incentives, which in cryptocurrencies are inextricably tied

to computational processes, can be used to recruit and reward participants.

In general, economic forces greatly reduce the scope of possible network errors. We can safely assume that errors which incur economic penalties are much less likely to occur than those which do not.

1.2. *Practical impact*

A market for work related to computationally-intensive smart contracts already exists. The Golem Project, which crowdfunded 820,000 ETH on the morning of November 11, 2016 (∼$8.6 million on that date), already cites TrueBit as a verification mechanism for their forthcoming outsourced computation network [4]. TrueBit has the potential to support many practical new applications. Here are some examples.

- *Outsourced computation.* TrueBit functions as a worldwide computation market. Anyone in the world can post a computational task, and anyone else can receive a reward for completing it. The system's incentive structure guarantees correctness of returned solutions.
- *Truly decentralized mining.* Centralized mining pools pose security threats. For any Nakamoto consensus-based cryptocurrency, one can use TrueBit to build a practical, efficient, and trustless mining pool managed by an Ethereum smart contract (see also [5]).
- *Trustless cryptocurrency exchange.* Currently, users must find trusted exchange partners to transfer currency (or messages) between blockchains. TrueBit can facilitate secure transfer, for example, of dogecoins to the Ethereum blockchain and back (modulo some new opcodes for dogecoin) [6].
- *Scalable blockchain.* By decoupling verification for miners into a separate protocol, we can achieve high transaction throughput without facing a Verifier's Dilemma.
- *Scalable "on-chain" storage.* Swarm [7] is developing a platform for incentivized, peer-to-peer storage. TrueBit can make Swarm data accessible to Ethereum smart contracts.

We discuss these ideas further in Sec. 7.

1.3. *Smart contracts*

Smart contracts, introduced in 1994 by Szabo [8] and first realized in the Ethereum [9] cryptocurrency in 2015, are uncensorable programs that live on Ethereum's blockchain and have their own executable code and internal states, including storage for variable values and ether currency balance. Proposed applications for smart contracts include outsourced computation and storage [1, 7, 10], prediction markets [11], decentralized venture funds [12], and Bitcoin mining pools [5, 13]. Ethereum's variable `gasLimit` restricts the number of computation steps that smart contracts can perform.

TrueBit itself is an Ethereum smart contract which allows users to call *TrueBit contracts* for trusted, computationally-intensive applications. Traditional Ethereum smart contracts can call a TrueBit contract as a subroutine, thereby effectively bypassing Ethereum's `gasLimit`. In other words, TrueBit increases the per block work that the Ethereum network can process correctly.

2. How TrueBit works

TrueBit's primary purpose is to realize correct, trustless computations despite miners' limited computation bandwidth. Intuitively, we wish to reward participants who correctly perform computational tasks, but who decides whether these tasks were done correctly? In absence of a dispute, the party who performs a computational task on behalf of a TrueBit contract simply receives a reward. On the other hand, if a dispute does occur, we must rely on the only trusted resource, the limited network of miners, to resolve it. Dispute resolution occurs as a "verification game" subroutine in TrueBit, and we will discuss its details in Sec. 3.

Who initiates a dispute in the case of an incorrect computation, and how can we ensure that such a dispute will actually occur? One option is to assume that each party with some stake in the contract brings their own trusted (although not necessarily mutually trusted) verifier, and flags for a challenge, if needed. This trusted verifier approach suffices for some applications, but in general, parties may need to rely on the results of a previously executed TrueBit contract *ex post facto* even if they themselves did not exist at the time of its creation. Like Ethereum smart contracts, TrueBit contracts must have universal validity.

Since there exist no trusted parties on Ethereum's network, by symmetry we must allow any party to be hired to solve any computational task, and similarly, anyone should be able to challenge a Solver's outcome of a computational task. The latter requirement ensures that TrueBit operates by unanimous consensus. Assuming that an honest and motivated Challenger exists, who will challenge any wrong answer given by the Solver, the system's incentives for the Solver are straightforward: reward correct solutions. But how can we make sure that such a motivated Challenger exists?

Idea 1: Offer rewards for checking computations.

While this strategy may encourage participation, it provides no incentive for correct verification. We have no *a priori* reason to trust that the Verifier will substantively inspect the computation task, and we gain no additional assurance by increasing her reward for checking.

Idea 2: Offer rewards for finding bugs.

While a bug bounty may incentivize a Verifier to correctly perform a check, it only does so when the Verifier has some reason to believe that a bug might actually exist. Unless a potential Verifier believes that she has some real chance to find a bug, in practice we cannot expect her to participate in the system. In summary, we need both an incentive to partake in verification and an incentive to perform checks correctly. This leads us to the following proposition.

Idea 3: Offer a bug bounty *and* provide expectation that bugs will exist.

TrueBit takes this third approach by occasionally forcing Solvers to submit incorrect solutions. During "forced errors", TrueBit reverses the normal system incentives: the Solver gets paid for submitting an incorrect solution and penalized for submitting a correct one. We discuss the details of TrueBit's forced error protocol in Sec. 4.

Let us return to our dispute resolution subroutine. At the heart of TrueBit's protocol lies an interactive "verification game", which decides whether or not a contested computational task was performed correctly (see Sec. 3). The verification game proceeds through a series of rounds, where each round recursively checks a smaller and smaller subset of the computation. A trusted network, in our case the Ethereum platform [9], merely enforces the rules of the game and therefore does not bear the bulk of the verification burden. Anyone on the network can generate tasks, compute, or verify in exchange for rewards. TrueBit does not require "honest" or "altruistic" nodes for correct operation but rather runs under the assumption that each node wishes to maximize its own profit. We shall discuss the details of TrueBit's incentives for participation in Sec. 4.

2.1. *System properties*

Any outsourced computation system should be *fair*, in the sense that parties who perform computational tasks indeed receive compensation for their work and *reliable*, in the sense that paid computations are performed correctly. In addition to these properties, TrueBit also ensures *accessibility* through easy joining or exiting of the verification ecosystem. Any *anonymous* party can play the role of Task Giver, Solver, or Verifier, where a Verifier is a party who checks solutions and becomes a Challenger whenever she reports an error. In particular, TrueBit does not trust or rely on the reputation of its participants. Everyone who puts a deposit into the system has a fair chance to be hired for a given computational task.

TrueBit offers several novel advantages over traditional cloud computing and verifiable computable models. Verifiable computing ensures a correct answer for an outsourced computation by forcing the cloud to provide a short proof, which witnesses the correctness of its computation. The idea is that this "short proof" should be much easier to check than performing the computation oneself. Researchers have achieved much progress on this method in recent years, however the cryptographic setup costs and computational overhead for the cloud in state-of-the-art systems make these methods unsuitable for most real-world applications. Moreover, many of the proof-based systems to-date, including Zaatar, Pinocchio, Ginger, and TinyRAM, require one to run thousands of instances of a single function before breaking even on the overall verification time for a 128×128 matrix multiplication puzzle [14]. The new cryptocurrency Zcash [15] successfully makes use of verifiable computing machinery, albeit only for a very restricted class of computation tasks. Below, we contrast TrueBit with verifiable computing and traditional cloud computing.

(1) *Incentives.* Unlike traditional cloud computing, where the user simply trusts the cloud to provide correct answers, TrueBit provides financial incentives to ensure correctness.
(2) *Transparency.* The entire inner workings of TrueBit's interpreter sit on the blockchain and are open for inspection (see Sec. 6). Furthermore, the user community can democratically update the interpreter as needed.
(3) *Efficiency.* Solvers in TrueBit have low computational overhead and minimal initial setup costs. The verification game (Sec. 3) does introduce some extra work, but in practice, due to high penalties for wrong answers and bogus challenges, we expect participants to appeal to the verification game only rarely, if at all.

(4) *Simplicity.* TrueBit's operation is relatively straightforward. Unlike traditional verifiable computing, TrueBit avoids deep probabilistically checkable proofs (PCPs), succinct non-interactive arguments of knowledge (SNARKs) [16], and exotic cryptographic assumptions (e.g. those used in zkSNARKs [17]). The standard cryptography used in TrueBit, namely hash functions and digital signatures, sits in the underlying blockchain network and does not surface in the TrueBit protocol itself.

(5) *Adaptability.* TrueBit runs on top of Ethereum's current protocol without impacting functionality.

(6) *Keyless entry.* Participants do not need to manage cryptographic keys beyond those used in Ethereum. TrueBit establishes identities through financial deposits alone, and therefore the system avoids risks from cryptographic trapdoors.

2.2. *Assumptions*

Traditional distributed systems models focus on tolerance against arbitrary, Byzantine errors. In Ethereum and other Nakamoto consensus-based cryptocurrencies, however, we have no reason to expect arbitrary errors—nodes generally deviate from protocol due to financial incentives. Our system model makes two basic assumptions about the underlying network.

(I) There exists a trusted network (i.e. Ethereum) that correctly performs very small computation tasks.

(II) Participants are rational, in the sense that they act to maximize their individual profits. In particular, we expect that CPUs will show up to do computational tasks if and only if they expect fair compensation for their work.

The consensus computer [1] shows how one can use the incentives in Ethereum to establish assumption (I) above, both in theory and in practice. The task of our new system will be to amplify this small amount of correct computation in order to handle larger tasks using assumption (II). Even though solo Ethereum miners rarely earn mining rewards, expected long-term returns suffice to merit their participation in the system (via mining pools). As we shall discuss in Sec. 3.1, anonymous parties may not necessarily perform correct computations when economic incentives, including conservation of CPU cycles, pressure them to do otherwise. This observation justifies assumption (II). We emphasize that we do not find the assumption that there exists a single honest (and non-lazy) participant [18] to be realistic. A person who loyally contributes CPU cycles without incentive to do so most likely does not exist in a large, general purpose, public system.

This chapter takes a somewhat simplified view of Nakamoto consensus. In some cryptocurrencies, such as Bitcoin, miners can selectively censor a class of transactions by deciding not to include its members in their blocks. In the case of TrueBit, censorship of challenges (see Sec. 4) or transactions in the verification game (Sec. 3.2) could result in catastrophic, bogus solutions being accepted on the blockchain. In Ethereum, however, censoring individual transactions is difficult as miners cannot easily see where an Ethereum transaction might call without executing it. Indeed, Ethereum users can obfuscate the function of their transactions, making it computationally costly for miners to hand-pick transactions for their blocks. Hence, we shall assume that censorship of individual transactions does not happen. In Sec. 5.1, we analyze an attack on TrueBit in which miners withhold entire blocks and show that it is not profitable.

2.3. *Attacker model*

TrueBit's security relies on two basic assumptions about the behavior and capabilities of potential attackers.

(I) *Attackers cannot subvert the underlying consensus computer.* The Judges in the verification game (Sec. 3) always adjudicate correctly, as do the Referees in the incentive layer (Sec. 4).

(II) *Attackers are rational.* We assume that attackers have limited financial resources and will not deviate from protocol, unless it is profitable for them to do so.

While an adversary with sufficient computational power can derail a Nakamoto consensus-based cryptocurrency [19], assumption (I) above neatly sweeps this issue outside our domain. TrueBit itself does not utilize proof-of-work. Although we have chosen to implement TrueBit in the Ethereum platform for convenience, we could easily replace the "Judges" from (I) with any other universally trusted computing mechanism without affecting the system's functionality.

Due to anonymity of all participants, *Sybil attacks*, in which a single party deviantly assumes multiple identities on the network, pose special threats to TrueBit (see Secs. 5.1 and 5.2). We will assume that an attacker may create as many identities as she likes, including simultaneously playing the roles of Task Giver, Solver, and Verifier, appearing as two distinct Verifiers, or managing multiple identities via pooled financial resources (see Sec. 5.3). Using assumption (II) above, we shall argue that TrueBit resists such attacks in Sec. 5.

The "forced errors" described in the beginning of this section (Sec. 2) pose a special challenge to TrueBit due to opportunistic attacks. As the prize for discovering forced errors is necessarily enormous, attackers may seek ways to obtain these large prizes without performing the intended verification tasks (Sec. 5.4). For this reason, the occurrence of forced errors must appear truly random to all prospective Verifiers. Moreover, Solvers must be incentivized from prematurely sharing the fact that their error was forced or claiming that an error was forced when it really wasn't. In general, we want consistent, not sporadic, verification. At least one Verifier should be present for every task. The attraction of forced errors represents the "Achilles' heel" of TrueBit, and we devote the incentive structure in Sec. 4 to careful defending of this potential attack surface.

3. Dispute resolution layer

In this section we present and analyze TrueBit's dispute resolution mechanism. TrueBit relies on network "Judges" with limited computational power who adjudicate all disputes, and we show how such Judges can correctly resolve disputes over complex computations.

3.1. *Bottleneck: The Verifier's Dilemma*

Ethereum's incentive structure severely restricts the amount of computation work that smart contracts can accurately enforce. Users interact with smart contracts by broadcasting *transactions* to the network. Such transactions contain program scripts, which require computational work to verify their validity prior to miners' acceptance on the blockchain. Each miner must not only locally verify but also remember the state of every single smart contract. On the surface, this redundant work limits smart contracts' computational scope due to inefficiency, but the true bottleneck lies deeper at the incentive level.

Let us recall miners' incentives for participating in Ethereum. In Nakamoto consensus-based cryptocurrencies [19], new transactions enter the blockchain through a process called *mining*. Participating *miners* maintain the integrity of the blockchain data by racing to solve computation-intensive, *proof-of-work* puzzles in exchange for rewards. The first miner who successfully broadcasts a solution to the current proof-of-work puzzle proves that she has spent the necessary computation power to merit appending her new set of transactions to the blockchain, and this step awards the miner a set of newly minted coins. Then, the race begins again on top of this new block.

When an Ethereum user broadcasts a transaction to the network, the miner who appends her valid transaction to the blockchain receives a *transaction fee* for the service. On the other hand, Ethereum's mining protocol dictates that other miners should verify this transaction *gratis*, for the "common good" so to speak. More specifically, the mining protocol dictates that the longest chain consisting of valid transactions is the correct one, and miners should mine on top of it. If the time to verify new blocks is negligible, miners may indeed agree to check them, but if the verification work becomes substantial, then miners risk falling behind in the mining race by following protocol and actually verifying. In order to save time, rational miners might simply skip the verification step, leaving the blockchain open to invalid transactions. On the other hand, by skipping verification, a miner might also waste her CPU cycles by mining on top of a stale chain, which the rest of the community deems invalid. Either way, the miner risks forfeiting rewards. A rational miner doesn't know what to do!

Any substantial verification burden placed on miners thus results in a *Verifier's Dilemma* [1]. Because of the Verifier's Dilemma, smart contracts, whose verifications require non-trivial computational efforts, will fail to execute correctly as rational miners, who may choose to deviate from the suggested protocol, stop donating their limited resources to thankless verification tasks. In short, Ethereum lacks a scalable verification mechanism and remains vulnerable to serious attacks [1] when miners have to verify more than a tiny amount.

TrueBit contracts securely execute computationally-intensive tasks for use in Ethereum smart contracts. The system's core consists of a smart contract interface, where a user can request a solution to a computational task or puzzle, and anyone can solve it in exchange for a reward. Our interactive verification game, described below, empowers users to both outsource arbitrary computations in a decentralized manner and receive back correct solutions. To achieve this goal, we will not only reduce the redundancy of verification work but also refine the incentive structure responsible for the Verifier's Dilemma.

3.2. *Solution: The verification game*

The goal of TrueBit's interactive verification game is to resolve a given dispute between a Solver, who has provided a solution to some computational task, and a Challenger, who disagrees with the solution and therefore calls the verification game to be played. The outer layer of TrueBit (Sec. 4) uses the verification game as a subroutine. The roles in the verification game include a Solver, who offers a solution to a given task, and a Challenger, who disagrees with the Solver's solution. The final party, the Judges, always perform computations correctly but possess extremely limited computational bandwidth. The Judges in TrueBit are the entire community of Ethereum miners, who reach verdicts through Nakamoto consensus.

The verification game proceeds in a series of rounds, each of which narrows down the portion of the computation in dispute. In the first round, the Challenger forces the Solver to commit to a pair of computation steps delimiting some time interval. In the next round, the Challenger iteratively challenges a subset of the Solver's computation over this time interval, and next she challengers over a subset of that subset, etc. until in the final round the final challenge is sufficiently trivial that the Judges can make a final ruling on whether the challenge was justified. The Judges also enforce that the Solver and Challenger follow the rules of the game. At the end of the verification game, either the cheating Solver will be discovered and punished in the outer layer of TrueBit (Sec. 4), or the Challenger will pay for the resources consumed by the false alarm.

Example 3.1: Matrix multiplication [18]
We give an illustrative example of a verification game. Suppose that the given task is to compute the product of two matrices A and B. The Solver claims that $A \times B = C$. In the first round, the Challenger must claim an error in one of the entries in C, and outputs coordinates i, j, such that

$$c_{i,j} \neq \sum_{k=1}^{n} a_{i,k} \cdot b_{k,j}$$

with corresponding evidence consisting of partial sums $d_0, d_1, \ldots, d_n$, where

$$d_m = \sum_{k=1}^{m} a_{i,k} \cdot b_{k,j}.$$

The Judges only verify that i, j are coordinates and that $d_n \neq c_{i,j}$ and that $d_0 = 0$. The Challenger loses if this fails to hold. In the second round, the Solver tries to defend himself by providing a k, such that $d_k \neq d_{k-1} +$

$a_{i,k} \cdot b_{k,j}$. If the Judges can verify this claim then the Solver wins the game, otherwise the Challenger wins.

3.3. *Detailed protocol*

We can use a variation of Example 3.1 above to check arbitrary computations by creating a binary search to pinpoint the erroneous (or not erroneous) step in a computation. The idea for the following verification game is essentially due to Canetti, Riva, and Rothblum [20, 21] and was later independently discovered by the authors of this chapter [18, 22]. Canetti, Riva, and Rothblum did not consider this game in the context of blockchains but rather a simpler scenario, in which a user outsources a computation to k different clouds under the assumption that at least 1 out of k of the clouds will perform the computation correctly. The assumption of at least one trusted Solver/Challenger cloud is too strong for our purposes since in a purely rational network, such a Solver/Challenger may not exist. Section 4 describes our workaround for the smart contract setting.

For purposes of reducing the number of rounds of interaction and space demands on the blockchain, our protocol roughly follows [20, 21], which combines the (parametrized) binary search optimization from [18, Proposition 9] with the probabilistic Merkle tree construction from [22]. Recall that a *Merkle tree* is a binary tree, in which each node is the hash of the concatenation of its children nodes. In general, the leaves of a Merkle tree will collectively contain some data of interest, and the root is a single hash value, which acts as a certificate commitment for the leaf values in the following sense. If one knows only the root of a Merkle tree and wants to confirm that some data x sits at one of the leaves, the holder of the original data can provide a path from the root to the leaf containing x together with the children of each node traversed in the Merkle tree. Such a path is difficult to fake because one needs to know the child preimages for each hash in the path, so with high likelihood the data holder will supply a correct path if and only if x actually sits at the designated leaf.

For clarity, we will present the verification game in the familiar and standard language of Turing machines, but in actuality TrueBit uses the Google Lanai architecture to simulate steps in the computation (see Sec. 6). The verification game consists of three parties:

- a *Solver* who has allegedly performed some computational task and claims some solution,
- a *Challenger* who disputes the correctness of the Solver's solution, and

- *Judges* with bounded computational power, who will rule on whether the output is correct.

TrueBit fills the roles of Solver and Challenger according to the incentive layer discussed in Sec. 4. The Judges are the set of miners in Ethereum who collectively record their decisions on the blockchain. The incentive layer (Sec. 4) determines the financial consequences of the Judges' verdict.

To start, the Solver and Challenger each privately compile a tableau of Turing configurations mapping each time step of the task to its full internal state. Assume that the task runs in time t steps and space s, where s bits suffice to include the complete state of the Turing machine at any given time, including tape contents, head position, and machine state. The game protocol includes a fixed, integer parameter $c > 1$, which says how many configurations the Solver broadcasts to the blockchain in each round of the verification game. More configurations per round mean fewer total rounds needed, however it also means that the total number of configurations sent to the blockchain increases. In particular, we reach diminishing returns as the data sent per round reaches Ethereum's per block capacity.

The verification game protocol proceeds as follows.

Main loop. The following steps are done iteratively to pinpoint the source of disagreement. The protocol determines timeout periods within which Solvers and Challengers must respond. Failure to respond within time bounds results in immediate loss for the non-responding party.

(1) The Solver selects c configurations equally spaced apart in time across the current range of the dispute. In the initial iteration, for example, the Solver selects configurations across the entire tableau at the following time steps:

$$\frac{t}{c}, \frac{2t}{c}, \frac{3t}{c}, \ldots, \frac{ct}{c}.$$

He then produces c Merkle trees, each with s leaves, where the leaves constitute the complete machine state at each of these times, and broadcasts each of the roots of these Merkle trees to the blockchain.

(2) The Challenger responds with a number $i \leq c$, indicating the first time step in this list that differs from her own, and broadcasts this number i to the blockchain.

(3) The Judges check that the Solver indeed posted c Merkle roots to the blockchain and that the Challenger's value i satisfies

$1 \leq i \leq c$. If either of these checks fails, the Solver, or respectively the Challenger, loses immediately.

(4) The next iteration of the game continues similarly, but restricted to configurations between the $i-1$-st and i-th indexed configurations. Here we interpret a 0 as the computation's initial configuration.

Final stage. After $\log t / \log c$ rounds, the loop above converges to the first, disputed computational step, and the Judges explicitly check this step of the computation as follows. Suppose that the first disagreement occurs at time e. The Solver provides paths from the Merkle root for time e to its leaves containing:

- the location of the machine head at time e,
- the content of the tape square under the machine head, the tape contents of its left and right neighbors, and
- the machine state at time e.

The Solver also provides the analogous paths for time $e-1$. The Solver loses the dispute if the Judges find that any of these paths fail to be valid. Finally, using this information and the original task code (which existed on the blockchain prior to the verification game), the Judges check the computational step at time e and rule whether the challenge was justified.

3.4. *Runtime and security analysis*

We now show that the work done by the Judges is small compared to the work required to perform the task, and that the number of interaction rounds is modest. Note that the majority of effort from the Judges simply involves recording data onto the blockchain. The Judges' work is split over many transactions and therefore avoids any chance of a Verifier's Dilemma.

We fix σ as the size of each hash in the Merkle trees, and p as the space required to store a single machine state for the computation. For computational tasks that run in time t and space s with c configuration roots posted per round, the protocol requires $\log t / \log c$ rounds of interaction on the main loop, writes $c\sigma + \log c$ bits to the blockchain during each iteration of the main loop, and writes $2 \cdot ((2\sigma + 1) \log s + 5 + p)$ bits in the final round (assuming that tape squares can contain either 0, 1, or be blank, so we need $3 \cdot \log 3 / \log 2 \approx 5$ bits to describe 3 of them). Summing these parts, the total time required for community-wide verification from the Judges is then

$$O\left[\frac{\log t}{\log c}\cdot(c\sigma+\log c)+2\cdot((2\sigma+1)\log s+5+p)\right]$$

where the hidden constant depends on the implementation of this protocol on the Judges' local machines. This estimate does not include the space and time required to store the instructions for the initial computation task. We assume that this data already exists on the blockchain before the verification game begins.

In comparison with other proposed blockchain-based computation systems, or even PCP-based verifiable computing systems [14], the work imposed on the Judges by the verification game is small [23], as is the computational overhead required to join the system [24]. Moreover, the economic costs to the Task Giver, who pays only a Solver and a Challenger and not all Judges to process the full task, is modest, as are the computational overheads for the Solver and Challenger.

Note that larger values for σ reduce the chances of hash collisions. Hash collisions might allow a dishonest Solver to substitute configurations in his computation tableau with fake ones sharing the same Merkle roots. Two consecutive configurations in the tableau whose Merkle roots happen to agree with the Merkle roots of two consecutive configurations that the dishonest Solver wishes to apply as substitute suffice to disrupt the integrity of the verification game protocol. Indeed, the Challenger will agree with both Merkle roots, and the Judges will confirm that the transition between them is valid and therefore incorrectly reject the challenge. For a fixed size σ, the chance that such collisions occurs by accident even becomes likely for sufficiently enormous tasks.

One could entirely eliminate the security risk discussed in the previous paragraph by posting complete machine states on the blockchain rather than just Merkle roots, but this makes the protocol more expensive. Alternatively, one could either increase the parameter σ or the number of states checked by the Judges in the final step. In this sense, the choice of σ bounds the maximum complexity of secure computations in TrueBit. We will also see in Sec. 4.1 that the capital held by the TrueBit contract as a "jackpot" for finding a forced error poses an additional constraint on computational capacity, but this value can scale as well. Finally, the effectiveness of TrueBit's verification game may degrade for extremely complex tasks due to the computational limitations of the Challenger. If the expected execution time of a verification game exceeds a human life expectancy, for example, then potential Challengers may lack motivation to participate despite financial incentives. Note that even when a task itself is feasible, its corresponding verification game may not be, as the verification game carries significant overhead. We anticipate that future versions of TrueBit may optimize the verification game for certain kinds of tasks, so as to reduce this discrepancy (see Sec. 7.4).

Remark 3.2: The verification game provides a significant privacy-protecting effect in that only disputed parts of a given computation must touch the public eye [25].

4. Incentive layer

We now discuss the financial incentives, which encourage anonymous parties to both contribute CPU cycles and perform requested computational tasks correctly. We use the dispute resolution layer from Sec. 3 as a subroutine and connect its role to the present construction. A *Verifier* is a potential Challenger, that is, someone who checks submitted solutions and calls for a challenge (see Sec. 3) if and when she detects an error. By assumption (II) in Sec. 2.2, Verifiers must receive adequate payment for verification tasks. As argued in the beginning of Sec. 2, however, Verifiers can only receive rewards when they actually find bugs. It follows that the reward for discovering errors must encompass amortized payment for all verification tasks checked, including those in which no bugs were found.

Any Ethereum user who offers a reward through TrueBit for performing a computational *task* is called a *Task Giver*. A party which offers a *solution* for performing these tasks in exchange for a reward is called a *Solver*, and, as indicated in the previous paragraph, Verifiers check that Solvers' solutions are correct. Solvers and Verifiers will play the roles of "Solvers" and "Challengers" respectively from the dispute resolution layer (Sec. 3) whenever a Verifier initiates a verification game. *Referees* enforce the rules of the TrueBit protocol in the incentive layer. While Ethereum miners enforce protocol rules both through the roles of "Judges" in Sec. 3 and the role of "Referees" in the present section, we use distinct labels for them because the layers and functions differ for these roles. The primary role of Judges is to interactively resolve a dispute, whereas Referees primarily enforce that Solvers and Verifiers timely submit appropriate data in the incentive layer.

In a nutshell. The main steps of a TrueBit contract are as follows. Items in quotes will be thoroughly explained in due course.

(1) A Task Giver announces a task and offers a reward for its solution.

(2) A Solver is elected by lottery, and he prepares both a "correct" and an "incorrect" solution to the task.
   (a) If a "forced error" is in effect, the Solver reveals the incorrect solution on the blockchain.
   (b) Otherwise, the Solver reveals the correct one.
(3) Verifiers check the Solver's solution. They win a large "jackpot" payout if both:
   (a) they correctly identify the solution as erroneous, and
   (b) a forced error was in effect.
(4) If no Verifier signals an error, then the system accepts the solution. Otherwise, acceptance depends on the outcome of a verification game.

In the rest of this section, we discuss the security measures in TrueBit, which ensure that participants comply with the skeletal procedure outlined above. A more complete description of the protocol appears in Sec. 4.6.

4.1. *Jackpots*

TrueBit periodically imposes *forced errors*, in which a Solver must offer a wrong solution to a task (see Sec. 2). This ensures that Verifiers who diligently search for errors will eventually find them. Accordingly, Verifiers who correctly report forced errors receive substantial *jackpot* payouts. By design, Verifiers cannot predict when forced errors will occur and therefore have incentive to check all tasks carefully. Forced errors occur only rarely, and we expect Solvers to make only few, if any, other errors. Rewards for identifying unforced errors may be modest compared to jackpot payouts, and we consider any such payments of these type incidental, rather than fundamental, to TrueBit's secure incentive structure.

The jackpot payout effectively bounds the complexity of computation tasks that TrueBit can perform securely. The Verifier must, on average, receive a payment, which fairly compensates her for the task at hand, which means that the jackpot payout should at least consist of fair compensation for the current task times the forced error rate. In this way, the jackpot cap bounds the Verifier's expected per task compensation which, by assumption (II) in Sec. 2.2, restricts Verifiers' available CPU cycles per task.

We fix a rate for forced errors among tasks. This fraction should not be so low so as to discourage Verifier participation through excessively infrequent rewards, but neither should it be so high so as to run the risk of

Referees bias (see Sec. 5.1). We set forced errors to occur, on average, once every thousand tasks.

4.2. *Taxes*

Given that a jackpot *repository* must exist, we now describe the mechanism for funding it. We assume that a generous philanthropist deposits some initial funds into the repository, but thereafter TrueBit will be self-sustaining through *taxes*. Any Task Giver who calls a TrueBit contract must pay not only the cost of computational work done by the Solver but also for the work done by the Verifier(s) (excluding unforced errors and bogus challenges), as well as the work done by Referees and Judges. We refer to the latter two costs as the *verification tax*. We ensure that the jackpot repository never disappears entirely by placing a cap on the jackpot size. To this end, we set the maximum jackpot payout for a forced error to be one third of the total repository size.

While a single, attentive Verifier suffices to ensure correctness and achieves ideal tax efficiency, in practice the verification tax requires a substantial cushion. We estimate the necessary verification tax to be 500%–5000% of the cost of performing the given task. As we shall see in Sec. 5.2, there is a quantitative tradeoff between tax efficiency, Solver deposits, and security of computations, so Solvers could potentially absorb some of this tax burden by contributing higher deposits. Our tax rate estimate incorporates Verifiers' incentives for participation and the fact that both the Task Giver and the Solver for a given task each have incentive to perform verification. Participation from these parties may necessitate higher taxes because the total jackpot payoff decreases exponentially as the number of challenges increases (see Sec. 5.3). Expected jackpot payoffs must be sufficiently high to consistently attract at least one Verifier per task. The required tax amounts may also depend on peculiarities of human behavior. Indeed, we may have to pay Verifiers more than Solvers per CPU cycle because Verifier rewards have the human-undesirable property of being sporadic, whereas Solvers always receive immediate rewards. Thus the optimal tax rate must, at least in part, be determined experimentally.

4.3. *Deposits*

TrueBit requires deposits from Solvers and Verifiers in order to thwart Sybil attacks (see Sec. 5.1) and incentivize correct computations. We set these deposits to be more than the expected jackpot payout for a given task plus the cost of playing a verification game. In particular, the deposits must be large enough to:

(1) pay for the (expensive) cost of a verification game, including all rewards and penalties for Solver and Challengers and work done by Judges,
(2) discourage Solvers and Verifiers from sacrificing deposits in order to obtain jackpots without performing verification (see Secs. 5.1 and 5.3),
(3) discourage Task Givers who collude with Solvers in effort to get bogus solutions onto the blockchain (see Sec. 5.2),
(4) refund taxes to the Task Giver in case the Solver causes an unforced error,
(5) deter Solvers from randomly guessing solutions to obtain task rewards instead of actually performing computations (as might be profitable for binary decision tasks with very high task rewards), and
(6) deter external, temporal pathologies.

Note that the currency used to pay Solvers and Verifiers need not be the same as the currency used to pay Judges and Referees, but for simplicity of presentation, we fix ether (ETH) as the unique underlying currency.

As an example of the second type, consider a situation where the Solver deposit is small (say 10 ETH) but the expected jackpot payout per task is high (say 1000 ETH). An individual playing both the role of the Solver and Verifier could offer a bogus solution and then challenge his own answer, hypothetically netting, on average, $1000 - 10 = 990$ ETH without providing any useful service. Such an action would degrade other Verifiers' incentive to participate.

As an example of the last case, if the outcome of a time-sensitive TrueBit contract controlled a 1000 ETH payout but only required a 10 ETH Solver deposit, then the party who stands to lose 1000 ETH from the TrueBit contract could attempt to cause a delay by posing as a Solver to the TrueBit contract and giving a bogus answer. We leave it to the Task Giver to determine appropriate minimum deposit values for such specific situations, as such contextual determinations lie outside of the scope of TrueBit itself.

4.4. *Generating forced errors*

In order to motivate verification of all tasks and to guard the jackpot repository against swindle, forced errors must appear unpredictably. TrueBit uses strings of random bits to determine whether or not a forced error occurs for a given task. The system derives its unpredictability via the following properties.

(1) The Task Giver does not know the random bits at the time that she announces a task.
(2) The Solver does not know the random bits until after he has committed his solution.
(3) Verifiers do not know the random bits until after they decide whether to challenge.

The first property makes it difficult for a Task Giver to create a task designed to swindle the jackpot repository, and the second discourages Solvers from being lazy and only volunteering to solve tasks which have forced errors. In order to satisfy these three properties, TrueBit combines random bits from the following two sources:

(a) "private" random bits from the Solver, and
(b) the hash of the block mined immediately after the block containing the Solver's solution.

By *hash of the block*, or *block hash*, we more precisely mean the hash of the block's *header*, a roughly 200-byte piece of data that contains the timestamp, nonce, previous block hash, and Merkle root hash of the transactions occurring in the block [26]. Source (b) achieves properties 1 and 2 above, and source (a) achieves property 3.

TrueBit generates Solvers' "private" random bits using a method reminiscent of RANDAO's random generator mechanism [27]. Before participating in a TrueBit contract, a Solver must privately generate a string of random bits r and publish its hash on the blockchain. This action commits the Solver to using r in the protocol without revealing the value r to others.

The Solver establishes property 2 above by committing both a hash of a "correct" solution and a hash of an "incorrect" solution prior to the broadcast of the block hash from item (b). At the time that this block hash is revealed, only the Solver, who has access to the value r, knows whether or not a forced error is in effect. He publicly designates one of his two hashed solutions for evaluation, however potential Verifiers do not know *a priori* whether the Solver intended his solution to be "correct" or "incorrect." Only after the timeout for challenges has elapsed do the Verifiers witness the Solver's private random bits in the clear and learn whether a forced error was in effect. In case the Solver's solution is challenged, the Solver must reveal r as well as his second "correct" solution in order to prove that the error was forced and avoid a penalty (see Sec. 5.4). Thus, in the case of a forced error, the Task Giver still obtains a correct solution. In case of an unforced error, i.e. when the Solver evidently fails to provide a correct solution, the Task Giver receives a full refund of her task reward and taxes (see Secs. 4.3 and 4.6).

Although in theory one could securely run this part of the protocol without hashing the Solver's solutions, hashing makes it easier for the Solver to provide a convincing "incorrect" solution which appears, upon casual inspection, indistinguishable from a "correct" one. The Solver can effectively use any randomly selected "incorrect" solution during a forced error because he never has to reveal its preimage.

4.5. *Solver and Verifier election*

The system effectively chooses Solvers by lottery. When a task is announced, Solvers broadcast their interest in solving it to the Referees in the form of Ethereum transactions. Referees, or more specifically miners, choose one of these transactions to include the next block, thereby electing the Solver for that given task. In case the miner who chooses the transactions for the current block happens to also hold a lottery ticket, he may bias his chances of winning the lottery. This bias does not affect the security of TrueBit, however, since the Solver must still provide a correct solution.

In contrast to the Solver selection, TrueBit does not impose a limit on the number of Verifiers, and we describe in Sec. 5.3 how multiple Verifiers who participate in a task must split their rewards after a successful challenge. Due to the computational work involved in verification, rational Verifiers will only verify tasks in case they expect to receive adequate compensation for their work (assumption (II), Sec. 2.2). Thus the number of Verifiers verifying each task will remain low due to the balance between computation costs and incentives (see Secs. 5.2 and 5.3).

4.6. *Protocol overview*

In this section we present an overview of the TrueBit protocol. We will discuss how our chosen parameters enable system security in Sec. 5.

Throughout the description below, Solvers and Verifiers, who deviate from protocol by broadcasting malformed data or failing to respond within timeout bounds, forfeit their security deposits to the jackpot repository.

Preprocessing steps. The following must be done prior to commencing TrueBit operation:

(1) A substantial jackpot repository must be established for the TrueBit contract prior to commencement of any interactions (see Sec. 4.1 for details).
(2) Solvers and Verifiers who wish to participate must commit deposits to the TrueBit smart contract (see Sec. 4.3). The deposits may be placed at any time during the protocol, so long as the underlying contract has sufficient time to confirm receipt of the funds.
(3) Solvers must also generate private random bits and commit their respective hashes to the blockchain. We denote the private random bits of the unique Solver selected to perform the `task` below by r (see Sec. 4.4 for more details).
(4) A universal tax rate T must be established (see Sec. 4.2).

Main algorithm. The protocol steps are as follows.

(1) A Task Giver provides the following:
    (a) a computational `task`,
    (b) the `timeOut` for accepting bids, performing the computation, and waiting for a challenge. In the protocol below, we do not distinguish between these various timeouts with distinct notation, and we colloquially use "`timeOut`" to refer to both events and lengths of time. In all cases, `timeOut` must be long enough to avoid microforks during which Referees temporarily disagree on current parameters.
    (c) a `reward` for a correct output, which must be at least the cash equivalent of the task difficulty d based on `timeOut` (see Sec. 5.5), plus a total tax of $T \cdot d$. The `reward` is held in escrow by the TrueBit contract while the taxes are immediately deposited into the jackpot repository.
    (d) the minimum deposit, `minDeposit`, needed to participate as a Solver or Verifier (see Secs. 4.3, 5.1–5.3).
(2) Solvers, who have the requisite `minDeposit` and random bits,

can bid to take on the task until the bidding `timeOut`. At most one Solver is selected (Sec. 4.5), and if no Solver takes on the task in the allotted time, the task request is canceled and the Task Giver receives a full refund.

(3) The Solver privately computes `task`. In case of a `timeOut`, Solver forfeits his deposit to the jackpot repository and the protocol terminates.

   (a) Solver commits two distinct hashes to the blockchain, thereby committing both a "correct" and an "incorrect" solution.
   (b) The hash of the next mined block is revealed, and then Solver knows whether or not there is a forced error (see Sec. 4.4).
   (c) Solver designates one of the two hashes as the hash of his `solution`.

(4) Verifiers who have posted `minDeposit` can challenge (the hash of) `solution` until `timeOut`. Prior to `timeOut`, the Verifier must broadcast the hash of an even integer to the blockchain in order to commit to a challenge. Hashing an odd number in case of no challenge is optional and may be used to camouflage real challenges from other Verifiers (see Sec. 5.3). After `timeOut`, the Verifier broadcasts to the blockchain this hashed number in the clear to reveal her action.

   (a) If no Verifier challenges `solution`, then
      i. Solver reveals r to prove that there was no forced error (i.e. the criteria in Step 4(b)i below fails),
      ii. Solver reveals `solution` in the clear on the blockchain,
      iii. Solver receives the task `reward`, and then
      iv. the protocol terminates.
   (b) Otherwise, some Verifier challenges `solution`.
      i. Solver reveals his private random string r on the blockchain, and Referees check it against his commitment from preprocessing Step 3. If the hash of the concatenation of r and the block hash following the `solution` announcement from 3(b) is small (as determined by the forced error rate, see Sec. 4.1), then a forced error is in effect (see Sec. 4.4).
      ii. If the value r reveals that Solver's error was forced,

then Solver must reveal his secondary solution in the clear (see Sec. 5.4).

A. If no Verifier challenges Solver's secondary solution before `timeOut`, then Verifier wins a fraction of the maximum jackpot amount J, scaled for task difficulty. In case of challenges from multiple Verifiers, the jackpot is split among them. In more detail, if there are k distinct challenges, then each participating Verifier receives $J/2^{k-1}$ times the ratio of the task's difficulty to the system's maximum task difficulty (see Secs. 4.1, 5.3, and 5.4 for further discussion).

B. Otherwise the Solver must play the verification game with the challenging Verifier(s). Verifier penalties, Verifier rewards, Solver penalties, and refunds to the Task Giver here are the same as described in the next step. In case the Solver successfully defends himself against all challenges, however, then the jackpot payouts proceed as in Step 4(b)i above.

iii. Otherwise the error was not forced. Solver reveals `solution` in the clear, and then Solver and Verifier must play a verification game (Sec. 3.2). In case of challenges from multiple Verifiers, the steps below are repeated until either a Verifier wins a challenge or Solver defeats all Verifier challenges. The Solver collects `reward` only once he wins all of the challenges, and if this does not happen, then the Task Giver receives a refund on his `reward`, and tax payments are reimbursed through Solver's deposit as described below.

A. If Solver wins, then Verifier forfeits half of her deposit to the jackpot repository and the other half to the Solver, where the latter suffices to compensate the Solver for playing the verification game (see Sec. 4.3).

B. Otherwise, Verifier wins. Solver pays at most half of his deposit to the Verifier(s), according to the distribution scheme in Step 4(b)ii above, pays back the Task Giver's taxes out of this amount, and forfeits the remaining funds to the jackpot repository (see Secs. 5.1 and 5.2).

End of protocol.

Note that Solver does not receive `reward` when a forced error occurs in Step 4(b)ii. In such situations, we expect the Solver to challenge his own "mistake" and receive a jackpot payout much greater than `reward`. TrueBit makes this payout automatic since the Solver would lose the verification game anyway if he and Verifier were to play it out.

4.7. *Sanity check*

Table 1 recaps the parameters used in the TrueBit protocol and hints at how to estimate their exact values. Each parameter is either a fixed constant, an input from a participant or something already on the blockchain, or can be computed from the elements, which appear above it in the table. By inspecting the list below from top to bottom, one can confirm that all notions are well-defined.

5. Defenses

We now analyze the security of the incentive layer from Sec. 4. TrueBit's security relies on the presence of at least one Verifier to check each task performed. In this section, we show that available incentives suffice to guarantee the existence of such Verifier(s) according to the network assumptions in Sec. 2.2 and the attacker model from Sec. 2.3. TrueBit defends against shortcuts which could divert verification incentives to non-verifying parties. In particular, we show that TrueBit resists Sybil attacks, collusion pools, opportunistic attacks related to jackpot payoffs, and certain external threats.

5.1. *Pairwise Sybil attacks*

In a Sybil attack, an adversary assumes multiple identities on the network in order to execute an exploit. Identities on TrueBit include Task Givers, Solvers, Verifiers, Judges, and Referees. By our assumption (Sec. 2.3), Judges and Referees always function as intended, and we provide additional justification for this axiom here. In this subsection, we consider all

Table 1. Relations between parameters used in TrueBit.

parameter	dependency
dispute layer parameters p, c, and σ (Sec. 3)	- fixed constants.
tax rate (Secs. 4.2, 5.2)	- fixed constant (500%–5000%).
forced error rate (Sec. 4.1)	- fixed constant (1/1000).
maximum jackpot payout (Sec. 4.2)	- 1/3 of current jackpot repository.
cash equivalent of CPU cycles (Sec. 5.5)	- based on external markets.
maximum task difficulty (Sec. 4.1)	- maximum jackpot payout divided by cash equivalent of a CPU cycle.
task	Parameters in this box are chosen by the Task Giver with minimums as described below:
timeouts (Sec. 4.6)	- long enough to avoid microforks.
task difficulty	- determined by timeouts.
Solver reward (Sec. 5.5)	- no less than the cash equivalent of the task difficulty.
expected jackpot payout (Secs. 4.1, 5.2, 5.4)	- cash equivalent of task difficulty, and number of active Verifier deposits.
Solver & Verifier deposits (Secs. 4.3, 5.1–5.3)	- more than the cost of verification game plus the expected jackpot payout. Also depends on tax rate.
actual number of Verifiers (Sec. 4.5)	- as many join (incentives limit overparticipation).
jackpot payout for challenging a forced error (Sec. 5.3)	- based on maximum jackpot payout, actual number of verifiers, and ratio of task difficulty to maximum task difficulty.
Payout for detecting an unforced error (Secs. 5.1, 5.2)	- at most half of Solver deposit is split among all Verifiers (rest goes to jackpot repository).

sets of pairs among the remaining identity types and show that pairwise cooperation does not harm operation of the system. While parties can freely join and leave TrueBit, each identity must make a deposit in order to participate. This deposit alone is a general deterrent against Sybil attacks, however as we shall see, multiple identities do not provide much cheating advantage.

Judges and Referees. Recall that Ethereum miners play the roles of Judges and Referees in TrueBit. Our analyses in Secs. 3.4 and 4.6 show that the work done by these parties is small, and hence they are not vulnerable to a Verifier's Dilemma (Sec. 3.1). Nakamoto consensus [19] therefore ensures that miners will not post bogus transactions, i.e. enforce rules incorrectly, lest other miners reject their block and cost them a block reward. Therefore the only threat from miners is block withholding, specifically with regard to random number generator bias (see Sec. 4.4).

While in theory miners could discard blocks, along with their mining reward and associated random bits for TrueBit, in practice this may never happen. A miner who drops a block must expect in exchange some income greater than the usual mining reward, in order to make this action worthwhile. If such income were to come as a bribe from a TrueBit participant, it would have to come from a Solver since only the Solver, who has unique access to his own private random bits, knows how to bias miners' random bits towards or away from a forced error. In short, miners cannot disturb randomness in TrueBit without a tip-off from a Solver. The Solver, however, has no reason to bias against a forced error because this would prevent him from challenging his own answer and winning the jackpot, or at the very least would invite others to share it with him, thereby decreasing the total jackpot payout (see Sec. 5.3). Moreover, the Solver is unlikely to succeed in biasing towards a forced error since miners have little control over their own block hashes. This "one-sided" effect of block withholding, which can lock away jackpot funds but (almost) never release them, makes block hashes a safe source of randomness in TrueBit. Hence the Solver's potential reward for challenging his own solution under a forced error is not merely an artifact of TrueBit's incentive structure — it guarantees unbiased Referees.

Task Giver and Solver. A Task Giver, who also acts as a Solver, does not benefit from solving her own task. One idea for an attack vector would be to create a task such that the Solver's private random bits force an error for the given task. Since the Task Giver cannot predict the random bits from the block hash at the time the task is created (Sec. 4.4, part (b)), the Task Giver cannot create such a task. Moreover, the Task Giver cannot afford to flood TrueBit with useless tasks and solve them herself in the hopes of eventually, by chance, encountering a forced error. The Task Giver would pay more taxes to the jackpot repository than she would expect to win from the jackpot payout (Sec. 4.2), even taking into consideration any rewards she might win by correctly solving her own tasks. Her losses are amplified through payments to other Solvers, Verifiers, and Judges who choose to participate in the protocol.

Solver and Verifier. The Solver's burned deposit always exceeds the Verifier's income for successfully challenging an unforced error (see Sec. 4.3). Hence, the Solver has no incentive to challenge himself as a Verifier in such cases. Similarly, a Verifier can only lose her deposit by posing bogus challenges. In certain situations, it is conceivable that a Solver–Verifier pair could benefit from submitting a false solution and then challenging it due to temporal constraints external to TrueBit itself (as mentioned in Sec. 4.3), and in such cases, the Task Giver must determine the necessary deposits to deter such actions.

In the case of an forced error, we expect the Solver will challenge himself in order to win the jackpot. Nevertheless, TrueBit's tax rate (Sec. 4.2), and hence its jackpot payout (Sec. 4.1), suffices to incentivize an independent Verifier to also check the solution. As we shall see in Sec. 5.3, the Solver lacks incentive to claim the same jackpot more than once, hence the Solver's self-verification does not spoil other Verifiers' motivation to participate.

Task Giver and Verifier. A Task giver can certainly verify the solution she receives. This checking can only improve the reliability of the system! We cannot, however, assume that the Task Giver will always perform such checks due to the Task Giver's possible computational and financial resource constraints or lack of motivation to involve herself in the details of a TrueBit contract.

5.2. *The trifecta*

Up until this point, we have implicitly assumed that the attacker's goal was to extract money from the jackpot without performing verification. However, there is another possibility: the attacker wishes to get a bogus solution accepted onto the blockchain. No penalty scheme can deter all such attacks since TrueBit, as a closed system, has no way to estimate the true economic impact of a bogus solution on Ethereum's blockchain. Hence, we now consider scenarios in which a party is willing to sacrifice a jackpot payout (or more) in order to skew computational results.

Scaring off Verifiers. Suppose that a Task Giver wishes to obtain an incorrect solution for a task. Posing as a Solver, she gives a wrong solution to her own task. Now the only thing that could stand in the way of her success would be a pesky Verifier. In order to dodge this obstacle, the attacker poses as a regular Verifier prior to posting her malicious task. She checks every task that comes along, until eventually she encounters a forced error. At this point, the attacker challenges not once but a hundred times, which according to Step 4(b)ii in Sec. 4.6, essentially wipes out the jackpot payout for herself and any other Verifiers who also challenged this solution. The upshot is that legitimate Verifiers no longer wish to participate because TrueBit failed to deliver their well-deserved jackpot payout, and more to the point, they have no way to estimate their income from future jackpots. So when the attacker finally broadcasts her malicious task, legitimate Verifiers simply ignore it.

As Verifier participation, and hence security, depends entirely on the expected value of jackpot payouts (assumption (ii) in Sec. 2.2), TrueBit must ensure sufficient expected value for these payouts. We therefore offer the following default strategy for all Verifiers. New tasks are submitted to TrueBit at various and overlapping times. A prospective Verifier, who wishes to earn rewards for verification, should count the number of active Verifier deposits and compare this to the average number of simultaneous tasks occurring on the network. If the tax rate and task flow profitably support an additional Verifier, then the prospective Verifier submits a deposit, randomly chooses an active task, and begins verifying. As soon as she finishes verifying this task, she randomly selects another active task and begins verifying that next one. Thus a Verifier with either more powerful hardware or better algorithms can gain a market advantage in TrueBit. Such a Verifier can check more solutions per unit time and therefore increase her expected jackpot rewards per unit time. Verifiers are free to leave the network at any time, for any reason. If the current number of active Verifiers exceeds the number supported by the tax rate and task flow, then the probabilistic default strategy tells each Verifier to exit the system with the minimal probability that one would expect to restore equilibrium if all Verifiers were following the default strategy.

Verifiers who choose tasks to verify at random according to the default strategy, as opposed to changing their behavior based on guesses as to who-solved-what before, can reasonably estimate their compensation for participation. When a significant number of Verifiers (not necessarily a majority) follow the default strategy, the adversary wishing to establish a bogus solution can neither predict which tasks will be verified, nor can he influence which tasks default strategy Verifiers will verify. According to assumption (ii) in Sec. 2.2, there exists some bound on the adversary's financial means. If not, he could simply flood the Verifier pool with deposits and begin verifying, which in turn would incentivize legitimate Verifiers to leave the system. Therefore the adversary can only post a bogus solution and hope that, by chance, it gets ignored. In this way, the default Verifier strategy spoils the scare-off strategy above.

Let us now estimate the chances that the adversary succeeds to post a bogus solution, which all legitimate Verifiers fail to notice. We make the worst-case assumption that the adversary Solver always gets elected to solve his own task. Suppose that the tax rate (see Sec. 4.2) supports at least 6 Verifiers per task, and that the adversary holds at most 1/6 of all Verifier deposits. Then effectively there are an average of 5 verifiers per task. Suppose n is the average number of simultaneous tasks on the network. Then the number of Verifiers is approximately $5n$, and so the chance that none of them check a particular solution is the chance that each of them independently and randomly choose other tasks, or $[(n-1)/n]^{5n}$. Therefore the probability that the adversary gets caught, regardless of n, is

$$1 - \left(1 - \frac{1}{n}\right)^{5n} > 1 - e^{-5} > 99\%.$$

The leftmost inequality follows from the standard limit definition for the Calculus constant e. We conclude that the adversary will most likely lose his Solver deposit by attempting such an attack.

By raising taxes further, we can make such an attack even more costly. In this sense the tax rate, together with the minimums for Solver deposits, bounds the mandatory "honesty" of the network. If bogus solutions never occur, then there would be no need for taxes. On the other hand, higher taxes make it more expensive to cheat. There is a tradeoff between overhead tax expenses for Task Givers, minimum deposits for Solvers, and security of computations.

High-stake tasks. For some high-stakes applications, it is possible to entirely fall back on the dispute resolution layer of Sec. 3. In a high-stakes situation, it is reasonable to expect that other parties, who have a significant financial stake in the outcome of a given task, will make an effort to verify the solution regardless of expected jackpot payouts. This motivation acts as a secondary deterrent against the "Scaring off Verifiers" attack above, as does the fact that the Verifier receives a fraction of the Solver's deposit in case of positive error detection (TrueBit burns part of the Solver's deposit in order to avoid collusion between Solvers and Verifiers).

5.3. *Collusion pools*

In this section, we analyze the potential effects of pooling computational, informational, and financial resources among Verifiers and Solvers. Participants in TrueBit may voluntarily form a *Cartel*, which may in turn use external smart contracts to enforce cooperation among its (mutually distrusting) members, however, Solver and Verifier deposits in TrueBit alone prevent many potentially harmful financial collusions designed to extract jackpot funds without exerting computational effort. Secondly, carefully chosen jackpot payout amounts, together with Solver and Verifier incentives and private commitment channels, prevent financial gain from unintended signaling of private random bits or challenges.

Rich but powerless. First, consider a Cartel of Solvers with limited CPU bandwidth but deep pockets which volunteers to solve each and every task that comes along. The Cartel makes no attempt to provide correct solutions to any of these tasks, but instead intends to absorb lost Solver deposits until a forced error comes along, at which point it splits the jackpot among its members. By construction (see Sec. 4.3), Solver deposits exceed the expected jackpot payout per task. Therefore, in the long run, such a Cartel strategy loses money. If members of the Cartel instead decide to start producing correct solutions, this does not harm security because that's what they are supposed to do anyway. We remark that, also by construction (Sec. 4.4), a Solver cannot simply decide to accept only tasks which have forced errors because the Solver does not know whether or not a forced error is in effect until after he has committed his solution to the blockchain.

Similarly, a Cartel, which challenges every single task in the hopes of eventually obtaining a jackpot, sustains losses in the long run due to lost Verifier deposits.

A flood of Challengers. Timely information about forced errors has economic value. While Solvers can earn jackpots by challenging their own forced errors, they risk sharing jackpot payouts by divulging information about such errors. It follows that Solvers have incentive to keep their knowledge of forced errors private and their fake solutions convincing. Information about active challenges also has economic value because a challenge posted to the blockchain could be a tip-off about a forced error. For this reason, Verifiers have incentive to announce "fake" challenges rather than remaining silent on tasks in which they don't detect errors (see Step 5 in Sec. 4.6). "Fake" challenges serve as noise to mask real challenges and protect Verifiers' jackpot revenue from potential copycats who might challenge just because they see someone else doing it.

If the jackpot for a given task were a fixed amount and equally divided among challenging participants, then one could potentially flood the set of participants with aliases to obtain more than one's share of the jackpot. A Cartel, in which members inform each other about forced errors, could therefore have the economic means to flood the network with challenges and monopolize jackpot payouts. This in turn would dilute jackpots for legitimate Verifiers, which would, in turn, degrade Verifier incentives for participation. While such a Cartel might successfully detect forced errors, it might also cause tasks with unforced errors to go unverified.

In order to dissuade a Cartel from flooding the network with harmful challenges, we must reduce the total jackpot payout based on the number of challenges that occur. The total jackpot paid out on a forced error with k challengers must not exceed

$$\frac{J}{2^{k-1}}. \tag{5.1}$$

The reason is as follows. We want a Challenger's net reward to decrease if she superfluously clones herself via the Cartel. Assume that there are n legitimate challenges plus a Cartel, which initially contributes k challenges to the pool and is considering adding another alias. We want the per challenge jackpot distribution among $n + k$ challenges to be less than it would be among $n + k + 1$ challenges, regardless of k, and n. Let J_i denote the total reward paid out when there are exactly i challenges. Then we want

$$\frac{k}{n} \cdot J_{n+k} > \frac{k+1}{n+1} \cdot J_{n+k+1},$$

or

$$J_{n+k+1} < J_{n+k} \cdot \underbrace{\frac{n+1}{n}}_{>1 \text{ for all } n \geq 0} \cdot \underbrace{\frac{k}{k+1}}_{\geq 1/2 \text{ for all } k \geq 1}.$$

Thus, it suffices to set $J_{n+k+1} = J_{n+k}/2$, or by induction $J_k \leq J_1/2^{k-1}$. In fact, we cannot do better than this because the case $k = 1$ asymptotically forces a matching bound.

The upper bound in (5.1) does not take into consideration the Verifier's cost of playing the verification game. In the case where the Solver's solution has been found to have an error through one of the challenges, there is no need to repeat the verification game with the other challenges. The Verifier, who actually plays the verification game with the Solver, receives a bonus from the Solver's deposit to compensate her for the CPU cycles spent during the challenge.

5.4. *On low-hanging fruit*

We analyze the security implications of Verifiers and Solvers, who opportunistically disappear for hard tasks and reappear for easy ones, or who take advantage of solution discrepancies.

Easy winners. If the jackpot for all tasks were equal, rational Verifiers might choose to verify only simple tasks and ignore complex ones. For this reason, the jackpot for each task scales proportionally with the task's complexity (Step 5 in Sec. 4.6). Scaling ensures that a Verifier's expected jackpot payout per CPU cycle remains constant across all tasks, and it equally incentivizes Verifiers to inspect simple and complex tasks. Solvers always appear to perform tasks since the minimum task reward suffices to compensate them for their work (see Sec. 4.6 and assumption (II) in Sec. 2.2).

Multiple solvers. For security reasons, TrueBit explicitly does not allow Task Givers to hire redundant Solvers for a single task (Main Algorithm, Step 2, Sec. 4.6). Suppose that two Solvers provided solutions to a single task and exactly one of them receives a forced error. Any Observer who notices a difference in the two solutions on the blockchain could, without verifying anything, challenge both solutions. By playing two verifications and sacrificing one deposit, such an Observer could potentially win a jackpot at negligible cost, thereby degrading the overall incentives for verification.

Forced error in disguise. What happens if a Solver has a forced error but gives a correct solution anyway? The Solver could then challenge his own solution while other Verifiers ignore it, resulting in both the Solver receiving a bigger share of the jackpot, as well as damage to the system's verification mechanism. Therefore, when the Solver reveals that a forced error was in effect, the protocol also forces him to reveal his committed "correct" solution, which must be distinct from the "incorrect" solution that he showed at first. If the second solution revealed by the Solver is correct and distinct from his first solution, then by uniqueness of deterministic processes, the first solution must have been incorrect (as desired). Verifiers have an opportunity to challenge this second solution. If an error is detected in it via a verification game, the Verifier(s) split the Solver's deposit rather than the jackpot payout according to Step 4(b)ii in Sec. 4.6. Since the Verifier(s) presumably already performed the task themselves when checking the first solution, no additional CPU cycles are required to do this second check. As the Solver loses both a jackpot payout and a deposit by following this course, by assumption (II) in Sec. 2.3, the situation described in this paragraph should never arise.

5.5. *A cash equivalence problem*

Consider the following scenario. A Task Giver wishes to get a bogus solution onto the blockchain. He offers a minimal reward for a difficult task, so as to ensure that no legitimate Solvers or Verifiers volunteer to inspect it. Acting as a Solver, he then provides a bogus solution, and the system accepts his solution because no one bothers to check it. It follows that for security purposes, the system must require that Task Givers compensate fairly based on the difficulty of the task. But how can TrueBit estimate the true cost of executing a given task in its native currency?

While one could potentially establish a long-term lower bound on the cost of a CPU cycle relative to a stable currency like the US dollar or Euro, calculating a lower bound relative to the value of a cryptocurrency token is another matter. Cryptocurrencies are extremely volatile. Moreover, TrueBit lives on the blockchain and does not have access to a newspaper with current exchange rates.

In the first iteration of TrueBit, we will manually update the internal cash equivalent of a CPU cycle based on a live feed (e.g. [28–30]). Ultimately, however, we would like to input these prices in a decentralized way without relying on a third-party. Later versions of the protocol may make use of Augur [11], a decentralized prediction market which ports outside information sources onto the Ethereum blockchain. As of this writing, Augur is currently in beta testing. Alternatively, we may build an independent blockchain for reporting prices, whose "transactions" consist of exchange rate updates.

6. Implementation

Formally, TrueBit is a smart contract in Ethereum, which uses Ethereum's existing smart contract framework to bootstrap the creation of computationally-intensive TrueBit contracts. Tasks in TrueBit take the form of C, C++, or Rust code, but the user must pass this code as input to the Google Lanai architecture [31] (discussed below) prior to submission to TrueBit. This latter step guarantees consistency of simulated architecture and allows Judges to adjudicate fairly.

Google's Lanai interpreter. The theoretical framework for the verification game requires a fixed computer architecture to be used for all verification tasks. In [21], the authors used the regular Intel X86 architecture for this purpose, but we believe that this architecture is far too complicated to be used in TrueBit. All participants of the verification game, including the Judges, have to simulate the behavior of the entire architecture. Even a slight difference in these implementations could result in catastrophic loss of funds for one of the parties. Moreover, simpler architecture costs less to simulate on the blockchain, and because of its simplicity and the fact that there now exists an efficient and trusted LLVM compiler writer, we have chosen to use Google's Lanai architecture [31]. The full Google Lanai interpreter will be available as a smart contract in Ethereum, so that it can be used by Judges. In all TrueBit contracts, the Lanai bytecode will be authoritative.

TrueBit's on-chain interpreter runs in Solidity. For efficiency reasons, tasks will not be run using the interpreters, except in case of dispute. In general, users will run tasks written in native programming languages and running on regular hardware. In the majority of cases where tasks do not require the Lanai interpreter, Solvers and Verifiers can privately optimize implementation of task executions and potentially gain a market advantage over other participants.

7. Applications

TrueBit is more than just an outsourced computation system. It is designed for use in trustless smart contracts. We present some examples of possible use cases.

7.1. *Practical decentralized pooled mining*

Mining rewards are extremely competitive. A typical desktop computer might only mine on average one Bitcoin block every thousand years. To reduce income variance, miners often join *mining pools*, which share both computing resources and mining rewards. Each pool has an *operator*, who distributes computing tasks and rewards to pool *members*. This centralization degrades the security of the system by giving the operator undue influence to censor transactions [5]. In extreme cases, where an operator controls more than half of the network's hash rate, as has been the case with `DwarfPool` [32] in Ethereum, `GHash.io` [33] in Bitcoin, and could happen again with Bitmain's proposed gigantic mining center [34], the operator can even withdraw cleared transactions and double-spend money by way of a *51% attack* [35].

SmartPool [13] introduces mining pools, whose operators are Ethereum smart contracts. As decentralized pool operators, smart contracts have many virtues. In addition to counteracting the censorship issues described in the previous paragraph, they can operate at low cost relative to centralized pools and do not rely on a social contract to ensure fairness. Unlike other decentralized mining pools, which either have higher payout variance [36] that negate incentives for joining the pool or require a change in the proof-of-work protocol [37], SmartPool offers low variance payouts and retrofits existing cryptocurrencies, and it can handle a huge number of participants with a wide range of computing power.

SmartPool's proposed Ethereum mining pool minimizes verification work by precomputing the 1 GB data sets needed to check Ethereum's proof-of-work [5, 38]. While this shortcut may help in checking Ethereum's proof-of-work, not all cryptocurrencies have this lucky feature. Fortunately, TrueBit contracts can check *any* proof-of-work, which means that with TrueBit, we can build a smart contract-based mining pool for *any* Nakamoto consensus-based cryptocurrency. At the time of this writing, for example, the effort required to check a Zcash proof-of-work [39] appears to exceed Ethereum's `gasLimit` capacity by a factor of 50. TrueBit is an option for bringing this task within reach of Ethereum smart contracts. Project Alchemy [40], which aims to bring smart contract functionality to Zcash in the spirit of the Dogecoin–Ethereum bridge below, may also benefit from TrueBit's ability to check Zcash's proof-of-work.

7.2. *Dogecoin–Ethereum bridge*

We can use TrueBit to build a two-way peg between Dogecoin [41] and Ethereum, enabling Dogecoin users to move dogecoins between Dogecoin's blockchain and Ethereum's blockchain without transmitting currency through a third-party exchange and effectively adding smart contract functionality to Dogecoin. The Dogecoin community maintains an active interest in such a bridge [6], and currently offers a prize of more than 6000 ETH for its construction [42, 43].

TrueBit contracts can check Dogecoin's Scrypt-based proof-of-work, whereas traditional Ethereum smart contracts cannot. If Dogecoin were to enhance its scripting language with an operation indicating currency transfer to Ethereum addresses, a TrueBit contract could then confirm such transfers. The newly created dogecoin token could then be passed around Ethereum, and a final signed transaction in Ethereum could finally send the dogecoin back onto Dogecoin's blockchain, assuming that Dogecoin miners are willing to treat the Ethereum blockchain as authoritative for such transfers.

7.3. *Scalable transaction throughput*

Building a system that can securely meet even the modest transaction volume demands of current Bitcoin users remains a formidable challenge [44]. Proposed Byzantine "sharding" [45–51], and miner-based "serializing" [52,53] protocols exist, which aim to distribute verification work, but here we take a simple and fundamentally different approach to decouple the two tasks which miners perform, namely

(1) selecting which transactions to include in the blockchain, and
(2) verifying that blockchain transactions are valid.

Using TrueBit, one can construct a verification protocol, whose incentives guarantee that task 2 is correctly performed by off-chain Solvers and Verifiers (with some help from Judges and Referees), while miners continue to perform task 1. In this way, complex transactions can securely reach the blockchain without overburdening miners.

7.4. *Towards a big data system*

In order to perform as a truly scalable cloud computer, TrueBit must have access to a scalable data storage system. Ethereum's blockchain alone does not suffice, as storing even moderate amounts of data directly on Ethereum's blockchain is prohibitively expensive. TrueBit can securely access and use portions of massive data sets, so long as the data is stored somewhere publicly and permanently, for example in Swarm [7] or on another blockchain. Parties who wish to rely on such data in a TrueBit contract must be sure that Verifiers have access to the full data set.

To use TrueBit on external data, one needs only store a Merkle root of the massive data set on the blockchain and add non-deterministic steps in the verification game, in which the Solver can "guess" the original data set represented by the Merkle root. While Solvers and Verifiers must have access to the full data, Judges and Referees do not. Indeed, if we modify the verification game so as to permit tasks for nondeterministic Turing machines, then the Solver can nondeterministically guess the certificate data as a step in the TrueBit contract. Only in cases of disputes would the Solver have to reveal external certificate data to the Judges via the blockchain. In some applications, the Solver might even be able to reveal to the Judges a privacy-preserving zkSNARK rather than the data itself. zkSNARKs have the potential to enable privacy for many different kinds of systems on Ethereum [54].

While in theory TrueBit's scalable protocol can process arbitrarily complex tasks, in practice the verification game is inefficient and therefore security of TrueBit computations degrades as tasks reach the level of big data. For big data applications, TrueBit may not be able to rely on a one-size-fits-all verification game. Therefore, we anticipate optimizing the verification game for certain classes of tasks. For example, Sec. 3.2 gives an example of an efficient, specialized verification game for matrix multiplication. In future versions of TrueBit, Task Givers might broadcast not only tasks but also an indication of the corresponding verification game to be played in case of disputes.

Remark 7.1: We conclude with a caveat: TrueBit may expose latent security vulnerabilities in the underlying Ethereum network as a result of new kinds of interactions between smart contracts and miners. By allowing smart contracts to check proof-of-works, for example, TrueBit may facilitate 38.2% attacks [55].

Acknowledgments

We thank Vitalik Buterin and Vlad Zamfir for suggesting the use of forced errors in the TrueBit protocol and Eli Bendersky for introducing us to Google Lanai. We also thank Loi Luu and Julia Koch for useful discussions.

8. Addendum

In the months following the initial release of this whitepaper, new work and feedback informed TrueBit's development roadmap and helped refine the protocol itself. We discuss a few recent developments.

8.1. *Security patches*

A TrueBit adversary has either one of two goals:

(1) get bogus computations onto the blockchain, or
(2) extract jackpot funds without performing verification.

We investigate three attacks of the first type followed by two of the second.

Premature disclosure of random bits (Zack Lawrence). A Solver can dissuade Verifier participation by publicly broadcasting his private random bits prior to the designated reveal time. Verifiers then know immediately whether or not a forced error is in effect. Since Verifiers expect to gain little from checking solutions without forced errors, they may decide not to verify, thereby offering opportunity to get bogus solutions onto the blockchain.

1protocol's [56] random number generator protocol, Arbit, solves this problem by instituting penalties for Solvers who prematurely reveal private random bits and rewarding users who report them. When a user correctly reports a premature reveal to TrueBit, the following occurs.

(1) The Solver's solutions are discarded, and a new Solver lottery takes place. This re-incentivizes Verifiers to participate in the task, while voiding the Solver's incentive to reveal private information.
(2) Half of the Solver's deposit gets burned. This makes the above attack expensive for the Solver.
(3) The other half of the Solver's deposit goes to the party who reported the Solver's random bits. This incentivizes Observers to report the Solver's prematurely revealed random bits.

Incorrect secondary solution (Sina Habibian and Harley Swick). [57]

Suppose that a forced error is in effect and that the Solver submits two incorrect solutions. When the Solver reveals his "correct" secondary solution in Step 4(b)ii of the protocol (Sec. 4.6), Verifiers ignore it because there's no chance of a jackpot payout. Indeed, the only "reward" for correctly challenging this secondary solution is to play a verification game. Hence, one of the Solver's bogus solutions ends up on the blockchain.

We eliminate the incorrect secondary solution vulnerability as follows. Denote the Solver's two solutions by A and B. In the beginning of Step 4, rather than signaling for a challenge with the hash of an even integer, the Verifier hashes an integer, whose value mod 3 the protocol interprets as follows:

$$\begin{aligned} 0 \mod 3&: \quad \text{challenge solution } A,\\ 1 \mod 3&: \quad \text{challenge solution } B,\\ 2 \mod 3&: \quad \text{challenge both } A \text{ and } B. \end{aligned}$$

The Verifiers indicate their choice without knowing which of the two solutions the Solver puts forth as an answer. The protocol hides this information from Verifiers via the following process. The Solver commits to either solution A or solution B by hashing either A or B paired with his private random bits, where the Solver's private random bits serve as "noise", which prevent Verifiers from guessing which option the Solver chose. The Solver has incentive not to share his private random bits due to the "Premature disclosure of random bits" patch above, as well as the fact that the Solver risks reducing his jackpot share by exposing this private information. Finally, once the timeout for challenges has passed, the Solver reveals his random bits in the clear, thereby indicating his commitment to either solution A or solution B. Challenges and forced errors then proceed as usual. In case the protocol forces the Solver to reveal his second "correct" solution, Verifiers who earlier committed to a challenge against this second solution are obligated to play a verification game. In this way, Verifiers catch errant secondary solutions just as they catch errant primary ones.

In case a forced error is *not* in effect, broadcasting a pair of incorrect solutions poses a cost to the Solver in the form of a lost deposit. Indeed, Verifiers have proper incentives to check the Solver's primary answer. Since forced errors occur rarely and unpredictably, the Solver expects to sacrifice several deposits in order to mount an "incorrect secondary solution" attack. This loss offsets the Solver's eventual jackpot gain from challenging his own forced error solution. We implicitly assume that the chance to win a jackpot sufficiently motivates a Verifier to challenge whenever a Solver submits a pair of incorrect solutions; any Verifier who challenges both submitted solutions must play a verification game.

The fix above has a potentially useful side effect of publicly indicating how many Verifiers are monitoring a given task. Indeed, a Verifier broadcasts one of the three commitment forms above if and only if she is paying attention. The option to signal "no challenge" is no longer needed for camouflage because broadcasting challenges no longer indicates presence of a forced error. Moreover, if the Solver were to submit two correct solutions, the smart contract could immediately recognize them as identical and penalize the Solver accordingly.

An adversary could potentially exploit the monitoring feature in the previous paragraph by broadcasting multiple challenge commitments from Sybil identities, thereby reducing the total payout in case of a forced error and discouraging other rational Verifiers from participating. For this reason, the protocol must prevent Verifiers from revealing which task they are challenging until the final phase of the protocol. Since each Verifier performs the given computational task without distraction from others' commitments, an adversary cannot deter Verifier participation via Sybil attack.

Program abort (Yaron Velner). A task must explicitly specify an upper bound on the number of steps for which the verification game can run, and the Solver's solution should return an "error" if and only if the task computation exceeds this bound. Indeed, the computation steps for a given task must form a deterministic sequence in order for Judges to determine their correctness. Ideally, one would like to run tasks directly in a native language like C, C++, or Rust, however this approach requires accurate metering of the number of computation steps. We can reduce the metering overhead by processing steps with a compiler rather than an interpreter.

Jackpot balloon attack (Clément Lesaege). In the "flood of challengers" attack (Sec. 5.3), a Cartel could artificially inflate the jackpot repository by repeatedly challenging a single forced error solution and thereby reduce jackpot payout. Eventually, the Cartel could offset the cost of this attack by cashing in on an extra large jackpot. This action could recruit membership for the Cartel at low cost by removing incentives for Verifiers who do not belong to the Cartel.

We mitigate against the benefit of this attack by placing a fixed, absolute bound on the jackpot payout, regardless of the actual jackpot repository balance. Extra revenue accumulated in the jackpot repository only becomes available for payout after a protocol upgrade. Note that the maximum jackpot amount does not necessarily pay out at each forced error; the actual payout depends on the difficulty of the task.

Incentivizing block withholding (Clément Lesaege). The attacker, who is a Solver, deploys a smart contract which pays money to miners who withhold (or intentionally uncle) blocks whose random seed fails to yield a forced error. See [58] for possible implementations. This attack is profitable for tasks in which the expected jackpot payout times the probability of a forced error exceeds a block reward. Through repetitions, this attack could drain the jackpot repository.

Arbit's security mechanism, discussed in "Premature disclosure of random bits" above, defends against this attack. In order to execute the present "incentivizing block withholding" attack, the Solver either has to reveal his bits publicly or has to provide them to miners through private channels. If the Solver's penalty for a premature reveal exceeds the expected jackpot payout times the probability of a forced error, then this attack results in an expected net loss.

8.2. *The TrueBit Virtual Machine*

Section 3.3 describes a verification game which operates with respect to a Turing machine. In practice, however, no compiler exists which can transform C++ code into something as simple as Turing machine language. In Sec. 6, we proposed Google Lanai as a practical compromise in the absence of Turing machine architecture. Due to Lanai's complexity, the fact that Google controls its codebase, and that progress on its development appears to have slowed, we have since migrated development away from Google Lanai.

In addition to executing exactly the same computation steps regardless of hardware configuration, the compiler architecture underlying the verification game, or *TrueBit Virtual Machine (TVM)*, must satisfy the following *simplicity properties.*

(1) A single computation step on the TVM runs comfortably within Ethereum's gas limit, and
(2) the space required to describe a TVM state change fits inside a single Ethereum transaction.

WebAssembly architecture [59] comes sufficiently close to matching these properties so as to make TVM execution practical today. WebAssembly has become increasingly ready-to-use due to contributions and testing by Apple, Google, Microsoft, and Mozilla [60]. Several cryptocurrency projects have begun to develop on WebAssembly, including Dfinity, eWASM, and Parity, due to the platform's machine independence and relative simplicity.

The TVM consists of two parts:

(1) an off-chain *interpreter* which enumerates a list of states for a given computation, and
(2) an on-chain *stepper* which, given a state, can compute the next state.

Solvers and Challengers use the interpreter to create a Merklized list of states for a computation. Once the Solver and Challenger have determined the first step at which they disagree, the Judges use the stepper to run this step and rule on the outcome of the verification game.

Since Native WebAssembly does not entirely satisfy the simplicity properties above, the interpreter must either further compile input code into a simpler architecture, or it must divide each WebAssembly instruction into substeps. In order to reduce the chances of inducing compiler error, the

TVM follows the latter strategy. The TVM's off-chain interpreter parses WebAssembly instructions into OCaml execution, which in turn creates registers describing WebAssembly suboperations. Each Ocaml-based substep only accesses one dynamic data structure at a time.

8.3. *Additional applications*

Finally, we mention a few more illustrative applications.

Video broadcasting. Livepeer [61] offers a new platform for decentralized, live streaming video. Users can broadcast, watch, or get paid for performing the computationally intensive process of transcoding video into different codecs and formats. TrueBit ensures that transcoders perform this work correctly, while Swarm [7] guarantees that video data remains available to TrueBit during the necessary verification period.

Autonomous machine learning. McConaghy's ArtDAO [62] generates art, sells it, and then uses its revenue to improve its own code. A TrueBit-based ArtDAO would allow a financed piece of code on the blockchain to access computational resources in such a way that no one can "pull its plug." We may eventually see other blockchain-based machine learning applications, like computer vision, as well.

Data marketplace. Hedge fund Numerai [63] crowdsources modeling problems to data scientists and then executes trades based upon their work. Numerai rewards successful models, however contributing data scientists must trust Numerai to both test their work and compensate them fairly. TrueBit enables autonomous data markets. Open Mined [64], paired with TrueBit, opens the possibility of trustless renumeration based on streaming flows of models and data. The Ocean Protocol [65], which facilitates contribution and sharing of data, also requires a verification mechanism.

Staking and random numbers. 1protocol [56] allows users who have either computing resources or capital, but not necessarily both, to participate as TrueBit Solvers and Verifiers by decoupling security deposits from work done. In addition, 1protocol's Arbit protocol uses interactive verification to generate random numbers in a decentralized way.

Other applications. Please check the TrueBit website for other current ideas in progress! `https://truebit.io`.

References

1. L. Luu, J. Teutsch, R. Kulkarni, and P. Saxena. Demystifying incentives in the consensus computer. In *Proceedings of the 22nd ACM SIGSAC Conference on Computer and Communications Security (CCS 2015)*, pp. 706–719, ACM, New York, NY, USA, 2015. ISBN 978-1-4503-3832-5.
2. Some miners generating invalid blocks. `https://bitcoin.org/en/alert/2015-07-04-spv-mining`, July, 2015.
3. I thikn the attacker is this miner—today he made over $50k. `https://www.reddit.com/r/ethereum/comments/55xh2w/i_thikn_the_attacker_is_this_miner_today_he_made/`, October, 2016.
4. The Golem Project: crowdfunding whitepaper. `https://golem.network/doc/Golemwhitepaper.pdf`, November, 2016.
5. L. Luu, Y. Velner, J. Teutsch, and P. Saxena. SmartPool: Practical decentralized pooled mining. Cryptology ePrint Archive, Paper 2017/019. URL `https://eprint.iacr.org/2017/019`, 2017. `https://eprint.iacr.org/2017/019`.
6. Dogethereum. `https://www.reddit.com/r/dogethereum/`.
7. Swarm. `https://www.ethswarm.org/`.
8. N. Szabo. The idea of smart contracts. `http://szabo.best.vwh.net/smart_contracts_idea.html`, 1997.
9. Ethereum. `http://ethereum.org/`.
10. Golem. `https://www.golem.network/`.
11. Augur. `https://www.augur.net/`.
12. The DAO. `http://daohub.org/`, 2016.
13. SmartPool. `http://smartpool.io`.
14. M. Walfish and A. J. Blumberg, Verifying computations without reexecuting them, *Communications of the ACM*. **58** (2015), 74–84. doi: 10.1145/2641562. URL `http://doi.acm.org/10.1145/2641562`.
15. Zcash. `https://z.cash/`.
16. E. Ben-Sasson, A. Chiesa, D. Genkin, E. Tromer, and M. Virza. SNARKs for C: Verifying program executions succinctly and in zero knowledge. In eds. R. Canetti and J. A. Garay, *Advances in Cryptology – CRYPTO 2013: 33rd Annual Cryptology Conference, Santa Barbara, CA, USA, August 18–22, 2013. Proceedings, Part II*, pp. 90–108, Springer Berlin Heidelberg, Berlin, Heidelberg, 2013. ISBN 978-3-642-40084-1. doi: 10.1007/978-3-642-40084-1_6. URL `http://dx.doi.org/10.1007/978-3-642-40084-1_6`.
17. N. Bitansky, R. Canetti, O. Paneth, and A. Rosen. Indistinguishability obfuscation vs. auxiliary-input extractable functions: One must fall. Cryptology ePrint Archive, Paper 2013/468. URL `https://eprint.iacr.org/2013/468.pdf`.
18. S. Jain, P. Saxena, F. Stephan, and J. Teutsch. How to verify computation with a rational network. URL `http://arxiv.org/abs/1606.05917`, June, 2016.
19. S. Nakamoto. Bitcoin P2P e-cash paper. `http://www.mail-archive.com/cryptography@metzdowd.com/msg09959.html`, November, 2008.

20. R. Canetti, B. Riva, and G. N. Rothblum. Practical delegation of computation using multiple servers. In *Proceedings of the 18th ACM Conference on Computer and Communications Security*, CCS '11, pp. 445–454, ACM, New York, NY, USA, 2011. ISBN 978-1-4503-0948-6. doi: 10.1145/2046707.2046759. URL `http://doi.acm.org/10.1145/2046707.2046759`.
21. R. Canetti, B. Riva, and G. N. Rothblum, Refereed delegation of computation, *Information and Computation*. **226** (2013), 16–36. doi: http://dx.doi.org/10.1016/j.ic.2013.03.003. URL `http://www.sciencedirect.com/science/article/pii/S0890540113000217`. Special Issue: Information Security as a Resource.
22. C. Reitwiessner. From smart contracts to courts with not so smart judges. `https://blog.ethereum.org/2016/02/17/smart-contracts-courts-not-smart-judges/`, 2016.
23. Ethereum Computation Market. `http://www.ethereum-computation-market.com/`.
24. J. van den Hooff, M. F. Kaashoek, and N. Zeldovich. Versum: Verifiable computations over large public logs. In *Proceedings of the 2014 ACM SIGSAC Conference on Computer and Communications Security*, CCS '14, pp. 1304–1316, ACM, New York, NY, USA, 2014. ISBN 978-1-4503-2957-6. doi: 10.1145/2660267.2660327. URL `http://doi.acm.org/10.1145/2660267.2660327`.
25. pruby. Interactive Verification of C Programs (Comment). https://www.reddit.com/r/ethereum/comments/51qjz6/interactive_verification_of_c_programs/d7ey41n/, September, 2016.
26. V. Buterin. Ethereum: A next-generation smart contract and decentralized application platform. `https://ethereum.org/en/whitepaper/`, 2014.
27. RANDAO. `https://github.com/randao/randao`, 2016.
28. Crypto-currency market capitalizations. `http://coinmarketcap.com/`.
29. Oraclize. `http://www.oraclize.it/`.
30. Reality Keys. `https://www.realitykeys.com/`.
31. Google Lanai. `http://llvm.org/docs/CompilerWriterInfo.html#lanai`.
32. Dwarfpool is now 50.5%. `http://forum.ethereum.org/discussion/5244/dwarfpool-is-now-50-5`.
33. Warning: Ghash.io is nearing 51% – leave the pool. `http://www.cryptocoinsnews.com/warning-ghash-io-nearing-51-leave-pool/`.
34. Bitmain responds to controversy surrounding its upcoming 140,000 kw mining center. http://www.newsbtc.com/2016/11/04/bitmain-response-new-mining-center/, 2016.
35. J. A. Kroll, I. C. Davey, and E. W. Felten. The economics of Bitcoin mining, or Bitcoin in the presence of adversaries. `http://www.econinfosec.org/archive/weis2013/papers/KrollDaveyFeltenWEIS2013.pdf`, June, 2013.
36. P2Pool. `https://web.archive.org/web/20160707052612/http://p2pool.org/`, 2016.
37. A. Miller, A. Kosba, J. Katz, and E. Shi. Nonoutsourceable scratch-off puzzles to discourage Bitcoin mining coalitions. In *Proceedings of the 22Nd ACM SIGSAC Conference on Computer and Communications Security*, CCS

'15, pp. 680–691, ACM, New York, NY, USA, 2015. ISBN 978-1-4503-3832-5. doi: 10.1145/2810103.2813621. URL http://doi.acm.org/10.1145/2810103.2813621.

38. Ethash: defining the seed hash. https://ethereum.org/en/developers/docs/consensus-mechanisms/pow/mining-algorithms/ethash#seed-hash.
39. Why Equihash? https://z.cash/blog/why-equihash.html.
40. Introducing Project Alchemy. https://z.cash/blog/project-alchemy.html.
41. Dogecoin. http://dogecoin.com/.
42. A. van de Sande. The Doge connection Bounty Dao is live and working. https://www.reddit.com/r/ethereum/comments/41ohhr/the_doge_connection_bounty_dao_is_live_and_working/, 2016.
43. Dogecoin–Ethereum bounty smart contract. https://etherscan.io/address/0xdbf03b407c01e7cd3cbea99509d93f8dddc8c6fb.
44. K. Croman, C. Decker, I. Eyal, A. E. Gencer, A. Juels, A. Kosba, A. Miller, P. Saxena, E. Shi, E. Gün Sirer, D. Song, and R. Wattenhofer. On scaling decentralized blockchains (a position paper). In *Financial Cryptography and Data Security 2016 BITCOIN Workshop*, vol. 9604, *Lecture Notes in Computer Science*, pp. 106–125, Springer Berlin Heidelberg, February, 2016. ISBN 978-3-662-53357-4. doi: 10.1007/978-3-662-53357-4_8. URL http://dx.doi.org/10.1007/978-3-662-53357-4_8.
45. I. Abraham, D. Malkhi, K. Nayak, L. Ren, and A. Spiegelman. Solidus: An incentive-compatible cryptocurrency based on permissionless Byzantine consensus. https://arxiv.org/abs/1612.02916, 2016.
46. C. Decker, J. Seidel, and R. Wattenhofer. Bitcoin meets strong consistency. In *Proceedings of the 17th International Conference on Distributed Computing and Networking*, ICDCN '16, pp. 13:1–13:10, ACM, New York, NY, USA, 2016. ISBN 978-1-4503-4032-8. doi: 10.1145/2833312.2833321. URL http://doi.acm.org/10.1145/2833312.2833321.
47. E. K. Kogias, P. Jovanovic, N. Gailly, I. Khoffi, L. Gasser, and B. Ford. Enhancing Bitcoin security and performance with strong consistency via collective signing. In *25th USENIX Security Symposium (USENIX Security 16)*, pp. 279–296, USENIX Association, Austin, TX, 2016. ISBN 978-1-931971-32-4. URL https://www.usenix.org/conference/usenixsecurity16/technical-sessions/presentation/kogias.
48. L. Luu, V. Narayanan, C. Zheng, K. Baweja, S. Gilbert, and P. Saxena. A secure sharding protocol for open blockchains. In *Proceedings of the 2016 ACM SIGSAC Conference on Computer and Communications Security*, CCS '16, pp. 17–30, ACM, New York, NY, USA, 2016. ISBN 978-1-4503-4139-4. doi: 10.1145/2976749.2978389. URL http://doi.acm.org/10.1145/2976749.2978389.
49. S. Micali. ALGORAND: the efficient and democratic ledger. http://arxiv.org/abs/1607.01341, 2016.
50. A. Miller, Y. Xia, K. Croman, E. Shi, and D. Song. The honey badger of BFT protocols. In *Proceedings of the 2016 ACM SIGSAC Conference on Computer and Communications Security*, CCS '16, pp. 31–42, ACM, New York, NY,

USA, 2016. ISBN 978-1-4503-4139-4. doi: 10.1145/2976749.2978399. URL http://doi.acm.org/10.1145/2976749.2978399.
51. R. Pass and E. Shi. Hybrid consensus: Efficient consensus in the permissionless model. https://eprint.iacr.org/2016/917.pdf, 2016.
52. I. Eyal, A. E. Gencer, E. G. Sirer, and R. V. Renesse. Bitcoin-NG: A scalable blockchain protocol. In *13th USENIX Symposium on Networked Systems Design and Implementation (NSDI 16)*, pp. 45–59, USENIX Association, Santa Clara, CA, Mar., 2016. ISBN 978-1-931971-29-4. URL https://www.usenix.org/conference/nsdi16/technical-sessions/presentation/eyal.
53. Y. Sompolinsky, Y. Lewenberg, and A. Zohar. SPECTRE: A fast and scalable cryptocurrency protocol. https://eprint.iacr.org/2016/1159.pdf, 2016.
54. C. Reitwiessner. zkSNARKs in a nutshell. https://blog.ethereum.org/2016/12/05/zksnarks-in-a-nutshell/, Dececmber, 2016.
55. J. Teutsch, S. Jain, and P. Saxena. When cryptocurrencies mine their own business. In *Financial Cryptography and Data Security: 20th International Conference (FC 2016) Christ Church, Barbados*, pp. 499–514, Springer Berlin / Heidelberg, 2017. ISBN 978-3-662-54970-4.
56. 1protocol. https://medium.com/vest/introducing-1protocol-d2f239285386, 2016.
57. Open problems. https://github.com/TrueBitFoundation/Developer-Resources/wiki/Open-Problems.
58. Y. Velner, J. Teutsch, and L. Luu. Smart contracts make Bitcoin mining pools vulnerable. In *Financial Cryptography and Data Security*, pp. 298–316, Springer International Publishing, Cham, 2017. ISBN 978-3-319-70278-0.
59. WebAssembly. http://webassembly.org/.
60. A. Haas, A. Rossberg, D. L. Schuff, B. L. Titzer, M. Holman, D. Gohman, L. Wagner, A. Zakai, and J. Bastien. Bringing the web up to speed with WebAssembly. In *Proceedings of the 38th ACM SIGPLAN Conference on Programming Language Design and Implementation*, PLDI 2017, pp. 185–200, ACM, New York, NY, USA, 2017. ISBN 978-1-4503-4988-8. doi: 10.1145/3062341.3062363. URL http://doi.acm.org/10.1145/3062341.3062363.
61. Livepeer. https://livepeer.org/.
62. T. McConaghy. Wild, wooly AI DAOs. https://blog.bigchaindb.com/wild-wooly-ai-daos-d1719e040956.
63. Numerai. https://numer.ai.
64. Open Mined. https://openmined.org/.
65. Ocean Protocol. https://oceanprotocol.com/.

https://doi.org/10.1142/9789811278631_0016

Weak Muller Conditions Make Delay Games Hard

Sarah Winter[a]
Université libre de Bruxelles
Brussels, Belgium
sarah.winter@ulb.ac.be

Martin Zimmermann[b]
Aalborg University
Aalborg, Denmark
mzi@cs.aau.dk

We show that solving delay games with winning conditions given by deterministic and non-deterministic weak Muller automata is 2EXPTIME-complete respectively 3EXPTIME-complete. Furthermore, doubly- and triply-exponential lookahead is necessary and sufficient to win such games. These results are the first that show that the succinctness of the automata types used to specify the winning conditions has an influence on the complexity of these problems.

1. Introduction

Is solving delay games with Muller conditions, i.e., determining its winner, harder than solving delay games with parity conditions? Is more lookahead required to win a delay game with a Muller condition than to win a delay game with a parity condition? Deterministic Muller automata are exponentially more succinct than deterministic parity automata [1][c] and solving

[a]This chapter is supported by the MIS project F451019F (F.R.S.-FNRS). Sarah Winter is a postdoctoral researcher at F.R.S.-FNRS.
[b]This chapter is supported by DIREC – Digital Research Centre Denmark.
[c]Strictly speaking, this is only true when the size of automata is measured in the number of states, which is a very crude measure for Muller automata. See the discussion about the representation of Muller conditions later in the introduction and also the works of Boker [1] and Hunter and Dawar [2, 3].

classical, delay-free Muller games is PSPACE-complete [2] while solving parity games is quasi-polynomial [4]. So, the answer to both questions should be yes, but surprisingly, and frustratingly, both questions are open. Here, we take a step towards resolving them by showing that the answer is indeed yes if we consider weak Muller conditions.

The games we consider here are Gale-Stewart games [5], arguably the simplest form of two-player games. In such a game, two players, called I and O, alternatingly pick letters, thereby constructing an infinite word α. The winner of such a play is determined by the winning condition of the game, a language of infinite words. Player O wins if α is in the winning condition, otherwise Player I wins.

Delay games were introduced by Hosch and Landweber [6] only three years after the seminal Büchi-Landweber Theorem [7], showing how to determine the winner of Gale-Stewart games with ω-regular winning conditions. Hosch and Landweber generalized the setting of Gale-Stewart games by allowing Player O to delay her moves to obtain a lookahead on Player I's moves. This advantage allows her to win games she cannot win without lookahead. These games capture the asynchronous interaction of two agents [8], the deterministic uniformization problem for relations over infinite words [9, 10], and are related to streaming transducers with delay [11].

In this chapter, we are interested in games with ω-regular winning conditions. These are typically represented by (deterministic or nondeterministic) ω-automata with various kinds of acceptance conditions, e.g., safety, parity or Muller. The choice between determinism and nondeterminism and succinctness of the acceptance condition can have an influence on the complexity of solving the game.

Hosch and Landweber showed that it is decidable whether Player O wins a given delay game with bounded lookahead [6]. Forty years later, Holtmann, Kaiser, and Thomas [10] revisited delay games and proved that if Player O wins a delay game then she wins it already with doubly-exponential lookahead (in the size of a given deterministic parity automaton recognizing the winning condition). Thus, unbounded lookahead does not offer any advantage over doubly-exponential lookahead in games with ω-regular winning conditions. Furthermore, they showed that the winner of a delay game, again with its winning condition given by a deterministic parity automaton, can be determined in doubly-exponential time.

Both upper bounds were improved and matching lower bounds were proven by Klein and Zimmermann: Solving delay games is EXPTIME-complete and exponential lookahead is both necessary to win some games

Table 1. The complexity of solving delay games. Results for reachability, safety, and parity (in gray) are from [12, 13]; results for weak Muller are proven here.

	deterministic	non-deterministic
Reachability	PSPACE-complete	PSPACE-complete
Safety	EXPTIME-complete	2EXPTIME-complete
Parity	EXPTIME-complete	2EXPTIME-complete
weak Muller	2EXPTIME-complete	3EXPTIME-complete

Table 2. The lookahead required to win delay games. All bounds are asymptotically tight. Results for reachability, safety, and parity (in gray) are from [12, 13]; results for weak Muller are proven here.

	deterministic	non-deterministic
Reachability	exponential	doubly-exponential
Safety	exponential	doubly-exponential
Parity	exponential	doubly-exponential
weak Muller	doubly-exponential	triply-exponential

and sufficient to win all games that can be won [12]. Both lower bounds already hold for winning conditions specified by deterministic safety automata, while the upper bounds hold for deterministic parity automata. They also considered reachability automata, for which there is no difference between the results for deterministic and non-deterministic automata.

For non-deterministic parity automata, solving delay games is 2EXPTIME-complete and doubly-exponential lookahead is both necessary and sufficient [13]. As before, both lower bounds already hold for non-deterministic safety automata, while the upper bounds hold for non-deterministic parity automata. See Tables 1 and 2 for an overview of the known results (in gray).

However, the parity condition is not very succinct, e.g., deterministic Rabin, Streett, and Muller automata can all be exponentially more succinct than deterministic parity automata, while they all recognize exactly the ω-regular languages.

Here, we need to discuss briefly how to measure the size of an ω-automaton, say with set Q of states. For all acceptance conditions induced by a subset $F \subseteq Q$ (e.g., reachability, safety, Büchi, and co-Büchi) or by a coloring $\Omega\colon Q \to \mathbb{N}$ (e.g., parity) the size of the automaton is captured by its number of states, as the acceptance condition can be encoded with a polynomial overhead. However, for more succinct automata, the situation is different. Rabin and Streett conditions are given by a finite family

$(R_i, G_i)_{i\in\mathcal{I}}$ of pairs of subsets $R_i, G_i \subseteq Q$. Here, $|\mathcal{I}|$ is called the index of the automaton and has to be taken into account when measuring its size.

Muller acceptance on the other hand is based on a set $\mathcal{F} \subseteq 2^Q$, which can be encoded in various representations. The simplest one is just listing all subsets in $\mathcal{F}$, the so-called explicit representation whose size is $|\mathcal{F}|$. In this chapter, we are mainly focused on representing $\mathcal{F}$ by a Boolean formula with variables in Q, so that the models of the formula are exactly the sets in $\mathcal{F}$. In this case, we measure the size of the acceptance condition by the size of the formula. Other representations include the use of circuits, trees, DAGs, and colorings. The relative succinctness of these representations and their influence on the complexity of solving arena-based (i.e., delay-free) games has been studied by Hunter and Dawar [2] and Horn [14].

In the setting of arena-based games, parity games can be solved in quasi-polynomial time [4], while solving Rabin games is NP-complete [15], solving Streett games is co-NP-complete (they are dual to Rabin games), and solving Muller games can be P-complete, co-NP-complete, or PSPACE-complete, depending on the representation of $\mathcal{F}$ [2, 14].

Thus, it is natural to ask whether winning conditions given by more succinct automata also make delay games harder to solve and increase the bounds on the necessary lookahead for Player O. However, no such results are known. Note that it is straightforward to obtain doubly-exponential (triply-exponential) upper bounds on the complexity and on the lookahead, as every deterministic (non-deterministic) Rabin, Streett, or Muller automaton can be turned into an equivalent exponentially larger deterministic (non-deterministic) parity automaton. This result is independent of the size and encoding of the acceptance condition of the automaton that is transformed.

Here, we are interested in Muller conditions, as they are the most succinct ones. As a first step towards answering our motivating questions, we show that for *weak* Muller conditions, there is indeed an exponential increase, both in the solution complexity and in the necessary lookahead. Thus, we provide matching lower bounds to the upper bounds obtained by transforming (weak) Muller automata into parity automata.

Recall that a Muller condition $\mathcal{F} \subseteq 2^Q$ represents all those runs $\rho \in Q^\omega$ whose infinity set, the set of states visited infinitely often by ρ, is in $\mathcal{F}$. In contrast, a weak Muller condition $\mathcal{F} \subseteq 2^Q$ represents all those runs ρ whose occurrence set, the set of states visited by ρ, is in $\mathcal{F}$. Note that weak Muller automata are strictly less expressive than (standard) Muller automata.

In this chapter, we settle the complexity of solving delay games with deterministic and non-deterministic weak Muller winning conditions and provide tight bounds on the necessary lookahead to win such games. More precisely, we show that solving delay games with deterministic weak Muller winning conditions is 2EXPTIME-complete and solving delay games with non-deterministic weak Muller winning conditions is 3EXPTIME-complete. Similarly, we show that doubly-exponential lookahead is necessary and sufficient to win delay games with deterministic weak Muller conditions, and triply-exponential lookahead is necessary and sufficient to win delay games with non-deterministic weak Muller conditions. All these results hold for $\mathcal{F}$ being represented by a Boolean formula. The lower bounds rely on the fact that in a weak Muller automaton, every visit to a state is relevant for acceptance. We show that this property allows to construct a small deterministic (non-deterministic) automaton that requires the players in a game to implement a cyclic counter with exponentially (doubly-exponentially) many values. This automaton is then used to construct games, in which Player O needs doubly-exponential (triply-exponential) lookahead and to construct games that simulate AEXPSPACE (A2EXPSPACE) Turing machines. Finally, we also give an exponential lower bound on the necessary lookahead in delay games with deterministic explicit weak Muller conditions.

We conclude this chapter by discussing the obstacles one faces when trying to take the next step, i.e., to raise the lower bounds for (standard) Muller conditions. There, we also return to the discussion of the different representations of Muller conditions, i.e., we discuss whether our lower bounds, which hold for the representation by Boolean formulas, can be lifted to less succinct representations, thereby giving stronger results.

2. Preliminaries

We introduce our notation. We denote the non-negative integers by $\mathbb{N}$, and the set $\{0, 1\}$ by $\mathbb{B}$. The size of a set X is denoted by $|X|$.

Words, languages. An *alphabet* Σ is a non-empty finite set of *letters* or *symbols*. A *word* over Σ is a sequence of letters of Σ. The set of finite resp. infinite words over Σ is denoted by Σ^* resp. Σ^ω. The set of non-empty finite words over Σ is denoted by Σ^+. The *empty word* is denoted by ε. Given a finite or infinite word α, we denote by $\alpha(i)$ the i-th letter of α, starting with 0, i.e., $\alpha = \alpha(0)\alpha(1)\alpha(2)\cdots$. The length of α is denoted by $|\alpha|$. A subset $L \subseteq \Sigma^*$ resp. $L \subseteq \Sigma^\omega$ is a *language* resp. an *ω-language* over Σ. We drop the prefix ω when it is clear from the context.

Given two infinite words $\alpha \in (\Sigma_0)^\omega$ and $\beta \in (\Sigma_1)^\omega$, we define $\binom{\alpha}{\beta} = \binom{\alpha(0)}{\beta(0)}\binom{\alpha(1)}{\beta(1)}\binom{\alpha(2)}{\beta(2)} \cdots \in (\Sigma_0 \times \Sigma_1)^\omega$.

Automata. An *ω-automaton* is a tuple $\mathfrak{A} = (Q, \Sigma, q_I, \Delta, \mathrm{Acc})$, where Q is a finite set of states, Σ is an alphabet, $q_I \in Q$ is an initial state, $\Delta \subseteq Q \times \Sigma \times Q$ is a transition relation, and $\mathrm{Acc} \subseteq \Delta^\omega$ is a set of accepting runs.

An infinite *run* ρ of $\mathfrak{A}$ is a sequence

$$\rho = (q_0, a_0, q_1)(q_1, a_1, q_2)(q_2, a_2, q_3) \cdots \in \Delta^\omega.$$

As usual, we say that ρ is initial if $q_0 = q_I$ and we say that ρ processes $a_0 a_1 a_2 \cdots \in \Sigma^\omega$. Given a run ρ, let $\mathrm{Inf}(\rho)$ be the set of states visited infinitely often by ρ, and let $\mathrm{Occ}(\rho)$ be the set of states visited by ρ.

The language *recognized* by $\mathfrak{A}$ is the set $L(\mathfrak{A}) \subseteq \Sigma^\omega$ that contains all ω-words that have an initial run in Acc processing it.

An *ω-automaton* $\mathfrak{A} = (Q, \Sigma, q_I, \Delta, \mathrm{Acc})$ is *deterministic* (and complete) if Δ is given as a total function $\delta \colon Q \times \Sigma \to Q$. If we speak of *the* run of $\mathfrak{A}$ on $\alpha \in \Sigma^\omega$, then we mean the unique initial run of $\mathfrak{A}$ processing α.

Acceptance conditions. We recall safety, reachability, parity, and (weak) Muller acceptance conditions.

An ω-automaton $\mathfrak{A} = (Q, \Sigma, q_I, \Delta, \mathrm{Acc})$ is a *safety automaton*, if there is a set $F \subseteq Q$ of accepting states such that

$$\mathrm{Acc} = \{(q_0, a_0, q_1)(q_1, a_1, q_2)(q_2, a_2, q_3) \cdots \in \Delta^\omega \mid q_i \in F \text{ for every } i\}.$$

Moreover, an ω-automaton $\mathfrak{A} = (Q, \Sigma, q_I, \Delta, \mathrm{Acc})$ is a *reachability automaton*, if there is a set $F \subseteq Q$ of accepting states such that

$$\mathrm{Acc} = \{(q_0, a_0, q_1)(q_1, a_1, q_2)(q_2, a_2, q_3) \cdots \in \Delta^\omega \mid q_i \in F \text{ for some } i\}.$$

Furthermore, an ω-automaton $\mathfrak{A} = (Q, \Sigma, q_I, \Delta, \mathrm{Acc})$ is a *parity automaton*, if there is some coloring $\Omega \colon Q \to \mathbb{N}$ such that

$$\mathrm{Acc} = \{\rho \in \Delta^\omega \mid \max\{\Omega(q) \mid q \in \mathrm{Inf}(\rho)\} \text{ is even}\}.$$

In the remainder, we use the acronyms DPA resp. NPA to refer to deterministic parity automata resp. non-deterministic parity automata.

Finally, an ω-automaton $\mathfrak{A} = (Q, \Sigma, q_I, \Delta, \mathrm{Acc})$ is a *Muller automaton* resp. a *weak Muller automaton*, if there is a family $\mathcal{F} \subseteq 2^Q$ of sets of states such that

$$\mathrm{Acc} = \{\rho \in \Delta^\omega \mid \mathrm{Inf}(\rho) \in \mathcal{F}\}$$

resp.

$$\mathrm{Acc} = \{\rho \in \Delta^\omega \mid \mathrm{Occ}(\rho) \in \mathcal{F}\}.$$

In the remainder, we use the acronyms wDMA resp. wNMA to refer to deterministic weak Muller automata resp. non-deterministic weak Muller automata.

The sets Acc are infinite objects, but they have finite representations. E.g., for safety conditions Acc is fully specified by a subset $F \subseteq Q$ of safe states; for (weak) Muller conditions Acc is fully specified by a set $\mathcal{F} \subseteq 2^Q$ of sets of states. While these representations are usually small in the size of the automaton, this is not guaranteed for Muller acceptance conditions. Working with Muller automata, the representations of the acceptance condition influence the complexity of decision problems [2, 3]. Hence, we recall different ways of representing Muller conditions which are more succinct than simply listing the elements of $\mathcal{F}$.

Representations of acceptance conditions. Hunter and Dawar studied the impact of the representation of Muller conditions on the complexity of solving delay-free Muller games [2, 3]. In this chapter, we follow their terminology and definitions for these representations.

The most straightforward representation of an acceptance condition of a Muller automaton is an *explicit condition*, i.e., an explicit list of the elements of $\mathcal{F}$. For short, we refer to a wDMA resp. wNMA, where the acceptance condition is represented as an explicit condition as explicit-wDMA resp. explicit-wNMA.

But there are more succinct representations of Muller automata acceptance conditions, such as *Muller conditions over colors*, *win-set conditions*, *Emerson-Lei conditions*, and *circuit conditions* (see [2, 3] definitions and for a comparison in the setting of delay-free games).

In this chapter, we consider Emerson-Lei conditions, which are Boolean formulas φ with variables from Q. The set $\mathcal{F}_\varphi$ specified by φ is the set of sets $F \subseteq Q$ such that the truth assignment that maps each element of F to true and each element of $Q \setminus F$ to false satisfies φ. The size of a formula φ is denoted $|\varphi|$ and is defined as usual as its length. For short, we refer to a wDMA resp. wNMA, where the acceptance condition is represented as an Emerson-Lei condition as an Emerson-Lei-wDMA resp. an Emerson-Lei-wNMA.

Let us conclude by mentioning that Emerson-Lei conditions are more succinct than Muller conditions over colors, which are more succinct than win-set conditions, which are more succinct than explicit conditions. Circuit conditions are at least as succinct as Emerson-Lei conditions, but it is open if circuit conditions are more succinct than Emerson-Lei conditions (meaning it is open whether circuits can always be translated into small formulas). These (and further) results can be found in [2, 3].

Size of automata. As discussed in the previous paragraph, the representation of the acceptance condition of a (weak) Muller automaton can be more or less succinct. This should be reflected when defining the size of a (weak) Muller automaton. Hence, we define the size of a (weak) Muller automaton with Emerson-Lei condition φ as $|Q| + |\varphi|$; the size of a (weak) Muller automaton with explicit condition $\mathcal{F}$ as $|Q| + |\mathcal{F}|$.

For other ω-automata acceptance conditions such as, e.g., reachability, safety or parity, we define the size of the automaton as its number of states, as the acceptance condition can be encoded with a polynomial overhead.

Delay games. A *delay function* is a mapping $f : \mathbb{N} \to \mathbb{N}_{\geq 1}$, which is said to be constant if $f(i) = 1$ for all $i > 0$. A *delay game* $\Gamma_f(L)$ consists of a delay function f and a *winning condition* $L \subseteq (\Sigma_I \times \Sigma_O)^\omega$ for some alphabets Σ_I and Σ_O. Such a game is played in rounds $i = 0, 1, 2, \ldots$ as follows: in round i, first Player I picks a word $u_i \in \Sigma_I^{f(i)}$, then Player O picks a letter $v_i \in \Sigma_O$. Player O *wins* a play $(u_0, v_0)(u_1, v_1)(u_2, v_2) \cdots$ if the outcome $\binom{u_0 u_1 u_2 \cdots}{v_0 v_1 v_2 \cdots}$ is in L; otherwise, Player I wins. Note that if f is a constant delay function, then Player O has a constant lookahead of $f(0)$ letters on her opponent's moves. Hence, the moniker constant refers to the lookahead induced by f, not to f itself.

A *strategy* for Player I in $\Gamma_f(L)$ is a mapping $\tau_I : \Sigma_O^* \to \Sigma_I^*$ satisfying $|\tau_I(w)| = f(|w|)$, while a strategy for Player O is a mapping $\tau_O : \Sigma_I^+ \to \Sigma_O$. A play $(u_0, v_0)(u_1, v_1)(u_2, v_2) \cdots$ is *consistent* with τ_I if $u_i = \tau_I(v_0 \cdots v_{i-1})$ for all i, and it is consistent with τ_O if $v_i = \tau_O(u_0 \cdots u_i)$ for all i. A strategy for Player $P \in \{I, O\}$ is *winning*, if every play that is consistent with the strategy is won by Player P. We say that Player $P \in \{I, O\}$ *wins* a game $\Gamma_f(L)$ if Player P has a winning strategy in $\Gamma_f(L)$.

Problem statement. In this chapter, we investigate delay games with weak Muller automata winning conditions. We are interested in two aspects.

Firstly, what is the computational complexity of deciding whether Player O wins such a game? Secondly, how much lookahead does Player O need to win such a game if possible?

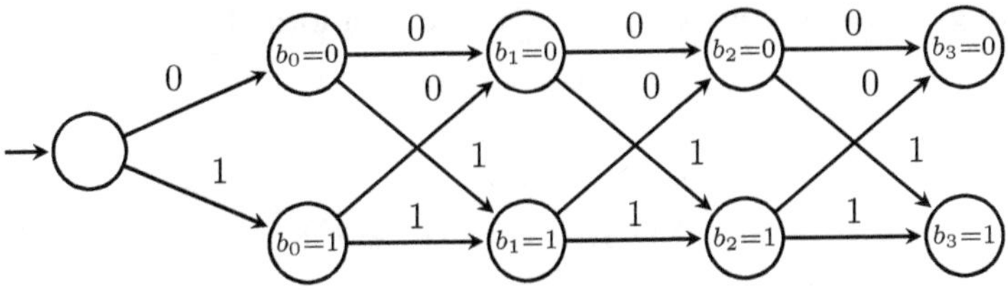

Fig. 1. A gadget to store a 4-bit sequence.

Note that delay games with wDMA resp. wNMA winning conditions are determined, because delay games with Borel winning conditions are determined [16]. Hence, we also speak of deciding whether Player O wins such a game as solving such a game.

We end this section with some introductory examples.

Example 2.1: In Fig. 1, we present a deterministic automaton gadget whose structure is suitable to store a 4-bit sequence. What we mean by that is, for a word starting with $a_0a_1a_2a_3 \in \mathbb{B}^4$, the corresponding run $(p_0, a_0, p_1)(p_1, a_1, p_2)(p_2, a_2, p_3)(p_3, a_3, p_4)\cdots$ of the automaton yields that the state p_{i+1} is equal to $b_i{=}a_i$ for $0 \leq i \leq 3$. This idea is easily scaled to an n-bit sequence for arbitrary $n \in \mathbb{N}_{\geq 1}$. Note that the gadget is of linear size in n.

Now, we present a delay game which makes use of such a gadget.

Example 2.2: Let $L \subseteq (\mathbb{B}^2)^\omega$ be the ω-language that contains all words of the form (α, β) such that $\beta(i) = \alpha(4 + i)$ for $0 \leq i \leq 3$. In words, the second four-bit segment of the input component is equal to the first four-bit segment of the output component.

We describe a wDMA $\mathfrak{A}$ that recognizes L. Basically, such an automaton is composed of two successive gadgets similar as in Fig. 1 and a sink state reached after the second gadget. In the first gadget, the output component of a letter determines the target state, and in the second gadget, the input component of a letter determines the target state. We assume that the states of the first gadget are called $G_1(b_0{=}0)$, $G_1(b_0{=}1)$, etc., and in the second gadget $G_2(b_0{=}0)$, $G_2(b_0{=}1)$, etc.

We give the acceptance condition of $\mathfrak{A}$ as a formula φ:

$$\begin{aligned} &(G_1(b_0{=}0) \leftrightarrow G_2(b_0{=}0)) \\ \wedge\, &(G_1(b_1{=}0) \leftrightarrow G_2(b_1{=}0)) \\ \wedge\, &(G_1(b_2{=}0) \leftrightarrow G_2(b_2{=}0)) \\ \wedge\, &(G_1(b_3{=}0) \leftrightarrow G_2(b_3{=}0)). \end{aligned}$$

Clearly, $L(\mathfrak{A}) = L$.

Now, let us consider a delay game with L as winning condition. Player O can win such a game if she is aware of the $(4+k)$-th letter that Player I plays before she has to give her k-th letter (for $1 \leq k \leq 4$). The constant delay function f with $f(0) = 5$ gives enough lookahead to ensure this.

We show that Player O wins $\Gamma_f(L)$. Due to the choice of $f(0)$, in each round i, Player I has produced a prefix $\alpha(0) \cdots \alpha(i+4)$ that Player O can base her move in that round on. Hence, she picks $\beta(i) = \alpha(i+4)$ in round i. This strategy is winning for her, as every consistent outcome is in L.

As for Example 2.1, this example is easily scaled to an n-bit sequence for arbitrary $n \in \mathbb{N}_{\geq 1}$. Note that a formula φ_n specifying the acceptance condition is of size $\mathcal{O}(n)$, while an explicit representation of $\mathcal{F}_{\varphi_n}$ is of size $\mathcal{O}(2^n)$ as all possible n-bit sequences must be explicitly represented.

In the above example, we presented a delay game where a small lookahead is sufficient for Player O to win. We now give an example of a family of ω-automata where, when used as a winning condition for a delay game, Player O needs exponential lookahead (in the size of the automaton) to win.

Example 2.3: Pick some $n \in \mathbb{N}_{\geq 1}$, and let $\Sigma_n = \{0, \ldots, n-1\}$. A sequence $\alpha \in (\Sigma_n)^* \cup (\Sigma_n)^\omega$ is said to contain a so-called *bad j-pair* if there are two positions $p < p'$ such that $\alpha(p) = \alpha(p') = j$ and $\alpha(q) < j$ for all $p < q < p'$.[d]

Now consider the language $P_n \subseteq (\Sigma_n \times \Sigma_n)^\omega$ that contains a word $\binom{\alpha}{\beta}$ if, and only if, $\alpha(1)\alpha(2)\cdots$ contains a bad j-pair where $j = \beta(0)$. In words, if the first output letter is j, then there is a bad j-pair in the input letter stream starting from the second letter.

We have that $P_n = L(\mathfrak{P}_n)$, where $\mathfrak{P}_n$ is the automaton specified in Fig. 2 when it is, for example, interpreted as a reachability automaton, where q_f has to be reached. Note that $\mathfrak{P}_n$ is deterministic and of size $\mathcal{O}(n)$.

[d]Note that we could equivalently require $\alpha(q) \leq j$ for all $p < q < p'$. The latter allows occurrences of j between positions p and p'. However, one can then pick two such occurrences without any j's in between, which satisfy the stricter definition with $<$.

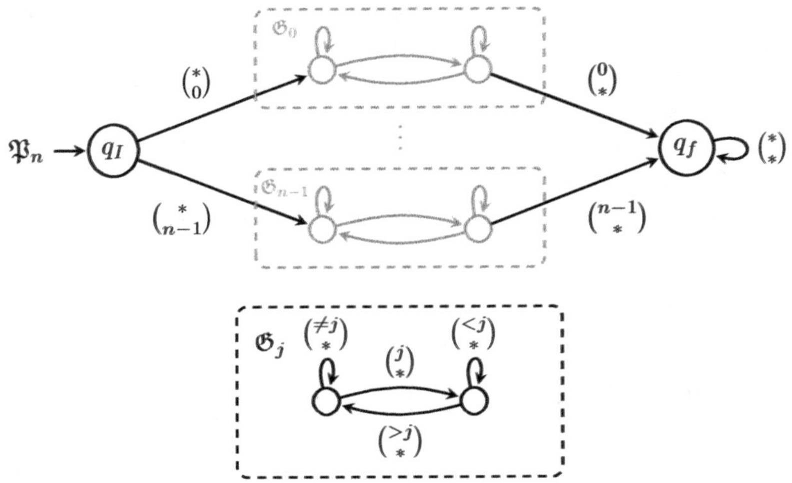

Fig. 2. The automaton $\mathfrak{P}_n$ (top) contains gadgets $\mathfrak{G}_0, \ldots, \mathfrak{G}_{n-1}$ (bottom). Transitions not depicted lead to a sink state, which is not drawn. Here, $*$ denotes an arbitrary letter from the respective alphabet.

Klein and Zimmermann [12] have proven that in order to be guaranteed the existence of a bad j-pair for $0 \le j \le n-1$, a sequence over Σ_n of length at least 2^n is necessary.

Proposition 2.4: *Let $n \in \mathbb{N}_{\geq 1}$, and let $\Sigma_n = \{0, \ldots, n-1\}$.*

- *Every word $w \in (\Sigma_n)^*$ with $|w| \ge 2^n$ contains a bad j-pair for some $0 \le j \le n-1$.*
- *There is a word $w \in (\Sigma_n)^*$ with $|w| = 2^n - 1$ that does not contain a bad j-pair for every $0 \le j \le n-1$.*

With this in mind, it is intuitively true that in a delay game with winning condition P_n as given in Example 2.3, Player O can always win if she has at least a lookahead of 2^n on Player I's moves, but not with less lookahead. This was formally proven in [12]. In our upcoming proofs, we will incorporate variations of Example 2.3.

To conclude, we show that the game presented in Example 2.3 implies an exponential lower bound result on the necessary lookahead for Player O to win delay games with deterministic weak Muller automata winning conditions which are stated explicitly.

The following results is obtained by picking $L_n = P_n$ and showing that $\mathfrak{P}_n$ is an explicit-wDMA with set $\mathcal{F} = \{F_0, \ldots, F_{n-1}\}$, where the set F_i

contains q_I and q_f, as well as all (that is, the two states) from gadget $\mathfrak{G}_i$. To see this, consider a run that starts in q_I and reaches the sink q_f: only one of the gadgets $\mathfrak{G}_i$ is visited. Also, both states of $\mathfrak{G}_i$ must be visited.

Corollary 2.5: *For every $n \in \mathbb{N}_{>1}$, there exists a language L_n recognized by an* explicit-wDMA $\mathfrak{A}_n$ *of size $\mathcal{O}(n)$ such that*

- *Player O wins $\Gamma_f(L_n)$ for some constant delay function f, but*
- *Player I wins $\Gamma_g(L_n)$ for every delay function g with $g(0) \leq 2^n$.*

3. Upper Bounds for Weak Muller Delay Games

We begin this section with a recap of known results used to prove our upper bounds on complexity and necessary lookahead. Klein and Zimmermann [12, 13] have shown the following complexity results about parity delay games.

Proposition 3.1:

(1) Solving delay games with DPA *winning conditions is* ExpTime-*complete.*

(2) Solving delay games with NPA *winning conditions is* 2ExpTime-*complete.*

Furthermore, Klein and Zimmermann [12, 13] have shown tight bounds on the necessary lookahead needed to win parity delay games. Here, we are interested in the upper bound results.

Proposition 3.2:

(1) For every DPA $\mathfrak{A}$, *the following are equivalent:*

- *Player O wins $\Gamma_f(L(\mathfrak{A}))$ for some constant delay function f, where $f(0)$ is exponential in the size of $\mathfrak{A}$.*
- *Player O wins $\Gamma_g(L(\mathfrak{A}))$ for some delay function g.*

(2) For every NPA $\mathfrak{A}$, *the following are equivalent:*

- *Player O wins $\Gamma_f(L(\mathfrak{A}))$ for some constant delay function f, where $f(0)$ is doubly-exponential in the size of $\mathfrak{A}$.*
- *Player O wins $\Gamma_g(L(\mathfrak{A}))$ for some delay function g.*

To obtain our upper bounds on weak Muller delay games stated below (see Theorems 3.4 and 3.5), we explicitly state the following easy result on the translation from weak Muller into parity automata.

Lemma 3.3:

(1) For every wDMA *(in any representation), there exists an equivalent exponentially-sized* DPA.
(2) For every wNMA *(in any representation), there exists an equivalent exponentially-sized* NPA.

Proof: Given a weak Muller automaton $\mathfrak{A}$, it suffices to construct a parity automaton $\mathfrak{A}'$ that additionally tracks the occurrence set of a run of $\mathfrak{A}$. To this end, the parity automaton uses the state set $Q \times 2^Q$, where Q is the set of states of $\mathfrak{A}$. The first component simulates a run of $\mathfrak{A}$, while the second one accumulates the occurrence set of the run prefix simulated thus far.

States of the form (q, O) with $O \in \mathcal{F}$ are assigned color 2 and those with $O \notin \mathcal{F}$ are assigned color 1. The resulting parity automaton is equivalent to the original weak Muller automaton, as the occurrence set eventually stabilizes. □

We are ready to prove our upper bounds (regarding complexity and necessary lookahead) for delay games with weak Muller automata winning conditions.

The combination of Proposition 3.1 and Lemma 3.3 immediately yields our first theorem, which states upper bound complexity results.

Theorem 3.4:

(1) Solving delay games with wDMA *winning conditions (in any representation) is in* 2EXPTIME.
(2) Solving delay games with wNMA *winning conditions (in any representation) is in* 3EXPTIME.

The combination of Proposition 3.2 and Lemma 3.3 immediately yields our next theorem, which gives upper bounds on the necessary lookahead for Player O to win delay games.

Theorem 3.5:

(1) For every wDMA $\mathfrak{A}$ *(in any representation), the following are equivalent:*

- *Player O wins $\Gamma_f(L(\mathfrak{A}))$ for some constant delay function f, where $f(0)$ is doubly-exponential in the number of states of $\mathfrak{A}$.*
- *Player O wins $\Gamma_g(L(\mathfrak{A}))$ for some delay function g.*

(2) *For every* wNMA $\mathfrak{A}$ *(in any representation), the following are equivalent:*

- *Player* O *wins* $\Gamma_f(L(\mathfrak{A}))$ *for some constant delay function* f*, where* $f(0)$ *is triply-exponential in the number of states of* $\mathfrak{A}$*.*
- *Player* O *wins* $\Gamma_g(L(\mathfrak{A}))$ *for some delay function* g*.*

In Secs. 4 and 5, we show matching lower bounds for Theorems 3.4 and 3.5.

4. Lower Bounds for Deterministic Weak Muller Delay Games

This section is devoted to showing lower bounds for delay games with deterministic weak Muller automata winning conditions.

Lookahead. We begin this section by showing a doubly-exponential lower bound on the necessary lookahead for Player O to win, which yields a tight bound in combination with Theorem 3.5.

Recall Example 2.3, where the concept of bad j-pairs was introduced, and an automaton of size $\mathcal{O}(n)$ was given, where the first letter, say $i_0 \in [0, n-1]$ of the output component indicates the existence of a bad i_0-pair in the input component (starting from the second letter), which is a word over $[0, n-1]$. Recall that according to Proposition 2.4, one needs exponential lookahead in n to correctly identify a bad j-pair in a word over $[0, n-1]$.

We are going to design a variant where a doubly-exponential lookahead is necessary. To this end, we encode the numbers in the range $[0, 2^n - 1]$ in binary and show how to construct a small automaton that checks whether the input component has a bad i_0-pair, where i_0 is the first number in the output component. This results in a game that Player O can only win with doubly-exponential lookahead.

In Example 2.3, the automaton stores i_0 in its state space, as there are only n possibilities. However, this is no longer feasible with the range $[0, 2^n - 1]$ and a polynomially-sized automaton, as there are 2^n possible values for i_0. Instead, Player O has to mark two numbers she claims to form a bad i_0-pair and the automaton then checks whether the two marked numbers are equal to i_0 and whether all numbers in between are strictly smaller than i_0. Using the succinctness of the Emerson-Lei condition, this can be achieved with a linearly-sized automaton.

Theorem 4.1: *For every* $n \in \mathbb{N}_{\geq 1}$*, there exists a language* L_n *recognized* *by an* Emerson-Lei-wDMA $\mathfrak{A}_n$ *of size* $\mathcal{O}(n)$ *such that*

- *Player O wins* $\Gamma_f(L_n)$ *for some constant delay function* f*, but*
- *Player I wins* $\Gamma_g(L_n)$ *for every delay function* g *with* $\sum_{i=0}^{n-1} g(i) \leq 2^{2^n}$.

Proof: Pick some $n \in \mathbb{N}_{\geq 1}$. Our goal is to give a language L_n recognized by an Emerson-Lei-wDMA $\mathfrak{A}_n$ of size $\mathcal{O}(n)$ such that the input component encodes a sequence $x_0, x_1, x_2, \ldots$ of numbers $x_i \in [0, 2^n - 1]$, and the output component encodes a sequence $y_0, y_1, y_2, \ldots$ of numbers $y_i \in [0, 2^n - 1]$. Each x_i and y_i is encoded as an n-bit sequence. For acceptance, we require that $x_1 x_2 \cdots$ contains a bad y_0-pair. According to Proposition 2.4, finding a bad j-pair in a word over $[0, 2^n - 1]$ requires, in the worst case, a word of length 2^{2^n}. Consequently, using L_n as the winning condition for a delay game will ensure that a doubly-exponential lookahead (in n) is necessary.

We go into details. The automaton we construct uses the product alphabet $\{0, 1, \#\} \times \{0, 1, \mathsf{X}, \#\}$. We first describe what we call valid input resp. output encodings:

- An input sequence α is a valid encoding if it is of the form
$$((0+1)^n \#)^\omega.$$
The i-th n-bit sequence encodes the number $x_i \in [0, 2^n - 1]$, the least significant bit is the leftmost one.
- An output sequence β is a valid encoding if it is of the form
$$((0+1)^n (\mathsf{X} + \#))^\omega$$
such that X occurs exactly twice. The i-th n-bit sequence encodes the number $y_i \in [0, 2^n - 1]$, the least significant bit is the leftmost one.

The behavior of the automaton is described below, where we distinguish three cases. The first two cases simply state that (viewed as a game) the player who first violates the correct encoding format loses.

(1) If there exists some i such that $\alpha(0)\alpha(1) \cdots \alpha(i)$ cannot be completed to a valid input encoding, but $\beta(0)\beta(1) \cdots \beta(i-1)$ can be completed to a valid output encoding, the automaton accepts.
(2) If there exists some i such that $\beta(0)\beta(1) \cdots \beta(i)$ cannot be completed to a valid output encoding, but $\alpha(0)\alpha(1) \cdots \alpha(i)$ can be completed to a valid input encoding, the automaton rejects.
(3) If α and β are valid encodings, the automaton accepts if a bad y_0-pair is correctly marked.

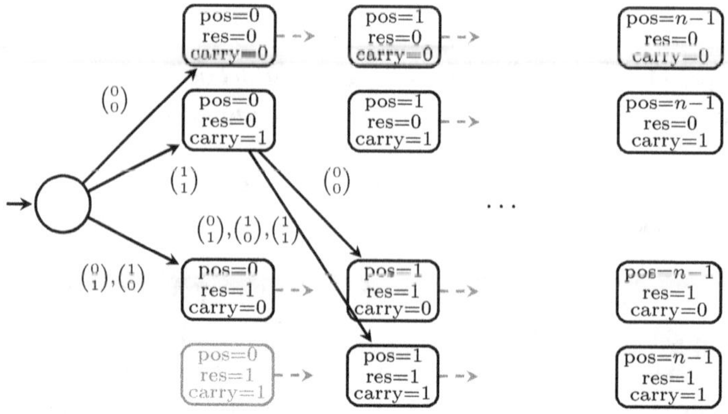

Fig. 3. A gadget to add two n-bit sequences in little endian notation. Missing transitions are implied by gray dashed arrows. States for positions 2 to $n-2$ are not drawn. The state label $\substack{\text{pos}=k\\ \text{res}=b\\ \text{carry}=b'}$ indicates that adding the bits at position k yields the result b and b' is the carry bit relevant for the addition at position $k+1$. The state $\substack{\text{pos}=0\\ \text{res}=1\\ \text{carry}=1}$ is not reachable and can be omitted.

We first explain the idea of Item 3. The intention of the marker $\times$ is to be placed before two numbers x_j and $x_{j'}$ such that $x_j = x_{j'} = y_0$ and they enclose a bad y_0-pair, i.e., $x_i < y_0$ for $j < i < j'$.

In order to check whether $x_j = x_{j'} = y_0$, it suffices to have three gadgets similar as in Example 2.1 for n-bit sequences. The first gadget stores y_0, we index its states with S for start. The second and third gadgets are entered when the first resp. second $\times$ is seen, and thus store x_j and $x_{j'}$, we index its states with M_1 resp. M_2 for marked numbers. Validating whether $x_j = x_{j'} = y_0$ is done via the acceptance condition by a formula similar to the formula presented in Example 2.2, which we present further below. Our formula also enforces that there are two marked numbers.

Instead of checking whether $x_i < y_0$ for $j < i < j'$, the automaton checks for another condition which is expressible with a small enough automaton and formula. Namely, whether $x_i + y_i = y_0$ for $j < i < j'$. Note that this is only equivalent to $x_i \leq y_0$. However, recall Footnote d. The intention is that, during a play, Player O can make this condition true if, and only if, x_j and $x_{j'}$ indeed enclose a bad y_0-pair.

In order to check whether $x_i + y_i = y_0$ for all $j < i < j'$, one gadget is used that computes the addition of two n-bit sequences. Such a gadget is depicted in Fig. 3. We index its states by A for addition. This gadget is entered for every n-bit sequence between the first two marked sequences.

Expressing that the automaton accepts if $x_i + y_i = y_0$ for all $j < i < j'$ is expressed via a formula, which we give further below.

Now, we briefly explain the overall automaton structure. We construct a deterministic automaton, thus, Items 1 to 3 must be handled in parallel. Note that the gadgets we introduced for Item 3 can easily be extended to check whether either sequence is invalid (i.e., whether Item 1 or Item 2 holds) without introducing new states for the gadgets. The automaton needs two additional sink states q_{acc} resp. q_{rej} that are entered when the input resp. output sequence is not a valid encoding.

After the S-gadget and before the M_1-gadget, and after the M_2-gadget, it suffices to check whether Item 1 or Item 2 holds and go to the corresponding sink state. This can be done with $\mathcal{O}(n)$ many states.

All in all, $\mathfrak{A}_n$ is a sequence of six different gadgets, which are all of size $\mathcal{O}(n)$, so $\mathfrak{A}_n$ is of size $\mathcal{O}(n)$:

(1) The first gadget, S, stores y_0 the first number picked by Player O.
(2) The second gadget checks that the format of the input and output is correct.
(3) The third gadget, M_1, is entered when the first $\times$ occurs, say at the beginning of the j-th block. It stores x_j.
(4) The fourth gadget, A, is entered immediately after the j-th block and only left when the second $\times$ occurs, say at the beginning of the j'-th block. It checks that $x_i + y_i = y_0$ holds for all $j < i < j'$.
(5) The fifth gadget, M_2, is entered when the second $\times$ occurs. It stores $x_{j'}$, the number marked by the second $\times$.
(6) The last gadget again checks that the format of the input and output is correct.

Note that the gadgets S, M_1, M_2, and A also check the format of the input and output is correct. As explained earlier, this does not increase their size.

Finally, we are ready to present the acceptance condition φ_n of the automaton $\mathfrak{A}_n$. We state its intended meaning in words first:

- the rejecting sink is not seen, and
- the accepting sink is seen, or
- for every position $k \in [0, n-1]$ and every bit value $b \in \mathbb{B}$:
  - If $y_0(k) = b$, then $x_j(k) = b$ and $x_{j'}(k) = b$, i.e., the two marked numbers are equal to the first one in the output.
  - If $y_0(k) = b$, then the k-th bit of $x_i + y_i$ is not $1 - b$ for every $j < i < j'$, i.e., all pairs of numbers between the marked ones add up to the first one in the output.

We define $\varphi_n = \neg q_{rej} \wedge (q_{acc} \vee \psi_n)$ with

$$\psi_n = \bigwedge_{k\in[0,n-1]} \bigwedge_{b\in\mathbb{B}} S(b_k{=}b) \to \left[M_1(b_k{=}b) \wedge M_2(b_k{=}b) \wedge \bigwedge_{b'\in\mathbb{B}} \neg A \begin{pmatrix} \text{pos}{=}k \\ \text{res}{=}1{-}b \\ \text{carry}{=}b' \end{pmatrix} \right].$$

Note that φ_n is of size $\mathcal{O}(n)$.

Now, we show that

- Player O wins $\Gamma_f(L_n)$ for some constant delay function f, but
- Player I wins $\Gamma_g(L_n)$ for every delay function g with $\sum_{i=0}^{n-1} g(i) \leq 2^{2^n}$.

We start by giving a constant delay function f such that Player O has a winning strategy in $\Gamma_f(L(\mathfrak{A}_n))$. According to Proposition 2.4, the sequence $x_1 \cdots x_{2^{2^n}}$ of numbers is guaranteed to contain a bad j-pair for some $j \in [0, 2^n - 1]$. Hence, we let $f(0) = m$, where m is some number large enough to ensure that $x_1, \ldots, x_{2^{2^n}}$ is included in the lookahead (assuming that Player I plays a valid input encoding). If Player I plays a valid input encoding, Player O's strategy is to output some y_0 such that $x_1 \cdots x_{2^{2^n}}$ contains a bad y_0-pair. She marks the pair correctly with X, and can ensure that she produces output y_i such that $x_i + y_i = y_0$ in between the marked numbers. At other positions, she plays arbitrary numbers, and no other marks.

We argue that φ_n is satisfied. The state q_{rej} is never reached, as if Player I does not play a valid input encoding, the state q_{acc} is reached independently of Player O's moves and the formula is trivially satisfied. Assume that Player I plays a valid input encoding, and Player O behaves as described. Clearly, $S(b_k{=}b) \to (M_1(b_k{=}b) \wedge M_2(b_k{=}b))$ for all $k \in [0, n-1]$ and $b \in \mathbb{B}$. Since for all numbers y_i between the marked numbers $x_i + y_i = y_0$ holds, we also have that if $S(b_k{=}b)$ has been seen, then the k-th bit of $x_i + y_i$ is equal to b. Hence, only the states $A\begin{pmatrix} \text{pos}{=}k \\ \text{res}{=}b \\ \text{carry}{=}0 \end{pmatrix}$ and $A\begin{pmatrix} \text{pos}{=}k \\ \text{res}{=}b \\ \text{carry}{=}1 \end{pmatrix}$ can be seen. This implies that $\neg A\begin{pmatrix} \text{pos}{=}k \\ \text{res}{=}1{-}b \\ \text{carry}{=}0 \end{pmatrix} \wedge \neg A\begin{pmatrix} \text{pos}{=}k \\ \text{res}{=}1{-}b \\ \text{carry}{=}1 \end{pmatrix}$ is true.

For the other direction, assume that g is a delay function such that $\sum_{i=0}^{n-1} g(i) \leq 2^{2^n}$. We show that Player I wins $\Gamma_g(L_n)$. According to Proposition 2.4, there exists a word w over $[0, 2^n - 1]$ with $|w| = 2^{2^n} - 1$ that contains no bad j-pair for every $j \in [0, 2^n - 1]$. Player I's strategy is to play a valid input encoding such that x_i is the n-bit sequence that encodes $w(i)$ for all $0 \leq i \leq 2^{2^n} - 1$. After round $n - 1$, Player O has finished y_0. Furthermore, the condition $\sum_{i=0}^{n-1} g(i) \leq 2^{2^n}$ ensures that Player I has not yet begun to spell the n-bit sequence that encodes $x_{2^{2^n}}$. Hence, since y_0 is known to Player I, he can pick a number $x \neq y_0$ and continues to produce a valid input encoding such that every x_i encodes x for all $i \geq 2^{2^n}$.

We show that Player I wins playing as described. Player O's goal is to satisfy the formula φ_n. Clearly, she loses if her moves do not produce a valid output encoding (as q_{rej} would be reached), so we assume that she produces a valid output encoding. The state q_{acc} is not reached, as it is only reached if the input sequence is invalid, hence, the other part of the formula must be satisfied. Note that the implication $S(b_k{=}b) \rightarrow (M_1(b_k{=}b) \wedge M_2(b_k{=}b))$ for all $k \in [0, n-1]$ and $b \in \mathbb{B}$ can only be satisfied if Player O marks two numbers with X, because the S-gadget stores y_0 making the left-hand side of the implication true. However, since Player I's moves ensured that the input sequence does not contain a bad y_0-pair, the right-hand side of the implication

$$\bigwedge_{k\in[0,n-1]} \bigwedge_{b\in\mathbb{B}} S(b_k{=}b) \rightarrow \left[M_1(b_k{=}b) \wedge M_2(b_k{=}b) \wedge \bigwedge_{b'\in\mathbb{B}} \neg A \begin{pmatrix} \text{pos}{=}k \\ \text{res}{=}1{-}b \\ \text{carry}{=}b' \end{pmatrix} \right]$$

is not satisfied. We have shown that Player I wins $\Gamma_g(L(\mathfrak{A}_n))$. □

Complexity. Next, we settle the complexity of solving delay games with Emerson-Lei-wDMA winning conditions by proving a 2EXPTIME lower bound. Our construction is a generalization of the EXPTIME lower bound for solving delay games with winning conditions given by deterministic parity automata [12].

Intuitively, Player I produces a sequence of configurations of an alternating exponential-space Turing machine. Each of these is of exponential length, i.e., the cells of each configuration can be addressed with polynomially many bits. Now, Player I picks the successor configuration of universal configurations, while Player O picks the successor configuration of existential configurations (by picking a transition to apply). Relying on the lookahead, Player O always has access to the full current configuration picked by Player I before she has to pick a transition to apply. Dually, to account for the lookahead, Player I is allowed to copy a configuration to fill the lookahead while he waits for Player O to pick a transition. Hence, two successive configurations played by Player I should either be equal or the second one should be a successor configuration of the first one.

To force Player I to faithfully simulate the Turing machine, i.e., to only copy configurations or to pick a valid successor configuration (in particular the one determined by Player O in case of existential configurations), Player O can mark cells for the automaton to check for correctness: This is sufficient as an error by Player I manifests itself in a single wrongly updated or copied cell. Using the addresses of the cells and the markers used by Player O, a small automaton can check whether there is an error or not. Finally, Player O is able to correctly apply the markers, as she has enough lookahead to always observe the next two configurations, enough to spot errors introduced by Player I.

Altogether, we obtain a delay game simulating an alternating exponential space Turing machine, and AExpSpace = 2ExpTime yields the desired result.

Theorem 4.2: *Solving delay games with* Emerson-Lei-wDMA *winning conditions is* 2ExpTime*-complete.*

Proof: The upper bound was shown in Theorem 3.4, we show the corresponding lower bound.

We give a reduction from the word problem for alternating exponential space Turing machines. Since 2ExpTime = AExpSpace [17], we obtain the desired result.

Let $\mathcal{M} = (Q = Q_\exists \uplus Q_\forall, \Sigma, q_I, \Delta, q_A, q_R)$ be an alternating exponential space Turing machine, where $\Delta \subseteq Q \times \Sigma \times \Sigma \times Q \times \{-1, 0, 1\}$. Furthermore, let p be a polynomial such that 2^p bounds the space consumption of $\mathcal{M}$ and let $w \in \Sigma^*$ be an input word. We fix $n = p(|w|)$. Wlog., we assume that the accepting state q_A and the rejecting state q_R are equipped with a self-loop. Furthermore, wlog., we assume that either q_A or q_R is reached in every run of $\mathcal{M}$ on w. An alternating exponential space Turing machine that does not have this property can be turned into an equivalent TM $\mathcal{M}'$ that satisfies the property as follows: The number of configurations that can be reached on w is bounded (doubly-exponentially in n). Hence, $\mathcal{M}$ can be equipped with an exponentially-sized counter in order to detect that a configuration repetition has occurred (when the counter has surpassed the maximum) and reject if so. Then, $\mathcal{M}'$ is used as the starting point for the reduction instead of $\mathcal{M}$.

We construct an Emerson-Lei-wDMA $\mathfrak{A}$ of polynomial size in $(|\Delta| + n)$ such that $\mathcal{M}$ accepts w if, and only if, Player O wins $\Gamma_f(L(\mathfrak{A}))$ for some delay function f.

The idea is that, in the delay game, the players build a run of $\mathcal{M}$ on w in the form of a configuration sequence. Player I controls the universal states and Player O the existential ones. Furthermore, Player I spells all configurations with his moves. The configurations are of exponential size in n, so each cell of $\mathcal{M}$'s tape is addressed by an n-bit counter. In between configurations there is a delimiter, either C or N, to denote if the next configuration is a *copy* of the previous one or a *new* one. The need to copy configurations arises since, in the delay game, Player O is behind with her moves, so Player I needs to wait for Player O's pick if the current configuration is existential.

We go into details. The automaton we construct uses the product alphabet

$$(\{0,1,C,N,\text{␣}\} \cup Q \cup \Sigma) \times (\{\checkmark, \text{✗}, \ast\} \cup \Delta)\,.$$

Recall that Q, Δ, and Σ are components of the Turing machine $\mathcal{M}$ and are used to encode configurations and to pick transitions to apply. We first describe what we call valid input resp. output formats:

- An input sequence α is valid if it is of the form

$$\Big((C+N)\,\big[\,(0+1)^n\,(\Sigma+Q+\text{␣})\,\big]^+\Big)^\omega,$$

  to be interpreted as a sequence of configurations $(C{+}N)c_0(C{+}N)c_1\cdots$ where each c_i is a configuration delimited by C or N. A configuration is encoded by a sequence $\big[(0+1)^n(\Sigma+Q+\text{␣})\big]^+$, where each n-bit sequence encodes an address $a \in [0, 2^n-1]$, the least significant bit is the rightmost one, and the letter after the address encoding is either the tape cell content (that is, a letter $\sigma \in \Sigma$, or a blank (denoted as ␣)) or a state $q \in Q$. Note that the format does not imply that a sequence in $\big[(0+1)^n(\Sigma+Q+\text{␣})\big]^+$ encodes a configuration of length 2^n. This will be later enforced by the rules of the game.
- An output sequence β is valid format if it is of the form

$$\Big(\Delta\big[(\checkmark+\text{✗}+\ast)^{n+1}\big]^+\Big)^\omega,$$

  and if for every position i, $\alpha(i) \in \{C,N\}$ if, and only if, $\beta(i) \in \Delta$, i.e., the letters $\tau \in \Delta$ occur at the same positions as the configuration delimiters C and N. We give the intended interpretation of this format, which is explained in more detail below: The markers ✓ resp. ✗ and ✱ are intended to indicate the absence of resp. the existence of an error. If ✗ marks a bit in an address block, it should indicate an error in updating the address block at this bit, if ✱ marks the first bit of an address block, then it should indicate that there is a copy or update mistake in the successive configuration, which manifests itself at the cell with the marked address.

The behavior of the automaton is as follows:

(1) If there exists some i such that $\alpha(0)\alpha(1)\cdots\alpha(i)$ cannot be completed to have a valid format, but $\beta(0)\beta(1)\cdots\beta(i-1)$ can be completed to have a valid format, the automaton accepts.
(2) If a configuration encoding c contains not exactly one state or the first address block is not zero, the automaton accepts.
(3) If c_0 does not encode the initial configuration on w, it accepts.
(4) If there exists some i such that $\beta(0)\beta(1)\cdots\beta(i)$ cannot be completed to have a valid format, but $\alpha(0)\alpha(1)\cdots\alpha(i)$ can be completed to have a valid format, the automaton rejects.

We refer to Items 1 to 3 as simple input errors, and to Item 4 as simple output error. These conditions can be easily checked using the automaton structure with $\mathcal{O}(|\Delta|+n)$ states. We use designated sink states q_{acc} and q_{rej}.

(5) If there are no simple errors, and there is a position marked by ✗ in some address block a_i (say $a_i(k)$), the automaton checks if $a_{i+1}(k)$ (assuming there is no $a_i(k')$ for some $k' < k$, which is also marked by ✗) is wrongly updated in the next address block a_{i+1}. We call this an *address update error*. If the address is indeed wrongly updated, then the automaton accepts, otherwise it rejects. We explain further below how to achieve this behavior.
(6) If there are no simple errors, and there is an address block a whose first position is marked by ✱, the following is checked:

    - The next configuration must contain an address block a' whose first position is marked by ✱, otherwise the automaton rejects. (This is easily checkable via the automaton structure.)
    - If $a \neq a'$, the automaton rejects.
    - If $a = a'$, check whether the tape cell addressed by a contains an error:

        - If the delimiter between the configurations is C: The cell content is not correctly copied. We call this *copy error*.
        - If the delimiter between the configurations is N: The cell content is not correctly updated according to the picked transitions. We call this *update error*.

    If an error is claimed correctly, the automaton accepts, otherwise it rejects.

We explain further below how to achieve this behavior.

(7) Lastly, if none of the above cases occurs, if there is a configuration that contains q_A, then the automaton accepts, if there is a configuration that contains q_R, then the automaton rejects. Therefore, when a rejecting resp. accepting configuration has been seen, the automaton goes into designated new sink states q_R resp. q_A.

Now, we explain how to handle Item 5, i.e., the detection of an address update error. When a bit is marked with ✗, its value is stored in the state space. Using a modulo counter, it is easy to find the same bit in the next address block. It is not difficult to see that the value of a bit only flips if all remaining bits (that is, all bits to the right) are one, because then adding one to the address counter causes an overflow of all less significant bits (which are denoted to the right of the marked bit). Hence, an automaton can check with $\mathcal{O}(n)$ states whether a marked address bit has been updated correctly.

Regarding Item 6, checking whether the two marked (with ✱) n-bit sequences are the same can be done by using two gadgets and a formula as explained in Examples 2.1 and 2.2. Such a formula is of size $\mathcal{O}(n)$.

In order to check whether there is a copy error between successive configurations, it suffices to remember whether the delimiter between them is C and the cell content of the marked cell.

Checking whether there is an update error between successive configurations c and c' is more involved:

- The delimiter between them has to be N.
- A window of three cell contents around the cell whose address is marked must be remembered.
- If c is a universal configuration, it has to be verified (using the stored three-cell window) whether the update in c' is achievable by applying a transition from Δ. (It is not specified in the input sequence which transition should be applied.)
- If c is an existential configuration, the transition to be applied is specified in the output sequence. However, the transition to be applied has been given (possibly much) earlier: When the delimiter before a configuration is N, the configuration is new. If this configuration is existential, in the output sequence, the letter after the configuration specifies the transition to be applied.
- Transitions that are given after copied configurations as well as universal configurations are irrelevant.
- Thus, the automaton stores a specified transition only if it occurs after

a new existential configuration. This is the transition to be applied to obtain the next new configuration.

An automaton needs a polynomial number of states (in $(|\Delta| + n)$) to store the required pieces of information and verify that there is an update error.

Since we are constructing a wDMA, all of Items 1 to 7 need to be checked simultaneously. Therefore, we build a product automaton, which we denote by $\mathfrak{A}$ with one special feature, namely, the introduced sink states $q_{acc}, q_{rej}, q_A, q_R$ are global sink states (as opposed to product states).

This product automaton has a number of states polynomial in $(|\Delta| + n)$ since it is constructed from gadgets whose number of states are polynomial in $(|\Delta| + n)$. As mentioned before, regarding Item 6, the formula needed to verify that two n-bit addresses are the same is of size $\mathcal{O}(n)$. Since $\mathfrak{A}$ is a product automaton, the formula describing that two addresses are equal needs to be adjusted to incorporate the product structure, which causes a polynomial in $(|\Delta| + n)$ blow-up of the formula. We briefly explain how to take a product structure into account by a small example. Let $\mathfrak{C}_1, \mathfrak{C}_2, \mathfrak{C}_3$ be some gadgets with state sets Q_1, Q_2, Q_3, respectively, and the product automaton is of the form $\mathfrak{C}_1 \times \mathfrak{C}_2 \times \mathfrak{C}_3$. Say a formula only refers to states in $\mathfrak{C}_1$, which contains the state p. Then the updated formula must replace all occurrences of p with $\bigvee_{q \in Q_2} \bigvee_{r \in Q_3} (p, q, r)$.

All in all, the final formula φ looks as follows:

$$\varphi = \neg(q_{rej} \vee q_R) \wedge (q_{acc} \vee q_A \vee \varphi_{error?}),$$

where $\varphi_{error?}$ describes that either the gadgets to store the n-bit address sequence have not been entered (meaning no copy/update error needs to be verified), or that they store the same n-bit address sequence (the claimed copy/update error is verified via the automaton structure). The formula $\varphi_{error?}$ is adjusted to the product automaton structure.

Recall that $n = p(|w|)$, hence, φ is of polynomial size in $(|\Delta| + p(|w|))$. All in all, $\mathfrak{A}$ is of polynomial size in $(|\Delta| + p(|w|))$.

It is left to show that $\mathcal{M}$ accepts w if, and only if, Player O wins $\Gamma_f(L(\mathfrak{A}))$ for some delay function f.

To begin with, assume $\mathcal{M}$ accepts w. Let f be a constant delay function such that $f(0) = m$, where m is chosen such that Player O has enough lookahead to see a full configuration before she has to start with its output. To be more concrete, $m = 2^n(1 + n) + 1$ suffices to capture a whole configuration including the address blocks and the delimiter.

We show that Player O wins $\Gamma_f(L(\mathfrak{A}))$. We assume that either player does not make simple errors (recall that this means playing in the wrong format). Firstly, assume that Player I either does not update an address correctly, or introduces a copy or update error. The lookahead is large enough to correctly claim such an error. The formula is then satisfied. Secondly, assume that Player I does not introduce such errors. Since $\mathcal{M}$ accepts w, Player O can pick existential transitions such that if Player I applies the picked transitions an accepting configuration will be reached. Note that, in order to pick an existential transition such that an accepting configuration will be reached, it is only necessary to have the knowledge of the current configuration. This fact can be easily seen when visualizing the run tree of $\mathcal{M}$ on w. If an accepting configuration is seen, the formula φ is clearly satisfied. Player I can prevent that an accepting configuration is seen by always copying the current configuration. But then, the formula φ is also satisfied, because the sub-formula $\varphi_{error?}$ is satisfied if no copy/update error is claimed or a copy/update error is correctly claimed. Player O does not claim such an error, as we assumed that Player I does not introduce such an error.

For the other direction, assume $\mathcal{M}$ rejects w. We show that Player I wins $\Gamma_g(L(\mathfrak{A}))$ for every delay function g. Again, we assume that either player does not make simple format errors. No matter the existential transitions that Player O chooses, Player I can always pick universal transitions such that eventually a rejecting configuration is reached. Hence, we assume that Player I advances the run by giving new configurations according to his and Player O's choices.

We go into more detail. Whenever Player I has provided a new existential configuration, he has to wait for Player O to provide a transition to be applied since she is behind with her moves. While Player O has not provided her transition choice, Player I copies the current configuration until the choice of Player O is known. Then, Player I produces a new configuration obtained by applying Player O's choice. Whenever Player I has provided a new universal configuration, there is no need to copy the configuration, a new configuration can be provided directly after.

Furthermore, we assume that Player I does not introduce address update errors or configuration copy/update errors, as this is not beneficial for him. Player O loses a play where a rejecting configuration is reached. Hence, her only chance is to claim an error. Since there are no errors introduced by Player I, this approach is not fruitful as this consequently falsifies φ. Player O cannot not win a play, i.e., Player I wins $\Gamma_g(L(\mathfrak{A}))$. □

5. Lower Bounds for Non-Deterministic Weak Muller Delay Games

In this section, we show lower bounds for delay games with non-deterministic weak Muller automata winning conditions.

Lookahead. We show a triply-exponential lower bound on the necessary lookahead for Player O to win, which yields a tight bound in combination with Theorem 3.5.

To this end, we again implement the bad j-pair game described in Example 2.3, this time with numbers in the range $[0, 2^{2^n} - 1]$. As argued before, this allows Player O to win with triply-exponential lookahead, but not with less. Thus, we show that winning conditions specified by non-deterministic automata allow to implement the bad j-pair game with exponentially larger numbers than deterministic automata. This increase requires some additional mechanism to allow the automaton to check the existence of a bad j-pair, which we describe below.

The binary encoding of each number in $[0, 2^{2^n} - 1]$ is of exponential length, i.e., each bit can be addressed with n additional address bits. Player I produces a sequence of such encodings of numbers and Player O picks a number i_0 in her first move, claiming that the sequence picked by Player I contains a bad i_0-pair. To verify this claim, Player O is required to repeat i_0 ad infinitum.

Thus, both players can now cheat during their moves, Player I by not updating the addresses correctly and Player O by not copying i_0. Thus, both players need to use markers to claim such errors, unlike in the analogous result for deterministic automata.

However, due to the lookahead, Player I cannot mark the position where Player O has wrongly copied i_0. However, such an error manifests itself in a single bit and Player I is able to mark its address at a later position. As there are exponentially many addresses, we rely on non-determinism to guess for each address whether it is equal to the marked one, i.e., it needs to be checked for a copy error, or whether it is not equal to the marked address. It is this exponential succinctness, that leads to the increase in required lookahead from doubly-exponential to triply-exponential when allowing non-determinism.

Theorem 5.1: *For every $n \in \mathbb{N}_{\geq 1}$, there exists a language L_n recognized by an* Emerson-Lei-wNMA $\mathfrak{A}_n$ *of size $\mathcal{O}(n^3)$ such that*

- *Player O wins $\Gamma_f(L_n)$ for some constant delay function f, but*
- *Player I wins $\Gamma_g(L_n)$ for every delay function g with $\sum_{i=0}^{(n+2)\cdot 2^n} g(i) \leq 2^{2^{2^n}}$.*

Proof: Pick some $n \in \mathbb{N}_{\geq 1}$. Our goal is to give an Emerson-Lei-wNMA $\mathfrak{A}_n$ of polynomial size in n such that Player O needs triply-exponential lookahead in n to win the corresponding delay game. Recall Example 2.3 and Proposition 2.4. We are constructing a bad j-pair automaton over numbers ranging between 0 and $2^{2^n} - 1$, which enforces the need of a word composed of at least $2^{2^{2^n}}$ numbers to be guaranteed the existence of a bad j-pair.

To construct an automaton of polynomial size in n, we encode a number $x \in R = [0, 2^{2^n} - 1]$ as a word over $\mathbb{B}$ as follows: To represent a number $x \in R$ in binary, one needs 2^n bits, we address each bit by an n-bit sequence.

We go into more detail. A sequence of the form $a_0 b_0 \cdots a_{2^n-1} b_{2^n-1}$ such that $a_i \in \mathbb{B}^n$ and $b_i \in \mathbb{B}$ for $0 \leq i \leq 2^n - 1$ is called a *superblock*. A block $a_i \in \mathbb{B}^n$ is interpreted as an *address block* whose least significant bit is the rightmost one. The bit sequence $b_0 \cdots b_{2^n-1}$ is interpreted as a number $x \in R$, the least significant bit is the rightmost one. We refer to these bits as *special bits*.

In order to explain the idea, assume that in a play, Player I plays an input sequence that encodes a sequence of numbers $x_0, x_1, \ldots \in R$. We give a construction such that Player O's moves must produce an output sequence that encodes a sequence of numbers $y_0, y_1, \ldots \in R$ such that $x_1 x_2 \cdots$ contains a bad y_0-pair and it holds that $y_0 = y_1 = y_2 = \cdots$. In words, we enforce that Player O has to copy her first number ad infinitum. The technique to enforce Player O to copy her first number ad infinitum is as follows: When Player I realizes that she has not faithfully copied her number, then there exists an address a such that there are two superblocks whose values of the special bits addressed by a differ. Player I marks an address block that encodes a, and the winning condition (i.e., the automaton) enforces that all special bits which are addressed by a must have the same value. This is then violated, so Player O loses if she does not copy her choice of y_0 ad infinitum.

Furthermore, the automaton we construct verifies that the input sequence indeed contains a bad y_0-pair. Since it is enforced that $y_0 = y_1 = y_2 = \cdots$, verifying whether a bad y_0-pair exists is doable with a small enough automaton, which we explain further below.

We go into details. The automaton we construct uses the product alphabet $\{0, 1, \times, \#\} \times \{0, 1, \#, \checkmark, \text{✗}\}$. We first describe what we call valid input resp. output format:

- An input sequence α has a valid format if it is of the form
$$\left((\times + \#)(0+1)^{n+1}\right)^{\omega},$$
where $\times$ occurs exactly once.
- An output sequence β has a valid format if it is of the form
$$\left(\#(\checkmark + \text{✗})^{n}(0+1)\right)^{\omega}.$$
The intention is that ✓ resp. ✗ are used to mark the absence resp. the presence of errors in updating the address blocks.

We now describe how the automaton $\mathfrak{A}_n$ behaves, then how to achieve this behavior.

(1) If there exists some i such that $\alpha(0)\alpha(1)\cdots\alpha(i)$ cannot be completed to have a valid format, but $\beta(0)\beta(1)\cdots\beta(i-1)$ can be completed to have a valid format, the automaton accepts.
(2) If there exists some i such that $\beta(0)\beta(1)\cdots\beta(i)$ cannot be completed to have a valid format, but $\alpha(0)\alpha(1)\cdots\alpha(i)$ can be completed to have a valid format, the automaton rejects.
(3) If the very first address is not zero, it accepts.
(4) If the k-th bit of an address block is marked with ✗ (and no earlier ✗ occurs), then it is verified if the k-th bit of the next address block is correctly updated. If the update is faulty, the automaton accepts, if the update is correct, the automaton rejects.
(5) If $\times$ marks an address a, then the special bits picked by Player O that are addressed with a either have all value zero or all have value one. If this condition is violated, the automaton rejects.
(6) If none of the conditions before have led to acceptance or rejection, the automaton verifies whether $x_1x_2\cdots$ contains a bad y_0-pair. If the answer is yes, it accepts, otherwise it rejects.

We now describe how to achieve the desired behavior. Note that Items 1 to 3 can be easily checked in parallel via the automaton structure of an automaton with $\mathcal{O}(n)$ states. We use designated sink states q_{acc} resp. q_{rej} that are reached when the automaton accepts resp. rejects.

Note that Item 4 can be handled as Item 5 introduced in the proof of Theorem 4.2, where we re-use the sink states q_{acc}, q_{rej}. Again, $\mathcal{O}(n)$ states are used.

It is left to handle Items 5 and 6. Regarding Item 5, we need one gadget that stores the n-bit sequence that has been marked by $\times$. We refer to this sequence as address a. Such a gadget uses $\mathcal{O}(n)$ states and was first introduced (and used) in Examples 2.1 and 2.2. We index its states with A for address.

The automaton needs to verify that all special bits that are addressed by a have the same value. Therefore, the automaton behaves as follows:

- Before a new address block begins, the automaton guesses whether this address will be equal to a.
- If the automaton guesses that the addresses are equal:
  - It must be verified that the guess is right. If the guess is wrong, the automaton rejects.
  - In order to do the verification, we need one additional gadget that stores an n-bit sequence such as in Example 2.1. We index its states by $A_=$ for equal address. The automaton enters this gadget every time it guesses that the address will be equal to a. Further below, we give a formula that is satisfied if, and only if, the guess is always right.
  - Additionally, the automaton must store the values (as a set) of the special bits picked by Player O that occur as the next letter.
- If the automaton guesses that the addresses are not equal:
  - It must be verified that the guess is right. If the guess is wrong, the automaton rejects.
  - To verify, guess some $i \in [0, n-1]$ and check that the value of the i-th address bit is different from the i-th bit of the address a (recall, which is marked by $\times$, and stored in gadget A).
  - In Fig. 4, we give a gadget that implements the guessing and storing of a bit. We index its states by G for guessing. The automaton enters this gadget every time it guesses that the address will be not equal to a.
  - Further below, we give a formula that is satisfied if, and only if, always a bit could be found such that the values at this bit in the inspected address and in address a differ.

Such an automaton needs $\mathcal{O}(n)$ states.

Lastly, regarding Item 6, the automaton guesses before the beginning of some superblock that its encoded number x_j is such that $x_j = y_0$. Whether a position is the beginning of a superblock can be guessed and verified by the

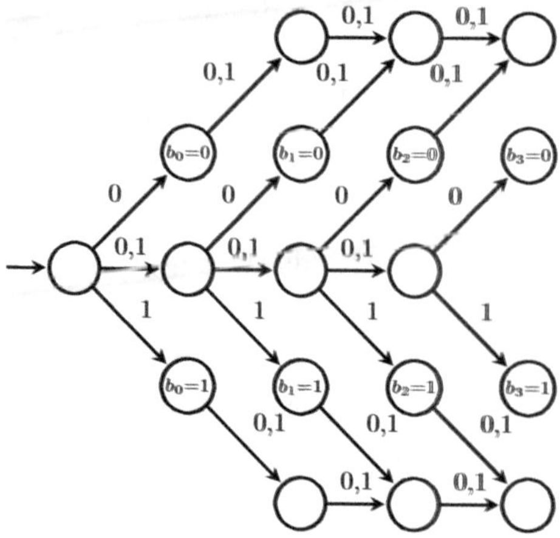

Fig. 4. A gadget to (non-deterministically) store a single bit from a 4-bit sequence. The top and bottom row of states serve the same purpose (namely, a bit has been stored) so they could be merged, but it is easier to draw the gadget in this way. This gadget is easily generalized to an n-bit sequence.

automaton, as all its special bits are zero. Since Item 5 ensures that $y_j = y_0$ and y_j occurs below x_j, it is easy to verify that the guess is correct (if the guess is wrong, the automaton rejects). Then, for the following numbers $x_{j+1}, x_{j+2}, \ldots$ it has to verify that $x_{j+1} < y_{j+1}, x_{j+2} < y_{j+2}, \ldots$ until some $x_{j'}$ is seen such that $x_{j'} = y_{j'}$. To verify that some number x is smaller than y, first, recall that x and y are encoded by their sequence of special bits such that the least significant bit is the rightmost one. If $x < y$, there exists a special bit at some position i such that all previous special bits had the same value for x and y, the special bit at position i has value zero in x and value one in y. The values of the special bits to the right play no role. If the verification fails, the automaton rejects by going to some new sink $q_{verif-fail}$. If some $x_{j'} = y_{j'}$ has been seen, the automaton goes into a new accepting sink $q_{verif-ok}$. Such an automaton needs a constant number of states, as the positions of special bits can be easily identified in a valid output sequence.

Recall that we construct a non-deterministic automaton, hence, the desired automaton $\mathfrak{A}_n$ is given as

$$(\mathfrak{A}_{format} \times \mathfrak{A}_{address}) \cup (\mathfrak{A}_{format} \times \mathfrak{A}_{copy} \times \mathfrak{A}_{pair})$$

where $\mathfrak{A}_{format}$ handles Items 1 to 3, $\mathfrak{A}_{address}$ handles Item 4, $\mathfrak{A}_{copy}$ handles Item 5, and $\mathfrak{A}_{pair}$ handles Item 6. The automaton has $\mathcal{O}(n^3)$ states.

The acceptance condition is given by the formula φ_n defined as

$$\neg q_{rej} \wedge \neg q_{verif-fail} \wedge (q_{acc} \vee (q_{verif-ok} \wedge \varphi_{copy})),$$

where φ_{copy} is defined as

$$\begin{aligned} & \bigwedge_{k\in[0,n-1]} \bigwedge_{b\in\mathbb{B}} A_{=}(b_k{=}b) \rightarrow A(b_k{=}b) \\ \wedge & \bigwedge_{k\in[0,n-1]} \bigwedge_{b\in\mathbb{B}} A(b_k{=}b) \rightarrow \neg G(b_k{=}b) \\ \wedge & \ \varphi_{same-value}. \end{aligned}$$

The first line of φ_{copy} is satisfied if, and only if, the automaton whenever it has guessed "current address equals marked address" is right. If such a guess is wrong, there is some $k \in [0, n-1]$ and $b \in \mathbb{B}$ such that $A(b_k{=}b)$ and $A_{=}(b_k{=}1-b)$ have been visited, falsifying the formula.

The second line of φ_{copy} is satisfied if, and only if, the automaton whenever it has guessed "current address is not equal to marked address" is right. Since the current address and the marked address differ in the value of at least one bit, the automaton can always guess to store this different bit in the gadget G. Hence, the second line of the formula is satisfiable in this way.

The formula $\varphi_{same-value}$ expresses that the relevant special bits either all have value zero, or all have value one.

Note that the formula φ_n needs to be adjusted to talk about the (product-like) structure of $\mathfrak{A}_n$ (which can be done as explained in the proof of Theorem 4.2), hence, its final size is $\mathcal{O}(n^3)$.

It is left to prove that

- Player O wins $\Gamma_f(L(\mathfrak{A}_n))$ for some constant delay function f, but
- Player I wins $\Gamma_g(L(\mathfrak{A}_n))$ for every delay function g with $\sum_{i=0}^{(n+2)\cdot 2^n} g(i) \leq 2^{2^{2^n}}$.

We start by giving a constant delay function f such that Player O has a winning strategy in $\Gamma_f(L(\mathfrak{A}_n))$. According to Proposition 2.4, the sequence $x_1 \cdots x_{2^{2^n}}$ of numbers is guaranteed to contain a bad j-pair for some $j \in [0, 2^{2^n} - 1]$. Hence, we let $f(0) = m$, where m is some number large enough to ensure that $x_1, \ldots, x_{2^{2^n}}$ is included in the lookahead (assuming that Player I plays valid superblocks with either # or $\times$ (used once) in front of an address block). We call playing valid superblocks with either # or $\times$ in front of an address block playing valid for short.

If Player I plays valid, Player O's strategy is to output some y_0 such that $x_1 \cdots x_{2^{2^n}}$ contains a bad y_0-pair. The number y_0 is repeated ad infinitum.

If Player I does not play valid, Player I either violates the format, or he introduces an error in updating the address counter. In the former case, q_{acc} is reached and the formula φ_n is trivially satisfied. In the latter case, Player O has enough lookahead to realize this and mark the error. Then, also q_{acc} is reached. The state q_{rec} is never reached as she does not violate the desired format.

The sub-formula φ_{copy} is also satisfied as she faithfully copies her first number y_0 ad infinitum. It is left to explain why $q_{verif-fail}$ is not reached and $q_{verif-ok}$ is reached. Since the sequence $x_1x_2\cdots$ contains a bad y_0-pair, it is possible for the automaton to pick a run that correctly verifies this, hence, only $q_{verif-ok}$ is reached. Thus, we have shown that the formula φ_n is satisfied when Player O plays according to her strategy.

We turn to the other direction. Let g be a delay function with $\sum_{i=0}^{(n+2)\cdot 2^n} g(i) \leq 2^{2^{2^n}}$. We show that Player I wins $\Gamma_g(L(\mathfrak{A}_n))$.

According to Proposition 2.4, there exists a word w over $[0, 2^{2^n}-1]$ with $|w| = 2^{2^{2^n}}-1$ that contains no bad j-pair for every $j \in [0, 2^{2^n}-1]$. Player I's strategy is to play a valid input sequence such that x_i is the bit sequence that encodes $w(i)$ for all $0 \leq i \leq 2^{2^{2^n}}-1$.

Recall, a superblock is of length $(n+2)\cdot 2^n$ as we have 2^n address blocks of length n, 2^n special bits, and, additionally, 2^n address "markers" (# signifying unmarked, ✕ signifying marked). After round $(n+2)\cdot 2^n$, Player O has finished y_0. Furthermore, the condition $\sum_{i=0}^{(n+2)\cdot 2^n} g(i) \leq 2^{2^{2^n}}$ ensures that Player I has not yet begun to spell the superblock that encodes $x_{2^{2^{2^n}}}$. Hence, since y_0 is known to Player I, he can pick a number $x \neq y_0$ and continue to produce a valid input sequence such that every x_i encodes x for all $i \geq 2^{2^{2^n}}$.

We show that Player I wins playing as described. Player O's goal is to satisfy the formula φ_n. Clearly, she loses if her moves have the wrong format (as q_{rej} would be reached), so we assume that she produces an output sequence that has the right format. The state q_{acc} is not reached, as it is only reached if the input sequence is invalid, hence, $\neg q_{verif-fail} \wedge (q_{verif-ok} \wedge \varphi_{copy})$ must be satisfied. We can assume that Player O faithfully copies y_0 to satisfy φ_{copy}. But, since the input sequence does not contain a bad y_0-pair, no matter which run $\mathfrak{A}_n$ takes, the verification of a bad y_0-pair is never successful. Hence, $q_{verif-fail}$ is reached, falsifying φ_n. Player I wins. □

Complexity. Finally, we turn to complexity showing a 3EXPTIME lower bound on solving delay games with Emerson-Lei-wNMA winning conditions.

To this end, we generalize the analogous construction for Emerson-Lei-wDMA presented in the proof of Theorem 4.2: Here, the players simulate a doubly-exponential space Turing machine. Thus, the use of non-deterministic automata for the winning condition allows us to add an additional exponential blowup in the space consumption of the Turing machine. Again, this requires some additional mechanisms, e.g., both players have to check their opponent's moves for correctness. We describe these changes in comparison to the construction for deterministic automata below.

As before, Player I is in charge of producing the sequence of configurations (with repetitions to fill the lookahead) and of picking successors of universal configurations, while Player O picks successors for existential configurations (by picking a transition for Player I to apply). As configurations are now of doubly-exponential length, we address their cells by exponentially long addresses. Hence, the bits of these addresses have to be addressed using linearly-sized addresses as well, i.e., there are two levels of addresses. Player O has again markers to force Player I to faithfully simulate the Turing machine, while Player I also has markers to ensure that Player O correctly marks errors. Altogether, we obtain a delay game that simulates an alternating doubly-exponential space Turing machine, and A2ExpSpace = 3ExpTime yields the desired lower bound.

Theorem 5.2: *Solving delay games with* Emerson-Lei-wNMA *winning conditions is* 3ExpTime*-complete.*

Proof: The upper bound was shown in Theorem 3.4, we show the corresponding lower bound.

We give a reduction from the word problem for alternating doubly-exponential space Turing machines. Since 3ExpTime = A2ExpSpace [17], we obtain the desired result.

Let $\mathcal{M} = (Q = Q_\exists \uplus Q_\forall, \Sigma, q_I, \Delta, q_A, q_R)$ be an alternating doubly-exponential space Turing machine, where $\Delta \subseteq Q \times \Sigma \times \Sigma \times Q \times \{-1, 0, 1\}$. Furthermore, let p be a polynomial such that 2^{2^p} bounds the space consumption of $\mathcal{M}$ and let $w \in \Sigma^*$ be an input word. We fix $n = p(|w|)$. Wlog., we assume that the accepting state q_A and the rejecting state q_R are equipped with a self-loop. Furthermore, wlog., we assume that either q_A or q_R is reached in every run of $\mathcal{M}$ on w. We have explained in the proof of Theorem 4.2 that such an assumption can be made.

We construct an Emerson-Lei-wNMA $\mathfrak{A}$ of polynomial size in $(|\Delta| + n)$ such that $\mathcal{M}$ accepts w if, and only if, Player O wins $\Gamma_f(L(\mathfrak{A}))$ for some delay function f.

As in the proof of Theorem 4.2, the idea is that, in the delay game, the players build a run of $\mathcal{M}$ on w in the form of a configuration sequence. Player I controls the universal states and Player O the existential ones. Furthermore, Player I spells all configurations with his moves.

The configurations are of doubly-exponential size in n, so each cell of $\mathcal{M}$'s tape is addressed by a superblock as introduced in the proof of Theorem 5.1.

In between configurations there is a delimiter, either C or N, to denote if the next configuration is a *copy* of the previous one or a *new* one. The need to copy configurations arises since, in the delay game, Player O is behind with her moves, so Player I needs to wait for Player O's pick if the current configuration is existential.

We go into details. The automaton we construct uses the product alphabet

$$(\{0, 1, C, N, \llcorner\!\lrcorner, \times, \#\} \cup Q \cup \Sigma) \times (\{0, 1, \#, \ast, \text{✖}, \checkmark, \text{✗}\} \cup \Delta).$$

Recall that Q, Δ, and Σ are components of the Turing machine $\mathcal{M}$ and are used to encode configurations and to pick transitions to apply. We first describe what we call valid input resp. output formats:

- An input sequence α is valid if it is of the form

$$\left((C+N)\left[S(\Sigma+Q+\llcorner\!\lrcorner)\right]^{+}\right)^{\omega},$$

where S stands for superblock and is defined as

$$S = \left((\times + \#)(0+1)^{n+1}\right)^{+}.$$

  - The sequence α is to be interpreted as a sequence of configurations $(C + N)c_0(C + N)c_1 \cdots$, where each c_i is a configuration delimited by C or N. Note that the correct encoding of such configurations is enforced by the rules of the game.
  - In a configuration encoding, a superblock defines the number of a cell, the letter after the superblock is either the tape content (that is, a letter $\sigma \in \Sigma$, or a blank (denoted as $\llcorner\!\lrcorner$)) or a state $q \in Q$.
  - Recall that a superblock S is to be interpreted as first a marker $\times$ or $\#$ to mark (the position before) an address block inside a superblock, then an n-bit address block encoding an address a, followed by a special bit b. Then again, a marker, an address block, a special bit, a marker, and so on.
  - The sequence of special bits of a superblock defines the number of the cell.

- The marker X can be used once in the input sequence to mark (the position before) an address inside a superblock. The purpose is explained further below.
- The letter # is to be interpreted as unmarked.

- An output sequence β is valid if it is of the form

$$\left(\Delta\left[(\ast + ✖ + \#)(✓ + ✗)^n(0+1)\right]^+\right)^\omega,$$

and if for every position i, $\alpha(i) \in \{C, N\}$ if, and only if, $\beta(i) \in \Delta$, i.e., the letters $\tau \in \Delta$ occur at the same positions as the configuration delimiters C and N. We give the intended interpretation of this format, which is detailed below.

  - The markers ✓ resp. ✗ are intended to indicate the absence of resp. the existence of an error an address block.
  - If ✗ marks a bit in an address block, it should indicate an error in updating the address block at this bit.
  - If ✱ marks (the position before) an address block, say the block encodes address a, then it should indicate that there is an error in updating the special bit with address a in the next superblock.
  - Finally, if ✖ marks (the position before) an address block, then it should indicate that there is a copy or update mistake in the successive configuration.

The behavior of the automaton is as follows. We first describe the occurrence of simple format errors that lead to acceptance resp. rejection:

(1) If there exists some i such that $\alpha(0)\alpha(1)\cdots\alpha(i)$ cannot be completed to have a valid format, but $\beta(0)\beta(1)\cdots\beta(i-1)$ can be completed to have a valid format, the automaton accepts.
(2) If a configuration encoding c contains not exactly one state or the first address block in a superblock is not zero, or the first superblock of a configuration does not encode the number zero, the automaton accepts.
(3) If c_0 does not encode the initial configuration on w, it accepts.
(4) If there exists some i such that $\beta(0)\beta(1)\cdots\beta(i)$ cannot be completed to have a valid format, but $\alpha(0)\alpha(1)\cdots\alpha(i)$ can be completed to have a valid format, the automaton rejects.

Items 1 to 4 can be handled similar to Items 1 to 4 in the proof of Theorem 4.2. One needs automata of polynomial size in $(|\Delta| + n)$ to do so.

We now describe different kinds of (more involved) errors and how they are handled:

(5) If the k-th bit of an address block is marked with ✗, then it is verified if the k-th bit of the next address block is correctly updated. If the update is faulty, the automaton accepts, if the update is correct, the automaton rejects.

This item can be handled as Item 5 in the proof of Theorem 4.2 with $\mathcal{O}(n)$ states.

(6) If ✱ marks an address a in a superblock, then the automaton checks whether the special bit that is addressed by a in the next superblock has been correctly updated. If an error has been correctly claimed, the automaton accepts, otherwise it rejects.

To achieve this behavior, a gadget as introduced in Example 2.1 is needed to store the address marked by ✱. To determine how the value of the special bit addressed by the marked address changes, it suffices to check whether all special bits (in the same superblock) to the right are one. If at least one zero is seen, the value of the special bit does not flip. Otherwise, it does. In the next superblock, the automaton has to find the special bit that has the same address. It has to guess when it has advanced to this address block and verify that the guess is correct. Therefore, we need another gadget as introduced in Example 2.1 to store the guessed bit and compare via a formula that the two stored addresses are equal. The gadgets need $\mathcal{O}(n)$ many states, checking whether the special bit is correctly updated can be done with a constant number of states.

(7) If the ✖ marker occurs in a superblock S, then the automaton checks whether there is a *copy* or *update error* between successive configurations that manifests in the cell whose number is encoded by S. If an error had been correctly claimed, the automaton accepts, otherwise it rejects.

We have detailed in the proof of Theorem 4.2 (regarding Item 6) how a copy or update error is checked when the relevant pieces of information are known (which requires $\mathcal{O}(|\Delta| + n)$ states). The difficulty lies in finding the same cell in the current and in the successor configuration. Therefore, the automaton guesses at the beginning of a superblock, say S, whether it has some address that is marked with ✖. (A wrong guess leads to rejection.) If the superblock S contains a ✖, then the automaton enforces that, from then on, the output must copy the special bits of S ad infinitum. This condition makes it easy to find the cell addressed by S in the next configuration, because it suffices to find the cell addressed by a superblock such that the special bits in the input and output sequence are equal.

We have described in detail in the proof of Theorem 5.1 (see Item 5) how to enforce that a bit sequence is faithfully copied. We use the same technique here, which needs $\mathcal{O}(n)$ states. To recap, in case of an error, the input sequence marks an address of a special bit with ✕. This address is where a copy error has manifested.

The last condition is concerned with whether w is accepted by $\mathcal{M}$:

(8) If the format is valid and no error occurred so far: If a configuration with q_A resp. q_R is seen, the automaton accepts resp. rejects.

This can be handled as Item 7 in the proof of Theorem 4.2 with a constant number of states.

All in all, the automaton $\mathfrak{A}$ needs a polynomial number of states (in $|\Delta| + n$).

Finally, we are ready to describe the acceptance condition of $\mathfrak{A}$ as a formula φ defined as

$$\neg q_{rej} \wedge \neg q_R \wedge (q_{acc} \vee q_A \vee \varphi_{*} \vee \varphi_{✖}) .$$

The formula φ_{*} is satisfied if the gadgets to store two addresses needed to find an update error regarding a special bit have not been entered (no special bit error is claimed) or the stored addresses are equal.

The formula $\varphi_{✖}$ is satisfied if the gadget to check faithful copying is either not entered or the gadget is entered and indicates that copying is successful. There are two reasons why the copy gadget is not entered. The marker ✖ is not used in the output, i.e., no configuration copy or update error is claimed. Or, ✖ is used, but the marker ✕ is not used in the input, i.e., no error in copying the special bits is claimed.

The size of φ is polynomial in $(|\Delta| + n)$.

It is left to show that $\mathcal{M}$ accepts w if, and only if, Player O wins $\Gamma_f(L(\mathfrak{A}))$ for some delay function f. To begin with, assume $\mathcal{M}$ accepts w. Let f be a constant delay function such that $f(0) = m$, where m is chosen such that Player O has enough lookahead to see a full configuration before she has to start with her output. This means m is triply-exponential in n. We show that Player O wins $\Gamma_f(L(\mathfrak{A}))$. We assume that either player does not make simple errors (recall that means playing in the wrong format). Firstly, assume that Player I either does not update an address block correctly, does not update a special bit in a superblock correctly, or introduces a configuration copy or update error. The lookahead is large enough for

Player O to correctly claim such an error. The formula is then satisfied. Secondly, assume that Player I does not introduce such errors. Since $\mathcal{M}$ accepts w, Player O can pick existential transitions such that if Player I applies the picked transitions an accepting configuration will be reached. If an accepting configuration is seen, the formula φ is clearly satisfied. Player I can prevent that an accepting configuration is seen by always copying the current configuration. But then, the formula φ is also satisfied, because the sub-formulas $\varphi_{\ast}$ and $\varphi_{\times}$ are satisfied if no error of any kind is claimed. Player O does need to claim an error, as we assumed that Player I does not introduce an error.

For the other direction, assume $\mathcal{M}$ rejects w. We show that Player I wins $\Gamma_g(L(\mathfrak{A}))$ for every delay function g. Again, we assume that either player does not make simple format errors. No matter the existential transitions that Player O chooses, Player I can always pick universal transitions such that eventually a rejecting configuration is reached. Hence, we assume that Player I advances the run by giving new configurations according to his and Player O's choices. A more detailed description on what it means to advance a run is given in the proof of Theorem 4.2.

Furthermore, we assume that Player I does not introduce any kind of errors as this is not beneficial for him. Player O loses a play where a rejecting configuration is reached. Hence, her only chance is to claim an error. Since there are no errors introduced by Player I, this approach is not fruitful as this consequently falsifies φ. Player O cannot win a play, i.e., Player I wins $\Gamma_g(L(\mathfrak{A}))$. □

6. Conclusion

We took the first step towards investigating delay games with Muller conditions by showing doubly- and triply-exponential bounds for deterministic and non-deterministic delay games with weak Muller conditions. More specifically, we have shown that solving such games is 2EXPTIME-complete (for deterministic automata) and 3EXPTIME-complete (for non-deterministic automata) and that doubly-exponential (for deterministic automata) and triply-exponential lookahead (for non-deterministic automata) is necessary and sufficient. These results have to be compared to those for deterministic safety and parity automata, for which solving delay games is EXPTIME-complete and exponential lookahead is necessary and sufficient. Similarly, for non-deterministic safety and parity automata solving delay games is 2EXPTIME-complete and doubly-exponential lookahead is necessary and sufficient. Thus, the succinctness of weak Muller conditions (encoded by formulas) yields an exponential increase in comparison to both safety and parity conditions.

There are two immediate directions for further research: Try to lift our results to (standard) Muller conditions and to strengthen the lower bounds by proving them for less succinct representations of $\mathcal{F}$.

Recall that our lower bounds for weak Muller conditions rely on gadgets comparing n-bit strings for some fixed n. This is possible for a fixed number of comparisons by having one gadget for each string to be compared and then using the formula defining $\mathcal{F}$ to implement the actual comparison. A natural way to lift, say, the bad j-pair lower bound game is to just play this game infinitely often. However, this requires to reuse the gadgets infinitely often, which then means one cannot compare strings from one iteration of the bad j-pair game, but compares strings from all iterations.

Let us stress again that all our upper bounds are independent of the representation of the acceptance condition. However, our lower bounds only hold for Emerson-Lei conditions, which is the most succinct representation. Strengthening the lower bounds via less succinct representations most likely requires a new approach, as all lower bounds we have proven require the comparison of n-bit strings. This can be done with a small formula, as shown here, but not with the less succinct presentations considered by Hunter and Dawar for delay-free games [2, 3]. In particular, it is even open whether delay games with explicit Muller conditions (the least succinct representation) are harder than delay games with parity or safety conditions. Note that in the delay-free case, explicit Muller games can be solved in polynomial time [14].

References

1. U. Boker. On the (in)succinctness of Muller automata. In eds. V. Goranko and M. Dam, *CSL 2017*, vol. 82, *LIPIcs*, pp. 12:1–12:16, Schloss Dagstuhl - Leibniz-Zentrum für Informatik, 2017. doi: 10.4230/LIPIcs.CSL.2017.12.
2. P. Hunter and A. Dawar. Complexity bounds for regular games. In eds. J. Jedrzejowicz and A. Szepietowski, *MFCS 2005*, vol. 3618, *LNCS*, pp. 495–506, Springer, 2005. doi: 10.1007/11549345_43.
3. P. Hunter. *Complexity and Infinite Games on Finite Graphs*. PhD thesis, Computer Laboratory, University of Cambridge, 2007. URL `http://web.comlab.ox.ac.uk/oucl/work/paul.hunter/papers/thesis.pdf`.
4. C. S. Calude, S. Jain, B. Khoussainov, W. Li, and F. Stephan. Deciding parity games in quasipolynomial time. In eds. H. Hatami, P. McKenzie, and V. King, *STOC 2017*, pp. 252–263, ACM, 2017. doi: 10.1145/3055399.3055409.
5. D. Gale and F. M. Stewart, Infinite games with perfect information, *Annals of Mathematics*. **28** (1953), 245–266. doi: 10.1515/9781400881970-014.

6. F. A. Hosch and L. H. Landweber. Finite delay solutions for sequential conditions. In ed. M. Nivat, *ICALP 1972*, pp. 45–60, North-Holland, Amsterdam, 1972.
7. J. R. Büchi and L. H. Landweber, Solving sequential conditions by finite-state strategies, *Trans. Amer. Math. Soc.* **138** (1969), pp. 295–311. doi: 10.2307/1994916.
8. M. Chen, M. Fränzle, Y. Li, P. N. Mosaad, and N. Zhan, Indecision and delays are the parents of failure - taming them algorithmically by synthesizing delay-resilient control, *Acta Informatica*. **58** (2021), 497–528. doi: 10.1007/s00236-020-00374-7.
9. A. Carayol and C. Löding. Uniformization in automata theory. In *Logic, Methodology and Philosophy of Science - Proceedings of the 14th International Congress*. College Publications, 2015.
10. M. Holtmann, L. Kaiser, and W. Thomas, Degrees of lookahead in regular infinite games, *Log. Methods Comput. Sci.* **8** (2012). doi: 10.2168/LMCS-8(3:24)2012.
11. E. Filiot and S. Winter. Synthesizing computable functions from rational specifications over infinite words. In eds. M. Bojanczyk and C. Chekuri, *FSTTCS 2021*, vol. 213, *LIPIcs*, pp. 43:1–43:16, Schloss Dagstuhl - Leibniz-Zentrum für Informatik, 2021. doi: 10.4230/LIPIcs.FSTTCS.2021.43.
12. F. Klein and M. Zimmermann, How much lookahead is needed to win infinite games?, *Log. Methods Comput. Sci.* **12** (2016). doi: 10.2168/LMCS-12(3:4)2016.
13. F. Klein and M. Zimmermann. Prompt delay. In eds. A. Lal, S. Akshay, S. Saurabh, and S. Sen, *FSTTCS 2016*, vol. 65, *LIPIcs*, pp. 43:1–43:14, Schloss Dagstuhl - Leibniz-Zentrum für Informatik, 2016. doi: 10.4230/LIPIcs.FSTTCS.2016.43.
14. F. Horn. Explicit Muller games are PTIME. In eds. R. Hariharan, M. Mukund, and V. Vinay, *FSTTCS 2008*, vol. 2, *LIPIcs*, pp. 235–243, Schloss Dagstuhl - Leibniz-Zentrum für Informatik, 2008. doi: 10.4230/LIPIcs.FSTTCS.2008.1756.
15. E. A. Emerson and C. S. Jutla, The complexity of tree automata and logics of programs, *SIAM J. Comput.* **29** (1999), 132–158. doi: 10.1137/S0097539793304741.
16. F. Klein and M. Zimmermann. What are strategies in delay games? Borel determinacy for games with lookahead. In ed. S. Kreutzer, *CSL 2015*, vol. 41, *LIPIcs*, pp. 519–533, Schloss Dagstuhl - Leibniz-Zentrum für Informatik, 2015. doi: 10.4230/LIPIcs.CSL.2015.519.
17. A. K. Chandra, D. Kozen, and L. J. Stockmeyer, Alternation, *J. ACM.* **28** (1981), 114–133. doi: 10.1145/322234.322243.

https://doi.org/10.1142/9789811278631_0017

A Turing Machine like Model for Computation on Real Numbers

Yue Yang

Department of Mathematics,
National University of Singapore,
Singapore, 119076
matyangy@nus.edu.sg

It is commonly accepted that Turing program formalizes the concept *algorithm*, as long as the domain is the set of natural numbers $\mathbb{N}$. It is still an open problem whether one can define algorithms beyond the domain $\mathbb{N}$. In this chapter, we study computability on the set of real numbers. We introduce a new Turing machine like computation model over real numbers, which is a natural generalization of the standard Turing machine. The basic properties of the new model are explored.

Contents

1. Introduction

It is well-known that the informal concept of algorithms was formalized in the 1930s; in fact, several equivalent formalizations were proposed, among which the most intuitive one is the notion of Turing machines. Over the domain of natural numbers, or any domain which can be coded into natural numbers, it is commonly accepted that the formal notion of Turing computability formalizes the informal concept of computability.

However, when we look at domains beyond natural numbers, for example, real or complex numbers, the views of formal computability become divergent. The investigation on computation over reals can be traced back to the beginning of recursion theory. It is Alan Turing who first defined the notion of computable real numbers [8]. A large number of formalizations were introduced (see Longley's survey paper [4]; Weihrauch's book [9], in particular, Chapter 9; and Blum, Shub and Smale [1]) but most of them were no longer compatible. Each individual formalization produces its own class of computable functions. On the other hand, working mathematicians seem to have a unique informal notion of algorithm, regardless whatever its domain is. This leads to the fundamental question whether the notion of effectiveness on arbitrary domains can be formalized, if so, what it is.

Even a quick reflection could tell us why the question is so challenging. Let us just state two of the most obvious difficulties. The first immediate difficulty is to balance the finite and infinite aspects. On the one hand, algorithms, no matter which domain they apply to, should always have a finite (or finitary) character. On the other hand, objects like real numbers are truly infinite. If one insists that all computations must be done within finitely many steps, then the computation would have to be finished before even reading the full input. The second difficulty is more relevant to computations on real numbers. On the one hand, the real line is continuous, working mathematicians often design algorithms using approximations. On the other hand, computation, as intuitively understood, is discrete. A related question is to what extent we should consider the topological structure of the real line. One can identify real number either as continuum many scattered points, or as Baire space ω^ω with product topology, or as the usual real line with the usual topology (i.e., rational centered intervals with rational radius are the basic open sets). Different identifications may lead to different notions of computability.

In this chapter, we introduce a Turing machine like computation model over real numbers, that has the following features:

(a) It is Turing machine like because both its hardware and program are natural extensions of the standard Turing machine (abbr. *STM*) over natural numbers. It includes STMs as natural components. However, it computes more functions than the STM when restricted to natural numbers, but not too much more.
(b) To a large extent, it agrees with the intuition of working mathematicians. On the one hand, it includes the functions that often occurred in calculus or scientific computing. Two typical examples are: The

exponential function e^x, and the sign function $\mathrm{sgn}(x)$ indicating x is negative, 0 or positive. On the other hand, it does not compute functions which has no obvious effective way to compute. For example, the characteristic function of rational numbers.

(c) Its theory of computation can be developed within arithmetic, or roughly in the same rank as the objects it computes. Each computable function is fully determined by its values on some fixed countable set (see Theorem 3.9). Even though problems involving real numbers can easily go beyond Zermelo-Fraenkel set theory, we want to stay at second order arithmetic for the moment. Involving Axiom of Determinacy or large cardinal axioms would be viewed as a sign of noneffectiveness.

(d) Its theory of computation depends on the representations of real numbers, but in a robust way. Essentially, there is only one system of representations needed, which is the equivalence class of Cauchy sequences with computable convergence rate,[a] for example, a real number x is represented by a sequence $\langle r_i : i \in \omega \rangle$ of rational numbers, such that $\lim_i r_i = x$ and for all $m, n > i$, $|r_m - r_n| < 2^{-i}$.

2. Description of the Computation Model

2.1. *The Machine*

We now describe a model for computing real numbers and will refer to it as *Master-Slave* machines. The physical device of a master-slave machine consists of the following parts:

(1) There is a standard Turing machine M, which is the main component of the device. We will refer to M as *the master*. M is able to handle all computations on natural numbers and has its own working tape(s). We can also assume M has input and output tapes for natural numbers. M also has channels to communicate with the slaves, which we will describe in item (5).

(2) There is an infinite sequence $\langle S_i : i \in \omega \rangle$ of universal (standard) oracle Turing machines, we shall call them *slaves*. Each slave S_i has its own label i, indicating its relative position to other slaves. There is a one-way infinite read-only tape, which is used for real number input. All slaves use this real number input tape as oracle, in particular, they can read as many cells as necessary. (In the case that one slave reads infinitely

[a]In this chapter, we view a real number as a Cauchy sequence with computable convergence rate. The equivalence class issue will be handled in a follow up paper [5].

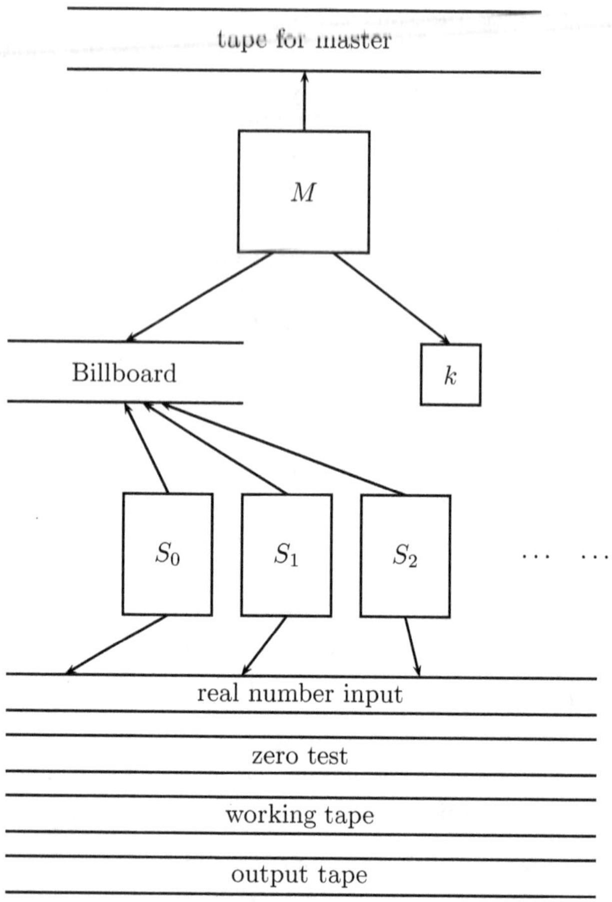

Fig. 1. A Master-Slave machine.

many cells, the computation will be defined as divergent. See item 2 on p. 470 below.) There are also a working tape and an output tape for real numbers (which is drawn at the bottom of Fig. 1 with connecting arrows omitted); all slaves are allowed to write on the working tape and output tape (but they cannot change other people's writing). Since we are only interested in the situation when every slave's computation halts, we may assume that the slaves write on the output tape in a sequential way, namely, if $i < j$, then S_j must wait until S_i finishes writing its output.

(3) Similar to the STMs, each tape is further divided into cells and each cell

holds one symbol from a fixed finite alphabet. For individual machines, regardless if it is master M or a slave S_i, their tapes are exactly the same as in the STM. For the input tape, which is used by all slaves as oracle, its content can be viewed as infinitely many finite binary strings separated by blank symbols $\square$ or by some new symbol like $; where the binary strings are codes for rational numbers and the very first bit is the sign bit "+" or "−". In the case when the input contains more than one real, we may assume that the input tape has several layers or tracks. Since our main target is computation on reals, we would like to fully employ the Church Thesis about computation on natural numbers, which says that any intuitively computable function on natural numbers is Turing computable.

(4) There is a special tape, called *zero-test tape*, which is used for testing whether a real number is zero. Initially the tape content is empty. The slaves will collectively produce a real number on the zero-test tape, testing if it is zero and write the result by a single boolean bit k for the master to read. After the test, the content on the zero-test tape is supposed to be cleaned and to become empty tape again. This zero-test will be viewed as a single (master) step, even if in reality it requires infinite amount of time. (We did not draw the connections on Fig. 1 for the sake of cleanness.)

(5) The communication between the master and slaves is done in two ways. The master can write a natural number p on a special tape (called the *billboard* tape). The slaves will take this p as input of their program; recall that the slaves are universal oracle Turing machines, which can simulate the p-th oracle Turing machine when reading the input p. The feedback from slaves to master is only the boolean bit k from (4), which M will take as an input.

The content p written in billboard catches one of the most important aspects of this computation model. Firstly, by letting all slaves execute the same program p, they only differ by precision preassigned to them; in some sense, one could even view the infinitely many slaves as a single one. Secondly, it also explains why we do not distinguish whether the input is a recursive real or not; although the recursive reals are philosophically appealing as they can be coded by a single natural number, when we do computation the program p is all we need, the program will treat recursive and non-recursive reals in the same way.

People might ask: Since the master machine itself can do the job p

assigned to the slaves and we have made the device infinitary anyway, why do we not give the master infinitely amount of time to finish the job? Indeed people had explored the idea of infinite time Turing machines (see for example, Hamkins and Lewis [2]). However the problem is that once one allows infinite time, the computation models often become too powerful. For example, Hamkins and Lewis machine can compute all Borel functions. The second point for keeping the master finitary is more philosophical. Like Turing machines, the machine model for computation is actually mimicking the human mind. The key idea behind this master-slave model is the trade-off between the infinitary hardware (slaves) and the finitary software (the master).

We now turn to describe the program of a master-slave machine. The master M has a finite set of states $Q = Q_0 \cup \{S, E\}$, where Q_0 consists of the usual STM states, S (for calling the action of slaves) and E (for the zero-test action) are two special new state. We ignore the states of slaves, as they all have the same fixed set of states of a fixed standard universal oracle Turing machine. We also have a fixed finite alphabet Σ, usually binary and with some auxiliary symbols for convenience. The master-slave machine will have a finite set of quadruples as its program. It has to be said that when multi-tapes (e.g. billboard and boolean bit) are used, the tuples in the program would be more than quadruples; we sacrifice a bit accuracy here for the sake of simplicity. However, as it is well-known that any multi-tape STM can be simulated by one with a single tape (see, for example, Chapter 7 of Hopcroft and Ullman [3]), this inaccuracy can be corrected easily. The quadruples are of the following three types:

1. First the standard ones, which are of the form $qaa'q'$ or $qaLq'$ or $qaRq'$ where $q \in Q_0$ and $q' \in Q$, a, a' are symbols in Σ, and L (left) or R (right) are symbols for the movement of the head. We also singled out two states q_s and q_h for starting and halting, respectively, just like in the STMs.
2. Commands of the form $Saaq$, where q is a state, a is a symbol in the alphabet, and S is one of the special new state symbol for slave calling. The intention is to trigger the action of the slaves; they will read the program input from the billboard tape and process the computation, when every slave machine halts (note that this is a theoretical step because the collective time taken by all slaves might be truly infinite), the state of the master becomes q. Note that symbol a is used only to make a quadruple.

3. Commands of the form $Eaaq$, where E is another new special symbol. The intention is to detect if the content on the zero-test tape is 0 (as a real number), if yes then change the content of the boolean bit to 1, otherwise change it to 0, and change the current state from E to q. [The content on the zero-test tape was prepared by slaves beforehand perhaps by another slave calling command. One can think of that as i-th slave produces the i-th digit (or the i-th member of the rational Cauchy sequence) of the real number.] Note again that symbol a is used only to make a quadruple.

We make the formal definition as a summary of the description above. For notational simplicity, we assume that both the master M and the slaves are working on their single tape respectively, and furthermore, we have merged the billboard and the boolean counter into the single tape of M.

Definition 2.1: A *master-slave machine* M is a quintuple

$$(Q, \Sigma, \delta, q_0, q_h),$$

where $Q = Q_0 \cup \{S, E\}$ is a finite set of symbols called *states* with two special symbols S and E, not in Q_0; $\Sigma \supseteq \{0,1\}$ is a finite set of symbols called *tape alphabets*; $q_0 \in Q_0$ is called *the initial state*; $q_h \in Q_0$ is called *the halting state*; and δ (called the *transition function*) is a partial function from $Q \times \Sigma \to (\Sigma \cup \{L, R\}) \times Q$, where L and R are two symbols not in Σ. We often identify δ as a finite set of quadruples by writing $\delta(q, a) = (b, r)$ as quadruple $qabr$.

This finishes the description of the master-slave model of computation. We pause to explain some motivation of the new features. The number of slaves is indeed infinite, and thus to execute the command involving S requires infinite amount of time. This is one place where we step into infinity; another place is the execution of zero-test command, which requires us to read infinitely many digits within one step. But we can justify it by observing that the infinitary features are actually driven by the input, which is the only source of infinity given to us by our opponent. If the input x is a finite string σ, no matter how long it is, we probably only need its length $|\sigma|$ many slaves or even just one master without slaves at all. To a true believer of finitism, the computation of real number does not make sense at all. If our opponent asked us to do computation on real numbers, he or she has already allowed certain kind of infinity, perhaps in the form of countably infinite sequences. What we did is merely accepting the same amount of infinity that our opponent enjoys, namely we have countably infinite slaves to handle the countably infinite sequences. For the zero-test command E, there is another fundamental justification: it is essentially the branching command "if... then ..., else ...", which must be admitted in any formalization of algorithms. Thus the balance of finite and infinite is achieved by the separation of master and slaves: The slaves collectively handle the infinite part, whereas the master remains as a standard Turing machine. Finally, the two new type of commands S and E allow the information to flow back and forth between the master and slaves, therefore the jobs of handling natural numbers and the real numbers are separately done by master and slaves.

2.2. *The Computation*

We now describe configurations and computation process of a master-slave machine. Note that the assumption of single tape as stated before Definition 2.1 still applies.

As in the classical Turing machine, a configuration (or an instantaneous description) of a master-slave machine simply records the current status of a computation process. Formally, an *MS-configuration* (or simply a configuration) C is defined as a pair $(n; x)$, where n is a natural number coding the configuration of the master machine as did for the STM and x is an infinite binary or Cauchy sequence currently handled by the slaves. More precisely, assuming the current state of the master is q, the symbol it is reading is a and the tape contents to the left and to the right are finite strings σ and τ respectively, then n is the code of the quadruple ((code of) σ, (code of) q, (code of) a, (code of) τ).

Given a master-slave machine M, "a configuration $C = (n; x)$ *yields* another configuration D by M" is defined as follows.

We decode from n in C to get the current state parameter q and the symbol a currently reading by the master, then check which quadruple in M starts with qa (in the case when q is S or E, we do not check the second component a). If no such quadruple exists, then C does not yield anything; otherwise, we have the following cases, depending on which quadruple it is, the resulting configuration $D = (n'; x')$ will be the one *yielded* by M.

- If the quadruple is $qaa'q'$ or $qaLq'$ or $qaRq'$, where $q \in Q_0$, then let the master perform one step computation just like we did in STM and code the new configuration of the affected part as n'. Then we say C yields $(n'; x)$.
- If the quadruple is $Saaq$, where S is the special slave calling state,

then for all i we let the i-th slave perform its own task by reading the program p from the billboard tape, using its own index i as natural number input and using x in C as oracle. If for every i, the (standard) oracle Turing machine $\Phi_p^x(i) \downarrow = b_i$, let x' be the infinite sequence $\langle b_i :\in \omega\rangle$ and let n' record the state change from S to q of the master, then we say that C yields $(n'; x')$; otherwise, C does not yield anything.

- If the command is $Eaaq$, where E is the special zero-test state, then we (let the slaves) test whether the content on zero-test tape is zero; if so write the boolean bit 1; otherwise, write the boolean bit 0 and change the state of the master to q. We may assume that the content on zero-test tape are cleared to become empty. Thus the only net change for the master is its state and the boolean bit; and the only net change for the slave part is the content of zero-test tape, which becomes empty. We let n' and x' denote the integer code of the configuration of the master and the infinite sequence written on the slave part respectively; and we say that C yields $(n'; x')$.

A master-slave machine always starts from its starting state q_0, with the real input x written on the tape shared by the slaves. To be more precise on the input, let us fix the alphabet Σ to be $\{\Box, 0, 1, +, -\}$, where $\Box$ is the symbol for blank, and $+$ and $-$ are the sign bit for real numbers. A natural number is identified as a finite binary string; an integer is either a natural number or a natural number with a $+$ or a $-$ sign in front; a rational number r can be coded also by a single binary string σ with a $+$ or $-$ sign in front and we read σ as a pair of natural numbers (a, b), where $b \neq 0$. The real number x is represented as a sequence $\langle r_i : i \in \omega\rangle$ of rational numbers. The input $x = \langle r_i\rangle$ is written on the oracle tape sequentially with a blank $\Box$ separating them, like $r_0\Box r_1\Box r_2\Box\ldots$. If n_0 is the code of the configuration of a standard Turing machine with state q_0 and empty tape content, then $(n_0; x)$ is called the *initial configuration*. We say that a configuration $(n; x)$ is *halting*, if its state component coded in n is the halting state q_h.

Definition 2.2: A *master-slave computation* or an MS computation on input x is either an infinite sequence of MS configurations $\langle C_i : i \in \omega\rangle$, or a finite sequence of configurations $\langle C_i : i \leq N\rangle$ for some natural number N, where C_0 is the initial MS-configuration $(n_0; x)$, and for all i, C_i yields C_{i+1} unless the sequence is finite and $i = N$.

If the sequence is finite and $C_N = (n_h; y)$ is a halting configuration then we say the MS-computation on input x *converges* or *halts*; and the output is y. Otherwise, that is, the sequence is infinite or the sequence is finite and C_N is not halting but does not yield anything, then we say the computation with input x diverges.

3. Computable Functions and Sets in the New Model

This computing device applies to any rational sequences (or any members of Baire space or Cantor space). However for computation on real numbers, we want to fix a reasonable representation, so that the usual topology comes into play. Let us fix a representation of a real number x by any Cauchy sequence $\langle r_i : i \in \omega \rangle$ of rational numbers, such that $\lim_i r_i = x$ and for all $m, n > i$, $|r_m - r_n| < 2^{-i}$. Note we do not require that r_i is a recursive sequence of rationals. Also note that one can choose other Cauchy sequences with known recursive convergence rate $k(i)$ instead of 2^{-i}. It is easy to see that these representations can be converted one another by master-slave machines.

Definition 3.1: We say that a partial function $f : \mathbb{R} \to \mathbb{R}$ is *master-slave computable* (or *MS-computable* or simply *computable*) if there is a master-slave machine M, such that $f(x) = y$ if and only if M halts whenever the input is a representation of x, with an output that is a representation of y, while $f(x)$ is undefined if and only if M does not halt for any representation of x.

Note that the definition can be easily generalized to functions with more than one input.

Remark: Notice the machine has a uniform procedure for all representing Cauchy sequences of an input x, information that was coded by a special representation has to be ignored by the machine. For example, for any subset A of natural numbers, A can be coded into the sequence that is 2^{-n} if n in A and 0 if n not in A. Such a sequence is a representation of 0, however, the MS-machine is unable to use this information, because it has to give the same output when input the constant 0 sequence.

Definition 3.2: We say that a subset S of real numbers is *master-slave recursive* or simply *recursive* or *computable*, if its characteristic function χ_S is master-slave computable.

We now establish some simple properties of master-slave computable functions and sets, and verify some features listed on page 466. As we are aiming for computation on real numbers, we make full use of Church thesis for natural numbers. Thus, we will just write an algorithm of how the master and slaves working on rational numbers as working mathematicians usually do, instead of writing codes for Turing machines. We begin with the fact that most functions occurring in scientific computing are MS-computable.

Theorem 3.3: *Let R be a positive real number and $f : (-R, R) \to \mathbb{R}$ be a function satisfying the following conditions: (1) Within the radius of convergence R, f has Taylor expansion $f(x) = \sum_{n=0}^{\infty} a_n x^n$ with recursive coefficient a_n, i.e., $\langle a_n : n \in \mathbb{N}\rangle$ is a (standard) recursive sequence of rational numbers; (2) the remainder term can be bounded uniformly and recursively, i.e., there is a (standard) recursive function $k(n)$, such that for any $x \in (-R, R)$, $|\sum_{i=k(n)}^{\infty} a_i x^i| < 2^{-n}$. Then $f(x)$ is MS-computable. In particular, e^x, $\ln(1+x)$, $\sin x$, and $\arctan x$ are all MS-computable on their respectively appropriate domains. Also, the binary functions $x + y$, $x - y$, $x \cdot y$ are MS-computable.*

Proof: Given any input $\langle r_i : i \in \omega\rangle$ representing x, the instruction for the n-th slave is: Calculate the number $N_1 = k(n+2)$. Once we know N_1, we know that for any x, r such that $-R < x, r < R$, we have

$$\left|\sum_{i=N_1}^{\infty} a_i(x^i - r^i)\right| < \left|\sum_{i=N_1}^{\infty} a_i x^i\right| + \left|\sum_{i=N_1}^{\infty} a_i r^i\right| \le 2^{-n-1}.$$

Next, the n-th slave searches for some N_2 so large that $|\sum_{i=0}^{N_1-1} a_i(x^i - r_{N_2}^i)| < 2^{-n-1}$, then calculate $\sum_{i=0}^{N_1-1} a_i r_{N_2}^i$ as the output. The master can write the code for this instruction on the billboard tape. The whole computation takes one single master step. □

Theorem 3.4: *The subsets of reals $\{0\}$, $\mathbb{N}$ and $\mathbb{Z}$ are MS-computable.*

Proof: First, given a Cauchy sequence $\langle r_n\rangle$ with convergence rate 2^{-n}, we just let the slave S_n check if for all $i \le n$, $-2^{-i} \le r_i \le 2^{-i}$; write a 0 on the zero-test tape if yes, otherwise, find the least such $j \le n$, which violates the condition and write $2^{-j} - 2^{-n}$ on the zero-test tape. This takes one master step. If x is 0, then the slaves will collectively write the constant 0 sequence on the zero-test tape; otherwise, they write a Cauchy sequence converging to 2^{-j}, where r_j is the first term in the sequence $\langle r_n\rangle$ showing that $r_n \not\to 0$. (Note that we are required to write a real number on the zero-test tape, that is why we take the extra care here.) Applying zero test, we get the answer.

Next, we compute the set of natural numbers: Given a Cauchy sequence $\langle r_n \rangle$ with convergence rate 2^{-n} representing x, the instructions for the slave S_n is: Check if for all $1 < i \leq n$ there is an integer m, such that $r_i - 2^{-i} \leq m \leq r_i + 2^{-i}$; if so write a 0 on the zero-test tape, otherwise find the $j \leq n$, which violates the condition and write $2^{-j} - 2^{-n}$ on the zero-test tape. This takes one master step. Applying zero test, we get the answer. Note here we exclude $i = 0, 1$ to ensure that there is at most one integer in the open interval $[r_i - 2^{-i}, r_i + 2^{-i}]$. A slave can decide if there is an integer between two rational numbers a/b and c/d, for example, it can simply calculate the (integral) quotient q of a/b and r of c/d and see if $r = q + 1$. This shows that the set of natural numbers is MS-recursive.

A similar argument shows that $\mathbb{Z}$ is also MS-recursive. □

Corollary 3.5: *The equality and less than relation between real numbers are MS-computable; and the sign function is MS-computable.*

Theorem 3.6: *Any recursively enumerable set (r.e.) of natural numbers in the standard sense, in particular, the halting set of standard Turing machines $K = \{e \in \mathbb{N} : \varphi_e(e) \downarrow\}$ is MS-recursive. (Here we use φ_e to denote the e-th partial recursive function over natural numbers.) Consequently, all co-r.e. set of natural numbers are also MS-recursive.*

Proof: Given an input x represented by a Cauchy sequence $\langle r_n \rangle$ with convergence rate 2^{-n}, first decide if it is in $\mathbb{N}$, if not, reject; if yes, use a similar argument as in the proof of Theorem 3.4 to find which integer e it is and convert $\langle r_n \rangle$ into the constant sequence $\langle e, e, \ldots \rangle$. Ask slave S_i to check if $\varphi_e(e)$ converges within i steps, if so write $2^{-j} - 2^{-i}$, where j is the least step that $\varphi_e(e)$ converges, otherwise write a zero. Then apply the zero test, the boolean bit is 1 if e is not in K; so we accept e if and only if the boolean bit is 0. □

We identify an effectively closed set C with a Π^0_1-class $[T]$, where T is a recursive binary tree and $[T]$ denote the class of all infinite paths of T.

Proposition 3.7: *Every effectively closed set, in particular, the Cantor middle-third set, is MS-computable.*

Proof: Fix a recursive binary tree T, we may assume that each node σ on T is a rational number; and if σ_1 and σ_2 are node on T of length i, then $|\sigma_1 - \sigma_2| \geq 2^{-i}$. Given a Cauchy representation $\langle r_n \rangle$ of x, use the

n-th slave S_n to check if there is a node σ on T of length $i \leq n$, such that $|r_{i+1} - \sigma| \leq 2^{-i-1}$, if yes write a 0 on the zero test tape, otherwise find the $j \leq n$, which violates the condition and write $2^{-j} - 2^{-n}$ on the zero-test tape. (This informally tells us if x passes through level i of T.) Then, use zero test to see if x is in the effectively closed set. □

In Definition 2.1, one can further fix a finite alphabet Σ and identify a master-slave machine with its transition function δ, in other words, each MS machine is essentially a finite set of quadruples over Σ. As did for STMs, the programs can be effectively coded, therefore there is a universal master-slave machine (see for example, Chapter 8 of Hopcroft and Ullman [3] or Chapter 1 in Rogers [6]). We fix an effective list of MS-computable functions $\varphi_0^2, \varphi_1^2, \ldots$, here the superscript 2 indicates it is of type two for computing real numbers.

Theorem 3.8: *The MS-halting set $K^2 = \{e : \varphi_e^2(e) \downarrow\}$ is not MS-computable.*

Proof: We show that the characteristic function χ of its complement is not MS-computable. Suppose χ is computed by the e-th MS-machine, then $\chi(e) = 1$ if and only if $e \notin K^2$ if and only if $\varphi_e^2(e) \uparrow$ if and only if $\chi(e) \uparrow$, a contradiction. □

Theorem 3.9: *There is a countable set X of reals (e.g. reals recursive in Kleene's $\mathcal{O}$), such that if two MS-machines M and M' agrees on every points in the set X, then they agree on every real numbers.*

Proof: (Sketch) We begin with a remark on the code of configurations of MS-computation. Fix a real number input x. For any MS-machine M, the computation $M(x)$ is a sequence of configurations $\langle C_i = (n_i, x_i) : i \leq \omega \rangle$. Notice that the real part of the configuration only changes when we apply the slave calling and zero-test commands, each x_i can actually be coded by a natural number e_i and the original input x, so that $\Phi_{e_i}^x = x_i$. Thus, the sequence of configurations $\langle C_i, x_i \rangle$ can be identified with a function $f(i, x)$. Observe that $f(i, x)$ is arithmetic in x: The relation C_i yields C_{i+1} is Π_2^0 relative to i and x, because it only requires the knowledge of if the procedure written on billboard is total, which is Π_2^0; and the zero test command is Π_1^0 relative to x. Thus, if M and M' are different, then there exists an x, such that either one and only one of $M(x)$ and $M'(x)$ is defined or both of them are defined but $M(x) \neq M'(x)$. Since $M(x) \neq M'(x)$ is arithmetic in

x, the whole predicate is Σ_1^1. By Kleene Basis Theorem (see Theorem 1.3 on Chapter 3 in Sacks [7]), we know that if such an x exists, then there is an x recursive in $\mathcal{O}$. □

This theorem illustrates the point (c) on page 466. Its proof also hints there is kind of normal form theorem in the style of Kleene. In the follow up of this chapter, Ng, Tavana and Yang [5] will formally establish the Normal Form Theorem and prove some of the noncomputability result stated on page 466. We will also clarify the connection of master-slave machines with the computation models of Blum, Shub and Smale [1] and the Type-two theory of effectivity (see for example, [9]). Moreover, we will define a notion of partial recursive functions from real numbers to real numbers and show that is equivalent to the master-slave computable ones.

Acknowledgments

The author has given several talks on this topic in seminars and in conferences since November, 2013; and is extremely grateful for the discussions with the audiences. In particular, the author wishes to thank Longyun Ding from Nankai University, China, Kengmeng Ng from Nanyang Technological University, Singapore and Nazanin Tavana from Amirkabir University of Technology, Iran for the discussions that cut the final shape of the computation model.

This work was partially supported by Singapore MOE grant MOE-2019-t2-2-121.

References

1. Blum, L., Shub, M. and Smale, S. On a theory of computation over the real numbers: NP completeness, recursive functions and universal machines. *Bull. Amer. Math. Soc.* (N.S.) **21**, no. 1, 1–46, 1989.
2. Hamkins, Joel D., and Lewis, A. Infinite time Turing machines. *Jour. Sym. Log.* **65**, no. 2, 567–604, 2000.
3. Hopcroft, John E., and Ullman, J. Introduction to Automata Theory, Languages and Computation. *Addison-Wesley*, 1979.
4. Longley, John R. Notions of computability at higher types. I. *Logic Colloquium 2000*, 19, 32–142, Assoc. Symbol. Logic. Urbana, IL, 2005.
5. Ng, Keng Meng, Tavana, Nazanin R. and Yang, Yue. A recursion theoretic foundation of computation over real numbers. *J. Logic Comput.* **31** (2021), no. 7, 1660–1689.
6. Rogers, Hartley Jr. Theory of Recursive Functions and Effective Computability. *McGraw-Hill*, 1967.

7. Sacks, Gerald E. Higher Recursion Theory. *Springer-Verlag*, Berlin, 1990.
8. Turing, A. M. On Computable Numbers, with an Application to the Entscheidungsproblem. *Proc. London Math. Soc.*, S2-42 (1): 230–265, 1936.
9. Weihrauch, Klaus. Computable Analysis, an introduction. *Springer-Verlag*, Berlin, 2000.

Printed in the USA
CPSIA information can be obtained
at www.ICGtesting.com
JSHW060316011123
51137JS00001B/6

9 789811 278624